Seed Production in Field and Horticulture Crops

NIPA® GENX ELECTRONIC RESOURCES & SOLUTIONS P. LTD.
New Delhi-110 034

About the Authors

Dr. S.R. Sree Rangasamy has been teaching the subject of Plant Breeding & Genetics to Under Graduate students for seven years & Post Graduate students for 20 years at AC & Rl Coimbatore. He was pursuing very active research on Cytogenetics in rice, millets and few fruit crops and Mutation Breeding for over 10 years in rice & on stabilizing CMS sources in rice. He has organized Plant Tissue Culture research to support crop improvement especially in rice and sorghum in production of Haploid and Doubled haploid. As a plant breeder he was involved in the release of more than 25 crop varieties in agri and horticultural crops. He has taken pioneering efforts to organize, develop & to establish Research on Plant Molecular Biology in 1983 to 95 which blossom to the Centre for Plant Molecular Biology and Biotechnology in Tamil Nadu Agricultural University.

After retirement from the TNAU he had been invited to be the GM (R&D) to organize Hybrid Research in Manjushree Plantation Ltd., Hozur and later in Vibha Agro Tech Ltd, Hyderabad for a period of 5 years as the GM (R&D). In both the respective leadership, the main focus was in grain, oilseeds and vegetable crops and in seed production in large scale. He was invited to be consultant Geneticist in Agroaxon SA, Almyros, Greece dealing on seeds and seed plants and to organize research on cotton for High Density Planting and too processing in tomato. The Nucleus and Breeder Seed is prime important in crop production and is his special interest. The Nucleus and Breeder seed is the credibility on the quality of seeds. Devotion, dedication and active involvement in true to type generated pure Nucleus and Breeder seed production are the goals. The Nucleus and Breeder seed are developed in a wide range of crops that are diverse in their genetic and breeding modes. To recognize his contribution he was awarded with Fellow of National Academy of Agricultural Sciences and Fellow of Indian Society of Genetics and Plant Breeding. He has been awarded with 13 academic awards including a best researcher award.

Dr. S. Manonmani graduated B.Sc., (Agri.) in 1989, M.Sc., (Agri.) and Ph.D. in Plant Breeding and Genetics in 1991 and 2000, respectively from Agricultural College and Research Institute, Tamil Nadu Agricultural University, Coimbatore. She was awarded with Senior Research Fellowship for doctoral research. She joined as Assistant Professor in TNAU during 1995. She worked in rice breeding from 2000 to 2017. Seventeen years of rice research resulted in the development of two rice hybrids (CORH 3 and CORH 4) and seven high yielding rice varieties. She attended training programme on "Marker Assisted Back Cross breeding" in 1RR1 Philippines. She has got 25 years of experience in teaching and handled 10 courses in Plant Breeding and Genetics for post graduate students. She is involved in the development and characterization of herbicide tolerant rice (Imazethapyr) through EMS mutagenesis. She moved to Hybrid Rice Evaluation Centre, Gudalur and worked for four and half years. During that period she developed Temperature sensitive Genie Male Sterile (TGMS) lines in rice. She also developed many parental lines, out of which six parental lines were registered at NPBGR, New Delhi. Presently she is working as Professor and Head, Department of Rice, Centre for Plant Breeding and Genetics, Tamil Nadu Agricultural University, Coimbatore.

Seed Production in Field and Horticulture Crops Nucleus and Breeder

S.R. Sree Rangasamy
Former Director
School of Genetics, Centre for Plant Molecular Biology
Professor, (Retd)
Tamil Nadu Agricultural University
Coimbatore, Tamil Nadu

S. Manonmani
Professor and Head
Department of Rice
Centre for Plant Breeding and Genetics
Tamil Nadu Agricultural University
Coimbatore, Tamil Nadu

NIPA® GENX ELECTRONIC RESOURCES & SOLUTIONS P. LTD.
New Delhi-110 034

NIPA® GENX ELECTRONIC RESOURCES & SOLUTIONS P. LTD.

101,103, Vikas Surya Plaza, CU Block
L.S.C. Market, Pitam Pura, New Delhi-110 034
Ph : +91 11 27341616, 27341717, 27341718
E-mail: newindiapublishingagency@gmail.com
www: www.nipabooks.com

For customer assistance, please contact
Phone: + 91-11-27 34 17 17
Fax: + 91-11-27 34 16 16
E-Mail: feedbacks@nipabooks.com

ISBN: 978-93-91383-03-9

Composed and Designed by NIPA®.

Acknowledgement

It is with great pleasure and appreciation, I acknowledge the immense inspiration, help and logistics kindly extended by Prof. Dr. (Mrs.) S. Manonmani, Prof. and Head, Department of Rice, TNAU, in the day to day progress and in shaping this Book. I am happy to mention the timely help and useful discussions by Professors Dr. R. Ravikesavan, Dr. N. Meenakshiganesan and Dr. (Mrs.) P. Jayamani. The subject matter Nucleus and Breeder Seed Production in a range of diverse crops is many sided. Prof. Dr. (Mrs.) R. Gnanam and Prof. Dr. Senthil Natesan have provided inputs respectively on the current research in Plant Cell Technology and in vitro clones and application of molecular markers in genetic purity testing. I am thankful to them. Prof. Dr. K. Rajamani helped me with details on medicinal tuber crops and I acknowledge his help. I have received frequent help and facilities from the Center for Plant Breeding and Genetics, TNAU. In this regard, I thank Dr. K. Ganesamurthy, the former Director and Prof. Dr. (Mrs.) S.Geetha, the present Director of the Center for Plant Breeding and Genetics. I express my fond appreciations to my wife Mrs. S. Dhanalakshmi for her encouragement and support.

Preface

Seed is associated with human civilization and culture and played important role in the successful crop production and sustaining prosperous agriculture. The philosophy based on the virtues of the seed has eulogized the seed as the important initial capital and the germinating seed as the real indicator of the success of the crop and the quality seed itself ensuring half the success of the crop. Seed is a living thing. Maintenance of genetic characteristics in self pollinated and cross pollinated crops through seeds demand well defined procedures. The quality assured seeds are the resultant products of breeding improved crop varieties and clonal crops and helped the mankind from food scarcity to achieve food security and added nutritional security as well. Quality assured seeds of the improved crop helped to defend environmental and climate challenges and biotic and abiotic stresses.

Improved varieties and the parental lines need to be faithfully perpetuated through generations in a true to type and truesome manner empowering the genetic potential intact by crop appropriate and specific propagules such as seeds, vegetative propagules and clones.

In majority of the crops, the seed is the main source of propagation from generation to generation and it is essential to maintain the trueness of characteristics and the genetic potential and purity of the original parent which is the primary source of seed. This important step necessitates the maintenance breeding procedures prior to the nucleus seed constitution stage to be followed by breeder seed production. Fidelity in clones and micropropagated plants and finger printing are included in adjudging the true to type selection and maintenance.

Quality assured nucleus and breeder seed system backstop and ensure the yield and stability and food and nutritional security, protein, oil and micronutrient contents and other qualities. These concerns are addressed while formatting protocols for constituting the nucleus and breeder seed. Innovative methods of evaluation could be introduced and adapted with the technological advancements.

In this Book on quality assured Nucleus and Breeder Seed Production in Field and Horticultural Crops, systematic organization and execution of the nucleus and breeder seed production protocols and procedures have been presented objectively in crop specific manner. The crops included are different grain, pulses, fibers and oilseeds crops, vegetable crops, fodder crops, green manure crops, fruit crops and few medicinal plants. The utilization of seed, clones, micropropagation, artificial seeds, grafts and *in vitro* plants and the application of DNA technologies in assessing the genetic purity fingerprinting, fidelity, uniformity, molecular GOT of the varieties and clones have been presented.

It is hoped that this Book on Nucleus and Breeder Seed Production of different categories of crops such as field and horticultural crops will be an useful source book to the breeders, academics, researchers, seed producers, research students and to help them to expand their innovativeness conjoining the genetics, breeding and molecular breeding perspectives to produce quality assured, true to type seeds. It is essential to retain/ maintain the prime quality of genetic potential of the original parent seed with dedicated pursuit which is the primary source seed material. It need to be ensured that the mean yield potential/ performance is maintained necessitating the introduction of systematic maintenance breeding procedure prior to nucleus seed constitution. While constituting nucleus and breeder seed, innovative and rapid methods for maintenance of nutritional qualities and evaluation could be introduced with the technological advancements like non destructive and other methods.

In this Book on Nucleus and Breeder Seed production in field, vegetable and other clonal crops, systematic organization of the seed chain and protocols of producing Nucleus to Breeders seed generations are presented in a comprehensive manner. This necessitated to organize and format the nucleus and breeder seed production of each crop, either by seed or by clonal mode by presenting a brief account of botany, floral architecture, pollination and reproduction mode, extent of natural cross pollination and other related aspects of genetics and breeding domains enabling the design and appropriate protocol development in maintenance breeding, nucleus and breeder seed production. At this juncture I recall the requests by my young breeder colleagues of the Private Seed R&D company to prepare a book on the merits, genetic purity and procedure for Nucleus and Breeder Seed Production which initiated to write this book.

It is my cherished hope that this book inclusive of all categories of crop plants that are either seed or clonally propagated will be an useful source book to the researchers, students and others involved in quality assured seed production. This book with help to promote or expand the seed innovations like futuristic seeds such as in vitro clones, micropropagation, artificial/synthetic seeds and organic seeds conjoining with the upcoming genetics, breeding and molecular breeding perspectives related to the Nucleus and Breeder Seed production in annual and perennial inclusive of grain, legume pulses, fibers, oil, vegetables, fruit and other horticultural crops seed and clonal crops.

Authors

Contents

Acknowledgement .... *vii*
Preface .... *ix*

1. Introduction 1
2. Importance of Seeds in Agriculture 11
3. Genetic Purity of Parents and Natural Cross Pollination 15
4. Clonal Seeds 25
5. *In vitro* Grafting 41
6. *In vitro* Clones in Plants 43
7. Organic Seed Production 59
8. Synthetic Seeds 65
9. Nucleus and Breeder Seed Production 69
10. Maintenance Breeding 83
11. Molecular Markers for Seed and Clonal Purity and Grow Out Testing (Molecular GOT) 87
12. Nucleus and Breeder Seed Production of Field Crops 105

Cereals and Millets

1. Barley (*Hordeum vulgare* L.) 105
2. Maize (*Zea mays* L.) 108
3. Oats (*Avena sativa* L.) 112
4. Pearl Millet (*Pennisetum glaucum* L.) 115
5. Rice (*Oryza sativa* L.) 122
6. Sorghum (*Sorghum bicolour* (L.) Moench) 142
7. Wheat (*Triticum aestivum* L. em. Thell) 148

Small Millets

8. Barnyard Millet (*Echinochloa frumentacea* Roxb. (Linn) 155
9. Finger Millet (*Eleusine corocana* (L.) Gaertn) 156

10. Foxtail Millet (*Setaria italica* Beauv) 158
11. Kodo Millet (*Paspalum scrobiculatum* L.) 159
12. Little Millet (*Panicum sumatranse* Roth ex Roem.Schult) 160
13. Proso Millet (*Panicum miliaceum* L.) 161

Pulses

14. Black Gram (*Vigna mungo* (L.) Hepper) 163
Green Gram (*Vigna radiata* (L.) Wilczek) 163
15. Broad Bean (*Vicia faba* L.) 166
16. Chickpea (*Cicer arietinum* L.) 168
17. Cowpea (*Cajanus cajan* L. Mill sp.) 171
18. Field Pea/Garden Pea (*Pisum sativum* L.) 175
19. French Bean (*Phaseolus vulgaris* L.) 178
20. Guar (*Cyamopsis tetragonoloba* L.) 181
21. Horse Gram (*Macrotyloma uniflorum* Lam.Verd) 184
22. Indian Bean (*LabLab purpureus* (L.) 186
23. Lentil (*Lens culinaris* Medik.) 190
24. Lima Bean (*Phaseolus lunatus* L.) 192
25. Pigeon Pea (*Cajanus cajan* L. Mill sp.) 193
26. Soybean (*Glycine max* (L.) Merrill) 206

Oilseeds

27. Castor (*Ricinus communis* L.) 208
28. Linseed (*Linum usitatissimum* L.) 215
29. Mustard (*Brassica nigra* L.) 217
30. Niger (*Guizotia abyssinica* L.) 222
31. Peanut/Groundnut (*Arachis hypogea* L.) 225
32. Safflower (*Carthamus tinctorius* L.) 230
33. Sesamum (*Sesamum indicum* L.) 234
34. Sunflower (*Helianthus annuus* L.) 238

Fibre Crops

35. Cotton (*Gossypium* sp.) 245
36. Jute (*Corchorus capsularis* L.) 255

37. Kapok (*Ceiba pentandra* L. Gaertn) 258

38. Mesta (*Hibiscus cannabinus* L. and *Hibiscus sabdariffa* L.) 260

39. Tobacco (*Nicotiana tabacum* L) 262

Green Manure Crops

40. Daincha *(Sesbania aculeata* L.) 266

41. Sesbania (*Sesbania speciosa* L.) 266

42. Stem nodulating Thailand Agathi (*Sesbania rostrata* L. *Bremek. & Oberm*) 268

43. Sunn hemp (*Crotalaria juncea* L.) 270

Forage Crops

44. Berseem (*Trifolium alexandrinum* L.) 273

45. Guinea Grass (*Panicum maximum Jacq.*) 276

46. Lucerne (*Medicago sativa* L.) 278

Sugar Crops

47. Sugarcane (*Saccharum officianarum* L.) 281

13. Nucleus and Breeder Seed Production in Horticultural Crops 287

Vegetables

48. Agathi (*Sesbania grandiflora* L.) 287

49. Amaranth (*Amaranthus* sp. L.) 288

50. Ashgourd (*Benincasa hispida*) (Thunb.) Cogn.) 293

51. Bathua Keerai (*Chenopodium album* L.) 294

52. Beet Root (*Beta vulgaris* ssp.*vulgaris* L.) 299

53. Bitter Gourd (*Momordica charantia.* L.) 301

54. Bottle Gourd (*Lagenaria siceraria* (Mol.) Standl. 305

55. Brinjal (*Solanum melongena* L.) 307

56. Cabbage (*Brassica oleracea* L.var. *capitata*) 310

57. Carrot (*Daucus carota* L. subsp. *sativus*) 315

58. Cauliflower (*Brassica oleracea* L. var *botrytis* L.) 318

59. Chilli Hot Pepper (*Capsicum annuum* L.) 322
60. Chow Chow (*Sechium edule* (Jacq) SW.) 327
61. Coleus/Chinese Potato (*Plectranthus rotundifolius* (Poir.) Spreng) 330
62. Coriander (*Coriandrum sativum* L.) 333
63. Cucumber (*Cucumis sativus* L.) 335
64. Curry Leaf (*Murraya koenigii* (L.) Sprengel) 338
65. Drumstick (*Moringa oleifera* Lam) 340
66. Elephant Yam (*Amorphophallus campanulatus* Roxb.) 344
67. Greater Yam (*Dioscorea alata* L.) 345
68. Ivy Gourd (*Coccinea grandis* (L) Voigt. (C. indica L) 346
69. Knol Khol (*Brassica oleracea* var.gongylode*s* L.) 347
70. Lettuce (*Lactuca sativa* L.) 350
71. Manathakkali – Makoi/Black Night (*Solanum nigrum* L.) 354
72. Mini Tuber Vegetables (*Colacasia sps.)* 357
73. Okra Bhendi (*Abelmoschus esculentus* L.) 359
74. Onion (*Allium cepa* L.) 361
75. Palak (Indian Spinach) (*Spinacea oleracea* var. *bengalensis*) 370
76. Pointed Gourd (*Trichosanthes dioica* Roxb.) 374
77. Potato (*Solanum tuberosum* L.) 376
78. Pumpkin and Squash (*Cucurbita* sp.) 382
79. Radish (*Raphanus sativus* L.) 386
80. Ridge Gourd (*Luffa acutangula* L. Roxb) 388
Sponge Gourd (*Luffa cylindrical Roem*) 388
81. Snakegourd (*Trichosanthes cucumerina* L.) 392
82. Sweet Potato (*Ipomoea batatas* (L.) Lam) 394
83. Tapioca (*Manihot esculenta* Grantz) 397
84. Taro Sembua (*Colacasia esculenta*. L.) 400
85. Tomato (*Lycopersicon esculentum* Miller) 402
86. Turnip (*Brassica campestris* var. *rapa* L.) 405

Fruits

87. Acid Lime, Sweet Orange and Mandarin Orange (*Citrus aurantifolia* Swingle) (*Citrus sinensis* (L) Osbeck), (*Citrus reticulate* Blanco) 408

88. Amla (*Phyllanthus emblica* L.) 411

89. Ber (Indian Jujube) (*Zizyphus mauritiana Lam.*) 414

90. Custard Apple (*Annona squamosa* L.) 418

91. Fig (*Ficus carica* L.) 420

92. Grapevine (*Vitis vinifera* L.) 422

93. Guava (*Psidium guajava* L.) 426

94. Jackfruit (*Artocarpus heterophyllus* Lamk) 429

95. Mango (*Mangifera indica* L.) 431

96. Mulberry (*Morus alba* L.) 435

97. Musk Melon (*Cucumis melo* L.) 438

98. Papaya (*Carica papaya* L.) 441

199. Pineapple (*Ananas comosus* L. Merr.) 449

100. Pomegranate (*Punica granatum* L.) 451

101. Sapota (*Manilkara achras* (L.) P. Royen) 453

102. Tinda (*Praecitrullus fistulosus* (Stocks) Pangale) 455

103. Watermelon (*Citrullus lanatus* (Thunb.) Matsum and Nakai) 457

Spices and Plantation Crop

104. Ajawain (*Carum copticum* [L.] Benth.et Hook.) 462

105. Anise (*Pimpinella anisum* Linn.) 464

106. Arecanut (*Areca catechu* L.) 466

107. Banana (*Musa* species L.) 470

108. Betelvine (*Piper betle* L.) 476

109. Black Pepper (*Piper nigrum* L.) 481

110. Caraway (*Carum carvi* Linn.) 483

111. Cardamom (*Elettaria cardamomum* Maton) 484

112. Cashew (*Anacardium occidentale* L.) 485

113. Cocoa (*Theobroma cacao* L.) 489

114. Coconut (*Cocosnucifera* Linn) ... 491
115. Coffee (*Coffea arabica* L. (Arabica) ... 497
116. Cumin (*Cuminum cyminum* L. Syn. *Cuminum odorum* Salisb) ... 502
117. Dill Seed Spice (*Anethum graveolens* L.) ... 504
118. Fennel (*Foeniculum vulgare* Mill.) ... 507
119. Fenugreek (*Trigonella foenum graceum* L.) ... 510
120. Garlic (*Allium sativum* L.) ... 512
121. Nigella (*Nigella sativa* L.) ... 515
122. Nutmeg (*Myristica fragrams*. Hutt.) ... 517
123. Palmyra palm (*Borassus flabellifer* L.) ... 519
124. Periwinkle or Sadhapahuli (*Catharanthus roseus* Linn) ... 523
125. Rubber (*Hevea brasiliensis* Muell) ... 526
126. Stevia (*Stevia rebaudiana* Berton) ... 529
127. Tamarind (*Tamarindus indica* L.) ... 531
128. Tea (*Camellia sinensis* (L) O.Kuntz) ... 534
129. Turmeric (*Curcuma longa* L.) ... 539
130. Zinger (*Zingiberan officinale* Rosc.) ... 542

Medicinal and Aromatic Plants

131. Aloe Vera (*Aloe barbadensis* Linn Burmf) ... 545
132. Avaram (*Cassia auriculata* L.) ... 548
133. Cassia Senna (*Cassia angustifolia* Vahl) ... 550
134. Glory Lily (*Gloriosa superba* Linn.) ... 525
135. Mentha Mint (*Mentha arvensis* L.) ... 554
136. Noni (*Morinda citrifolia* L.) ... 556

14. Long-term Conservation of Crop Varieties ... 561
15. Protection of Released Varieties ... 567

References ... 571

Subject Index ... 585

1

Introduction

India had fourfold increase in population from 0.3 billion to 1.34 billion and registered a fivefold increase in food grain production from 50 million tons to 285 million tons in 2018 (Paroda, 2020).

Agricultural production is keeping pace to meet the needs of the increasing population, which is growing at the rate of one percent, with sustained supplies of food, vegetables and fruits. The agricultural production inclusive of food grains, pulses and oilseeds is registering a steady growth of about 5% over recent years assuring food and nutritional security. The production of horticultural components such as vegetables, fruits, plantation crops and spices is ever increasing at more than 4.5% to 5.0% growth rate and supplements the nutritional, vitamin, neutroceutrical vegetables and greens to the society. Seed is the key promoter of all technologies associated and passed through seed that holds together all the characteristics present in the cultivar. Responses of all other inputs depends on the quality of the seed. Thus the seed is a dynamic and living biological instrument.

The familiar adage is that good seed showing vigorous germination is the best indicator of the promising crop and its yield. Good seed is half the way to the good yield. Seed is the basic and vital input and is the highest contributor to the yield. Quality assured seeds sustain high productivity besides assuring food and nutritional security. Seeds are the genetic powerhouse for increased yield and performance insulated with tolerance or resistance to the biotic and abiotic stresses and climate change. Seeds facilitate the genetic transfer of vital agrotechnologies, innovations, genetic fortification with micronutrients and pharmaceutical properties. Improved quality assured seeds are the prime importance for the public and private sectors doing agricultural services. The growth in the seed industry is 15 to 18%. The seed industries in India is regulated by suitable legislations and framework of protocols and practices. In the recent past inclusion of nutritive qualities, taste, micronutrient content biofortification and cooking qualities have become part of the quality concerns. Seeds effectively help to reach the benefits to the unreached.

In India, the productions of quality assured seeds of different generations such as nucleus, breeder, foundation and certified seeds has been scientifically organized with quality seed production practices, quality verification, regulation, quality assurance and control system.

Basic seeds available with the breeders or the institutions are maintained with assured best qualities by maintenance breeding. In the nucleus and breeder seeds, true to type trueness is replicated faithfully for maintaining genetic purity and seed qualities standards.

In our Nation, the Indian Seed Sector is formally and highly organized. The seed sector evolution in India has been facilitated by the following policies intervention.

1. Royal Commission of Agriculture (1925) recommending spread of improved varieties and seeds.
2. Enactment of Seed Act (1966)
3. Seed Review Team (1968)
4. National Commission on Agriculture (1972)
5. National Seed Programs (NSP) in three phases (1975-1985). (Creating State Seed Corporation, State Seed Testing laboratories and Breeder Seed Programs).
6. Seed Control Order (1983)
7. Technology Mission on Oilseeds and Pulses (TMOP)
8. Production of Seed Minikits
9. New Policy on Seed Development (1988)
10. Indian Minimum Seed Certification Standards (1988), (2013)
11. Seed Bank Scheme (2000)
12. Protection of Plant Varieties and Farmers Right Act (2001)
13. National Seed Policy (2002)
14. The Seeds Bill (2004)
15. Formulation of National Seed Plan (2005)
16. National Food Security Mission (2007)
17. National Biotech Regulatory Authority (2009)
18. Rastriya Krishi Vikas Yojna (2017)
19. Liberalising the Seed Trade.

The formal seed production system of nucleus, breeder, foundation, certified and truthful level seeds is taken care of by the State Agricultural Universities, Indian Council of Agricultural Research and National Seeds Corporation, and by private sector seed companies and their retailers. In the informal system of seed production, it is only in respect of farm/farmers saved seeds. The annual production of quality certified seed is around 40 million tons (Directorate of Economics and Statistics, GOI; Paroda 2019) of which vegetable seed is around 20 to 25% with 34% in varieties and 66% in hybrids. In terms of trade, the total of the seed industry accounts for 25000 Crore Rupees of which 25% is the share of vegetables that comprise low volume and high value vegetable seeds.

The food grain production in India in 2020-2021 is 305 million tons while the horticultural production in 2020-2021 is 329 million tons. The horticultural production started out spacing the food production since 2011-2012, thereby signaling the surging surplus (Fig. 1). The share of vegetable production to the total horticultural production is 30 to 32% (Fig. 3). The importance of surging horticultural production emphasis the need for concentrating on the production of quality assured seeds and clones. Large quantity of quality assured clonally propagated planting material of fruit trees and plantation crops such as grafts, rooted cuttings, setts, micropropagated plants, cloned true to type propagules and other kinds of planting materials are in high demand. There is an increasing needs of Nursery planting quality assured propagules in horticultural crops that remain to be solved (Chadha, 2006).

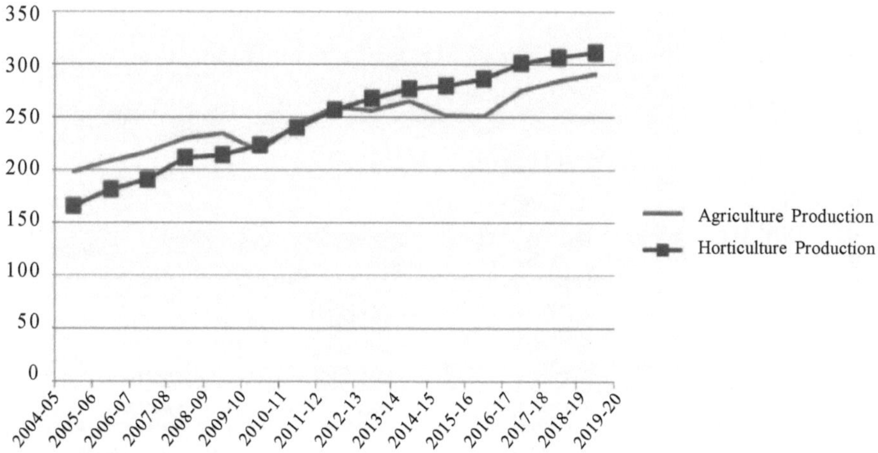

Fig. 1. Production in Agriculture and Horticulture (in Million Tons)

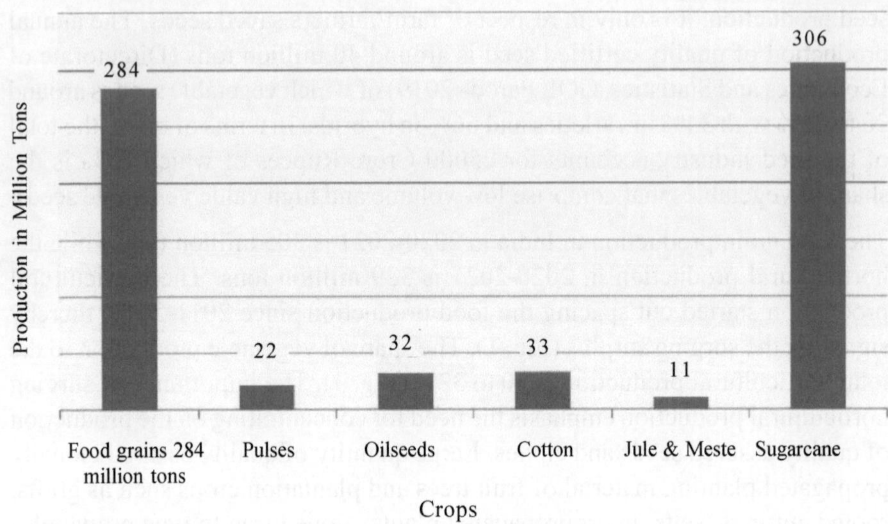

Fig. 2. Agricultural Production in India 2017-2018 (Million Tons)

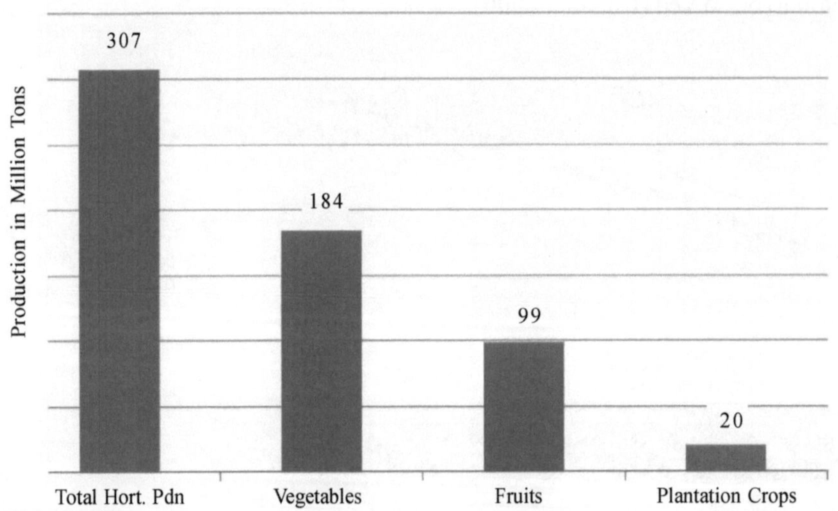

Fig. 3. Production of Vegetable, Fruits and Plantation Crops (2017-2018)

The breeder seed production has five fold increased in production in 2018-19 as compared to 2005-06 (Fig. 4).

The Tamil Nadu Agricultural University Coimbatore has established Breeder Seed Unit in 2013 in the Center for Plant Breeding and Genetics to actively monitor and look after the breeder seed production in agricultural and horticultural crops. The breeder seed production is monitored and seeds are tested for seed quality standards and physical properties.

The breeder seed production in India is about 1.75 million tons in 2019-2020. Systematically produced Nucleus Seed and Breeder Seed possess technically enhanced or superior qualities with high protein, oil content, minerals (Fe, Zn) and vitamins.

The production of quality assured 100% genetically pure and true to type seeds in Nucleus and Breeder stages requires sourcing of released varieties and clones from the respective Breeders/ Institutions concerned. The source base population is thoroughly studied and maintained ensuring genetic purity stability and true to type nature and other qualities. The NS and BS production activities respective to the crop among self pollinated inbreds, often cross pollinated, cross pollinated and clonally propagated crops are to be precisely understood and practiced and the quality assured seeds need to be produced. In many horticultural crops, this will include scions and root stocks that are compatible resulting in a perfect graft union. Focused attention of the breeder is to be bestowed on the genetic principles and protocols, maintaining genetic purity and true to type nature and conforming to the prescribed seed quality standards. In this Book, detailed production protocols and sound production practices have been critically presented for crops along with the DUS characteristics or characteristics prescribed in the guidelines by the PPV & FRA Act to guide and to facilitate the selection of the crop variety concerned while genetically pure and true to type production is being made.

Understanding of the breeding behaviour of the crop plants is very essential and to understand this, details of floral biology, anthesis and pollination and nature of incompatibilities prevailing and harnessing pollination are presented in sufficient details for each crops. Plant varieties present a set of characteristics that enable the distinction, uniformity and stability. For each crop, set of the characteristics are identified and adopted in the Protection of Plant Varieties and Farmers Right Act.

In the methodologies and protocols, the genetic structures, breeding behaviours, isolation distance needed, the size of the samples and number of plants or progenies to be carried followed, flowering and maturity, inbreeding depression if present, combination of uniformity, health vigour, assessing true to type nature in individual plants and progenies, harvesting and single plant seed collection and laboratory examination and quality determination and assurance are presented in detail. The overall effort is to maintain the minimum prescribed genetic purity and other seed standards in the seed chain of NS, BS, FS and CS. Nucleus and Breeder Seed, Foundation Seed and certified seed in respect of clonal seed, the prescribed standards of planting material is adopted.

The technological achievements in seed production utilizing different kinds of genetically controlled male sterile parent lines, application of technologies of scion and root stock grafts, clonal multiplication and in vitro grafting technologies, synthetic seeds and related seed innovations will greatly enable the fast, true to type propagules and planting material production with quality assurance.

The nucleus and breeder seed should be 100% genetic purity and true to type and necessitates the stepped up systematic maintenance breeding activity to backstop the genetic purity standards of nucleus and breeder seed. The nucleus and breeder seed production technology is of very precise protocols specific to each kind of crop such as self pollinated, cross pollinated and clonal crops and the Nucleus Seed-Breeder Seed cycle need to be periodically practised.

The genetic purity of the Quality Assured seed is further confirmed by the Grow Out Test that is prescribed for each crop. The application of DNA marker technology further facilitates rapid and fast molecular test to adjudge the genetic purity. The detailed account of protocols and procedures for seed production in field crops, vegetable crops and horticultural crops are presented. The repeatability and superior performance of yield and traits of varieties need to be maintained over generations of crop growing. Published versions on standards and production of nucleus and breeder seeds are available in seed propagated and vegetables propagated field crops. This Book further elucidates latest protocols and additions. Advances in plant cell technologies such as in vitro plants clones, synthetic seeds, micropropagtion, rapid fast cloning of true to type nature are included.

The Central Seed Certification Board in the Seed Division of the Ministry of Agriculture adds new crops and species and notified varieties are certified to fulfill the Seed certification Standards as per the guidelines of the Seed Act (1966) and the seed control order (1983).

Some of the important commercial crops such as arecanut and coconut and plantation crops like cardamom, coffee, tea, rubber that are dealt in detail in this Book, Nucelus and Breeder Seed Production have not been included in the Indian minimum Seed Certification (2013). In addition, detailed protocols of Nucleus and Breeder Seed Production are also not presented for many other field and horticultural crops. In this Book many such crops have been included with detailed protocols and procedures for Nucleus and Breeder Seed production. In view of the importance of breeder seed which assures high yield, augment the crop diversity, adds to the resistance levels to the pests, diseases of crop varieties and abiotic stresses and quality seed supplies to the niche environment areas and to manage the climatic change challenges, need to be fully reached.

The protocols of Indian Minimums Seed Certification Standards (2013) of the central Seed Certification Board. Department of Agriculture. Govt. of India has been prepared and presented for guidelines adoption by Trivedi and Gunasekharan (2013) for foundation and certified seeds. A review by Rajendra Prasad *et al.* (2017) explains the National and International Seed Quality Assurance system and strategies for seed production chain in field crops in India. They have recommended in the perspectives.

1. Need to introduce Bar/QR code for the traceability of the breeder seed in the multiplication chain for quality seed production.
2. Variety specific molecular markers, rapid genetic purity testing for nucleus and breeder seed.
3. Suggested a national datas base of crop varieties, DUS guidelines, DNA finger printing to facilitate quality breeder seed production as foremost steps. Varieties in the seed chain in India in 2015-2016 are 563 in cereals, 203 in pulses, 225 in oilseeds, 28 in forages and 46 in fibers.

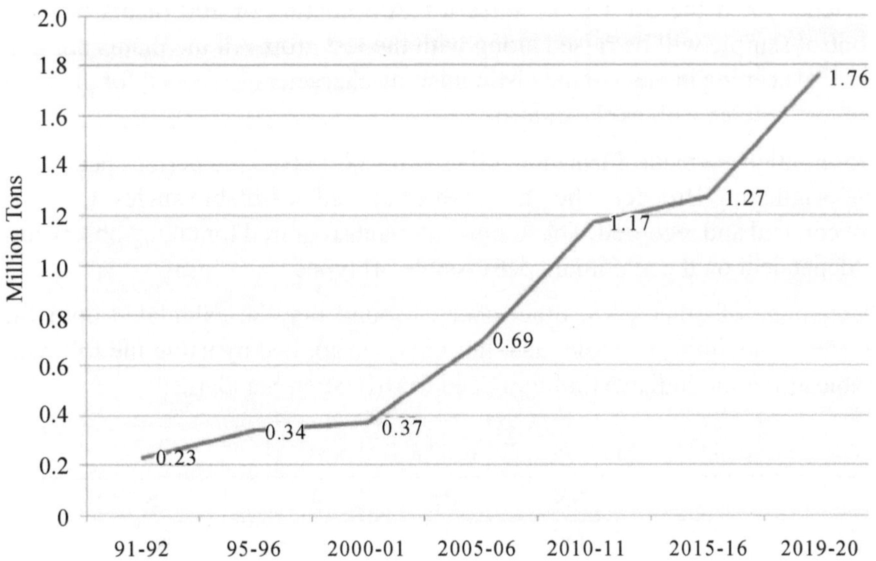

Fig. 4. Breeder Seed Production in India (Paroda, 2020)

Grow Out Test (GOT)

Basic qualification of GOT is prescribed under Seed Rules 1968. GOT is conducted to determine the genetic purity of a given Seed lot of a released variety and the extent to which the submitted sample confirm to the prescribed seed standards. The size of the seed sample for conducting the GOT is prescribed and it varies from 100 g. to 1000g. per seed sample depending upon the crop.

It is essential to sow the various samples of the same CV. in succession and standard samples are sown at suitable intervals, one standard sample for every ten samples to be tested. Row length, plant to plant distance, spacing between rows and between plots are specified for each crop for adherence while sowing the seeds.

Usually the cultivar differences are more distinct if the growing conditions are favourable so that the genetical differences express as clearly as possible. Grow out test is based on the heredity characteristics of the plants. In self fertilised crop, the individuals of the CV may be theoretically identical and it is easy to determine CV purity. In the cross pollinated species, the greater part of examination is made on the mutual comparison between the samples to be tested and the standard sample.

The seed ratio is adjusted depending on the germination percentage of individual samples and the sowing is done by dibbling. Subsequent thinning is not recommended. The test crop may be raised in the areas recommended for the varieties or in the off season nurseries. A minimum of 200 plants from the control sample will be raised along with the test crop. All the plants are to be studied keeping in view of the distinguishing characters described for the CV in the test crop as well as the control.

Observation are made during the full growing period or for a period specified by the originating Breeder. The plants are examined at suitable stages. Off types are counted and recorded. The number of plants required for taking observation is dependent on the maximum permissible off types.

Percentage of other CVs, other species found may be calculated upto first degree of decimal. The tolerance limit may be applied by using the following Table as per the Indian Minimum Seed Certification Standards.

Table. Genetic and Physical Purity in different kinds of seeds

%	NS	BS	FS	CS
Genetic Purity	100	100	97-98	98
Physical Purity	100	100	>95	>90
Minimum germination	100	100	70 to 80	60 to 85
Maximum inert matter	0	0	2	2

SEED

Seeds or propagating materials directly controlled by the originating Breeder or Institutions or in some case sponsored breeder concerned have the status of Breeder seed. The seed sources, be it a seed or propagating material are of highly diversified origin and structure with the ultimate function of reproduction.

True seeds are highly differentiated and specialized structures arise from zygotic or adventitious embryos. Seeds are classified as orthodox and unorthodox or recalcitrant, the former being capable of surviving or freezing within certain temperature limit while the later do not survive drying or freezing.

Seeds exhibit dormancy to various degrees and time duration. Seeds with non dormancy or no dormancy nature are also frequent. In some plants seeds that are just matured are germinating while being inside the fruit known as viviparous germination. In a few cases, seeds develop from nucellar tissues which are of maternal tissues. Nucellar seedlings are maternal in nature and find favour in citrus, lime and other crops to maintain the genetic uniformity either as scion source or root stock. Apomictic seeds are the other kind of seeds that develop by parthenogenesis without fertilization. Apomictic seeds are similar to sexual seeds in their structure.

Asexual propagation predominates the vegetative propagation. Some plant species have sexual and vegetative reproduction mode or both. Vegetative reproduction replicates its genetic constitution and structure faithfully leading to cloning process with the clonality as the primary feature.

Clonal seeds, bulbils, bulblets, clonal cuttings, setts, settlings, layers, air layerings and suckers are the means of vegetative propagation. Plants produce storage organs and such organs are also endowed with propagation potential. Storage organs like bulbs, corms, tubers, root stolons, root suckers, rhizomes and runners are participating in clonal propagation.

Advances in cell and tissue culture technologies intervention enables in vitro production and propagation of plants at will and the process is known as micropropagation. Micropropagated plantlets will develop into full-fledged plant which are actually clones that could be designated as in vitro clones. Ex vitro clones are another combination of propagation technologies for clonal multiplication with rooting of micro cutting using mainly green house and mist chambers/tunnels.

Somatic embryo induction techniques in vitro generates large number of true to type of differentiated embryoids of somatic cell origin. Artificial seeds or synthetic seeds are developed with somatic embryoids. Artificial seeds technology for encapsulated somatic embryoids is offering promise to propagate in primary nurseries in controlled conditions. In some fruit trees, shoot meristems are excised and encapsulated and used as propagules.

2

Importance of Seeds in Agriculture

Seed is the culture and heritage. Seed is associated as an integral component in human civilization and culture

1. Seed is the key ingredient in establishment and expansion, diversification and improvement of crops/and crop husbandry.
2. Seed is the most effective and efficient means of crop propagation.
3. Seed in small quantity is convenient to store, has longevity to transport, easy to sow, relatively free of diseases and pests.
4. Seed is the important carrier and transmitter in crop improvement
5. Seed is the repository of heritabilities of the progeny. Seed has helped to accelerate the science of Genetics.
6. Seeds are the mysterious Genetic capsules; Genetics and Genomics endowments are stored, maintained, transmitted and multiplied through seeds.
7. Seed coating or dressing of agricultural chemicals is required in small quantities for protection, regulating plant growth and development.
8. Seed is a catalyst for exploitation of agricultural technical improvements in crop production. Seeds played a pivotal role in Green Revolution.
9. Seeds are living and alive. Seeds replicate. Highly nutritious, carries nutrients and micronutrients.
10. Seeds need to maintain the propagation functions. Needs to plan for production in advance of the time and seasons to be used. Seed is not a shelf item like fertilizers or agricultural chemicals.
11. The Royal Commission of Agriculture (1925) recommended the spread of improved varieties and seeds distribution to improve food production.
12. Good quality assured seeds can contribute about 20 to 25% increase in yield over farm saved seeds/clones
13. Seed is the basic tool for Food supply, Food Security and Nutrition Security.
14. Seed is a powerful medium for rapid rehabilation after natural disasters.

15. Less infestation with weed seeds.
16. Less disease and insect problem.
17. Quality seed helps to increase the input use efficiently.
18. High produce value and marketability.
19. Sustaining high yields productivity, production and high crop quality.
20. To address critical gaps in specific problem solving strategies.
21. Seeds in pursuit of Excellence.
22. Simple ways to share benefits and/socialize genetic/breeding technologies and gains.
23. Sustainability in nutritional security.
24. Single window delivery of all virtues.
25. Seed is a wealth. Save the good seed as your treasure.
26. Seed is the first determinant of the future plant development.
27. Seed is the master key to success with cultivation.
28. It is desirable/advisable to make seed replacement once in 4 years in respect of self pollinated crops and once in 3 years in cross pollinated crops.
29. Seeds ensure future with sustainability of agricultural and horticultural production and food and nutritional security.
30. Seed is a vehicle to carry all genetic technical innovations in genomics. Helps to exploit genetic potentialities of new varieties. Seeds and clone play catalystic roles and alleviate rural poverty and help resource poor farmers.
31. Seed is Culture and Heritage and Civilisation. We nurture seeds and Seeds nurture us.
32. Good Quality seeds assure good yield.
33. Good sprouting seeds assure Good stand of the crop
34. Heirloom seeds are our heritage with sustainability.
35. Tryst with quality seeds. Seed is a means of dispersal for plant population in space (spatial) and time (temporal) representing continuity and change
36. The plant genome embedded with all virtue is present in the seed which is the single window delivery.
37. Scientific and technological information and mechanisms for students, scientists, researchers, seed producers and various stakeholders to achieve high quality assured seeds/planting materials/propagules for sustainable development.

38. Quality assured seeds provide/confer prime factor leadership in increasing and sustaining higher crop yield and in backstopping genetic purity.
39. Seed security is inclusive of quality assured seeds of appropriate crop varieties, availability and accessibility to the farmers supported by quality seed production system fulfilling seed standards and quality norms. The authentic seed security system is of paramount importance in the food and nutritional security.

3

Genetic Purity of Parents and Natural Cross Pollination

Genetic purity is the vital determinant in continuing the genetic integrity in seed from generation to generation. This is the basis of variety purity maintenance. The genetic purity depends upon the purity of the seeds planted. The genetic purity of the seeds planted must exceed or equal to the final product purity standards required as the purity generally decreases with subsequent generations of propagation. The improved varieties loose their identity and healthiness unless properly maintained. Maintenance selection can prevent the seed quality deterioration.

Natural cross pollination of a given genotype from different genotypes, genetic resources, land races, semi wild and related taxa results in genetic contamination. The genetic structure of the plant population in a variety or inbred is influenced in part by the rate of natured cross pollination/out crossing.

Genetic purity of parent lines and varieties are affected to various degrees due to the natural cross pollination influenced by the following causes.

1. Adventitious pollination/outcrossing
2. Vicinisim
3. Pollen drift
4. Absence of pollen barrier
5. Protracted or extended anthesis duration
6. Extended protogyny
7. Flowering in multitillering multi branching over long period resulting in cross pollination
8. Outcrossing due to wind drift
9. Insect flight increase in frequency
10. Insufficient isolation distance. For nucleus and breeder seed, the isolation distance need to be 1600m. or more to maintain 100% genetic purity
11. Incompatibility in different crop varieties may express some intrinsic or unknown variation that encourages outcrossing.

12. Seed production field shape like narrow and long seed production plots across the wind velocity.

Natural cross pollination due to out crossing results in adventitious contamination and could be due to the floral structure, extent of stigmatic protrusion, extended stigma receptivity and protogynous nature. Genetic causes such as self incompatibility, genic male sterility, cytoplasmic genic male sterility and photoperiod and temperature sensitive genic male sterility are the other factors that increase the natural cross pollination. The out crossing or the vicinism between varieties is in general high to very high in the genetically regulated male sterile system. Growing the parental lines across the wind velocity and direction besides in close distance proximity and the prevailing activities of pollinators such as wide range of insects and bees, aiding cross pollination are the causes leading to natural cross pollination. Nature of genotype and plant spacing, dense or sparse, wide spaced planting, narrow stripes of crop and climatic factors like wind velocity, temperature, humidity, seasonal and location specific or region specific climatic feature interplay with harnessing the pollination and the seed production system. Insect pollination is strongly directional in activity moving to the nearest neighbouring flower in the same row of plants. In some crops, extent of staminate flowers and pollen production as affected by the genotype and the prevailing pollinator and environment conditions are the major factors in the natural out crossing (Table 1).

Table 1. Natural cross pollination (%) in different crop plants

S. No.	Crop	Natural Cross Pollination (%)
1.	Rice	0.1 to 40% Virmani and Athwal (1977)
		0.0 to 4% (TNAU portal) 3.14 to 36.16% in CMS lines 15.0 to 75.0% in CMS lines
		20-40% in CMS. (Subba Rao *et al.,* 2003)
		0.6 to 1.2% in autumn season 0.1% in winter season. and 0.1 to 0.4 (Mondal, 2009)
2.	Wheat	Upto 5% Rajkumar *et al.* (2003)
3.	Barley, Oats	0.4 to 1.0 Chowboy and Roy (2003)
4.	Maize	15.82% of pollen released by each plant participated in the dispersal (Ning *et al.,* 2014). Upto 95% (Singh *et al.,* 2003)
5.	Grain sorghum	2.5% (Sekar & Sarathy, 1975)
		2 to 10% (Kanna Babu *et al.,* 2003)
6.	Ragi	Less than 1% (Seetharam *et al.,* 2003)
7.	*Panicum miliaceum*	Less than 1%
8.	*Panicum scrobiculatum*	Cleistogamous
9.	*Panicum sumatranse*	Negligible (less than 1%)
10.	*Setaria italica*	0 to 0.4% (7 to 10 TNAU : Agri Portal)

11.	Cotton	Up to 5% (Sekar and Sarathy, 1975) Adjacent row – 0.52 to 1.70% Adjacent plants – 0.00 to 0.30% Adjacent rows – 12.9 to 13.30% Adjacent Plants – 8.93 to 8.10% Natural coloured cotton 5 to 10%
12.	Jute	Over 2% (Sekar & Sarathy (1975) (Kumar *et al.*, 2003) Capsularis (1.38% to 16.0%) and Olitorious (30.3 to 17.0%)
13.	*Hibiscus sabdariffa*	0.20 to 0.68 to 15.6 (Vaidya 2000)
14.	Soybean	0.4% at 9 m. distance less than 1% (Hussain *et al.*, 2003)
	Soybean with wild sp G.soja	0.03% at 5.4 m. distance. Hussain *et al.* (2003) 0.16 to 0.19% Doesn't not exceed 1% (TNAU : Agri Portal)
15.	Safflower	5 to 40% (although basically self pollinated) upto 20% Viswanathan *et al.* (2003)
16.	Groundnut	Less than 1% Sekhar and Sarathy (1975) Upto 20% Chuni Lal *et al.* (2003)
17.	Gingelly	Upto 6.5%, 4 to 5 % average (Duhoon *et al.*, 2003)
18.	Sunflower	Highly Cross pollinated
19.	Mustard *B. juncea* Rapeseed	5 to 15% Yadava and Singh (2003)
20.	Canola *B.napus*	0.70%
21.	Castor	Upto 14% Viswanathan and Rao (2003)
22.	Cicer chickpea	Less than 1%, 1.81% 1.58% TNAU Portal). (Dua and Sewak, 2003)
23.	Pigeonpea	Upto 28% (TNAU Portal) 5 to 40% Upto 20% 9.2 to 21.4% (mean 15.5) Vishnu Dhar *et al.* (2003) Saxena *et al.* (2018)
24.	Peas	Less than 5% (Dixit and Tripathi, 2003)
25.	Lentil	Less than 1% (Dixit and Tripathi, 2003)
26.	Faba bean (*Vicia faba*)	Above 50 to 60% (Suso *et al.*, 1996)
27.	*Lathyrus sativus* grass pea	Highly cleiostogamous (TNAU)
28.	Cowpea	Pollination before flower opening (TNAU)
29.	Blackgram	
30.	Greengram	Not above 5% (Dixit and Tripathy, 2003)
31.	Horsegram	Less than 1%
32.	Linseed	0.5 to 0.9 (Kherki *et al.*, 2003)
33.	Vegetables (S.No. 34 to 45)	(T.R. Gopalakrishnan, 2007)
34.	Broccoli	95%
35.	Kahlrobi	91%
36.	Kale	83%

37.	Cabbage	73%
38.	Brussels sprout	72%
39.	*Brascica oleracea*	Highly cross pollinated 3000m. isolation distance
40.	Cauliflowers	Highly self and cross pollinated
		72% self pollinated. Winter and early cauliflowers cross pollination very high.
41.	Okra	Maximum of 42%
42.	Egg Plant – brinjal	Upto 8% (Agarwal, 1980)
		6 to 20%
		6.4%
		30-40%
		0.14 to 1.99%
		0.00 to 82.0% (average 27%) 3 to 7% (AVRDC Ann. Rep.)
		0.7 to 15%
		11.6 to 24.0 (Kanwar & Gill, 2000)
43.	Tomato	0.5 to 40% (AVRDC) 5.51% 2 to 5%
44.	Hot pepper	8 to 37% (AVRDC)
45.	Watermelon	0.8% average for rows 3 to 6 m. apart
		0.0% between plants 10m. apart (Kumar and Wehner, 2011)
46.	Tobacco	1 to 5 % (Deo Singh *et al.* (2003))
47.	Berseem	Cross pollinated (Malaviya *et al.*, 2003)
48.	Lucerne	Cross Pollinated (Yadavendra *et al.*, 2003)

Reports on Natural Cross Pollination are also available in tomato (Currence and Jenkins, 1942 and Prok 1949), radish (Crane and Mather, 1943), sesamu (Pathivana, 1949) pigeon pea and lab lab (Veerasamy *et al.,* 1973), soybean (Ahrat and Cariness, 1984), cloves (Beevi *et al.,* 1984), okra (Gibbs, 1984), Voselle (Vaidya, 2000), rice (De silvoa *et al.,* and Pusaka *et al.,* 2018) and chilli (Pitough and Sykor, 2018).

The natural cross pollination differs from variety to variety in the same crop species and between cultivated and wild relatives as in rice (Marathi and Jena 2015) (Table 2 & 3). Growing the parent line with requisite isolation distance from other varieties of the same crop species will ensure in avoiding or excluding the natural cross pollination. Ning *et al.* (2014) experimented on avoiding natural cross pollination and adventitious contamination within the specified threshold levels with proper isolation distance. They have arrived at a model with-maize crop for the prediction of gene flow frequency equal to or lower than the threshold of 1.0% or 0.1% and recommended distance from 10 m. to 49 m. for 0.1% and from 17 m. to 25 m. for 1.0% outcrossing respectively. They have inferred that only 15.8% of the total amount of pollen released by each plant in maize participated in the dispersal program. In rice, maximum threshold distance (MTD) was estimated for gene flow in rice growing areas in Southern China by Jio *et al.* (2014). They found that MTD for 0.1% natural cross pollination in varieties of rice is basically less than 5 m. in the whole region despite climatic differences in the region. Flowering time isolation and 2 m. high cloth screen barrier offered effective protection in minimizing the gene flow. They have recommended maximum distance between CVs of rice from 30 to 40 m. and between male sterile line and pollinators 40 to 150 m. At parallel planting at 0.m zone, the gene flow in male sterile lines ranged from 3.14 to 36.16%.

Table 2. Extent of percentage outcrossing in cultivated and wild relatives of Rice (Balram Marathi and K. Jana, 2015)

Cultivated Rice	%
Australia	0.44
USA	0.00 to 3.40
Malaya	0.41 to 0.48
Srilanka	3.6
Africa	0.0 to 1.10
India	0.00 to 4.30
India	0.4
India	0.06 to 1.23 (Sahadevan and Namboodre) (1963)
India	0.11
India	0.01
Wild species of Oryza	
Thailand	44.00
Srilanka	26.50 to 22.40
Thailand	30.70
Thailand	55.90
Thailand	50.69
Africa	3.50 to 19.70
India	79.00
India	Upto 20%
Taiwan	30.70

Source: Balram Marathi and K Jena 2015, Euphytica 2015 Vol 201 : 1-14

Table 3. Frequency of cultivars in rice and wild relatives of rice in different Natural cross pollination (NCP) classes

Wild progenitors and wild relatives of cultivated rice		Cultivated rice NCP%	
NCP% (Range)	**Reported status**	**NCP % (Range)**	**Reported status**
10-20	1	0-1	7
21-30	3	2-3	2
31-40	2	3-4	1
41-50	-	4-5	2
51-60	2	4.07	1
55	1		
79%	1		
(Range 3.2% to 79%)		(Range 0.01% to 4.4.30%)	

Pollinators and floral adaptations play a crucial role in maintenance of genetic purity under natural conditions. Virmani and Athwal (1977) recorded variations in floral traits of rice varieties in respect of anther length, stigma length and breadth and stigma exertions in rice varieties and their role in influencing outcrossing. Virmani *et al.* (2003) reported natural cross pollination from 0.1 to 4.0% in cultivated rice varieties and 15 to 75% out crossing in male sterile lines. Isolation distance of 40 to 100 m. or time isolation of 21 days need to be maintained to avoid cross pollination in rice. Row ratio of 1:6 for A and R lines across the wind direction and 4:2 for A and B line are found adequate for out crossing for maximum seed sett.

Suso *et al.* (1996) have critically discussed the interplay of plant pollinator role, field situation locations and climatic factors in the pollination events in Vicia faba bean. Among the factors, the ability of the genotypes to outcross is very distinct.

Seed genetic purity threshold generally reflects a balance between the quality requirements of the market and the costs required to achieve a given purity standard. Mechanisms promoting self pollination and cross pollination ensure the self pollination and cross pollination respectively in crop specific manner. However, some amount of cross pollination in self pollinated crops and some amount of self pollination in cross pollinated crops is of common occurrence. Genetically controlled floral features such as CMS, GMS, TGMS and self incompatibility further influence with the naturally occurring pollination. Variability in the natural population depends upon the extent of cross pollination prevailing in nature. Pollen barriers, live or physical, growing buffer crops, raising adequate number of border rows, time isolation, space isolation, caging and bagging are recommended to provide physical isolation and avoiding pollen contamination.

Pollen movement in muskmelon is restricted to 3 m. from the donor plant and 2 to 3 meters in cucumber (Kumar and Wehner, 2011). In watermelon, they have found near zero natural cross pollination in rows separated by 6 m. and it was 0.8% for rows 3 m. to 6 m. apart. In andromonoecious families of watermelon, Ferreria *et al.* (2002) recorded an out crossing of 65% and inbreeding coefficient of 0.41. When averaged over monoeuious and andromonoecious families, the out crossing was 77%. In cucumber families, Wehner and Jenking (1985) recorded the natural out crossing rate of 36% (29 to 43%) in crosses, 17% (0-42%) in sibs and 47% (23 to 77) in self pollination. Thus 64% of pollinations were self or sib but not cross pollination among families. Watermelon and cucumber are similar in growth and sex expression and have mixed mating system. Self pollination can occur both in monoecious and andromonoecious populations (Kumar and Wehner, 2011).

Kanwar and Gill (2000) studied the natural cross pollination in egg plant in three different planting designs such as honey comb, alternate rows and concentric circle (Fig. 1) and found mean out crossing percentage of 13.5, 11.6 and 11.8% respectively with self fertilizing flowers whereas the natural cross pollination ranged from 18 to 24% with emasculated flowers. Natural cross pollinations decreased as the distance from the pollinator is increased. Chowdhuwry and Choomsai (1970) recorded the natural cross pollination percentage of 1.99, 1.47 and 1.36% respectively at the distance of 0.60,1.80 and 2.40m. from the pollinator. Natural cross pollination decreased at a distance and decreased as the distance is increased with a sharp cut off point. Adventitious contamination of the products below a specified threshold level is thus ensured with isolation distances specific for each crop species.

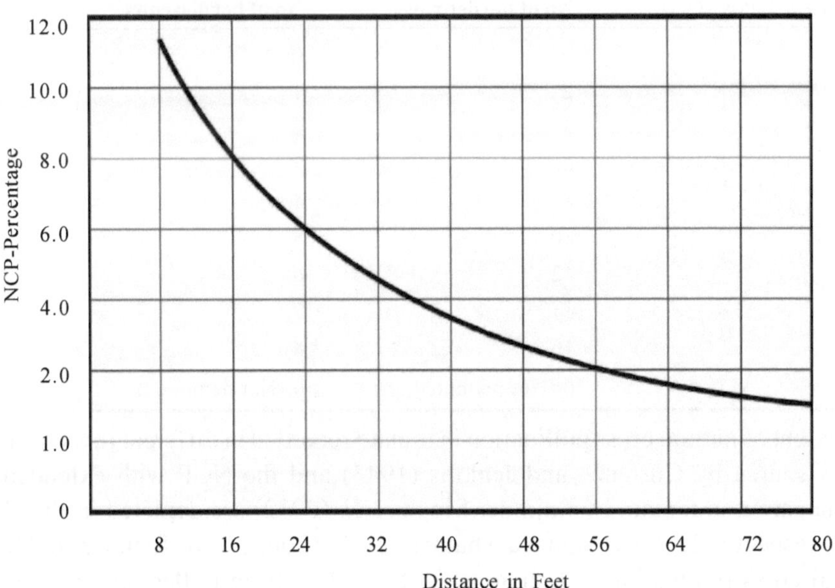

Fig. 1. Natural cross pollination (NCP) in Brinjal planted in concentric circle design based on data from J.S.Kanwar and B.S. Gill (2000)

Self pollinating crops like rice requires small isolation distance of 5 to 10 meters that would also primarily prevent mechanical mixtures during harvesting. Cross pollinating species require upto 1600 m. or sometimes more of isolation distance to prevent cross pollination from plants of the same species to prevent outcrossing depending upon the flower structure at male and female phase maturity, pollen production quantity, duration of pollen viability and the type of seed class such as Nucleus seed, Breeder seed, Foundation seed and Certified seeds being produced.

In general, large enough population is needed to maintain genetic composition and resiliencing. Too small a population will result in variety that no longer perform as well. Population size and isolation distance are functions of how a crop pollination itself. Border rows around the field that are not harvested with the seed crop can be grown to intercept pollen and to reduce contamination of the remainder of the field.

Minimum number of border rows required for isolation for hybrid corn seed production have been prescribed by the Association of Official Seed, USA (Table 4).

Table 4. Minimum number of border rows for isolation for hybrid corn seed production.

Minimum distance from other corn fields (feet)	Field size 1 to 20 acres: No of border rows	Field size 20 acres or more: No of border rows
410	0	0
370	2	1
330	4	2
290	6	3
245	8	4
205	10	5
165	12	6
125	14	7
85	16	8
0	0 (Not permitted)	0 (Not permitted)

The extent of natural cross pollination in tomato recorded in different rows from pollen source by Currence and Jenkins (1943) and the NCP with extended stigma parent in tomato by Rajinder Singh *et al.* (2012) are depicted in Fig. 2 and 3 respectively. Veeraraghava Thatham *et al.* (2005) have found 2 to 3% natural cross pollination in tomato at 25 to 30 feet from pollen source. The relationship between NCP and isolation distance in maize reported by Bateman (1947a) and the natural cross pollination by Trought (1930) in Egyptian cotton are presented in Fig 4 and 5 respectively.

Allogamy or cross pollination occurring in natural conditions (natural cross pollination) may be due to wind pollination (anemophily), water (hydrophily), insects (entomophily) and due to animal (zoophily).

The extent of natural cross pollination in tomato, cotton, brinjal and maize are presented in fig 1 to 5. In cross pollinated crops, natural self pollination occurs to the extent of 5 to 10%.

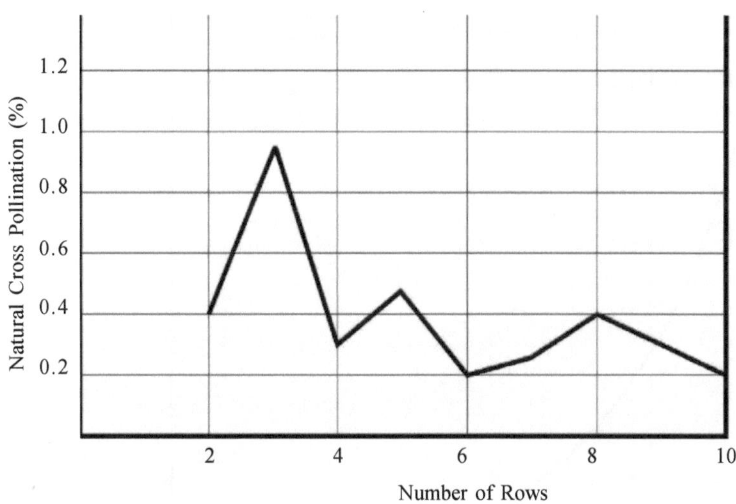

Fig. 2. Natural Cross Pollination and Number of Rows (6 feet apart) from pollen source in tomato
Source: Currence and Jenkins (1942)

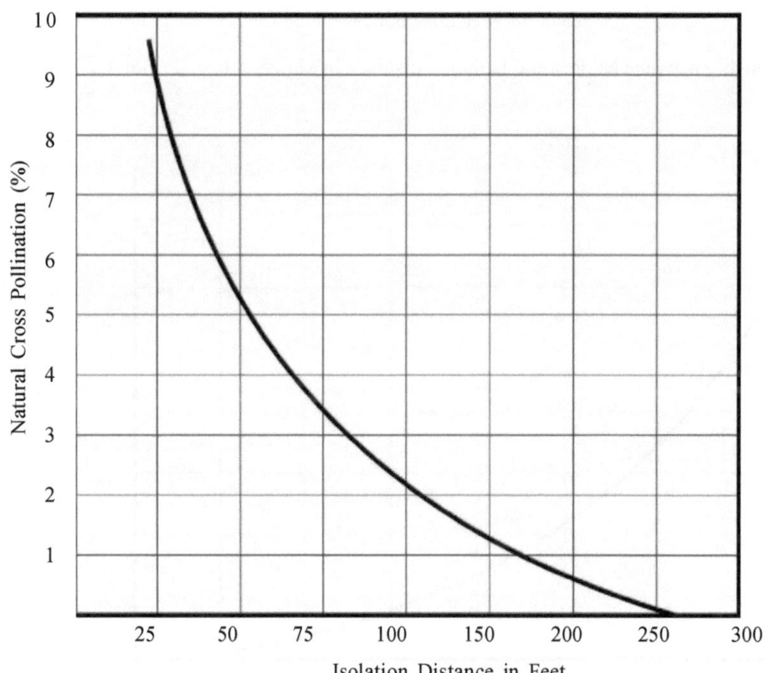

Fig. 3. Per cent of cross pollination in tomato in different isolation distance. Data obtained with exerted stigma type pistillate parent and cut flower type as pollinator parent.
Source: Rajendra Singh *et al.* 2012.

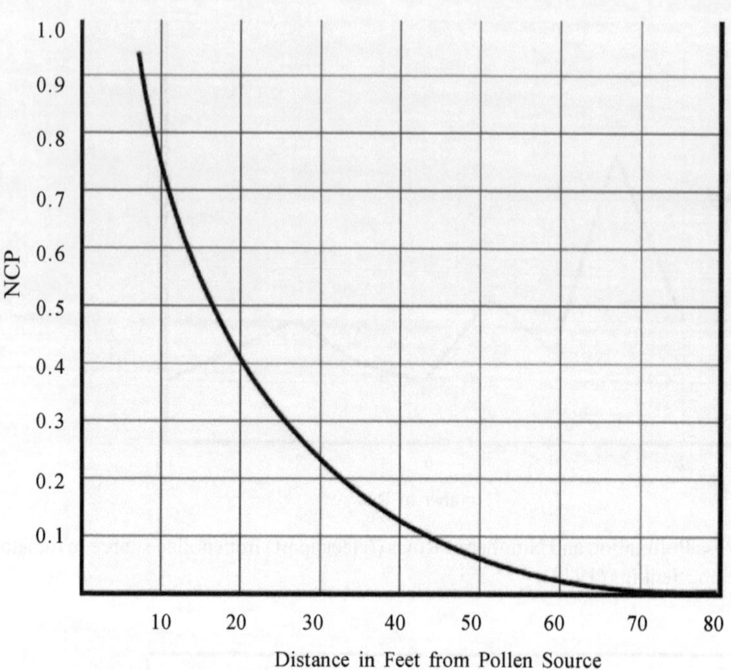

Fig. 4. Relationship between NCP % and Isolation distance in Maize, Adapted from A.J. Batemen (1947)

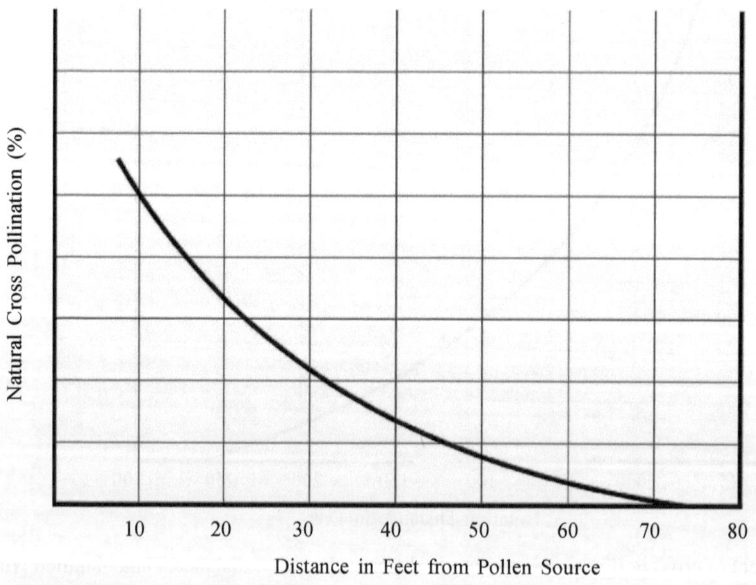

Fig. 5. Natural Cross Pollination in Cotton (Trought 1930)

In the flower architecture and sex expression, the cross pollination is enabled due to the conditions such as unisexuality, monoecy (maize), dioecy (papaya), dichogamy (anther and stigma of bisexual flower maturing at different time, protogyny (bajra), protandry (maize, cotton), heterostyly (different length of stigma and anther as in linseed), heterogamy (presence of physical barriers or mechanical obstacles between anther and stigma (lucerne waxy film on stigma), self incompatibility, male sterility, and cytoplasmic genic male sterility and temperature genetic mate sterility (TGMS). Often cross pollinated crops show 5% to 50% cross pollination.

Natural cross pollination (NCP): In mathematical terms the amount of crossing follows geometric curve and theoretically never approaches Zero at larger distance but seed growers are interested in biological significance. In point of fact, most pollinators do not fly more than 400m and the amount of cross pollination does approach Zero within 800 to 1600m distance (Fig. 6).

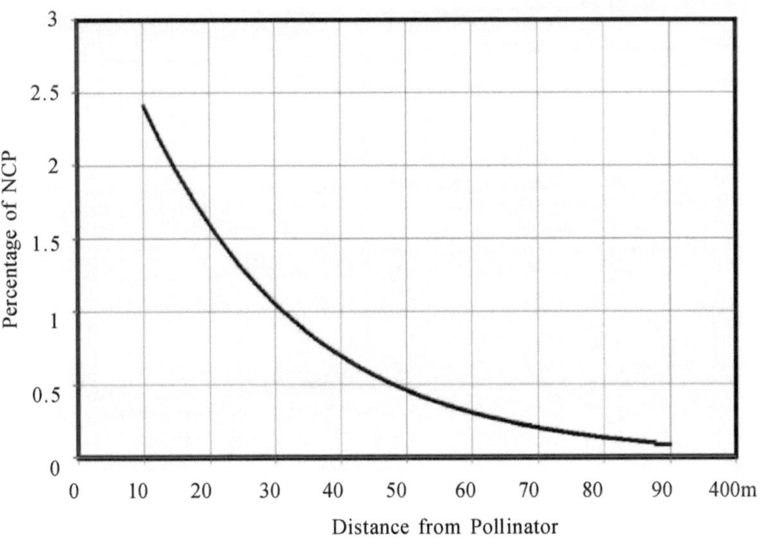

Fig. 6. Extent of Natural Cross pollination (NCP)

General Relationship between cross pollination and distance from pollinating source. Extent of natural cross pollination in mathematical term, the amount of NCP follows geometric curve (Jeffrey. H. Mc. Cormick (2004).

4

Clonal Seeds

Clonal crops belong to at least 34 families including herbs, shrubs, trees and vines. Clonal crop improvement dates back to 10,000 years ago. The diversity of clonal crops and their wild relatives are very significant but not yet fully understood and exploited. Many crops represent combination of sexual and clonal reproduction. Different parts of the clonal crops which may or may not be the same as consumed parts have been shaped as clonal propagules such as stem, setts, tuber, rhizome, corms and bulbs among others.

Clonal propagation helps in fixation of agronomically valuable genotypes, true to type propagules and prevention of sexual reproduction to avoid biparental inbreeding and inbreeding depression. Clonal preservation is practiced with very specific biochemical composition as in hops, grapevines and mangoes and easy propagation of favourable mutations and useful somatic mutations in meristems. In citrus, more than 1600 bud mutations have been recorded and propagated. Maintenance of plant genetic resources, control of gene flow from wild species to the cultivated species and ease of propagation are the other advantages in clonal propagation. Pit falls are : loss of some components of diversity, continuance of deleterious mutations, competition of plant parts as propagules and accumulation and build up of pathogens, viruses and pests. Clonality (clonal) and sexual reproduction interaction shapes the evolutionary dynamics in plants (Doyle Mackey *et al.,* 2010).

Clonal seeds are asexually propagated propagules, usually of any one of the vegetative organs or parts such as air layers, leaf cuttings, stem cultings, root suckers, tubers, corms, rhizomes, stolons and grafts of different kind. Apomictic seeds are yet another clonal seed source in Agriculture (Spillane *et al.,* 2001). Of late, micropropagated plantlets (*in vitro* plantlets) originating from explants of vegetative origin/organs are finding increasing use as clonal seeds. Tissue culture produced somatic embryo or its encapsulated form and synthetic seeds or artificial seeds are the useful additions to the clonal seeds.

There is a marked shift to the use of clonal seeds and area under clonal crop is increasing and grown with quality planting material and focused on adoption of disease and pest free material which encourages seed replacement rates. The rapid technological advancements *in vitro* techniques and diagnostic capabilities of disease and pests in clones further facilitate the upscaling of disease free micropropagules combined approach of *in vitro* and *ex vitro* methods now in vogue.

The advantage in clonal crop is its cloning of any selected superior genotype at different breeding stages and selection system without undergoing sexual reproduction. The sexual reproduction is circumvented and cloning faithfully replicates and reproduces genetically similar homogenous and uniform individuals with high genetic purity. The standard of clonal seeds should be equal to that of sexually reproduced seeds in terms of genetical, physiological and phytosanitary qualities besides their ability in field standards such as field establishment. The stability for the clonal seeds for the genetic fidelity is of paramount importance.

Clonal Seeds: Clonal seeds are true to type, homogenous uniform and ready to plant propagules in commercial planting. Clonal crop breeding results in improved and genetically enhanced clones. Ease of vegetative propagation like stem cuttings, root suckers, bud grafting, air layering and many other vegetative propagation methods are adopted in fruits trees and forest trees besides in many field crops. Consequences of inbreeding depression, natural cross pollination, self and cross incompatibilities, other kinds of pollination and stigma-pollen adverse interfaces, polyploidy and chromosomal instability, male and female sterility, poor seed production, high degree of seed inviability and high seed sterility are not experienced. In many perennial heterozygous trees, vegetative propagation maintains the identity and genetic fidelity. Vegetative propagation reduces the juvenile period and thereby resulting in earliness than usual pre flowering or bearing period. The Indian Minimum Seed Certification Standards shall include both true seeds and vegetative propagules as seed. Innovative approaches like Plant Tissue Culture propagules is used for propagation.

Clonal Seed Perspective and Application

1. Promote technology for commercialization in relevant ecological areas/ agro ecoproduction
2. Develop facilities for aeroponics, hydroponics, glass houses, mist chambers and hardening facilities.
3. Cost effectiveness
4. Efficient production distribution or delivery system to the farmers/ farm fields.
5. Reducing the cost of traditional bulk quality planting material as its share in the main crop yield is very high. The cost is highly reduced by micro-prapagated plants.

6. Sound technology for freeing the planting material and cleaning from virus, bacterial, fungal and insect contamination and infection.
7. Introducing stages of clonal multiplication in primary and secondary nurseries.
8. In case of micropropagation, utilizing facilities such as hardening, screen house, glass house, multiplication, sand beds and rooting beds in successive generations to make up required volumes of basic materials. Ensuring the clonal seed standards and norms and clonal health standards are prerequisites in this clonal seed system.
9. In clonal seeds, edible portions are used as seed material (potato, taro, yam). Other plant parts are used in banana, cassava and sweet potato. Markets are not the usual sources to get clonal seed material for planting. Sourcing clonal seed materials is more expensive than in crops like corn (Ref. clonal seed crop. Global science pdf.)
10. In clonal seeds, phytosanitary problems are transmitted by vegetative propagation. Access to improved crops in clonal crops is incipient. There is lack of systematic and intensified efforts in cultivar development combined with clean planting materials development to specific conditions for important agricultural production systems. A clonal crop system consists of all individuals, organizations and institutions which are linked to the genetic material, the quality and quantity and supply of clean planting material is to be necessarily initiated.
11. The clonal seed quality depends upon the genetic purity of heritable traits, physiological, germination, phytosanitary and sanitary aspects. Culling diseased plants, insect infested plants, off-types and deviant plants are important steps. Use of molecular markers to verify genetic fidelity, uniformity and purity is advocated. Clonal crops show low multiplication rate and large volume of seed material needed for planting. Most clonal seeds are edible material. Farmers cannot access that much planting material in village markets and transportation from source nursery is expensive and requires management.
12. Clonal seeds should have 100% varietal authenticity, genetic purity, zero pest and disease tolerance, freeing of virus, hitech multiplication methods and technologies, pest and pathogen free production fields, production in non traditional environment free of pests and diseases.
13. Variation may preexists in natural population. It may develop as a result of micropropagation (Forneck, 2005; Grunberg *et al.,* 2006). Genetic variation will arise in clonal or vegetative propagation due to multiplication of small mutations eg.in grapevine, clonal variation, retro-transpositions, chimerism and epigenetic changes (Forneck, 2005). These changes can be confirmed

by using molecular markers system. Most perennials combine sexual reproduction with some form of clonal propagation. The balance between these forms of reproduction affects propagule size, establishment dynamics, and transmission of genetic variation. Clonality occurs by stolons, root tubers, rhizomes, vegetative sprouting, corms, tubers and axis splitting.

14. Vegetative propagation helps to control growth phases to overcome juvenility to shorten the time for flowering, to rejuvenate the plantings and by grafting in combining in one plant with the superior genotypes. Zimmerman (1972) has reviewed, the juvenility and flowering in plants.
15. Direct in vitro regeneration of disease free planting material through tissue culture requires three crop seasons in the nursery to produce sufficient quantity of rhizome for planting. Microrhizome induction technique for large scale propagation material is now a reality. Ancymid is finding use to induce microrhizome (1.0 or 0.5 mg). It is a low cost technique avoiding gelling agent (Archana *et al.,* 2013).
16. Applications of high thoroughput technologies of in vitro system, micropropogation, micro and minituber, micro and minirhizomes can facilitate faster scaling up of disease free propagules. There is scope to integrate *in vitro* system with *ex vitro* system wherever possible to make the seed multiplication rate highly appreciable and commercialising.
17. There is a realistic scope to develop/formulate crop specific Clonal Regeneration System and to reduce the cost of usually adopted bulb vegetative propagules. On a global perspective most important clonally propagated crops are potato, cassava, sweet potato. Dioscorea yam, taro (*Colocasia esculenta*), sugarcane, banana, citrus fruits, grapes, apple and strawberry. In India, potato, tapioca, sweet potato, banana, colocasia, yams and sugarcane are most important clonal crops (FAO Stat., 2006).
18. Bisognin (2011) has published detailed procedures of breeding vegetatively propagated crops.

Precleaning the clonal seeds

1. Use of virus free meristem
2. Cryotherapy of shoot tips form efficient elimination of virus
3. Low cost Plant Tissue Culture protocols, production of quality planting material and laboratory facilities
4. Screening and selecting proper planting materials
5. Molecular tools to detect presence of virus
6. High throughput production of quality planting material

7. Quality seed multiplication technology in protected conditions
8. Model delivery system

Clean Seed System: The following steps in clean clonal system are followed in I.I.T.A. to clonal breeder seed production in Tapioca free from CMD and CBCD.

1. Clean seed production site
2. Clean and virus free seed
3. Clean seed indexing and sharing virus indexed germplasm.
4. Clean seed program through tissue culture production of high quality planting material.
5. Clean seed standard certification

QUALITY CLONES IN FRUIT TREES

Protocols and standards of vegetative propagation in Fruit trees: Continuous and increased demand for horticultural production necessitates increased production from fruit crops mainly by increased productivity level. This can be achieved by stepped up activity in (i) identifying and developing quality and superior mother trees (ii) producing and planting quality planting material that are healthy superior in yield and free from diseases and pests and (iii) producing quality fruits that are superior in nutritional quality (Chadha, 2006).

Fruit crop being mostly perennial in nature demand utmost care in selection of varieties, quality planting material and adoption of mass multiplication of planting material. There is acute shortage of quality planting material in the country and lack of proper mechanism of dissemination of production technology of mass multiplication of disease free, healthy and quality material in fruit crops. Most of the planting material made available are poor genetic value and health (Gorakh Singh, 2012; Chadha, 2014). Only 35 to 40% of the demand of quality planting material in India is being met and vegetative propagation has immense potential in ensuring true to type and clean planting material and in enhancing fruit production.

Mother Tree: The quality of mother tree is usually determined by the tree caliper (diameter of the trunk stem), number of feathers (branches) and angle of the feathers. The number of feathers on the tree is generally thought to complete with vigour (Gorakh Singh, 2012). High variations in the cultivar and degeneration of varieties are observed. Therefore it is necessary to select the best tree or tree of outstanding merit within each selected variety and using them as mother trees for future multiplication. The mother trees must have been tested for the performance over a number of years and should be free from transmittable diseases and be in a healthy condition. The fruit shape, size and quality must conform to the typical specifications of the variety. High yielding clones from the different locations should be selected and assembled in new Mother Block. The selected plants are planted in closer spacing in order to get continuous supply of scions and these plants are pruned periodically to keep them in vegetative phase and to produce enough shoots for supply of scions. Adequate care is taken to maintain the mother trees in healthy condition and free from pests and diseases. It is necessary to keep proper records of the individual trees in respect of fruiting, yield, quality and regularity in bearing. The mother tree records are maintained.

Clonal Root Stocks: Clonal root stocks are genetically identical clones of the parent root stock and they exhibit desirable characteristics such as adaptation to problem soil, problem water, drought of varying degree, tolerance to winter and cold, high temperature, tolerant to ill drained soil situations and heavy and sandy soils and resistance to pests, diseases, nematodes and wilt disease. Root stocks control and influence the ultimate tree size. It is a known fact that the interaction and influence of root stock on the growth, development, health, longevity and the productivity of the trees that originate from the scion grafted onto the root stock. The rootstock- scion somato genetic interaction (Tyson and Dhingre, 2013) and plant grafting new mechanisms (God schichimid, 2014) are of useful understanding. The right choice and selection of root stock suitable to the region and the scion is the prerequisite to enhance the productivity, quality of the fruits and the yield stability. It is highly necessary to organize and to establish a Root Stock Mother Stock beds for mass multiplication of virus and disease free root stocks by stooling and cuttings. In many fruit crops, there are well documented and identified root stocks being maintained for utilizing in grafting operations (Table 1).

Polyembroyonic root stocks are common in citrus propagation and they are derived from seeds that develop from the nucellar tissues without the help of male gametes in their formation implying that they originate from other than the sexual embryo or true seed. The nucellar seedlings are identified by similarity and uniformity of the seedlings which fall within one group and are more or less same size. Others which are either small or too tall than the commonly prevailing type are discarded and considered as off type and/or zygotic. Gorak Singh (2012) advocates periodical roguing of the zygotic seedlings from the rest when the seedlings are about 10 to 20 cm. height and ready for transplanting in the nursery and afterwards at the time of grafting or budding and for the last time at the time of transplanting in the field. It is desirable to establish separate root stock plots for each of the growing situations that require specific adaptation needs. The scion shoots from selected Mother Stock Plants and shoots for root stock plants from Mother Root Stock gardens are vegetatively propagated. In case of polyembryonic/nucellar seedlings for root stock nurseries are rasied to select nucellar seedlings. Recommended roots stocks for fruit crops are listed by Gorak Singh (2012) in his treatise on protocols and standards for Vegetative Production of Fruit crops. He has provided (i) a list on type of root stocks, and commercial propagation techniques for 37 individual fruit crops along with (ii) detailed procedures for different layering, budding and grafting techniques (iii) propagation techniques utilizing specialized organs like runners, suckers and cuttings and (iv) standards for planting material of fruit crops (Table 1).

Table 1. Recommended techniques and root stocks for fruit trees (Gorak Singh, 2012)

S.No.	Fruit crops	Commercial propagation techniques	Type of rootstocks
1.	Almond	T-budding and wedge grafting.	Open pollinated peach and bitter almond seedlings
2.	Aonla	Patch budding and wedge grafting (winter).	Open pollinated seedlings
3.	Apple	T-budding/tongue and wedge grafting.	Clonal root stocks
4.	Apricot	T-budding and wedge grafting.	Open pollinated apricot/peach/ plum seedlings
5.	Avocado	T-budding and wedge grafting.	Open pollinated seedlings
6.	Bael	Wedge grafting.	Open pollinated seedlings
7.	Banana	Suckers / corm.	An elite mother plant, free from diseases, producing quality fruits
8.	Cashew	Soft wood grating.	Open pollinated seedlings
9.	Cherry	Tongue and wedge grafting.	Clonal root stocks
10.	Custard apple	Wedge grafting	Open pollinated seedlings
11.	Date palm	Sucker / off shoot.	An elite mother plant, free from diseases, producing large sized quality fruits
12.	Fig	Hard wood/semi-hard wood cutting.	An elite mother plant, free from diseases, producing large sized quality fruits
13.	Grape	Hard wood cutting and wedge grafting (for in-situ/bench grafting).	An elite mother plant, free from diseases, producing large sized quality berries (for cuttings)
14.	Guava	Wedge grafting.	Open pollinated seedlings
15.	Gooseberry	Hardwood/semi-hard wood cutting.	An elite mother plant, free from diseases, producing large sized quality fruits
16.	Jackfruit	Patch budding and soft wood grafting.	Open pollinated seedlings
17.	Jamun	Soft wood grafting	Open pollinated seedlings

18.	Khimee	Soft wood grafting	Open pollinated seedlings
19.	Kiwi fruit	Hard/semi hard wood cutting and budding	An elite mother plant, free from diseases, producing large sized quality fruits. Open pollinated seedlings (for budding)
20.	Lemon /lime	Cutting and budding	An elite mother plant, free from diseases, producing large sized quality fruits (for cuttings). Open pollinated seedlings (for budding)
21.	Litchi	Air layering and wedge grafting	Open pollinated seedlings (for wedge grafting)
22.	Mahua	Soft wood grafting	Open pollinated seedlings
23.	Mandarin	T-budding and wedge grafting	Polyembryonic seedlings
24.	Mango	Soft wood, wedge and veneer grafting	Open pollinated seedlings and polyembryonic (13-1, Vellaikulumban)
25.	Peach	T-budding, and tongue grafting	Clonal rootstocks
26.	Pear	T-budding, wedge and tongue grafting	Clonal rootstocks
27.	Pecan nut	Patch budding and wedge grafting	Open pollinated seedlings
28.	Pineapple	Slip/sucker	An elite mother plant, free from diseases, producing large sized quality fruits
29.	Plum	T-budding, tongue and wedge grafting	Clonal rootstocks
30.	Pomegranate	Wedge grafting / air layering	Open pollinated seedlings (for wedge grafting) and an elite mother plant, free from diseases, producing large sized quality fruits (for air layering)
31.	Raspberry	Sucker	An elite mother plant, free from diseases, producing large sized quality fruits
32.	Blackberry	Sucker	An elite mother plant, free from diseases, producing large sized quality fruits
33.	Sapota	Wedge grafting	Khirnee seedlings
34.	Strawberry	Runner	An elite mother plant, free from diseases, producing large sized quality fruits
35.	Sweet Orange	T-budding / wedge grafting	Polyembryonic seedlings
36.	Wood apple	Soft wood grafting	Open pollinated seedlings
37.	Walnut	Patch budding and wedge grafting	Wild seedlings and Paradox

The propagation planting material should be labelled (12 cm. x 6 cm.) indicating the crop and variety, date and age of grafting, method of propagation source of root stock and budwood, treatment if any, date of treatment, freedom from diseases and pests and signature of the authorised person.

The XIth Five Year Plan focus on scion in fruit trees is (Chadha, 2006)

1. Establishment of Mother Blocks and Root Stock Blocks with Hitech Nurseries and Tissue Culture Units.
2. Accreditation of Nurseries for production and supply of quality planting material.
3. Producing Quality Planting Material. The guidelines for establishing the Mother Block for scion and rootstock are followed.

   The genetic purity standards and clone size specifications of the scion (clone), root stock and the graft are prescribed in the Indian Minimum Seed Certification standards (2013). The details are presented in Table 2.

Singh *et al.* (2012) and Malhotra (2016) have published in detail the production of quality planting material in horticultural and spice crop respectively. Parthipan (2017) has developed miniclone technology for tree crops that can be applied for fruit trees as well.

Methods of maintenance of Plant Genetic Resources in tree spices by Parthasarathy (2010) and conservation and utilization of Horticultural genetic resources by Rajasekharan and Ramanathan Rao (2019) and advances in quality planting material production in spice and horticulutral crops and testing are published by ICAR (2017 and 2018) are noteworthy publications.

Table 2. Genetic purity and clone size specifications

S. No.	Crop	Genetic Purity %	Diameter of the clone (cm.)	Diameter of the root stock (cm.)	Height of the grafting (range in cm.)	Diameter of the graft at 10 cm above the graft union (range in cm.)	Height of the graft (range in cm.)	Grafted clone should be free from suckers	Clones not conforming to the specified size in clone lot shall not exceed 5% by number	Genetic Purity Trueness, Uniformity and conformity to the variety characteristics (100 %)	Free from diseases and pests
1.	Acid lime	100	0.60 -1.00	(Layers / nucellar seedlings)			20-40	-	-	-	Virus canker greening
2.	Aonla	100	-	0.75- 1.00	10-15	1.00-1.25	50-100	-	-	-	-
3.	Apple	100	0.60 -0.70	-	10-20	0.75-1.00	50 - 120	-	-	-	-
4.	Banana	100	7.0 -10.0						-	-	Virus
5.	Ber	100	-	0.60 -1.00	10-15	0.75-1.00	20-30	-	-	-	-
6.	Cashewnut	100	-	0.60-0.75	10-20	0.75-1.00	50-120	-	-	-	Wooly aphid
7.	Custard apple	100	-	0.75-1.00	15-20	1.00-1.25	30-50	-	-	-	-
8.	Fig	100	1.00-1.50	Rooted cuttings		1.00-1.25	25-30	-	-	-	-
9.	Grapes	100	-	1.00-1.25	8.0-10.0	1.00-1.25	10-15	-	-	-	-
10.	Guava	100	-	0.75-1.00	10-15	1.00-1.25	50-75	-	-	-	-
11.	Jack Fruit	100		1.00-1.25	15-20	1.25-1.50	30-50	-	-	-	-

12.	Lemon	100	1.00-1.25	-	25-40	-	25-40	-	-	-	Virus
13.	Mandarin	100	-	0.60-1.00	15- 20	1.00-1.25	40-60	-	-	-	Virus
14.	Mango	100	-	1.00-1.25	15-25	1.2-1.50	25-40	-	-	-	Die back
15.	Pineapple	100	Suckers weight			400-500 (g)	-	-	-	-	-
			Slip weight			350-400 (g)	-	-	-	-	
16.	Pomegr anate	100	0.75-1.25	-	-	-	25-40	-	-	-	Nematode
17.	Sapota	100	-	1.00-1.25	15-25	1.00-1.25	50-70	-	-	-	-
18.	Sweet orange	100	-	0.60-1.00	15-20	1.00-1.25	30-50	-	-	-	Virus
19.	Tamarind	100	-	0.75-1.00	15-20	1.00-1.25	50-100	-	-	-	-

Note : Pure living clones should be 99.5% by number in foundation class and 98.0% by number in certified class.
(*Source:* Indian Minimum Seed Certification Standards (2013) Central Seed Certification Board MOA, GOI, New Delhi)

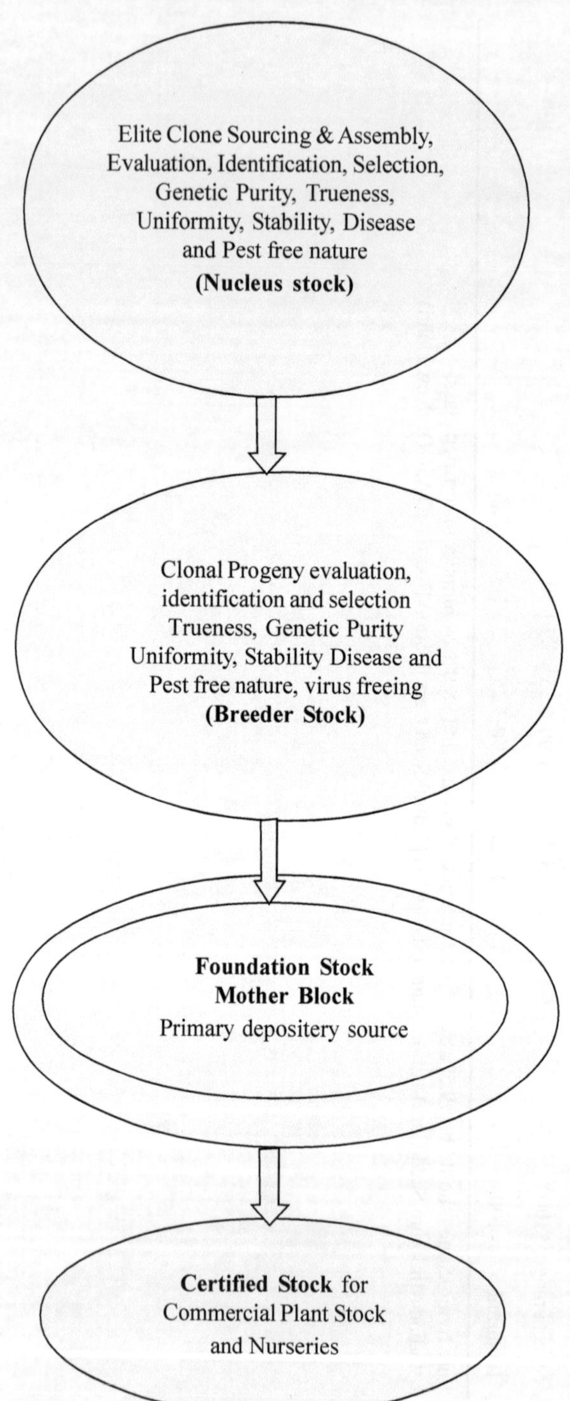

Fig. 1. Establishing Mother Blocks for Scion in Fruit Tree Crops

Guidelines for establishing Mother Blocks for scion in fruit tree crops are provided by Ram Chandra (2012) for Mother block establishment and planting material by Saroj (2012) and by Singh *et al.* (2012) on the quality planting material in horticultural crops. The different aspects are:

1. Only pedigreed fruit trees true to type are selected for scion sourcing. Each stage is supported by Primary and Secondary nurseries, Discard pest and disease affected plants
2. Healthy bud woods/buds selected from trees that are regular and prolific bearing observed for at least for two to three years/regular bearing seasons
3. Trees showing precocity/earliness in flowering and free from diseases and pests and free of physiological disorders are selected as mother trees for bud wood collection
4. Indexing of trees is done for the presence of viral diseases and pathogens. Biodiagnostic methods are applied.
5. Containerized nurseries with sterilized/fumigated/solarized soil compost, Farm Yard manure medium are raised to exclude recontamination, with root infecting pathogens like phythium, rhizoctonia, phytopthora etc and nematodes
6. Plants at each stage of nucleus, breeder, foundation and certified stocks are raised in isolation and indexed. When necessary disease freeing is done
7. Nursery multiplication techniques are carried out in phytosanitary hygienic conditions. To the extent needed, micropropogation, shoot tip culture, shoot tip grafting, and other *in vitro* techniques are adopted (adjunct measures). Protected high-tech nursery practices are adopted.
8. Minimum Standards of Plant Propagation protocols and quality are assured and maintained.
9. Mother plants in Mother Blocks are replanted in a phased manner with new cultivar
10. In case of root stocks that are clonally propagated, similar steps as in Scion stocks are to be followed
11. It is desirable to grow one to two rows planted in border with the same source stock as in the nucleus or breeder or mother block or in certified block
12. Bud woods selected or budlings developed should be of superior quality. Appropriate scion-root stock specific grafting techniques that are highly standardized to produce highly compatible scion-root stock union and ideal growth/development of scion shoot and root system development in the root stock are adopted. Maintenance of pedigreed root stock plants in mother block and certified stock nursery is necessary to continuously provide quality root stock plants resource seedlings.

13. DUS Testing is carried out as prescribed
14. In mother block, upto 20% reduced spacing per plant is advocated. Mother block is grown in isolation.
15. Root stock plants of seed origin, seedlings, nucellar seedlings, apomictic seedlings and somatic embryo seedlings are to be grown following accepted protocols for evaluation for genetic purity, uniformity and sound disease free/or preventing recontamination. Root stocks providing dwarfing or semi dwarfing nature, compatible to known varieties (scion stock) tolerant or resistant to diseases, pests, drought, physiological disorders and root system free from micronutrient disorders are separately maintained.
16. Indexing is done for Root Stock resources also
17. Genetic purity and clone size specifications are shown in Table 2.
18. Foundation Plant Services established. (for example Foundation Plant Services of U.C. Davis. USA).

Foundation Plant Services (University of California, Davis, USA)

1. To identify quality elite planting material testing and maintenance in Fruits, Nuts, Grapes, Propagation stocks
2. Plant importation, quarantine disease testing, virus elimination
3. DNA identification
4. Conduct Release of Univ. California patented horticultural varieties
5. Link Researchers, Nurseries and Producers

CANADIAN CERTIFICATION

1. Generation G.I
2. Generation GIA (Isolation, maintenance of trees, virus testing)
3. Generation G1 Mother Plants avoid contamination, reinfestation
4. Generation G2 Elite Primary stock increase block
5. Generation G3 Foundation seeds increase
6. Generation G4 Certified Nursery Stock preparation from G1, G2 and G3 mother plants

SAPO Services:

1. Establish Foundation and Mother Blocks
2. True to type control
3. Prune to make Bud woods and not fruits

5

In vitro Grafting

Micrografting is a method to produce virus free, viroid free and disease free grafts that are clonally propagated. This method is an aid to mass multiplication. *In vitro* grafting is applied to difficult to root clones. Micrografting is performed to understand incompatibility prediction between the two grafting partners, virus indexing, deduction of latent virus, safe germplasm exchange and to do grafting functions throughout the year. It is not season specific. Protocols for micrografting are available for almond, pistachia, olive, prunus, grape, mulberry, cashew, citrus, peach, pear, avocado, hazlenut, walnut, chestnut and grape fruit (Hussein *et al.*, 2014). Combination of *in vitro* produced scion, bud and *in vitro* produced root stock is known as full vegetative graft and the combination of *in vitro* produced scion but grafted on the seedling root stock is termed as half vegetative graft are comparable in their performance as experimented in rubber in respect to growth, physiological aspects, rubber yield and others agronomical traits. The advantages in the full vegetative grafts are better graft union, disease free nature, and uniform grafts that are true to type since the scion and root stock are *in vitro* clones. Hussein *et al.* (2014) have presented a detailed review on micrografting for fruit tree improvement. *In vitro* grafting in apple (*Malus domestica*) reported by Bhat *et al.* (2013) could be considered a typical protocol development involving LalAmri CV scion and M. 9 as root stock. Microshoots of root stock M 9 were developed and rooted in *in vitro* and used as root stock plants grafted with 10 mm. long microshoots of LalAmri apple. Micrografting technology being followed was an improvement by Murashige (1977) over the micrografting technology developed by Doorenbos (1953) in Ivy and Holmes (1956) in Chrysanthemum. There is increasing interest in developing this *in vitro* grafting or micrografting technique involving the placement of a meristem of shoot tip explants onto a decapitated root stock that has been grown aseptically from seed or micropropagated culture (Hartmann *et al.*, 2004). The micrografting stages include *in vitro* establishment and multiplication of scion and root stock and preparing them and performing micrografting. The conventional method of grafting is season dependent and failure of grafting means loss of one year for the production of grafted plants. The micrografting is performed under controlled conditions throughout year according to the market demand. The micrografting protocols have been evolved for many fruit and plantation crops and are applied commercially. Apple, citrus, almond and grapes are some of the crops receiving the Shoot Tip Grafting or micrografting technology. In citrus, nucellar seedlings are good source of clonal and virus free nature but the seedlings are juvenile and take many years to flower. Shoot tip grafting helps to overcome this problem. In almond, the shoot tip grafting has been profitably utilized in Israel and Morocco. The size of the shoot apex has a paramount influence on virus elimination from the plants. In many horticultural crops, this method could be standardized and applied to purify, renovate and rejuvenate the trees from viral disease. Graft incompatibility is expressed/deducted in less than 3 months by *in vitro* callus fusion techniques (Errea *et al.,* 2001) instead of waiting to know the compatibility or otherwise in the grafted trees planted in the field. In cashew which is difficult to root, this method is useful and success has been obtained by *in vitro* grafting (Thimmappaiah *et al.,* 2002). Walnut, pistachio and almond are the other crops that are difficult to root and the micrografts were highly successful.

6

In vitro Clones in Plants

In vitro cloning is a vegetative propagation performed in the Plant Tissue Culture system. *In vitro* micropropagation cloning propagation is mostly applied to the vegetatively propagated plants and in particular in ornamental, fruit and in a few orchard crops and in vegetable crops besides in many forest tree species. From 1980's cloning for commercial purposes is being adopted especially in ornamental crops. Many R&D laboratories have been established to perform micro-propagation on commercial scale. Steps in *in vitro* cloning are selection of elite superior disease free donor mother plants and harvesting the appropriate explants, such as meristem, axillary buds, tubers, rhizomes and setts. The explants are sterilized, then placed in the induction medium followed by multiple shoots production. The multiple shoots from the induction phase are subsequently multiplied in repeated transfer and passages in culture medium. This procedure is repeatedly made to achieve multiplication to the desired extent or scaling up. In the mass multiplication, the rate of multiplication of vegetative shoots varies from low to medium to high extent depending upon the genotype response and the crop species. Standardized *in vitro* multiplication protocols are available for many of the ornamental plants, fruit trees, plantation crops and forestry species. Superior root stocks of desirable genotypes are multiplied with this system so that benefits of superior uniform root stocks that are compatible with the scions are made available. *In vitro* multiplication of scion is also made in commercial plantings for making complete vegetative propagation method of graft as different from semi vegetative propagation involving *in vitro* produced scion buds grafted onto root stocks that are not from *in vitro* developed stock.

In the successive micropropagation multiplication cycle of shoots, extra care is bestowed to transfer the well differentiated shoots alone and to discard the adventitious shoots from passages. At the end of the optimum number of passages or multiplications, the shoots are transferred to the rooting medium. Depending on the crop system or plant system, single shoot in each culture tube or groups of 3 to 5 individual shoots in culture vessels are placed in the specific rooting medium. This cycle of repeated transferring to the rooting medium may be three to four times so as to get required number of rooted plantlets as stock or prebasic cultures.

The *in vitro* shoot to rooting cycles are respectively designated as G1, G2 and G3 denoting. Clonal Generation 1, Clonal Generation 2 and Clonal Generation 3 stages respectively that could be treated as prebasic or basic seeds. The subsequent two cycles are Nucleus and Breeder seeds respectively. This is the seeds class scheme or model followed / advocated in *in vitro* cloning propagation. Breeder seed class is extended to Breeder stage I and II as and when needed. Subsequent to the rooting in the medium, the rooted shoots are transferred to appropriate size cavities or plugs in the protrays to develop plug seedlings. Seedlings tray are also made use filled with peat-coir pith compost enriched medium. After allowing for strong root development and hardening, they are planted in primary nursery to grow enough to make successful field establishment. The long term crops like palms, dates, cardamom, bamboo and other plantation crops and fruit trees may require growing in the wide spaced secondary nurseries or polybags of required size filled with compost medium for adequate growth and development of shoot and root and in some case to produce suckers also. The plug seedlings and polybags are easily transported to the field for planting. The *in vitro* clones can be also grown in aeroponic system for good development of shoot and root of the plants. Hydroponic system is also practised depending on the adaptation nature and behavior of the plants. Well developed plants in *in vitro* system that have undergone hardening, rooting and primary nursery phases are healthy, free from pests and diseases and show very high uniformity and field establishment.

In vitro cloning facilitates multiplication of faithful, true to type, genetically uniform and stable plantlets or clones. Juvenile phase period is short and highly reduced in the clonal plants and early to flower as compared to seed plant or seedling raised plants.

Many studies comparing the performance of the *in vitro* clonal plants with the mother plants/varieties showed similar and comparable performances in physiological and agronomic endowments besides the yield and quality likewise. *In vitro* clones and vegetatively cloned plants showed comparable performances and in their genetical properties and quality parameters as in banana, cardamom, poplars, teak, eucalyptus and in other crops. Grafted rubber plants with buds from *in vitro* clone scion grafted onto *in vitro* clone root stock are comparable with the performance of grafts made with seedling root stock and buds from donor trees which are the usual grafts of scion and root stock combination.

In potato, *in vitro* clones are produced from seed potatos. *In vitro* plants are cultured *in vitro* to produce microtubers and from the microtubers, minitubers are produced in green houses. Minitubers are also produced from the *in vitro* plants grown in green house. Field planted microtubers, minitubers and seed potato seeds performed in comparable manner in respect of vegetative and tuber characteristics and in the physiological traits and yield. Micro and minitubers and their field performance are presented in detail by Naik and Karihaloo (2007). *In vitro* formation of storage organs such as bulb, rhizome, tuber and corms can be in favour of using them as propagules because these propagules can be directly planted in the primary nursery or in the well prepared field without hardening or acclimatization. These organs can be transported without culture medium. Further improvement in the protocols are needed to obtain larger sized and vigorous organs that can establish well in commercial plantings as compared to small rhizomes that have low survival and produce weak unhealthy plants.

The basis for *in vitro* cloning is micropropagation and faithful mass multiplication of true to type of the genotypes concerned. Mass cloning in the crops is a reliable and well defined approach for multiplication with genetic purity, genetic fidelity, genetic stability and uniformity. *In vitro* cloning with shoot tip culture helps in virus and viroids elimination and freedom from pathogens and faster multiplication.

Meristem culture is most reliable in many aspects. *In vitro* clone or *in vitro* plants (IVP) system is used to generate plants of same sex, male plant or female plant, male and female sterile lines and polyploids. There is no genetic drift or cross pollination in IVP which is known for quality assured fast multiplication of true to type clones at a low cost. This is a system by itself ensures genetic and phytosanitary qualities and is practiced in the controlled plant tissue culture facilities and hence not season dependent that is experienced in conventional methods.

The systematic planned manner of execution of micropropagation from the mother plant donor selection, explants selection, shoot induction onwards upto many passages in plant tissue culture and maintaining genetic fidelity , quality and uniformity has been now outlined by the Department of Biotechnology, Government of India in the DBT-NCS-TCP project. In this system formulated for the first time in India in 1996 and also globally, the DBT has incorporated monitoring, evaluation , quality testing, phytosanitary norms and identified referral laboratory and institutions for virus indexing and DNA testing to establish genetic fidelity. The regulatory mechanisms of the DBT are very enabling. Guidelines for National Certification System for tissue culture plants have been formulated for adoption in the following eight crops.

1. Potato
2. Apple
3. Banana
4. Blackpepper
5. Bamboo
6. Citrus
7. Sugarcane
8. Vanilla

The guidelines are crop specific. The guidelines include procedure, protocol and related regulations. Across the Nation, many Public and Private R&D facilities engaged in commercial micropropagation of plants follow the guidelines of DBT- NCS-TCP. Steps to identify off type, somaclones and epigenetic variation are necessary to eliminate them in successive passages. Infrastructure facilities such as laboratories, glass house, hardening facilities, nurseries and service space facilities are needed to participate in the DBT-NCS-TCP system. *In vitro* system for plant cloning micropropagation is executed in number of generations commencing from Generation 1 (G1) onwards upto G3 and continued upto 8 to 12 sub cultures or passages and certified. Clone only Breeder Seed production, minimum quality standards and disease and pest free nature are subjected to molecular GOT (DNA markers) and Monitoring guidelines by DBT. The Central Seeds Certification Guidelines of Government of India are extended to apply for micropropagated planting material. Role of Clonal Breeder Seed in vegetables, fruits trees, plantation crops and forest species is increasing.

In vitro cloning propagation offers high rate of multiplication. It is rather enormous. In respect of sugarcane crop, the *in vitro* multiplication system developed and the multiplication operated by the Sugarcane Breeding Institute, Coimbatore and the benefits are described hereunder (N.C. Jalaja *et al.,* 2008; TNAU Agri Portal, 2013).

The comparative performance of different source planting material from sugarcane CO 470 variety is presented below:

Particualrs	AST	MC	Check
Germination in 3 budded setts %	46	48	37
Number of tillers per plant	2.41	2.28	2.02
Cane yield t/ha	105.3	112.4	90.5
Sucrose in juice %	18.82	20.25	18.59
% of juice extraction	60.3	62.0	54.1

AST: Aerated Steam Theraphy; M.C: Micropropagated Clones; Check: Untreated Seed Cane Setts (*Source:* B. Sundara 1995)

In vitro micropropagation techniques are now often preferred to conventional practices of vegetative propagation not only from commercial point of view but also with regard to crop improvement program. The benefits are 1. the rate of true to type multiplication is enormous which ensure rapid spread of newly released varieties. 2. spread of disease and pest through indiscriminate movement of seed can be totally eliminated. 3. The vigour of the varieties is regulated to its original level, thus improving cane yield and quality. 4. an effective varietal scheduling is possible as the seed production program is under total control. 5. ratoon crops are excelling due to use of quality clonal seed and 6. old degenerative varieties can be replaced and new varieties can be introduced and fastly multiplied to ensure higher production.

Rate of *in vitro* multiplication in sugarcane

Shoot tip number	No. of shoots	Ratio	No. of days in each cycle	Total no. of days
1	1	-	15	15
1	1	-	21	36
1	1	-	21	57
1	6	1:6	45	102
6	36	1:6	45	147
36	216	1:6	45	192
216	2160	1:10	45	237
2160	21600	1:10	45	282
21600	216000	1:10	45	327
Rooting	200000	-	45	372
Nursery	180000	-	45	-

(*Source:* D. Neelamathi, 2013)

This system can be initially utilized to produce the breeder seed of sugarcane in each factory. Foundation and certification seed can be produced from plants through vegetative cutting. Considering various benefits of this system, several sugarcane varieties are multiplied at a considerable cost. It is possible to obtain improved cane yield and increased sugar production by introducing the newly identified varieties at shorter interval. Tissue culture laboratories have been established throughout India during last decade producing tissue culture plants and quantum of tissue culture plant would be increasing at the higher rate. The quality of plants will have to be ensured for getting desired effect (DBT-NCS-PTC).

National Organizations such as National Horticultural Board, National Horticultural Mission, APEDA, Biotechnology Consortium Ltds. (BCIL) besides the DBT, Ministry of Agriculture, ICAR and SAUs are converging to facilitate and to support the *in vitro* cloning protocols applied and extended to all possible crop plants. Season independent production of the *in vitro* plants is always an added advantage.

In banana the main steps involved *in vitro* system are

1. Inoculation
2. Initiation of culture
3. Shoot multiplication in different passages

4. Discarding the adventitious shoots in each passage
5. Rooting, hardening or acclimatization
6. Primary nursery, plug seedlings and poly bags nursery

Multiplication rate of Banana from a single explants under tissue culture system

Stage	Generation	Duration (days)	No of Plants
Initiation	-	25	1
	1	50	3
	2	75-80	12
	3	100-110	48
	4	120-130	192
	5	175-180	760
	6	200-210	3040
	7	220-230	12160
Rooting		255-260	11000 to 12000
Primary	Hardening	270-280	11000 to 11500
Secondary	Hardening	310-320	10000 – 10500 **

Note : ** major discards are off-types and somaclones

To ensure only quality banana clones are supplied, the banana *in vitro* plant standards has been prescribed as follows;

i) Height - 30 cm
ii) Pseudo stem girth - 5 to 6 cm, 60 days after hardening
iii) Number of leaves - 4 to 5, photosynthetically active leaves
iv) Inter foliar space - 5 cm or less
v) Number of active roots - 25 to 30
vi) Length of roots - 15 cm
vii) Free from visual symptoms of leaf spot, pseudo stem root, physical deformation, nematodes, root pathogens and virus disease.

Performance of banana TC plants

Bunches / ha	TC Plant banana	Sucker propagated
Mean yield – bunches /ha	2663	2416
Currently more than 10% of banana planted is from tissue culture plants.		

In banana the identification of suitable mother plants is an important step. Only true to type, vigours plus productive superior plants are selected and the conventional suckers from individual plants are tested for field performance, agronomic traits, yield, and quality of bunch and fruits. Mother Nursery Block is then established in fertile and clean field to select mother plants. Isolation of 500m is provided to maintain purity and to avoid virus infection/spread. Maintaining pedigree record, tagging individual plants, indexing the plants for virus, decapitating the selected indexed plants to facilitate development of axillary buds, decapitation of suckers and decortications of suckers are systematically followed. Paring all roots and scalping the rhizome surface are done to be followed by secondary decortications and hardening. *In situ* mass multiplication is possible by pouring 4 ml of 40 ppm BAP into the decorticated cavity and covering the cavity. This protocol is adopted from first generation to third generation.

The Indian Minimum Seed Certification Standards (2013) has included the National Certification System for Tissue Culture Raised plants (NCS-TCP) of the DBT. GOI. The document Plant Tissue Culture Techno Commercial easibility by NCS-TCP (BCIL-DBT) 2008 provides the detailed steps on the system inclusive of Virus-free and quality certification vis-a-vis Reference DNA Sample tissure culture raised plants, Certification, referral laboratories, Facilities involved in TCP and approval of the DBT. Tissue culture raised plants by micropropagation is now being practised in flower, fruit, vegetable, medicinal, tuber crops and in field crops for rapid multiplication and maintaining genetic purity and genetic fidelity, quality of planting material and free from disease, round the year production and storage besides improving the yield and quality.

Select 250 healthy disease and pest free, true to type plants from clonally raised crop. Selection of plants made after conformity for DUS characters

↓

Clone the selected 250 true to type plants. Prepare 3 to 5 setts (3 to 5 budded) from each matured stem

↓

Raise individual clonal progeny rows in disease free clean field. Clones planted after precleaning/disinfection

Examine individual plants in each clone progeny for true to type, uniformity and healthiness and virus free nature. At maturity, assess the performance and select the progenies performing above mean values. Select the superior progeny rows and clone the selected progeny rows.

Raise progeny rows (2 rows plots) 4m. long in 2 replications. Raise in clean field and follow phytosanintary measures. Off-types, deviating types of plants, disease, virus, insects affected plants are uprooted and burnt. Examine progeny rows for uniformity, conformity to the crop specific DUS characteristics. Early and late maturing plants are discarded. At maturity, progeny rows performing above mean + 2SE are selected.

Identify superior plants within each selected progeny row. Examine the plants for healthiness, virus, disease and insect free nature, true to type and conformity of DUS characteristics.

Selected plants are lifted up, underground portion washed and examined for true to type and disease, nematode, virus free nature. Clone the selected plants and bulk them to constatute the nucleus stock

Raise Breeder Crop

Visit of monitoring team at mid maturity / Pre harvest stage conduct GOT.

Fig. 1. Development of Nucleus and Breeder Stock in Clonal Crops. Selection in asexually propogated crops.

Note: 1. Minimum isolation distance 5m. is followed, 2. Crop rotation is adapted, 3. Select clean field only, 4. True to type/ descriptor are adhered, 5. Standard crop specific production schedules followed, 6. Removal of off-types and deviating type of plants carried out, 7. Clean equipments are used for field and cloning operations.

The protocols and standard parameters for the production of Apple, banana, black pepper, citrus, sugarcane and potato are shown in figures (2-9) as presented in the Indian Minimum Seed Certification Standards for Tissue culture raised plants.

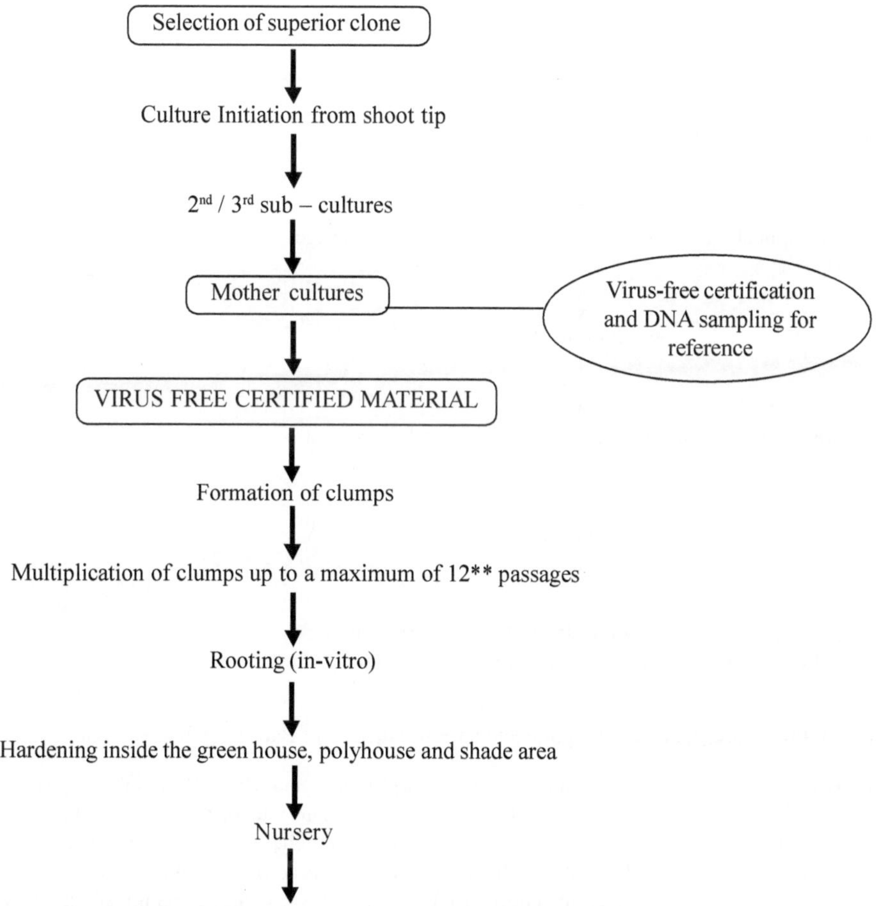

Fig. 2. Procedures and standard parameters for production of Apple by tissue culture

** In tissue culture it is well known that lesser the number of sub-cultures, lower will be the chances of somaclonal variation. However, it must also be realized that if the number of passages are far too small then the entire production process becomes economically unviable. Therefore, efforts should be made to optimise the shoot multiplication process and extend the number of passages only till the clonal uniformity of the progenies is maintained. This could be achieved through a strict monitoring of shoot multiplication process ensuring that adventitious shoots are not multiplied

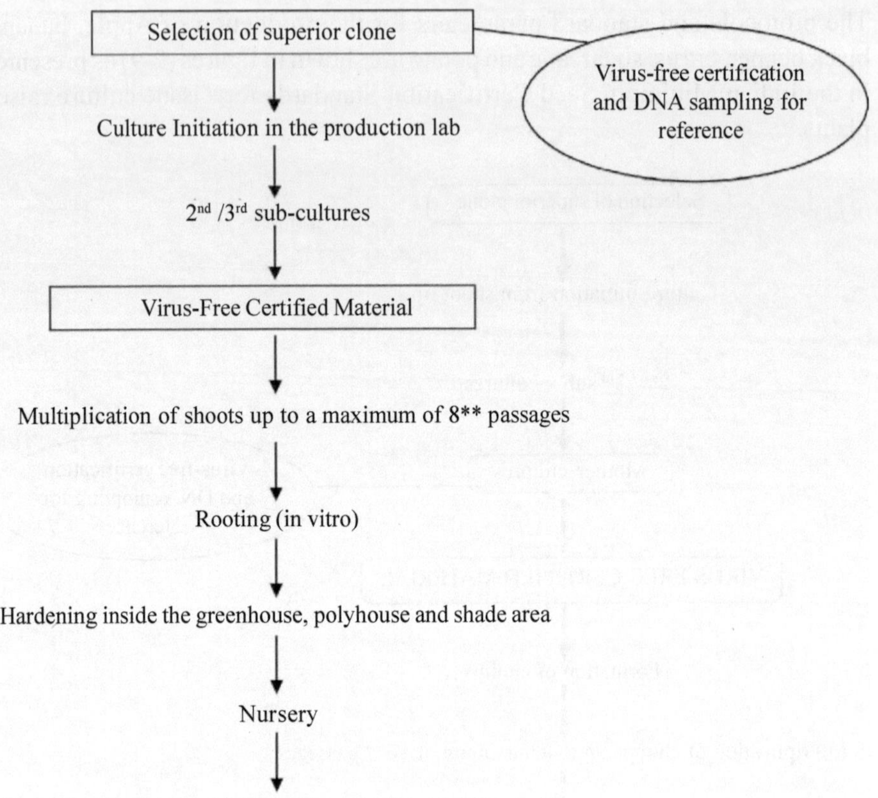

Fig. 3. Procedures and standard parameters for production of Banana by tissue culture

There is a difference of opinion among researchers and production labs on the number of passages that could be regarded as safe for shoot multiplication in banana with respect to clonal uniformity of plants. In tissue culture, it is well known that lesser the number of sub-cultures, lower will be the chances of somaclonal variation. However, it must also be realized that if the number of passages are for too small then the entire production process becomes economically unviable. Therefore, efforts should be made to optimize the shoot multiplication process and extend the number of passages only till the clonal uniformity of the progenies is maintained. This could be achieved though d) strict monitoring of shoot multiplication process ensuring that the adventitures shoots are not multiplied.

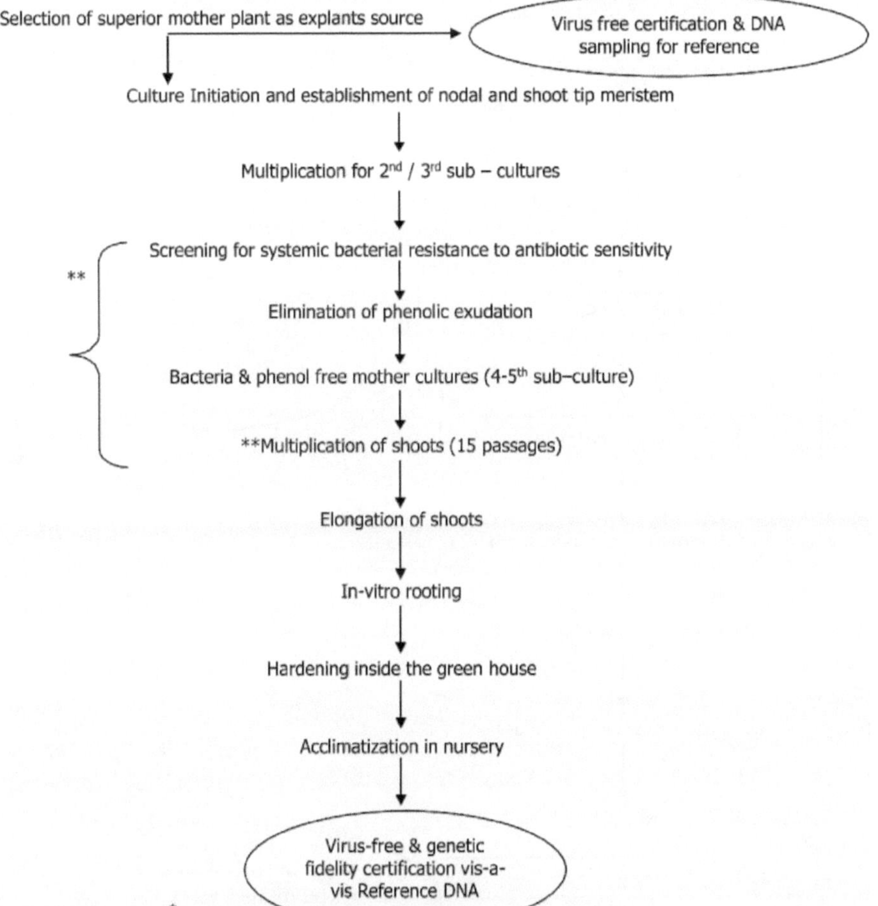

Fig. 4. Procedures and standard parameters for production of black pepper by tissue culture

*Plants should be of superior quality in terms of growth, disease/insect resistance, drought tolerance, constant bearing (confirmed for at least three consecutive years), high yield (fresh and dry weight), oil and oleoresin content. The explants should be healthy and free from microbial infections.

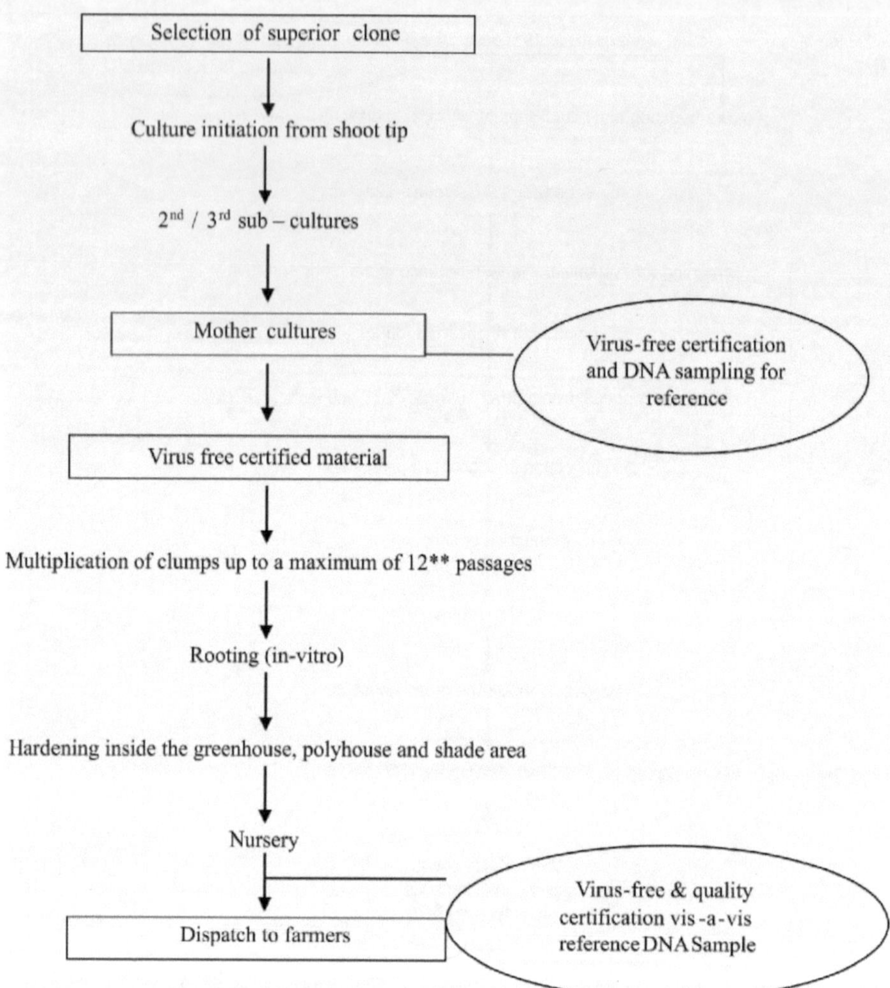

Fig. 5. Procedures and standard parameters for production of citrus by tissue culture

*In tissue culture it is well known that lesser the number of sub-cultures, lower will be the chances of somaclonal variation. However, it must also be realized that if the number of passages are far too small then the entire production process becomes economically unviable. Therefore, efforts should be made to optimize the shoot multiplication process and extend the number of passages only till the clonal uniformity of the progenies is maintained. This could be achieved through a) strict monitoring of shot multiplication process ensuring that adventitious shoots are not multiplied and b) confirming the clonal fidelity of tissue cultured plants using molecular markers in different passages. Citrus shoots have been subculturedupto 12 passages without any loss of clonal fidelity.

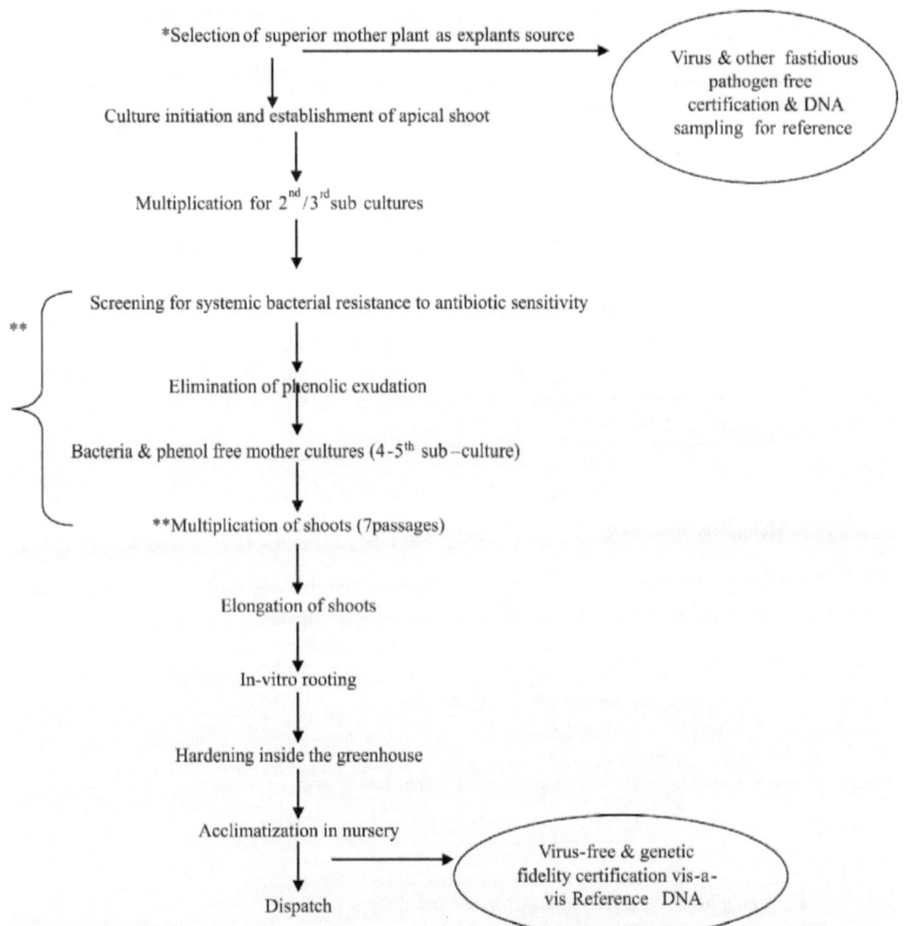

Fig. 6. Procedures and standard parameters for production of Sugarcane by tissue culture

*Plants should be of superior quality in terms of growth, disease/insect resistance, drought tolerance, high yield (fresh weight), sugar content etc. The explants should be healthy and free from microbial infections, smut and, grassy shoot. One set of mother plants must be maintained in the insect proof glass house as reference sample.

**Since sugarcane tissue culture is frequently confronted with endogenous bacterial contamination and phenolic exudation, these should be eliminated using appropriate method.

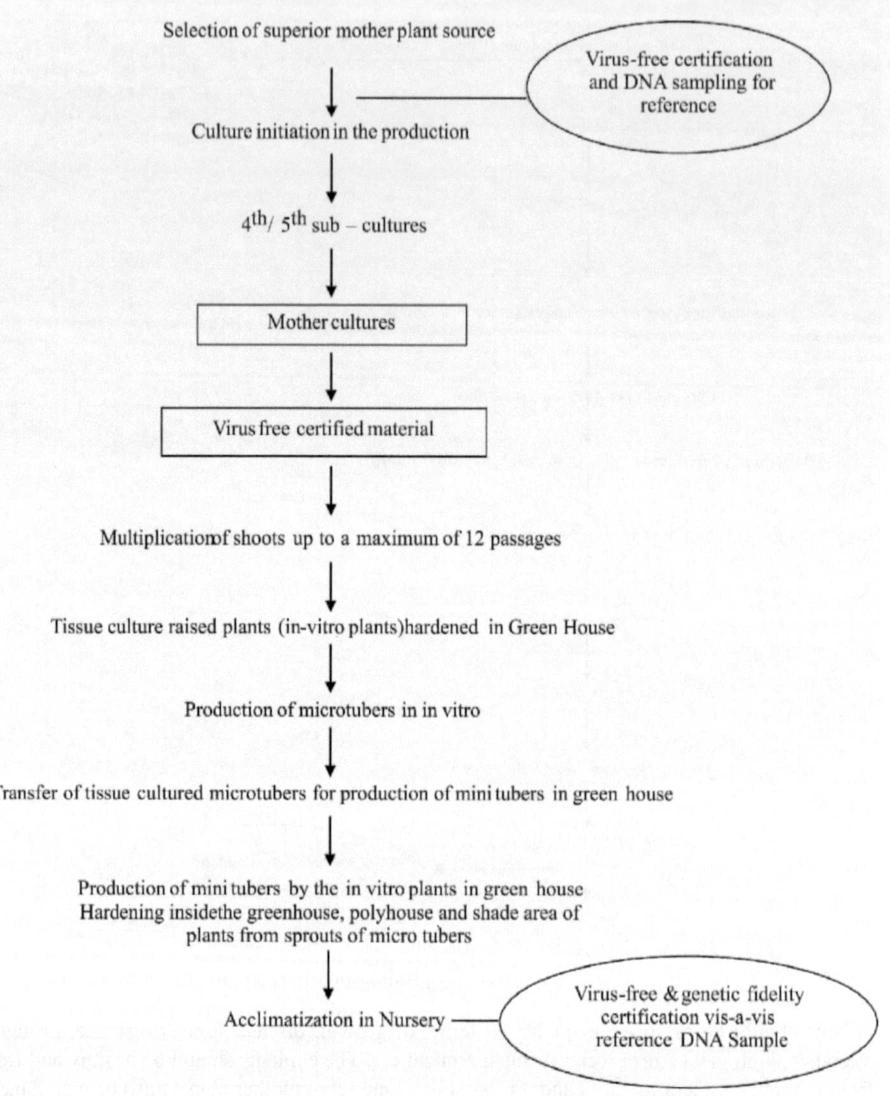

Fig. 7. Procedures and standard parameters for production of Potato by tissue culture

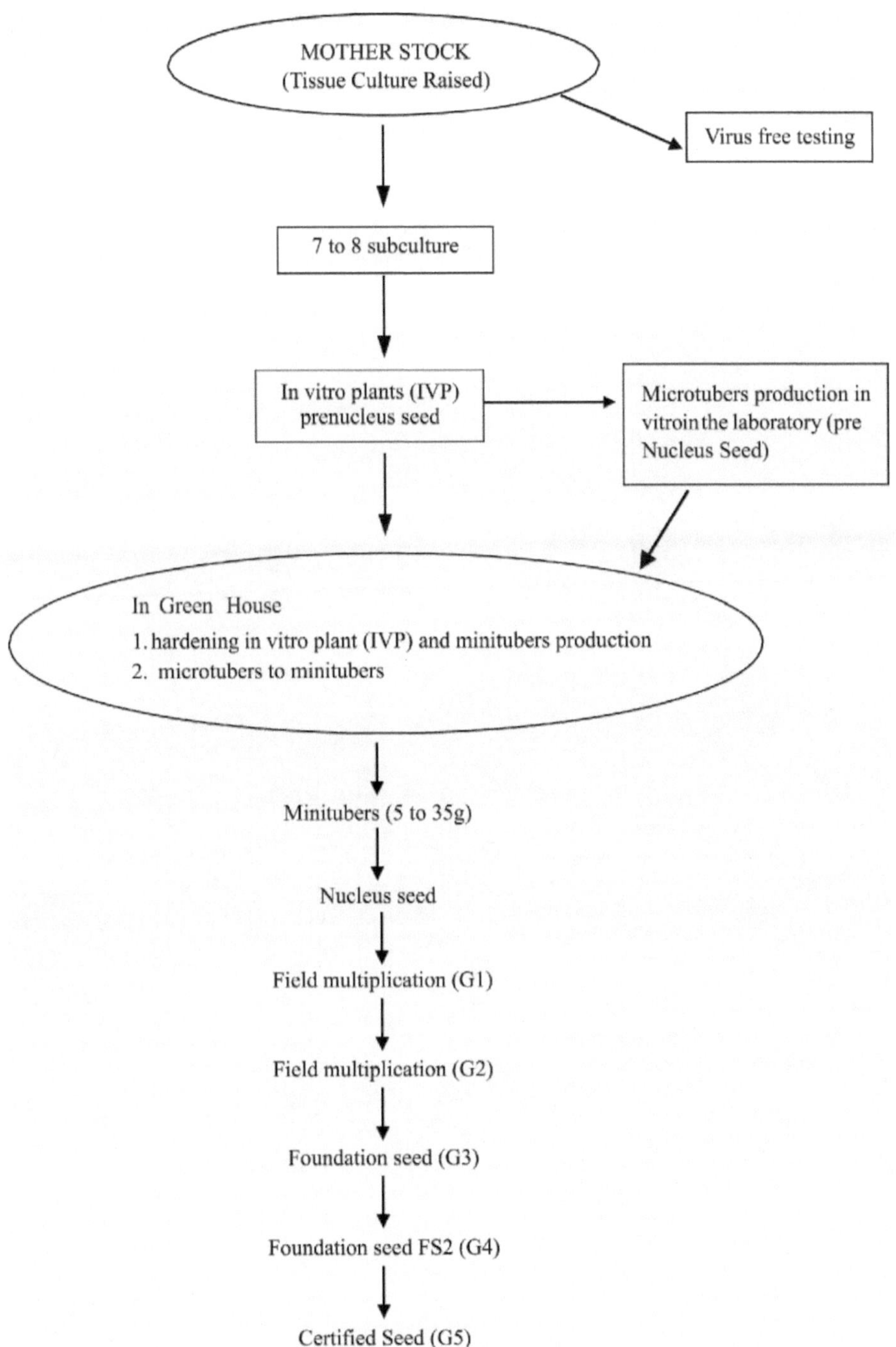

Fig. 8. Multiplication based system for potato seed production proposed by Naik and Karihalbo (2003)

7

Organic Seed Production

Organically produced seeds are obtained by adapting the organic crop production system. Organic seed is a critical input but specialised efforts to produce organic seeds are not adequately and systematically followed. Accredited organic farming standards recognises the organic seed production as a prerequisite to organic farming. Seed health is important in organic seeds since seeds are not treated with chemical fungicides. It is reported that 851 different varieties of organic seeds are commercially available to the organic farmers and growers and 90% of them are vegetables varieties and the rest are cereal grains and clones. Much of the problems associated with organic seed productions are weeds, diseases and pests. Varieties suitable for organic seed productions are yet to be evolved or identified. Research is needed on organic seed production and organic seed standards and maintenance protocols and procedures and varieties suitable for organic farming. Tolerance to select micronutrient deficiencies, low soil fertility and low soil moisture adaptation, high nutrient use efficiency, withstanding weed competition, fairly/fast field establishment and growth responsiveness and highly tolerance to organic growing integrity and ability, tolerance to resisting seed borne diseases are much required features. Lammerts *et al.* (2012) have described in detail the procedures in Organic Crop Breeding.

Availability of organically produced seeds to the farmers is practically absent and most of the farmers use conventional seeds that are non organic seeds but not chemically treated. NOP (National Organic Program USDA) and NPOP (National Program for Organic Production-APEDA) guideline documents on organic seed and also list out the seeds that need to be excluded. GM Seeds, seeds arising from cell fusion and artificial seeds and gene editing, cross pollinated seeds by pollinations from non organic seed crops, male sterile plants pollinated by non maintainers, crop not well isolated, highly seed borne disease affected fields, contaminated seeds and use of different origin of seeds are prohibited. Steen *et al.* (2004) and Schonten *et al.* (2006) have outlined the challenges for the production of high quality organic seeds. Madhya Pradesh State has launched Organic Seed Production program in 2015 with the involvement of Madhya Pradesh Department of Agriculture and Seed Development Corporation and associated with organic farmers.

In organic seed production, the field must not have been applied with prohibited input materials in the immediate past three years. Only compost manures, natural biofertilizers, manures, organic plant pesticides and fungicides and release of beneficial insects are practised. Crop rotation in the field not applied with prohibited materials and safe buffer distance from other crops are recommended. Raising of green manure crops, application of natural sources of manure that supply N, P, K and micronutrients, seed meal, mineral powders, rock phosphate, green sand, proper tillage, manual weed removal,organic sourced mulching, acetic acid containing formulations, corn gluten meal, essential oils, IPM, biocontrols, natural insecticides, neem, pyrethrums, rotono, legume root extracts, tea compost, vermin compost, FYM compost and organic seed treatments are permitted. Plant extracts like Thyme, lemongrass, cinnamon, garlic, thyme oil, as seed treatment, bleach disinfection, surface disinfection are to be made to protect the seeds from disease infection. Biopriming with cow dung, vermin wash, Trichoderma fluorescens at 20% for 6 hours, Hydropriming, Phosphate solubilizing bacteria (250C+), cow dung slurry + PSB, GA3 50 ppm, calchloride 1%, KNO3 1%, ABA 200 ppm, Pseudomonas and phospho bacterium 15% are also recommended for adoption for seed priming in the organic seed production practices.

In organic seed production practices, the key and prime component is the source seed which should be always from organically produced crop source. That means, the seeds are sourced from organically grown crop in the field that was already applied with organic inputs only at least for more than 3 years. In case of annual crops, the organic seeds are collected/harvested once a year and in biennial crops, seeds are collected once in 2 years. Plant propagules/planting materials from vegetatively propagated perennials are collected as setts or cuttings or suckers and saved.

Organic Seed Production: The fertile field that is under continuous organic culture at least for more than 3 years and free of soil borne disease is selected. Isolation distance of 1600m. from the same crops is adapted. All round the field enough space (3 to 5m) is provided for raising a border of the same crop or non cultivated buffer zone that will prevent drifts of sprays from neighbouring field. The required area in the field is prepared to fine tilth for seed sowing. Seeds are sown with the spacing recommended for the seed crop of the particular variety. Basal application of well decomposed organic compost or FMY or vermicompost is applied at 20t/ha. Well cleaned, graded organic seeds from previous season organic crop that are handpicked to discard disease infected seeds are dibbled at close spacing. Before sowing, the seeds are washed in clean water and then priming is done with biological agents specified for the crop. After the priming duration, the seeds are then dibbled with 2 to 3 seeds per hill. Twenty days after germination, only one healthy vigorous seedling is retained in each hill and other seedlings are uprooted. Recommended organic production packages in respect of organic fertilizers application are adopted. At different phases of development and growth, individual plants are examined for trueness and deviating types of plants and off-types are uprooted. At flowering stage, very early flowering and late flowering plants are uprooted. Plants are examined for uniformity and trueness and off-types and deviating type of plants are discarded. In case of nursery raising for organic seedlings production only organic inputs are used.

The plant population in the field is maintained highly adequately to facilitate interplant pollination in cross pollinated crops to avoid inbreeding depression. At preflowering stage and at postflowering phase, it is advisable to spray organic nutrients like Jeevamirtham, moringa leaf extract and micro nutrients. Plant protections measures are done by utilizing organic/ biopesticides as follows.

Natural fertilisers: Carriers based microorganisms, Rhizobium, Azotobacter, Azospirillum, BGA, Azolla, Microrhizae and Phosphobacterium.

Biological Agents: Protozoa, natural enemies like spider, insects, mites, nematodes, birds, fungi, bacteria and virus may be the biological agents.

Bio plant growth promoters: Plants extracts as liquid manures (eg: Eupatorium weed, glyricidia). Added to this, are Biodynamic Agriculture, the indigenous and biodynamic preparations such as cow horn manure, horn silica, compost preparations and BD 502-508 can be used in organic nutrient management. Organic manure, biocompost, vermicompost, biofertilisers, biological control agents, bioplant growth promoter are applied. Livestock keeping and dairy provide continuous supply of organic manure. In addition, green manure crops like Sesbania, Daincha all of which can fix 60 to 100kg of N/ha can be applied by growing and incorporation. Biofertilisers is an important component in organic agriculture which converts and fix unavailable source of elemental N, bound phosphate and decomposed plant residues into available forms (Sripathi *et al.*, 2012; Pathak and Ram, 2013).

The legume-rhizobium association can fix N ranging from 40 to 120 kg/ha. Under optimum conditions FYM, biogas sludge, compost, press mud, vermicompost coupled with inoculation of biofertilisers like Azospirillum, Azotobacter and green manuring can be applied. Based on the principles of organic farming certain inputs have been identified by IFOAM as permitted and some as restricted in usage as fertilizer and soil conditioners (International Federation of Organic Agriculture Movements) (IFOAM-2011).

Pest and disease management: Pest and disease management with reduced cost and without/free from chemical residues is an important step. Biocontrol organisms are: *Trichogramma* sp, *Bracon* sp, *Chrysopera* sp and biopesticides such as Nuclear polyhedrosis, virus (NPV), for control important pests like *Helicoverpa* and *Bacillus thurinjiensis*, *Trichoderma*, *Harzianum* etc. are used. Use of botanicals such as neem seed kernel extract, neem or margosa, pongamia powder and cake are effective against number of pests and application of them is encouraged. Organic seed treatment is made to control seed borne pathogens and neem seed kernel extract, garlic extract, vermin wash, compost treatment for potato late blight and tomato phytophthora infestans are applied. Cassia and clove oils, inhibit growth of seed borne infection of *Aspergillus* and *Curvaularia*. Aqueous extracts of nuxvomica, ginger rhizome, basil leaves or fruits of neemare used to control alternaria in rice seed. Physical treatments like solar treatment, water treatment, hot oven treatment, can be effectively used for management of seed priming and to control seed pathogens. Seed borne pathogens are controlled by neem seed kernel extract, garlic extract, vermin wash compost to control late blight of potato and in tomato *Phytophthora infestans*. Horse compost extract controls some of the seed borne disease. Cassia oil and clove oil inhibit seed borne infections of *Aspergillus*, *Curvularia*.

The harvesting and processing tools and equipments are well cleaned before use so as not to leave traces of agricultural chemicals residues. The harvested produce should be stored separately and free of contamination.

Quality control of organic seed certification: Main problem is lack of training to the farmers, problem with quality standards and control of diseases, pest and seed borne diseases. Strong organic breeding program is essential for development of varieties in various crops which must suit local conditions and climate change. Appropriate seed certification standard is needed for organic seed compared to conventional seed crop (Yadav, 2015).

Definition: Organic seed is seed (planting material) produced by the crop that is planted and raised organically for at least one generation in the case of annual crops, and two generations in the case of biennial crops and perennial crop. (Lammerts and Myers, 2012 and IFOAM, 2011). The criteria for organic seed productions are: initially the mother seed is obtained by purchasing conventional non treated seeds which are grown organically for one season (C1) and sold as organic seeds to the organic farmers (C2). The organic seeds also mean a cultivar that is to perform optionally in organic system. In EU, organic seed production has gained a momentum. In India, presently organic seed is mainly done by private seed companies. Organic seed production is taken up in vegetable crops. For clonally propagated crops, 4 to 6 months old growing shoots under organic culture nursery is utilized.

Organic seeds need to suit for the specific organic culture/cultivation system. The aim of organic plant breeding should be to develop plants that enhance the potential of organic farming and biodiversity and aiming to develop variety/hybrids with improved productivity under organic culture principles. The Organic plant breeding approach respects natural crossing barriers and is based on fertile plants that could establish harmonious viable relationship with living soil and maintain seed plant-seed cycle. The organic breeding should include selection improvement of competitiveness with weeds, nutritional quality, soil NUE, and support sustainable agriculture. It is better to attempt the organic breeding to develop a set of varieties that could adopt to the regional range/ local conditions, healthy flavours, tasty combined with crop resistance and crop resilience. Bosinti *et al.* (2017) has described the organic seed production in lettuce.

Breeder Seed of Organic Seed: Organic seed is certified seed. Breeder seed need to be raised organically in order for its progeny, foundation seed, registered seed and certified seed to be represented or sold as organic seed. At present the seed of any generation plants or the conventional untreated and produced under organic conditions can be called organic seed. Natural organic standards allow non organic seed can be used if equal variety of organic seeds is not produced or untreated variety is not commercially available (IFOAM).

Certified Organic Seed Production: Refers to seed or planting material that has been propagated under organic growing conditions. The propagation was certified for the compliance with organic guidelines by a third party. The definition is clearly distinct from the term certified seed which refers to the certification of plant materials for its compliance with National Seed Laws. Here the certification is issued or granted if minimum requirements with respect to purity, germination rate and seed health is achieved.

Double Certified Seed: The term double certified seed refers to the plant material that has passed the certification for purity, germination rate and seed health as well as the certification for organic seed propagation.

Methods like GM, CMS hybridization without restorer gene, protoplast fusion are not approved. When hybrids are developed, they should be fertile and able to adopt for organic culture conditions. Secured novel breeding techniques are developed eg. Cis or intragenesis and reverse breeding including genetic engineering in such a way that the end product in future might not be classified as genetic modification (GMO) by GMO regulation (Schonten *et al.,* 2006).

Derogation: Lack of availability of organically produced propagation material made the European countries to introduce a temporary derogation system under which the use of conventionally produced non chemical treated propagation material can be allowed in organic seed production and has a greater risk of weed, pest and disease infestations. Risk invasion by GMO contamination is emerging/on increase. New sanitation of treatment need to be developed as alternatives for the present use of fungicides. Such treatments should also maintain the viability of the seeds and should be in accordance with the organic farming.

Novel Seed Sorting Techniques may also be of use by sorting out diseased seeds from contaminated lots or discarding less vigorous seeds Biennial crops are exposed to various diseases during two subsequent seasons. Disease transmission was studied and measured in all stages of the reproductive chain. Organic seed production requires a high degree of sanitation and disease freedom of the basic seed. Physical treatments or hot water treatment is used. Compounds of natural origin are important to control disease and pests. Essential oil and other organic acids are used. Thyme oil exhibits high *in vitro* inhibition activity against Xanthomonas, Clavibactor, Botrytis, Alternaria, Ascorbic acid in 2.5% concentration are used as crop protectants.

Organic Seedlings: Organic seedlings are produced by sowing seeds collected from organic crop variety following organic seed production protocols (Tujela and Oztation, 2017). Well matured, disease and pest free hand picked organic seeds are subjected to biopriming with any of the recommended biological compounds or bio fertilizer and other biological agents. Biopriming duration is as per the recommended practices for the crop variety. The nursery or small seed beds or plugs(in the case of plug seedlings production) are prepared with sterilized soil-compost mixture added with biofertilizers and biological agents like Trichoderma, PSB, neem kernel extract or other biological agents. Seeds are sown after biopriming. Twenty days after germination, healthy and vigorous seedlings, one per hill is retained and others are discarded. Seedlings are allowed to grow to the prescribed duration with added biofertilizers and before planting the seedlings are given root dipping with biological agents comprising Trichoderma,neem oil/pongamia oil and azospirillum. The seedlings are planted in the mainfield with recommended spacing for the crop variety. The crop production is meticulously followed by adopting organic seed production protocols. Vijayan and Krishnasamy (2017) reported that nursery soil application of vermicompost and Panchakavya spray @ 3% produced organic vigorous seedling in rice variety with increased seedlings length, dry matter production, root volume, root ATpase activity and chlorophyll content.

8

Synthetic Seeds

Synthetic seeds are also known as somatic seeds, artificial seeds, clonal seeds and synseeds. They are analogue for botanical seeds. Somatic embryoids which are tissue culture produced embyoids with bipolar structures, apical and basal meristematic regions which are capable of forming shoot and root respectively. Somatic embryoids are encapsulated and are used as artificial seeds possessing the ability to convert into plants in *in vitro* and *ex vitro* conditions. Artificial seeds are broadly two types. Desiccated synthetic seeds are produced by encapsulating somatic embryos in polyoxy ethylene glycol (polyoxy) followed by desiccation and these types of synthetic seeds are provided only in plant species whose somatic embryos are desiccation tolerant. (i) Desiccated system in which somatic embryoids are hardened to withstand desiccation and then encapsulated in suitable coating material. Somatic embryos may be hardened either by treating or coating mature somatic embyoids with suitable polymer followed by treating with ABA to improve germination of somatic embryoids. Janick *et al.* (1989) first developed desiccated artificial seeds from carrot somatic embryoids by treating with 5% solution of polyethylene oxide. (ii) Hydrated artificial seeds are prepared by enclosing the somatic embryoids in gels which remains hydrated. Calcium alginate is most suitable. Hydrated artificial seeds are sticky and difficult to handle on large scale and rapidly dry in the open air. A waxy coating over the artificial seeds can solve the problem. Hydrated artificial seeds need to be planted soon after they are produced. Radenbangh (1990) developed hydrated seeds by mixing the somatic embryoids of alfalfa, celery and cauliflower with sodium alginate followed by dropping into a solution of calcium chloride or calcium nitrate to form calcium alginate. The effect of genotypes and culture medium on somatre embryogenesis is an important factor (Grunberg *et al.*, 2006).

Hydrated synthetic seeds are produced when the somatic embryoids are sensitive to desiccation and recalcitrant.

Artificial Seed Technology is useful in crop improvement and propagation of plants, non seeds producing plants, polyploidy with elite traits and in plants with problems in seed propagation. Being clone in nature, it is applied to propagate useful heterozygous plants and recombinants. Artificial seeds are adjuvant for micropropagation and conserving plant genetic resources. It will be useful in large scale monoculture.

Broadly the procedures to produce the artificial seeds are:

1. Establishing somatic embryoids
2. Maturing the somatic embryoids
3. Synchronized and singulate somatic embryoids
4. Mass production of somatic embryoids.
5. Standardizing encapsulation
6. Standardizing artificial seeds
7. Mass production of synthetic seeds
8. Greenhouse and field planting.

Alginate hydrogels help as a matrix of synthetic seeds because of its moderate viscosity and low toxicity for somatic embryoids, quick gellation and low cost. Alginate is suitable for long term storage and is better than agar and enhances capsule formation and protect the somatic embryoids against mechanical injury. Large scale production of artificial seeds will be useful in planting large scale monoculture. Seneratne (1992) has described methods to convert shoot buds of apple into artificial seeds by encapsulation technique. Buds of clonal root stocks in apple are converted by encapsulation as propagules after appropriate root induction treatment with IBA for 3 to 6 days. Artificial seeds confer advantage in scaling up of propagules production, genetic uniformity, easy delivery to the field and eliminates transplants. Chief limitations are limited production of somatic embryoids, asynchronous development of somatic embryoids, improper and lack of dormancy, maturation and stress tolerance. Details on the properties of artificial seeds, production protocols, and application aspects, have been reviewed and presented by Sai Prasad (2001) and Ravi and Anand (2012).

The scope for production and application of aritifical seeds are evident from the results on apple roots stock, indica rice, garlic bulblets, acacia, sugarcane, sweet potato, soybean and ornamental crops, Dendrocalamus, bamboo, turmeric and zinger and many other crops are reported by several authors. Rai *et al.* (2009) have described in detail the encapaulation technology in fruit crops. Adhitari *et al.* (2014) have established high genetic stability in encapsulated shoot it developed plants in Turmaria.

Procedure for Production of Artificial Seeds (Synthetic Seeds)

Establish Somatic Embryoids
↓
Mature Somatic Embryoids
↓
Synchrony and Singulate Somatic Embryoids
↓
Standardisation of Somatic Encapsulation
↓
Standardisation of Somatic Embryoids
↓
Mass Production of Somatic Embryoids
↓
Standardisation of Artificial Endosperm
↓
Mass Production of Synthetic Seeds
↓
Green House or Field growing

Source: (G.V.S. Prasad, 2001)

The benefits of synthetic seeds are particularly in recalcitrant seeds species (mango, cocoa coconut), non seed producing plants, polyploids, medicinal plants and plant species with problems in seed propagation, pathogen free nature, genetic uniformity, fast reproduction, low cost and direct to the field establishment.

Application of synthetic seeds are *in vitro* conservation, *ex situ* germplasm conservation and cryopreservation. Synthetic seed technology is emerging as a new trend in seed technology in a wide range of crops. Synthetic seeds do functional mimics of botanical seeds.

Many aspects of synthetic seed technology, productions, cryo preservation are presented by many authors in the latest book on synthetic seeds (Faisal. M and Alatar. A Eds, 2019) and review papers on synthetic seeds in medicinal, horticultural crops including. Zinger, turmeric, banana and other crops (Khatoon *et al.,* 2017; Saikat *et al.,* 2015; Rihan *et al.,* 2019) and for crop improvement (Chandra *et al.,* 2018; Ghosh and Hacque, 2019; Tripathy 2017). The genetic fidelity of plants derived from synthetic seeds was reported in Annanas, Cosmosus and Cineraria.

Technologies to develop somatic embryoids, encapsulated seed and enhancement of synthetic seed with synthetic endosperm and conversion to seedlings in hybrid rice have been developed with a success rate of 52% germination (Kumar *et al.,* 2005).

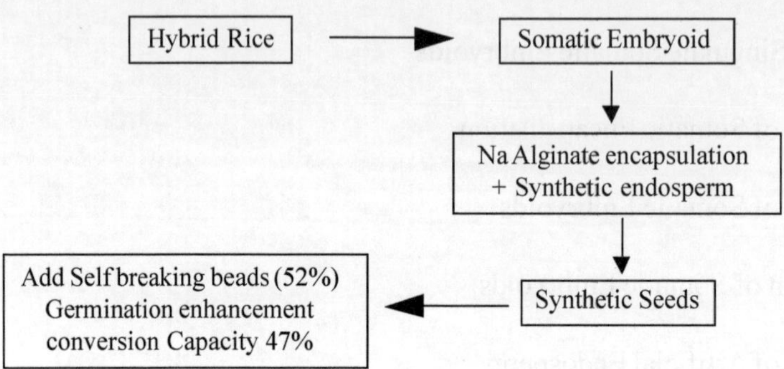

The utilization of the synthetic seed or artificial seed technology in the production of nucleus, breeder and basic seed is to explored.

Nair and Gupta (2007) have reported regeneration of plants from encapsulated somatic embryoids in pifipers.

Aspect of encapsulation and synthetic seed technology in crop plants, fruit and medicinal plants have been reviewed and presented by Rai *et al.* (2009), Sundararaj *et al.* (2010), Sharma *et al.* (2013) and Saikat *et al.* (2015).

9

Nucleus and Breeder Seed Production

Crop improvement is a continuous process. Crop improvement aims to genetically enhance, enrich and empower the crop plants with superior performance, enhanced yield level, higher quality of food, nutritional, fodder, energy produce and adaptability to climate change and crop growing environments. The intrinsic virtues of the improved cultivars could be sustained mainly through the production of the quality assured seeds endowed with genetic capabilities. The quality assured seed delivery system is through the seed chain from the basic seeds through Nucleus-Breeder-Foundation-Certified seed. Quality seed production system need to be based on genetic, breeding and seed technological protocols including field and laboratory testing to ensure very high genetic purity, true to type and superiority.

The crop plants are self-pollinated, cross pollinated or asexually propagated and each one of the crop plants has its own genetics and breeding system behavior. Every released crop variety is to be necessarily maintained by well-defined maintenance breeding, the principles of which are extended to the Nucleus seed and Breeder seed generations. The initial stages of seed chain, the Nucleus and Breeder seeds are of paramount importance to ensure complete genetic purity, true to type and physical standards of seeds. The seed quality in the other two later seed generations, Foundation seed and Certified Seed are very much dependent on the Breeder Seed quality.

It is well known established fact that the improved cultivar is higher yielding than the previously released variety. Mostly improved varieties have a yield advantage of 10% or more besides the increased resistance virtues and grain nutrition and fodder quality and input use efficiency as a result of genetically combining together all the essential needed agronomic traits. To exploit this advantage, it is necessary to start with quality assured clean Nucleus and Breeder seeds. Highest purity in Breeder seed will be reflected in the Foundation seed and Certified seed generations for the reasons these two stages are handled in many locations and by many agencies, formal and informal.

A comprehensive treatise is presented in this Book on the Nucleus and Breeder seed production in field and horticultural crops including clonal seeds in vegetatively propagated crops that are more than 30% of the crops. Good quality seeds are the prime movers in food production and enhance the nutritional security of our Nation. Seed is inclusive of virtues of such superior genotypes and enhanced performance for yield and quality of the crop, fortified with the genes for nutrition content, pest, disease and abiotic resistance. Seeds thus embody agricultural prosperity and livelihood and benefits of nutritional and consumer health. Improved and quality assured seeds are the carriers of new technologies in a single window system and thereby are the "change agents" and in revitalising the Agriculture and Nutrition. Improved seeds could be effectively delivered through well designed Genetic Delivery system through seed generations holding intact all the characteristics of the genotype per se without genetic deterioration and maintaining the original characteristics without deviation. The seed generation system is influenced by the breeding behavior of the crop to varying extent and environmental influence and exaggeration in the traits expression and also genetic drift. Essential safeguards such as adequate isolation distance to avoid cross pollination by wind and insect pollination, maintaining uniformity and true to type nature, removal of volunteers, off types and deviating type of plants as and when they occur and total efforts to avoid vicinism are to be adhered and followed. Maintenance Breeding of the improved lines, pre-released varieties, recently released varieties, established varieties and popular varieties is the prerequisite step in the framework of Nucleus-Breeder seed development and delivery chain. Application of diagnostic characteristics of the concerned variety or parental line for different characters in each generation and in each progeny and validating with true to type, uniformity and stability are the major steps to assess and adjudge the genetic purity.

Causes of deterioration in seed quality

1. Off type: a single off type rogue plant of the first season multiplies rapidly within four generations and cause total deterioration of the crop quality and crop seed degradation.
2. It is not always possible to identify the off type from the improved type due to genotypical basis without morphological differences.
3. Lack of inbuilt progeny testing of the selected superior plants for nucleus/ breeder seed production.
4. Residual genetic variability of the selected strain. Mather (1941) has pointed out that in homozygous types internal balancing and in the heterozygous types relational balancing throw out variation.
5. Premature release of variety still while segregation and the crop is not yet stabilized.
6. Recessive gene mutations.

7. Mechanical mixture.
8. Out crossing.
9. Volunteer plants.
10. Genetic drift of varieties bred for climatic conditions and grown in different environment conditions. Seeds multiplied outside its region of adapation for several generations and repeated growing in biotic and abotic stressed situation, adapt and shifts in adaptation selection. Loss of adaptation may start with first generation and become significant after three generations.
11. Always organise basic seed production in the favourable season in the region to which the variety has been bred to facilitate ideal expression of the traits including yield and free of diseases and pests.

Feistritzer (1975) reported yield increases in crops due to the quality assured seeds. Improved seeds increase the yield upto 10 to 12% in cereals, 14% in potato and 40% in sugar beet. Genetic purity refers to true to type determined by GOT, seed testing and molecular markers. Such tests help to confirm genetic purity of seeds and identifying sibs, determination of variants, segregation, seed mixtures in varieties, inbreds, hybrids, identification of outcrossing in seed lots and to determine if the Breeder seed is homozygous or not.

Diagnostic characters: Traits that are stable highly heritable and spread over the seedling, vegetative, foliar, stem, floral, spikelet, grain, glumes color, pollen and stigma color could function as diagnostic characteristics. DUS guidelines provide details on different characteristics in different growth phases in individual crops (UPOV) and Protection of Plant Variety and Farmers' Rights Authority (PPV & FRA) (Plant Authority).

The National Bureau of Plant Genetic Resources (NBPGR) has developed and presented Minimum descriptor for Agri-Hort Crops in four parts namely (1) Cereal and Other Crops (2) Vegetable Crops (3) Fruit Crops and (4) Medicinal and Aromatic Plants for guidance (NBPGR, 2001 and 2002).

Seed Generation system: India cultivates wide range of crop plants that produce food, grain legumes, oilseeds, fodder, fuel, vegetables, fruits, spices and scores of plantation crops such as coconut, arecanut, palmyrah, cocoa, pepper, cardamum, rubber, tea and coffee. Quality planting materials are essentially required which are key components. A scientifically designed seed or planting material delivery system is a prerequisite through Nucleus Seed-Breeder Seed-Foundation Seed-Certified Seed for various seeds, clones, grafts and other vegetative prapagules and *in vitro* plants for assured crop performance in generation after generation of growing. (Chowdhury and Lal, 2003). The quality planting materials are regulated by The Seeds Act (1966) dealing on the Foundation Seeds and Certified seeds. Certified seeds are produced from the

Foundation Seed which in turn is produced from Breeder Seed. Truth Full Seeds exists with declared scope within the Seed Act. Clonal clean seed is part of the seed system and is the key agent functioning in the production chain in the vegetatively propagated crop. Apart from the breeder seed and clonal seeds, micropropagation supplements the clonally propagated crops. The continued technological and R&D advancements in seed production and clonal seed production ensure quality assured seed and planting materials which support basically the maintenance of genetic purity and true to type and sustaining the increased yield and related traits performance. It is estimated that at least 8 to 10% increase in yield in seed sown crop and 25 to 30% yield increase in clonal raised crops as a result of maintaining the genetic purity and true to type nature besides repeatable yield levels. There are many field crops, vegetable and fruit crops being addressed and supported by R and D efforts on the quality assured seeds program through *in vitro* in conjunction with *ex vitro* protocols. Somatic embryoids and synthetic seeds find favour in the *in vitro* system to promote synthetic seeds.

Yield levels of crops getting gradually reduced as the crop generations advance from the time of release. The seeds of varieties show much decline in yield due to genetic deterioration besides other causes. This is evident from the intrinsic yield differences between the crops grown from Breeder seed and other generations such as certified and farm grown or farm saved seeds.

Dedicated and focussed Nucleus and Breeder seed support is an essential prerequisite to maintain the genetic potentials of the yield in the newly released varieties, established varieties, pre-released varieties, and parental lines.

A comprehensive presentation of authentic protocols and practices of the Nucleus seeds and Breeder seed inclusive of stock seeds is presented in this Book. The aim is to present a reference source Book to the agricultural scientists involved in the Nucleus and Breeder seed and Clonal seed maintenance and production in field, vegetable and perennial crops. Few crops that have not been previously focused for Breeder Seed Production protocols have been included.

The labelling colour and storage of the Breeder seeds are essential steps as given below:

The colour of the label for breeder seed is golden yellow and the size should be 12 cm. x 6 cm. The rear of the tag should be rubber stamped with proper signature, name and designation of the producing breeder. The colour of the tag for foundation seed is white tag, certified seed is azure blue and for Nucleus Seed, there is no specific tag (Fig. 1).

Grow Out Tests are prescribed to assess and adjudge the genetic purity and trait purity by actual growing of the crop with seed samples from Nucleus Seed and Breeder Seed. The GOT procedures and other guidelines are enumerated below as detailed and in the Indian Minimum Seed Certification standards (2013). The genetic purity of the nucleus and breeder seed is 100% and that of foundation seed is 99.00% and certified seed is 98%.

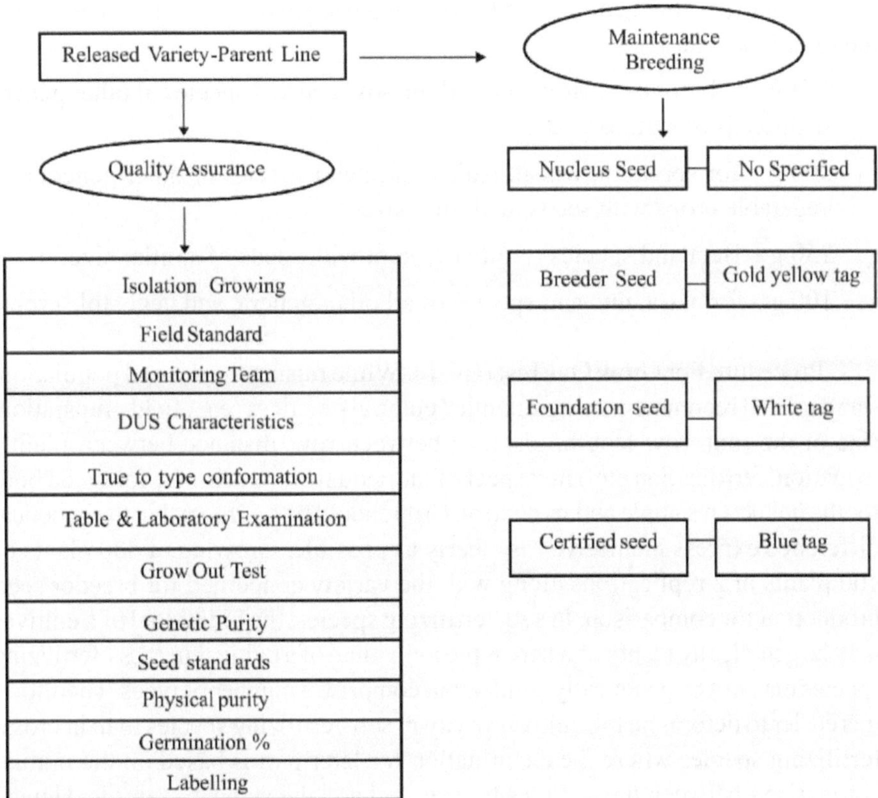

Fig. 1. Generation System of Seeds

Storage: Seed should be properly dried to 10 to 12 percent moisture content (or below 9 percent, if packed in moisture proof container) and stored in a proper place or site preferably dry and cool. During the storage period, care should be taken for control of storage insects and fungi.

Grow Out Test: (Indian Minimum Seed Certification Standard) The monitoring of genetic purity of nucleus and breeder seed lots is very important which can be done by grow out tests. For this a sample of appropriate size is drawn from the graded seed lots and grown as per the guidelines in the field to observe the genetic purity. The grow out test is conducted in the area where crop can maximum express the characters without any variation due to environmental influence. The crop is monitored throughout the growing season at different stages of crop growth and off type plants, if any, are tagged and counted at the maturity to assess the genetic purity. The lots which are not of desired genetic purity should be rejected.

Sampling: The samples for grow-out test are to be drawn simultaneously with the samples for other quality tests following the standard procedure. The size of the sample will be:

- 1,000 g – for maize, cotton, groundnut, soybean and species of other genera with seeds of similar size
- 500 g – for sorghum, wheat, paddy, okra and species of other genera and vegetable crops with seeds of similar size
- 250 g – Beet and species of other genera with seeds of similar size
- 100 g – for bajra, jute and species of all other genera, and vegetable crops.

Procedure For Grow Out Test (GOT): While raising the desired population, standard and recommended agronomic / cultural practices (e.g., field preparation, size of the plot, row length, distance between row, distance between plants, irrigation, fertilization etc.) in respect of individual crops are to be followed both for the unknown sample and its control. Crop should be so grown that the genetical difference express themselves as clearly as possible. Growing of 800 plants or 400 plants in 2 replications along with the variety concerned for breeder seed production for comparison. In self-fertilizing species, the individual of a cultivar may be genetically identical whereas the individual of a cultivar in cross-fertilizing species may not be genetically similar, but comprise a number of types. Therefore, it is easier to determine the cultivar purity in self-fertilizing species than in cross-fertilizing species where the examination for later part is based on the mutual comparison between the samples to be tested and the standard sample. Hence, it is essential to sow the various samples of the same cultivar in succession and standard samples are sown at suitable intervals (for example one standard sample for every ten samples to be tested). The size of plots, row length etc. will differ from crop to crop. However, the specifications for different crops are indicated in the following Table 1.

Table 1. Specifications for different crops for conducting GOT

S. No.	Crop	Row length (m.)	Distance (cm.) between plants	Distance (cm.) between rows	Distance (cm.) between plots	No. of replications
1.	Wheat, barley, oats	6	2	25	50	2
2.	Pea, cowpea	6	10	45	90	2
3.	Chickpea, greengram, blackgram	6	10	30	60	2
4.	Maize	10	25	60	90	2
5.	Hybrid cotton	5	10	45	45	2
6.	Paddy:					
	a) Very early to mediumduration	6	15	20	45	2
	b) Late and very late duration	6	25	30	60	2
7.	Pearl millet	6	10	60	90	2
8.	Sorghum	6	10	45	60	2

Source: Indian Minimum Seed Certification Standards (2013)

The seed rate may be adjusted depending on the germination percentage of individual samples and the sowing may be done by dibbling. Subsequent thinning is not recommended. The test crop may be raised along with the control either in the areas recommended for the variety or in off-season nurseries. The authentic control sample from the originating plant breeder/breeding Institute is to be maintained by the testing Station/Agency following standard procedures. A minimum of two hundred plants from control sample will be raised along with the test crop.

Observations: All plants are to be studied keeping in view the distinguishing characters described for the cultivar both in the test crop as well as the control. Necessary corrections may be incorporated if the control is found to be heterogeneous. Observations are made during the full growing period or for a period specified by originating breeding Institute and deviations from the standard sample of the same variety are recorded. At suitable development stage of crop growth, the plots are examined carefully and plants which are obviously of other cultivar are counted and recorded.

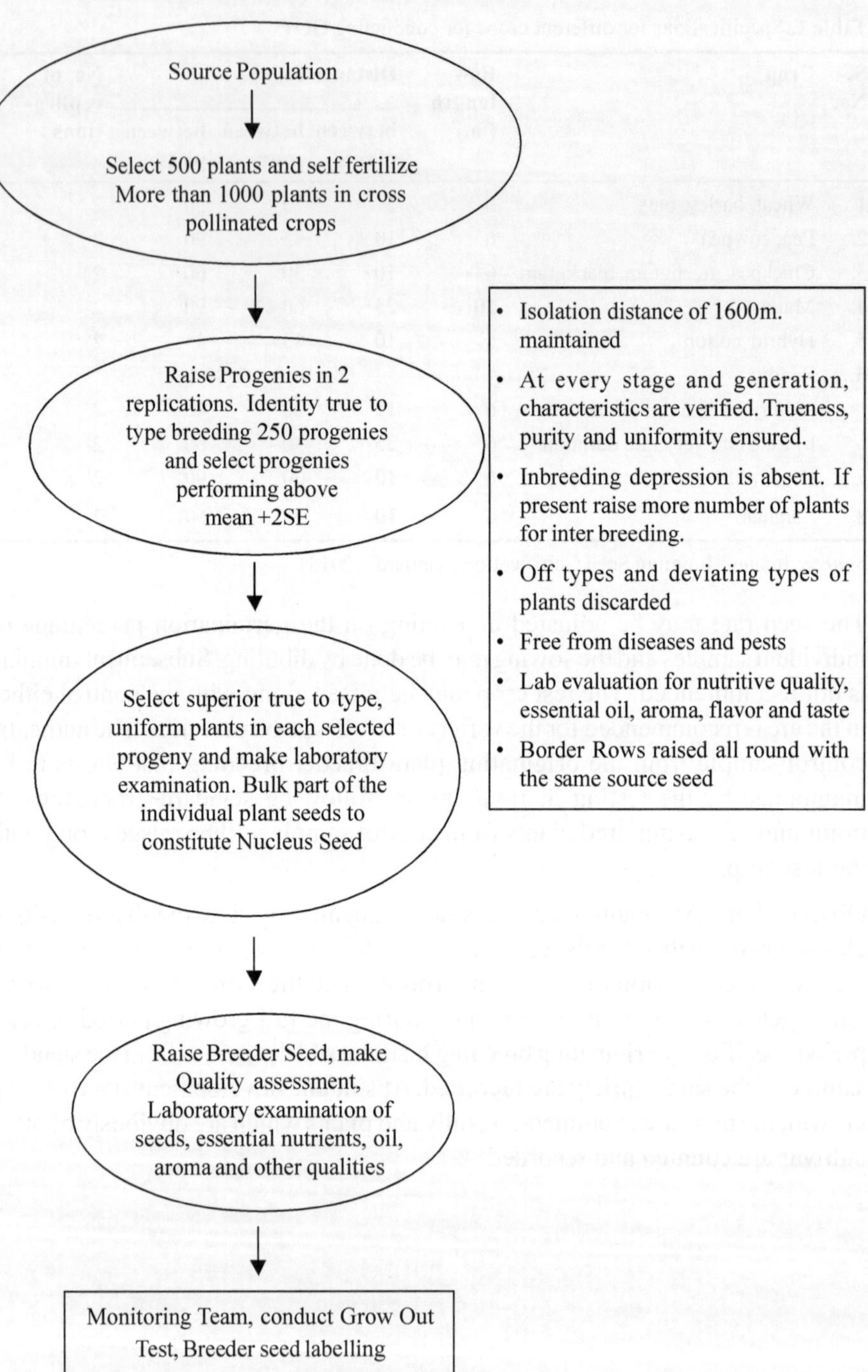

Fig. 2. Flow diagram for Nucleus and Breeder Seed Production

Clonal Seed Sourcing

↓

Raise in clean field

↓

Clean the clones/propagules from diseases, pests, virus and other contaminants

↓

BASIC SOURCE CLONES

↓

Clonal Progeny Evaluation

↓

BASIC CLONE – G0

↓

G1, G2, G2, G4 generations.

Fig. 3. Clonal Seed Production

The basis of the number of plants required for taking observations is dependent on maximum permissible off types which are as follows (Table 2).

Table 2. Minimum genetic purity (%) and maximum permissible off types

Maximum permissible off types (%)	Minimum genetic purity (%)	Number of plants required per sample for observation
0.10	99.9	4,000
0.20	99.8	2,000
0.30	99.7	1,350
0.50	99.5	800
1.00 and above	99.0 and below	400

Source: Indian Minimum Seed Certification Standards (2013)

Calculation, interpretation and reporting: Percentage of other cultivars, other species or aberrant plants found may be calculated upto first place of decimal. While interpreting the result, use of tolerance may be applied by using the reject number (Table 3).

Table 3. Reject numbers for sample size of standards (Indian Minimum Seed Certification Standard)

Reject number for prescribed standards and sample size	800	400
99.5 (1 in 200)	8	*
99.0 (1 in 100)	16	8
95.0 (5 in 100)	48	24
90.0 (10 in 100)	88	44
85.0 (15 in 100)	128	64

* Indicate that the sample size is too small for a valid test.

Source: Indian Minimum Seed Certification Standards (2013)

Progeny Testing: Progeny testing of selected true to type single plants is essential while forwarding the superior plants in each of the generations, such as maintenance breeding, nucleus seed and breeder seed. After visual evaluation for trueness in the vegetative and reproductive phases, progeny performance such as grain yield, biomass yield and fruit yield are recorded and progenies recording more than mean ± SE or mean ± 2 SE as the case may be selected to the next generation. Standard error is the measure of uncontrolled variation in a sample, estimated by dividing the standard deviation (SD) by the square root of the number of observations in the sample. It is a measure of variability among population mean.

Besides the field evaluation of the single plants or progenies, laboratory or table examination of ear heads, glume, and grain characteristics are examined in single plants and only those ears and grains in conformity alone are selected for next generation. The composite samples of the nucleus/ breeder seed are tested/ estimated for nutritive content, oil contents, micronutrient, to ensure that these contents are equal to or more than the values of the check entry.

GOT is performed by molecular testing using appropriate DNA markers. Seedling level genetic purity is designed to be conducted under controlled growing conditions in green houses and scoring and the seedling stage itself, prior to 30 DAS.

In India, Seed Production is being done in a formal system regulated by public policies, Seed Act, Seed Testing and controlled by regulations. In the informal system, seed production is being carried out by Farmers and by the private sector.

Classes of Seed (Indian Minimum Seed Certification Standards)

Nucleus Seed: The nucleus seed is the initial seed obtained from selected individual plants of a particular variety/ parental line for the purpose of purifying and maintaining that variety/parental line by the originating Breeder and its further multiplication under his own supervision or the supervision of a qualified breeder. It is used to produce breeder seed and forms the basis of the total seed production chain.

Breeder Seed: Breeder Seed is the seed or the vegetative propagating material directly controlled by the originating Breeder or sponsored plant breeder or institution and /or seed whose production is personally supervised by a qualified plant breeder and which provides the source for the initial and recurring increase of foundation seed. Breeder seed shall be genetically so pure as to guarantee that in the subsequent generations ie foundation and certified classes shall confirm to the prescribed standards of genetic purity. The other quality factors of breeder seed such as physical purity, inert matter and germination shall be indicated on the label on actual bases.

Foundation Seed: The foundation seed should be the progeny of breeder seed or be produced from foundation seed which can be clearly traced to the breeder seed. The production of foundation seed shall be supervised and approved by the certification agency and be handled as to maintain specific genetic identify and genetic purity and shall be required to confirm certification standard specified for the crop during variety being certified.

Certified Seed: Certified seed should be the progeny of foundation seed and its production should be handled as to maintain specific genetic identify and purity according to standards prescribed for the crop being certified. Certified seed may be the progeny of certified seed provided its reproduction does not exceed three generations beyond foundation seed stage I and it is determined by the certification agency that genetic identify and purity not altered significantly and the certification Agency is satisfied that there is genuine shortage of foundation seed.

The genetic identity and purity of the foundation and certified seeds very much depends on the quality of the nucleus and breeder seed used. The maintenance of original characteristics of the variety/parental line at the nucleus and breeder seed level is therefore very much important. It is absolutely necessary to select single plants possessing true to type characteristics of the particular variety or parent line from the base population and grow them in plant progeny rows for proper evaluation and rejection of undesirable types.

The breeder seed production is highly specialized procedure and demands special care and attention. Although the breeder seed does not come under the purview of Indian Seed Act (1966), the breeder has to follow much more stricter minimum seed certification standards for nucleus and breeder seed production than what the Indian Seeds Certification Board has prescribed for the foundation and certified seed production. Roguing is done at least three times. Variations if any, for DUS characteristics are noticed and the off-types are rogued. Each and every plant is examined through the different stages throughout the growing season with emphasis on the marker characters and time of their expression. The difference between certified seed and truthful seeds are given in Table 4.

Table 4. Differences between certified seed and truthful seeds

Certified Seed	Truthful seed
Certification is voluntary	Truthful labelling is compulsory for notified kind of varieties
Applied to notified kinds only	Applied to both notified and released varieties
Shall satisfy Seed quality standards	Tested for physical purity
Minimum field and seed standards	Purity and germination
Seed certification officials, seed inspectors can take samples for inspection	Seed inspector alone can take samples for checking

The plant characteristics which are highly heritable and least influenced by the environment are considered to check the purity and genuineness of a variety.

The groups of crops included for Nucleus and Breeder Seed Production are:

1. Self pollinated crops
2. Cross pollinated crops
3. Often cross pollinated crops
4. Composites and synthetics
5. Parental lines of hybrids
6. Vegetativey propagated crops
7. Apomictically reproduced crops

In addition, crop plants propagated by *in vitro* means like *in vitro* plants, synthetic seeds, mini and microtuber and rhizome can be also considered.

Table 5. Seed standards of genetic purity (Minimum Standards of Genetic Purity) for different classes of seeds. (Indian Minimum Seed Certification Standard)

Class	% Genetic purity
Nucleus Seed	100.00
Breeder Seed	100.00
Foundation Seed	99.00
Certified Seed	
i) Varieties, parents lines, composites, synthetics, multilines	98.00
ii) Hybrids	95.00
iii) Hybrids of cotton, musk melon, brinjal and TPS	90.00
iv) Hybrid castor	85.00

Table 5 a) Genetic Purity in different kinds of seeds

%	NS	BS	FS	CS
Genetic Purity	100	100	97-98	98
Physical Purity	100	100	>95	>90
Minimum germination	100	100	70 to 80	60 to 85
Maximum inert matter	0	0	2	2

Rajendra Prasad *et al.* (2017) have discussed and presented the National and International quality assurance system and strategies for emerging seed. production chain of field crops in India. The Notification and denotification system in varieties in India is published by Chand *et al.* (2020).

The Association of Official Seed Certifying Agency AOSCA is following the four classes of seed:

1. Breeder Seed: The Seed which is contributed directly by the organization or the sponsoring plant breeder.
2. Foundation Seed: The progeny of Breeder Seed or foundation seed handled to maintain specific genetic purity and identity.
3. Registered Seed: The progeny of breeder seed or foundation seed handled to maintain satisfactory genetic purity and identity.
4. Certified seed: Progeny of Breeder Seed or foundation seed or registered seed handled to maintain satisfactory genetic purity and identity.

Prerequisites or prior knowledge desirable for planning and execution of Nucleus and Breeder Seed Production

1. Annual or Biennial crop or perennial clonally propagated.
2. Breeding behavior of the crop
3. Pollination mode (self or cross pollinated crop)
4. Extent of Natural cross pollination
5. Pollinator / pollinating agent
6. Trait purity
7. Heritable nature of the traits
   - Low heritability
   - High heritability
8. Residual variation-extent
9. Inbreeding depression- Biparental, if, any
10. Diagnostic characteristics (DUS)
11. Impact / influence of the season and location
12. Photoperiod response
13. Maintenance Breeding
14. Number of plants forwarded from Base / Maintenance Breeding population
15. Number of Progenies and progeny size; Progeny selection above the mean ±2SE
16. Pests and diseases incidence, prone to micronutrient deficiency
17. Molecular GOT / DNA Finger Printing details / protocols

18. Isolation distance required
19. Clean, free from contamination of seed and planting material (Clean Seed)
20. In case of clonally propagated crops, the cleanliness and contamination free nature of propagules. Crop rotation is ideal for clonal seed multiplication nursery raising fields to avoid contamination build up.

10

Maintenance Breeding

Deterioration of Crop varieties in their genetic endowments/qualities has been explained by Kadam (1942). Publication on maintenance and management of breeder seed production (Laverack 1991), and maintennce of self pollinated crop varieies (De Hanns 1953; Parlevliet 2007) explain in detail various aspects in maintenance of the seed quality.

Maintenance breeding aims at preserving the genetic status of a variety or parent. It is a plant breeding activity involving steps to maintain the genetic structure, genetic purity, genetic integrity and true to type nature of the basic seed material of the variety or parent.

In the course of maintenance breeding, procedure of avoiding genetic deterioration, genetic drift, mutations, genetic segregation and inbreeding depression (biparental) (Allard *et al.,* 1968) are adapted so as to provide continuous flow of quality assured true to type seeds for developing nucleus and breeder seed. Maintenance breeding also improves the adaptation of the newly developed varieties to environment changes (Peng and Khush, 2010).

In the maintenance of crop varieties developmental variations due to place/ location effect, mechanical mixture, micromutations, cryptic variability, selective influence of pests and diseases and climate changes are overcome (Harrington, 1952).

Continued maintenance breeding efforts to preserve the yield potential, resistance to rapidly evolving biotic stresses and increased tolerance to bio physical and climate changes are recommended as a follow up after releasing a new cultivar.

In the self pollinated crops, 500 ears of healthy representative and true to type plants are selected at maturity from the base population raised from the seeds of breeder source. Off-types and deviating type of plants are carefully discarded from selection. In cross pollinated crops varieties parent lines, 2000 to 2500 representative healthy, true to type plants are selected. The earheads of selected single plants are verified for true to type nature and deviating types or variants are discarded. Seeds separated from the selected ear heads are studied in the laboratory for shape, grains setting and uniformity, true to type and free from seed borne diseases. The bulked seeds obtained from the cross pollinated crop will constitute basic seed.

In respect of inbreds, it is advisable that 500 true to type plants may be selected and properly self fertilised and seed set upon selfing could be collected and examined in the laboratory and saved plantwise. The seeds of the selected plants in the self pollinated true to type plants are bulked together to make the basic seed of the inbreds.

Mass selection is thus adopted based on phenotype expression in respect of traits and characteristics that are highly heritable and/or low heritable in nature. The steps to basic seed production by maintenance breeding are:

1. Fertile, disease free highly productive field is selected that is isolated by 1600 meters from other fields growing with the same crop. The location and season should be highly favourable for the variety to maximum expression of the characteristics, conforming to the descriptor of characteristics for the variety.
2. Seed are sown in hills in 1000 sq.m area and single plant per hill is retained at seedling stage. All recommended crop production/ agronomic practices are followed.
3. The field is kept free of weeds. Offt-ypes and volunteer plants, disease and pest affected plants are uprooted. The individual plants are examined periodically till harvest and deviating plants are identified and discarded. This exercise is regularly continued in each stage of growth from seedling to flowering and maturity in respect of inbreds and parents. Selected true to type plants are self fertilized in case of inbred lines.
4. Representative true to type and healthy plants from the middle of the field are selected and tagged, harvested plant wise in separate paper bags, dried, seeds extracted and saved for laboratory examination of the seeds. Based on the verification and conformity of seeds, selection is made. Samples of the seeds are bulked and tested for germination and seed purity norms and other seed characters. The quality assured seeds are saved as basic seeds for utilizing in nucleus-breeder seed production cycle.

In respect maintenance breeding in clonal crops selection against bud sports and somaclonal variation need to be practiced intensively. Production of disease free and other contamination fee is absolutely necessary.

It is desirable to estimate the nutritive, micronutrients, oil content and other qualities in the seeds, as may be the case, in order to assure nutritional qualities.

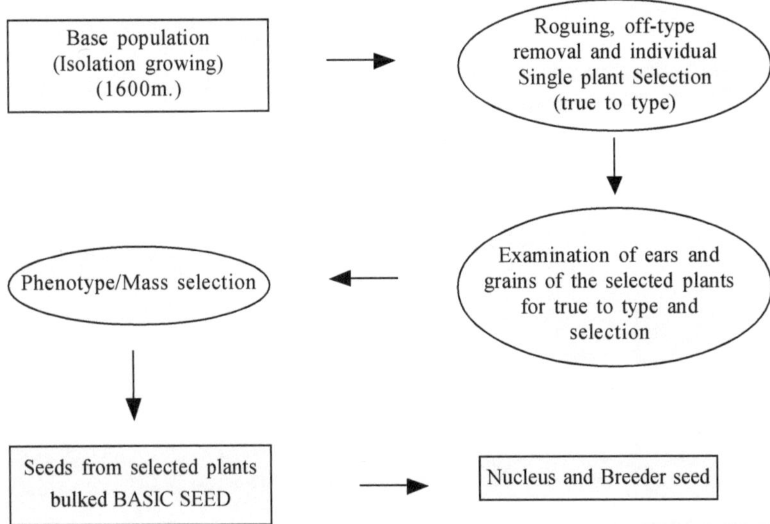

Fig. 1. Maintenance Breeding

Minor variation is caused by polygene complexes under changed environments. In fact, homozygous types posses internal balancing and heterozygous types in relational balancing effects (Mather, 1941). Both of them throw out variations in case of polygenes. Therefore compensating mechanisms in the form of balanced polygenes exist in nature (Mather, 1949). However, the defect will be lesser in selfers than in selections from hybrid programs (eg. in cotton).

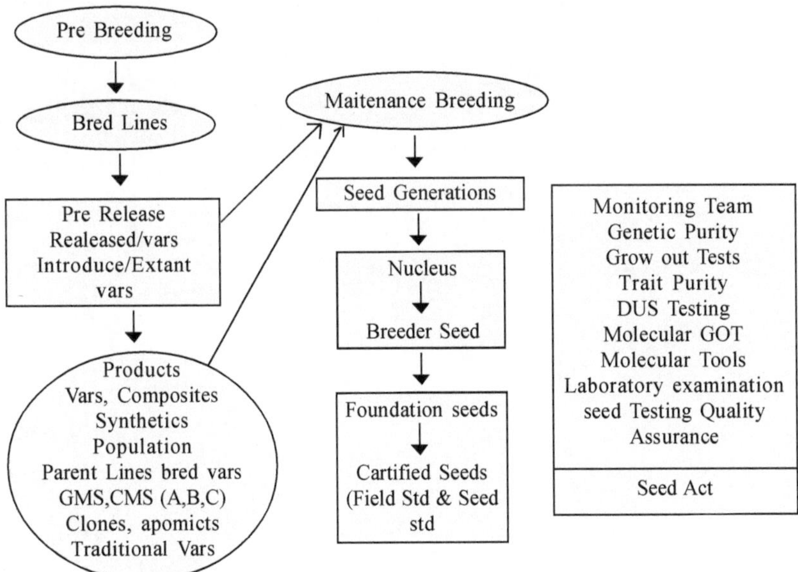

Fig. 2. Breeding Maintenance breeding Seed generations and Genetic purity

11

Molecular Markers for Seed and Clonal Purity and Grow Out Testing (Molecular GOT)

Application of molecular markers:

1. Molecular genotyping / finger printing profile
2. Genetic characterization of parents, inbreds and clones
3. Cultivar identification
4. Genetic diversity studies
5. Genetic purity testing
6. Clonal fidelity testing
7. Molecular GOT
8. DNA barcode for identification and IPR protection
9. Germplasm identification

Molecular markers may be a short DNA sequence that flanks base pairs like SNP, a Single Nucleotude Polymorphism, small fragment or SSR Single Sequence Repeat or an extended site as in the case of microsatellites. Schematic representation of the SSR assay is shown in Fig. 1. Genetic markers have to be easily identified, linked to a specific locus and polymorphic. Molecular markers are subdivided into codominant monolocus and polyallelic markers and the second group contains dominant, polyallelic and biallelic markers. The most wide spread codominant markers are the microsatellites. Codominant markers are utilized for the identification and registration of varieties and determination of genetic purity. With the advent of PCR, DNA marker technology has been revolutionized. PCR based markers like RAPD, SSR, and ISSR are faster and technically easier and the cost is less.

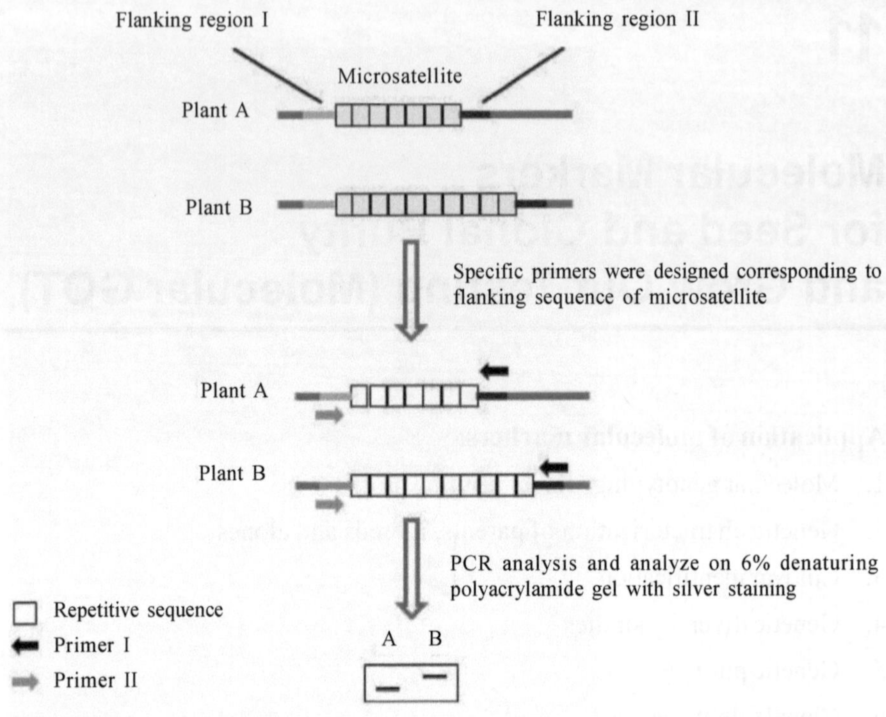

Fig. 1. Schematic representation of SSR assay

DNA markers reflect heritable differences in homologous sequence among individuals. DNA markers are abundant, codon neutrals and hence no effect on phenotype, no epistasis effect, no pleiotropic effect and not influenced by the environment. They are used for the identification and registration of varieties, seed purity and authenticity.

SSR based DNA barcodes were used as a tool for the identification of plant varieties. DNA barcode help in easy and accurate identification of genotypes in combination of morphological characters and also to address the IPR issues.

Microsatellite markers are short tandem repeats of simple 1 to 6 nucleotides as motifs. Their multiple alleles and codominant nature are inherited. They are spread over coding regions as well (Morgante *et al.,* 2002). Genotypes with at least one specific combination of alleles that allow accurate fingerprinting and easy detection of genotypic differences are useful.

Compared to other markers, SSR markers are widely utilized for studying and analyzing the genetic relationship between individuals, genetic divergence and genetic variability among plant populations (Fig. 2). SSR is frequently used due to its codominant nature, a high degree of polymorphism, multiple alleles for each marker, stably inherited, high reproducibility, transferability between related species and high-resolution capacity.

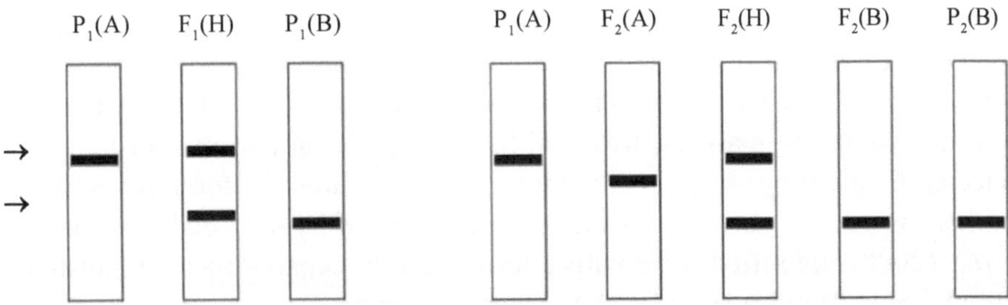

Fig. 2. Co-dominant SSR markers showing Mendelian segregation in parents and F2 population

The main restriction for their use is that microsatellite loci have to be selected and sequenced to promote specific amplification and that their polymorphism has to be tested in each sample (De Lucas *et al.,* 2008). Molecular barcoding using microsatellite markers for fingerprinting profile for CVs clearly shows unique patterns and represents standard or reference barcode profile. It is a visual representation of genotypic differences.

DNA markers are historically and conceptually an extension of morphometrics used in evolutionary, taxonomy, population studies and breeding. DNA markers are answering the same questions as to: (i) how similar are individuals and populations and (ii) how are the differences among the subjects associated with (Alizadeh *et al.,* 2015). The comparison of most used markers is presentrd in the Table 1.

Table 1. Comparison of most used markers

Feature	RAPD	SSR	SNP
DNA required µg	0.02	0.005	0.05
DNA quality	High	Moderate	High
PCR Based	Yes	Yes	Yes
No. of polymorphic loci analysed	1.5 to 50	1.0 to 3.0	1.0
Ease of use	Easy	Easy	Easy
Amenable to automation	Moderate	High	High
Reproducibility	Unreliable	High	High
Development Cost	Low	High	High
Cost per anaylysis	Low	Low	Low
Samples per day	50	50	Unlimited due toautomation

DNA fingerprinting utilizing microsatellites is now applied in many crops. In rice, Rahman *et al.* (2009) found discrimination among HYV and local rice

CVs and two wild sp with the aid of microsatellite that is more polymorphic. DNA fingerprinting was applied in aromatic rice vars. Short grain aromatic rice CVs formed different clusters with high similarity/local among themselves. All the aromatic rice CVs included in the study could be distinguished from each other on the basis of comparison over primers (Choudhary *et al.*, 2001). Variation within parent lines in rice was detected by Nandakumar *et al.* (2004) by tagging with microsatellite markers. Residual heterozygosity at two marker loci was detected by them by analysis of plant to plant variation. Restorer gene-linked markers were also used to test the genetic purity of hybrid seeds. Sundaram *et al.*, (2008) identified informative SSR markers capable of distinguishing parental lines of hybrid rice for testing genetic purity.

Recently Robin *et al.* (2019) reported DNA fingerprinting of rice variety CR1009sub 1 and CR 1009 produces distinguishable alleles with the allele size of 80bp and 90bp respectively for RM219 and also produces 185bp and 190bp respectively for RM23865 in the agarose gel electrophoresis (Fig. 3).

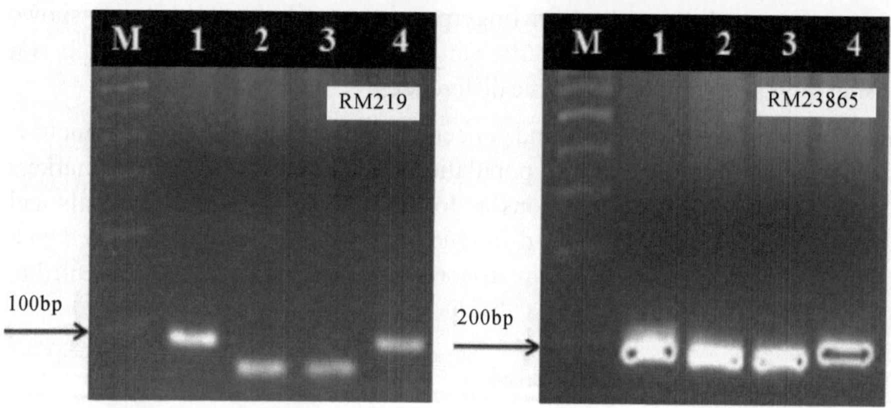

Fig. 3. DNA fingerprinting of rice variety CR1009sub 1 and CR 1009 produces distinguishable alleles with the allele size of 80bp and 90bp respectively for RM219 and also produces 185bp and 190bp respectively for RM23865 under the agarose gel electrophoresis.

Note: (M) Ladder (100bp); (1) SS1; (2) S; (3) CR 1009; (4) CR 1009sub 1.

DNA fingerprinting in rice was done with three varieties viz., CO53, PMK 3 and Norungan which revealed that the amplicons produced are with distinguishable base pair size. For instance 210bp in Norungan, 220bp in CO53 and 240bp in PMK 3 for the marker RM13819. Also 250bp in Norungan, 260bp in CO53 and 280bp in PMK3 for the marker RM 5883 (Fig. 4). Sandhiya *et al.* (2020) reported DNA finger printing in the eight sorghum varieties. All the vaieties showed distinct allele for the marker xtxp 312 (Fig. 5). Senthil *et al.* (2020) reported DNA finger printing of little millet (*Panicum sumatranse*) variety ATL 1 where the marker phi213984 scored distinct allelic pattern for the newly released variety ATL 1 from the other varieties (Fig. 6).

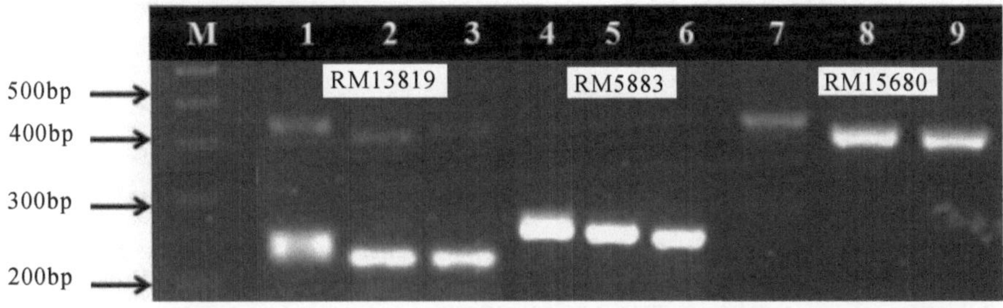

Fig. 4. DNA fingerprinting of rice CO53 variety with PMK 3 and Norungan produces distinguishable alleles with the allele size as follows. PMK 3 (240bp), CO53 (220bp), Norungan (210bp) for RM 13819; PMK 3 (280bp), CO53 (260bp), Norungan (250bp) for RM 5883; PMK 3 (450bp), CO53 (400bp), Norungan (420bp) for RM 15680.

Note: (M) Ladder (100bp); (1,4 and 7) PMK 3; (2,5 and 8) CO53; (3,6 and 9) Norungan.

Table 2. Allele size (bp) polymorphic (variation) in three rice varieties shown in Fig. 4.

Marker	Co53	PMK3	Norungan
RM 13819	220	240	210
RM 5883	260	280	250
RM 15680	400	450	430

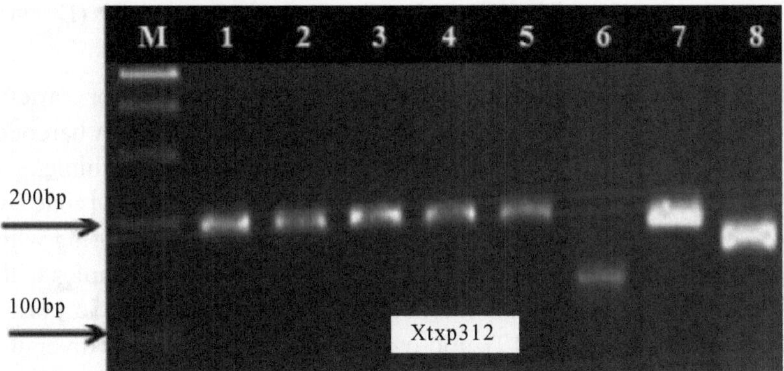

Fig. 5. DNA Fingerprinting of sorghum varieties/cultures using sorghum SSR markers. Allelic pattern clearly distinguishes the genotypes.

Note: M-Ladder (100 bp); (1) CO 30; (2) CO(S)28; (3) TNS 660; (4) TNS 661; (5) CO 32; (6) CSV 33 MF; (7) K8; (8) K12.

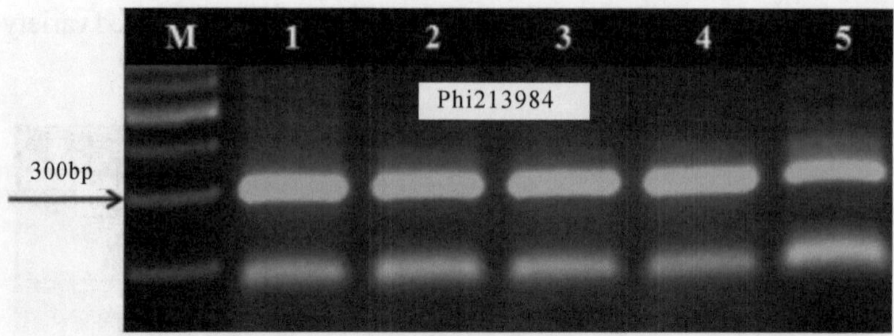

Fig. 6. DNA Fingerprinting of Little millets (*Panicum sumatrense*) using maize SSR marker (phi213984) of the newly released variety (5) ATL had a differential allele size of 310bp with all other varieties viz., (1) BL6; (2) TNSu176; (3) CO (Samai) 4; (4) Paiyur 2. (M) Ladder (100bp).

Genetic Purity testing by DNA markers

Microsatellites, DNA markers are used for patenting, Plant Variety Protection (PVP) and Essentialy Derived Variety (EDV). In DNA markers, the repeat region is variable between samples while the flanking regions are conserved. The alleles differ by the number of repeats. The repeats are made of motifs of various nucleotides lengths and highly polymorphic (2 to 15 alleles per locus) and Mendelian codominant.

Genetic purity of cotton hybrids was ascertained by using SSR markers by Selvakumar *et al.* (2010). Use of all three markers RAPD, ISSR, and microsatellites in combination was more reliable than using the three in isolation for identification and testing of genetic purity of parents of cotton hybrids (Dongre *et al.,* 2011).

In a range of vegetable crops, molecular markers were applied for variety identification and genetic purity by Liu *et al.*, (2004). SSR based DNA barcode as a tool for identification of eggplant and to address IPR issues by Chinnappa Reddy *et al.* (2012), identification of grapevine by utilizing DNA barcode by Gabbaes *et al.* (2015), CV identification in guava by DNA fingerprinting with microsatellites by Kanupriya *et al.* (2011), DNA profiling of chilli samples with ISSR by Lekha *et al.* (2001), SSR marker fingerprinting and genetic purity testing in onion (Cohen *et al.,* 2012), microsatellite marker for genetic diversity studies in onion by Khar *et al.* (2011) and assessment of genetic relations in Mango using RAPD markers by Ravishankar *et al.* (2000) are some of the examples for application of molecular markers. The DNA markers techniques for plant variety identification was reviewed and presented in cotton hybrids by Senthil *et al.* (2016).

Variety identification involves the DUS testing system based on the morphological and traditional phenotypic evaluation following crop-specific description (UPOV,

IPGRI and PPV & FRA). It becomes difficult to distinguish varieties solely on the morphological characters when all of them have been developed from few elite plant varieties. DNA fingerprinting is an authentic way to protect plant varieties by appropriate characters and documentation.

Genotyping benefits for breeding programs are 1. assessment of genetic diversity of breeding varieties, lines, clones, and cloned seeds 2. determination of similarities of breeding population 3. protection of valuable IPR and 4. identification and verification of unique genetic fingerprinting for PVP patent application.

According to genetics, a variety is a genotype or combination of genotypes which differs in its traits from another group of plants. A genotype is characterized by a set of alleles that uniquely differentiates an organism or group of the organism from others. Marker technologies promote a significant increase and the acceleration of the variety development. (Sivolop, 2013).

The morphological differences noticed may not be sufficient for discrimination and identification of old varieties, new varieties and extant varieties. This emphasizes the need for an alternate method to characterize CV. Molecular markers distributed over a large number of loci throughout the genome of plants are made use of to study the differences.

Identification of Plant varieties (UPOV system): is a plant grouping within a single botanical taxon of the lowest known rank. It can be defined by traits resulting from (a) given genotypes and distinguished from another plant grouping by at least one of the said traits. It is considered as a unit with regard to its suitability for being propagated unchanged. Identification of variety is by using diagnostic traits (markers) to recognize an individual as being part of a particular sole species division and as a separate one such sole species cluster from others.

Genetic purity is a population concept and is assessed by the identity of its members. DNA markers are used for genetic purity testing based on DNA fingerprint/ genotyping of markers. The diagnosis is done for genetic purity. Genetic purity testing is used by plant breeders and seed producers as a quality assurance tool to identify selves, outcrosses, seed mixes, seed swaps, and segregation within the parent lines. Steps involved in the fingerprinting for variety purity test is given in Fig. 7. It is a Regulatory compliance. In the DNA based methods to study genetic purity, SNP is being mostly used.

Genetic purity testing benefits are: 1 confirmation of genetic purity standards 2. identification of selfing and outcrossing in seed lots 3. determination of variants, segregants, seed mixes. and 4 variety verification.

Genetic purity testing benefits for breeding programs are the identification of outcrossing present in the parents, inbreds are pure, free of outcrossing and free of segregation and assuring purity of a line as Nucleus, Breeder, Foundation and Certified seeds.

Sampling from seed lots
Sub-Sampling
↓
DNA Extraction
↓
Purification and Quantification of DNA
PCR amplification using genome wide SSR markers
↓
Electrophoresis and Documentation
↓
Assessing seed genetic purity

Fig. 7. Steps involved in finger printing process for varietal purity test

Clonal Purity / Clonal Fidelity Testing: Extensive testing for indicating purity exactness, faithfulness and true to type nature is assessed in clonally propagated, micropropagated and kinds of other *in vitro* derived plants by using DNA molecular markers. Assessment of clonal fidelity by using molecular markers in an array of horticultural crops has been reported by Alizadeh *et al.* (2015) and Debnath (2017) in vegetable and fruits crops and by De Lucas *et al.* (2008) in commercial Populus clones. SSR markers helped to precisely identifying CV, to verify the botanical names and useful in cataloging, also help to avoid duplication and useful in certification protocols. While the morphological, physiological, phenological and floral characters are always not advisable since they are visual, the marker-based genotype profiling is precise and not influenced by the environment.

Kanupriya *et al.* (2011) made identification of CVs in guava by genetic fingerprinting by developing molecular barcodes (allelic barcode) for all the nine varieties studied. Molecular diversity analysis was reported in guava by Prakash *et al.* (2002). Identification of grapevines cultivars has been demonstrated by Gabbaes *et al.* (2015) by utilizing the microsatellites based DNA barcodes. DNA profiling with molecular markers has helped to solve the disputed chilli samples (*Capsicum annuum*) by using ISSR-PCR and FISSR-PCR marker arrays (Lekha *et al.,* 2001). Raghu *et al.* (2007) reported Cassava varieties viz., CO_1, CO_2, CO_3 and CO (TP) 4 obtained fron TNAU and CTCRI were allele specific to SSRY324. Thus, it indicates the existence of wide variation in the cassava accession while comparing with morphological data (Fig. 8).

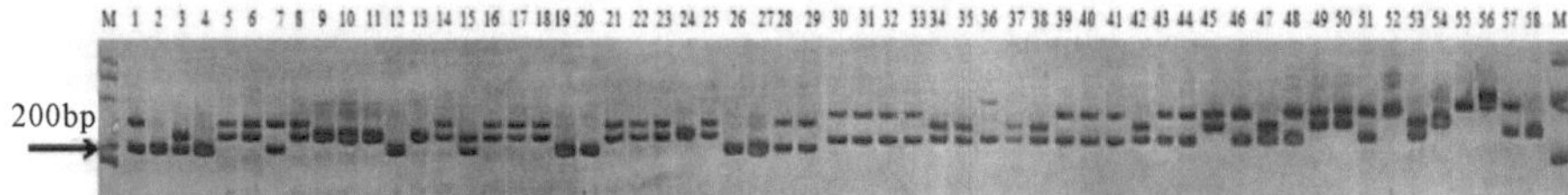

Fig. 8. Fingerprinting of South Indian cassava germplasm using SSR marker (SSRY324).

Note: (M) Ladder (100bp); (1) Me.1; (2) Me.2; (3) Me.4; (4) CO 1; (5) Me.11; (6) Me.12; (7) Me.14; (8) Me.16; (9) Me.25; (10) Me.37; (11) Me.41; (12) Me.48; (13) Me.50; (14) Me.57; (15) Me.59; (16) Me.62; (17) Me.73; (18) Me.75; (19) Me.77; (20) Me.80; (21) Me.81; (22) Me.86; (23) Mulluvadi; (24) Me.136; (25) Me.142; (26) Me.143; (27) Me.149; (28) Me.150; (29) Me.152; (30) H226 (Vellairose); (31) Me.156; (32) Sree Sahaya; (33) Me.158; (34) Me.162; (35) Me.163; (36) Me.164; (37) Me.165; (38) Me.166; (39) Me.167; (40) Me.168; (41) Me.169; (42) Me.170; (43) Me.171; (44) Me.173; (45) Me.175; (46) Me.178; (47) Me.179; (48) Me.681; (49) Me.690; (50) Me.695; (51) Me.725; (52) Sree badmanabha; (53) CO 2; (54) CO 3; (55) CO (TP)4; (56) Sree prakash; (57) Ambakadan; (58) Wild.

Tamilkumar *et al.* (2009) reporded the marker RM234 as a unique microsatellite marker which differentiates the rice parental line of CORH3 hybrid. Likewise markers viz., RM276, RM202 and RM204 for the hybrids KRH2, PRA10, AJAY and RAJALAKSHMI respectively (Fig. 9). Impurities are identified in the parental lines through the DNA markers. Two dimension bulked DNA strategy was followed by 20 x 20 growout matrix (Fig. 10). Using this technique the CMS line IR58025A was purified from the seed lot (Sundaram *et al.,* 2008). El Awady *et al.* (2012) screened 10 tomato varieties using SSR marker which revealed that 23 alleles out of 38 alleles amplified was polymorphic with the scorable fragment size ranging from 75bp to 275bp which covers 60.5% polymorphism (Fig. 11).

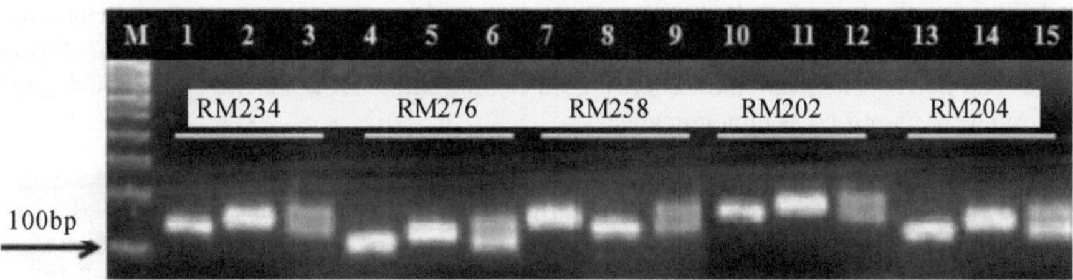

Fig. 9. SSR based markers based polymorphism between the parental lines and the hybrids of rice.

Note: (M); Ladder (100bp) (1) TNAUCMS2A; (2) CR 87R; (3) CORH 3; (4) IR58025A; (5) KMR3R; (6) KRH-2; (7)PUSA 6A; (8) PRR78; (9) PRR10; (10) CRMS31A; (11) IR-42266-29-3R; (12) AJAY; (13) CRMS-32A; (14) IR-42266-29-3R; (15) RAJALAKSHMI

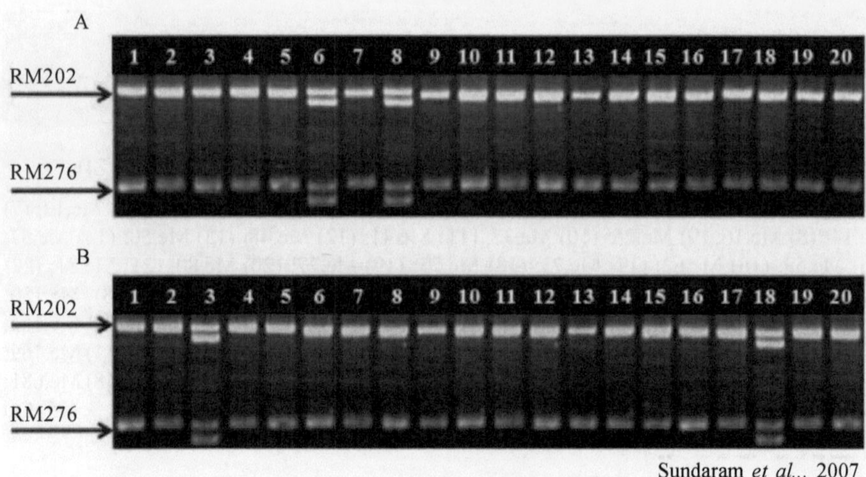

Sundaram *et al.*., 2007

20x20 MATRIX

Column

Row

Fig. 10. Two-dimension assay involving a 20 *20 grow-out matrix for assessment of purity of IR58025A with the help of SSR markers RM202 and RM276. **(Fig. 9) (a) Row-wise lanes 6 & 8** and Column-wise lanes 3 & 18 (indicated by arrows) represent contaminants. **(b) Schematic representation of the 20 *20 matrix** based method for rapid identification of contaminants in IR58025A. Plants at intersections of 6th row 18th column and 8th row and 3rd column (indicated by arrow) were identified as contaminants.

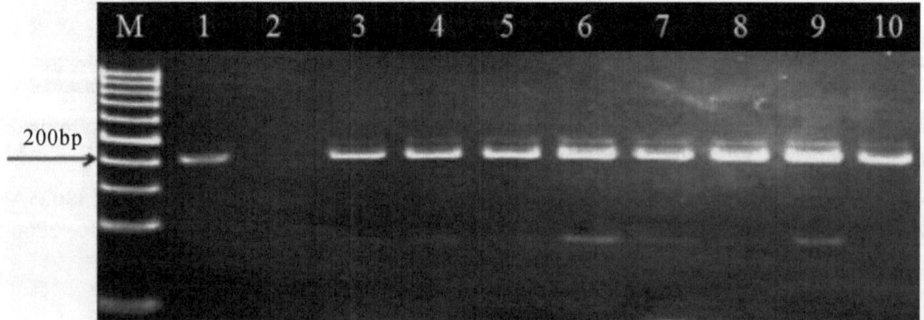

Fig. 11. SSR marker (Tom 57-58) profile for the 10 tomato varieties.

Note: (M) Ladder (50bp); (1) Edkawy; (2) Manapal; (3) Jefferson; (4) Meral; (5) Giza-80; (6) Roma-VF; (7) Pakmor; (8) Peto-86; (9) Flora-dad; (10) VG-151.

Genetic Fidelity: The degree of similarity between barcoding patterns provides information about genetic similarities and relationships between samples studied. A number of molecular markers such as RAPD, AP-PCE, DAF (DNA amplified fingerprinting), SSR, STR, SCAP STS, AFLP, ISSR, EST-PCR, and CAPS are utilized. Lakshman *et al.* (2007) produced *in vitro* true to type banana and verified genetic Fidelity, Biswas *et al.* (2009) in micro propagated strawberry and Pathak and Dhawan (2010) in apple have established genetic fidelity in the crops they have studied. Surabhi *et al.* (2015) screened 16 tissue culture banana variety using ISSR marker and reported clear, reproducible and scorable bands (Fig. 12A). Also tested the purity of the variety which scored monomorphic allele from the randomly selected tissue culture plants (Fig. 12B).

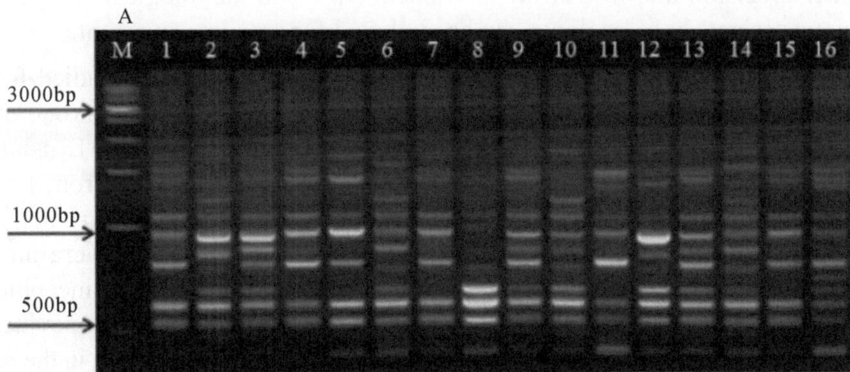

Fig. 12A. PCR amplification of 16 banana varieties using ISSR markar

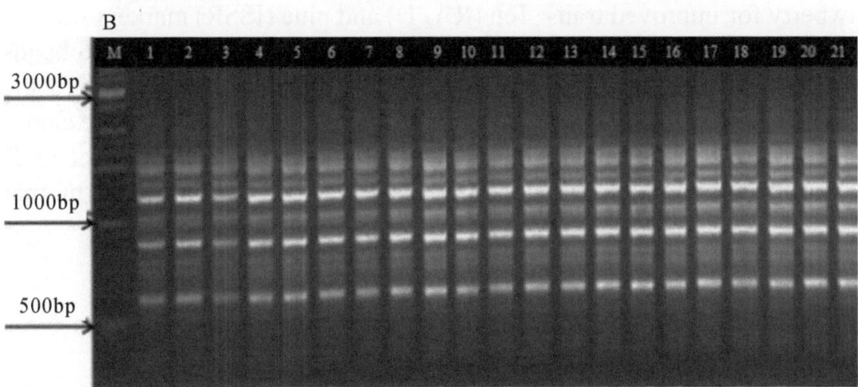

Fig. 12B. PCR amplification of 21 randomly selected tissue culture plants of Grand Naine variety using ISSR Markar

Note: (M) Ladder; (1) Cheni Champa; (2) Red Green Banana; (3) Hybrid H.531; (4) Ney Poovam; (5) Patkapura (Satasankha); (6) Robusta; (7) Karpuravalli; (8) Amrutapani; (9) Champa (Patia); (10) Martman; (11) Ganga tulasi; (12) Red Banana; (13) Champa; (14) Grand Naine; (15) Patkapura; (16) Bantala.

Somaclonal variation: *In vitro* micropropagation facilitates fast and large scale multiplication of true to type plants-that are to be hardened and matured till seedling stage ready for field planting. In the course of micropropagation, plants showing deviations or variations occur owing to the somaclonal variation as explained by Larkin and Scowcroft (1981). Somaclonal variation is explained by Debnath (2017) variously involving 1. late replication of heterochromatin 2. imbalance in the deoxyribonucleotide pool. 3. deregulation of the inherently hyperactive and unstable plant genome 4. accumulation of spontaneous mutations mainly a result of genomic shock. 5. mutagenic responses elicited from by-products of cell death and degradation and 6. preexisting variation in plant tissues. Molecular markers are powerful tools in identification of genetic stability with greater precision and less effort than phenotypic and karyological analysis. Genetic variants or somaclones can be exploited for crop improvement.

Alizadeh *et al.* (2015) have compiled the list of plants that were studied for fidelity and trueness or otherwise (somaclone) showing variations in one or more traits or character. The crops assessed for clonal fidelity are 1. using RAPD markers: apple rootstocks, banana, chrysanthemum. Clerodendron, date palm, Dioscorea, ginger, grapevine, Kiwi fruit, lemon, mentha, papaya, strawberry, pineapple, pyrus, sugarcane. 2. using ISSR marker: banana, Gerbera and grapevine, 3. using AFLP markers: Kiwi-fruit, Eucalyptus, Papaver, Pineapple and 4 using more than one marker Banana, citrus, date palm, ginger, grapevine, lemon, pineapple. Goyal *et al.* (2015) studied the somaclonal variation in the *in vitro* raised plantlets in *Dendrocalamus strictus* by RAPD and ISSR marker (Fig. 13). Dutt *et al.* (2009) evaluated *in vitro* generated somaclonal variation in strawberry for improved traits. Ten (RAPD) and nine (ISSR) markers showed amplification which resulted in 58 scorable bands from RAPD and 66 bands from ISSR respectively. The uniformity in the banding pattern showed the uniformity among the regenerants. Govindhan *et al.* (2010) screened the clones of *Hibiscus sabdhariffa* using RAPD marker and reported the marker OPB 01 showed polymorphism among the clones (Fig. 14) comparing to the mother plant.

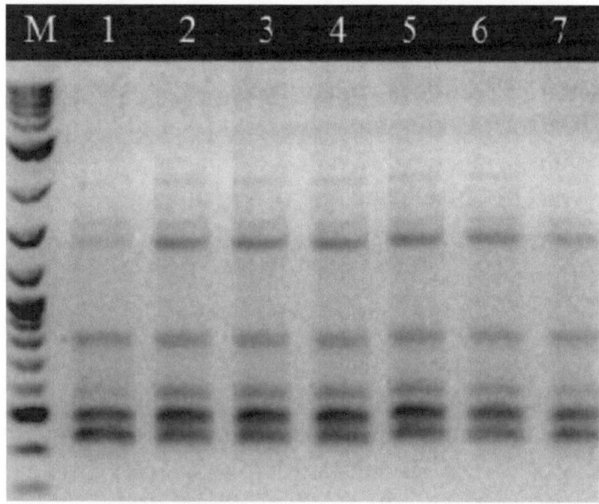

Fig. 13. DNA fingerprinting pattern of *in vitro* raised *D. strictus* regenerants (2 to 7) comparing with the mother donor parent using ISSR marker (UBC824), (Goyal *et al.*, 2015).

Note: (M) Ladder (0.1 – 10kb); **(1)** Donor

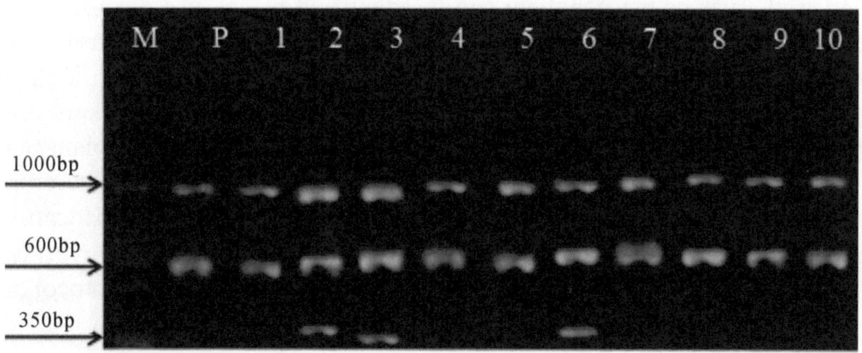

Fig. 14. Detection of somaclonal variation in the clones of *Hibiscus sabdhariffa* using RAPD marker (OPB 01).

Note: (M) Ladder; **(P)** mother plant; **(1 to 10)** clones of *Hibiscus sabdhariffa*.

Molecular Grow Out Test (GOT)

Morphological characteristics in the plants are phenotype expression and are the result of GXE interaction. In the morphological observations, there are difficulties or lack of precise observation and there will be ambiguities in visual observations. This could be overcome with the application of molecular marker tools that are more stable, reliable and objective.

Molecular fingerprinting to assess genetic purity in seed lots to ensure the minimum prescribed standards such as breeder seed (100%), foundation seed

(99%) and for certified seed (98%) is now practiced. The genetic purity of seeds is decided by conducting GOT based on morphological, physiological, and phenological traits assessment from seedling to maturity stage by observing the plant growth stage-specific character expression. This field GOT test requires times spanning over 4 to 6 months or more in different crops and the opportunity to test in adults that are influenced by macro and microenvironment.

Benefits of Molecular Grow Out Test are 1. quality control through verification of seed variety 2. assurance that the seed tag labelling is accurate for the stated variety 3. identification of questionable varieties, parents, or inbreds.

Types of Seed Material used are: 1. varieties that are true to type, genetically uniform, and stable. 2. seeds, clones, somatic embryoids,synthetic seeds and grafts 3. A, B and R lines, GMS lines and TGMS lines in crops like rice, pearl millet, sorghum, cotton, sunflower, pigeon pea, mustard and 4. sugarcane, banana, and other micropropagated crops.

The definition of plant variety is needed for the use of seed producers and for the protection of varieties. Traditionally DUS is applied for characterization of varieties by phenotype parameters. In terms of molecular biology or molecular genetics, the var or the genotype can be identified by the use of molecular markers. The GOT by phenotype is stage-specific of the crop and time-consuming. DNA profiling is more effective as compared to the traditional morphological physiological and proteinaceous enzyme markers. The DNA markers are independent of the environment. The presence of DNA in cells of the plant can be isolated at any time and tested at all stages of the ontogeny of the crop variety. Use of 12 to 15 SSR loci allows one to differentiate uniquely and identify the variety. The system of codominant markers is also able to identify hybrids. Time taken for Molecular GOT is short unlike DUS test. The broad protocol of DNA based GOT is presented stepwise in Figure 15.

Comparison of Traditional GOT and Molecular GOT: Arpita *et al.* (2018) made a comparative study of traditional grow out test and DNA based array to estimate the hybrid purity in the cauliflower hybrid. The basis for grow out test in the field growing is by identifying true to type plants. The study was conducted by Arpita *et al.* (2018) with the sample of hybrid seeds with added admixture of 5% of seeds of female parents. Six seed samples studied are 40, 80, 100, 200, 300 and 400 in number. One codominant SSR marker was utilized in DNA based array was adopted for genetic purity estimation. The results of estimation showed similarities in the tratitional and DNA marker based array GOT (Table 3) with all the six seed samples sizes for 99% purity, in comparision with 400 seeds minimum size of sample for 99% purity. The genetic purity estimation was based on a locus specific informatives allele marker SSR. The residual heterozygosity was also found.

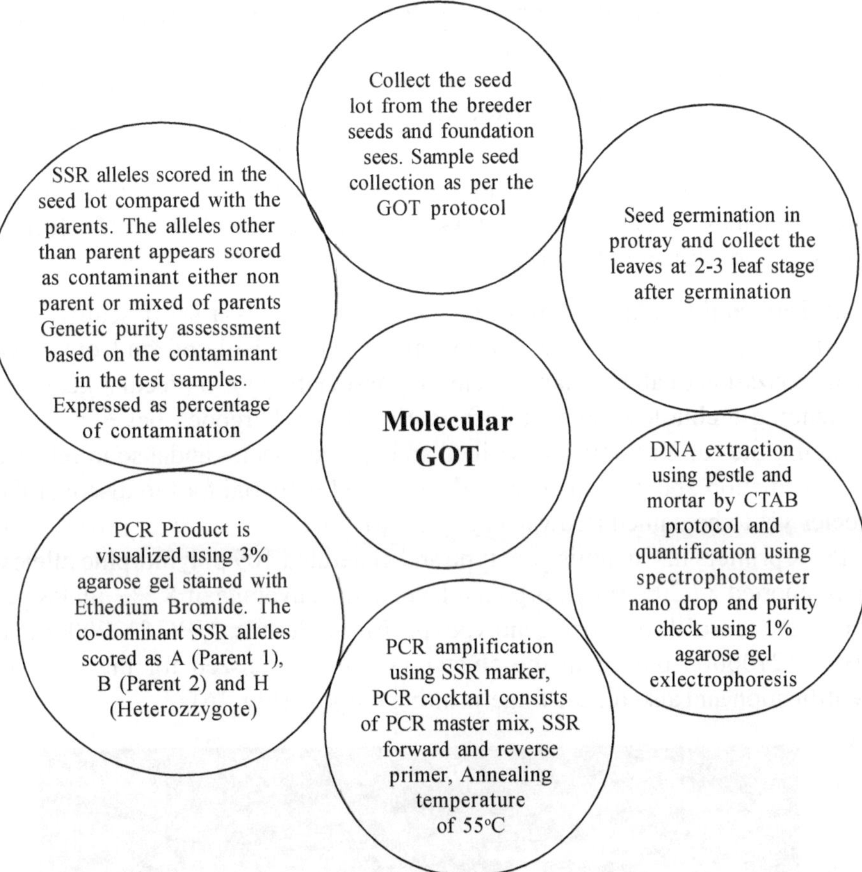

Fig. 15. Procedure involved in Grow Out Test (GOT)

Table 3. Comparision of Molecular Grow Out Test and Traditional Grow out Test is given below in the table. (mean values)

Seed Sample size (No. of seeds)	Field grown GOT purity %	SSR array purity % (Mol. GOT)	Calculated purity %
400	99.36	99.26	99.40
300	99.30	99.20	99.30
200	99.30	99.30	99.30
100	99.60	99.40	99.60
80	99.90	99.90	99.50
40	100.00	100.00	100.00

1. 400 seeds are recommended by the minimum seed standard. Sample seeds used in this study contain seeds of female parent (5%)
2. Mean values calculated based on the data published by Arpita *et al.* (2018).

The molecular GOT is effective substitute for the traditional field grown grow out test. In the field GOT, duration of test is around 75 to 90 days or more according to the crop duration till maturity while the molecular GOT takes 48 hours. The minimum seed standards has prescribed minimum of 400 seeds for the seed samles drawn and sown in 2 replications, 200 seeds in each replication. The study on optimization of the number of seeds to be used for genetic purity assessment indicated minimum of 100 seeds for Molecular GOT. The Molecular GoT steps are shown in Fig. 18.

DNA Bar coding (Allele barcode): DNA barcodes is a short sequence of about 400 and 800 bp long which can be easily isolated and effective characterization of all the plant species. By associating the molecular genetics, sequencing technology and bioinformatics, DNA barcodes paves ways to recognize the earlier known, described and named species and also to retrieve the information about them. This is the best and potential tool to discover the species yet to be named (Cowan *et al.,* 2006). Parvathaneni *et al.* (2011) used 15 ISSR primers in Cucumis genotype and generated 109 polymorphic alleles. Also reported 87.20% polymorphism level and an average of 8.3% alleles per primer. ISSR profile revealed genotype specific alleles viz., UBC 812700bp and UBC 8121000bp placed in the cluster IV can be utilized for the specific identification and also for the fingerprinting purpose (Fig. 16).

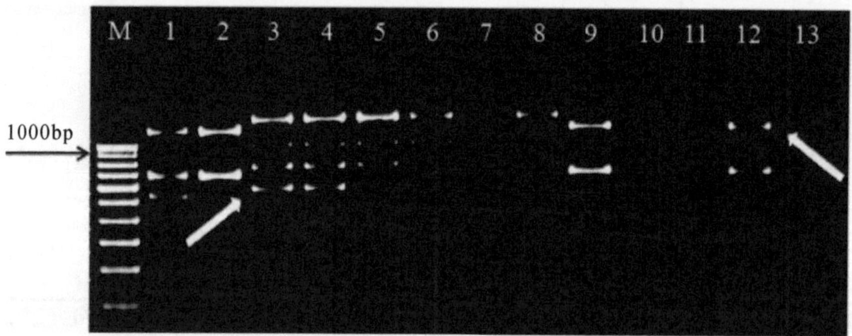

Fig. 16. ISSR marker (UBC 812) profile on cucumis genotypes.

Note: (M) Ladder (100bp); (1) Orissa; (2) Rupamail; (3) Edapaddi; (4) Kanivelari; (5) Sangiri; (6) Long melon; (7) Senpatti; (8) Perundurai; (9) Long green; (10) Andaman; (11) Karur; (12) AAUC-2 (13) N-78.

The allele barcode representation can be used as a finger printing profile for each CV as it clearly exhibit a visual pattern. At least one specific combination of alleles that allow accurate finger printing and easy detection of genetic differences and in the identification of the individual need to be identified. The allele barcode facilitate easy detection of genotypic difference thereby help in identification of the individual with certainity. This representation further help as a visual reference or standard (Fig. 17).

SSR pair represented loci and the DNA bands are alleles. Amplicon profiles generated by primer pairs help in identifying the differences. DNA barcoding construction through SSR markers in the Indian soybean varieties was accomplished by Rani *et al.* (2016). They have generated ten digital barcodes for providing unique identification code for each one of the 102 soynean varieties. Zhao *et al.* (2012) screened microsatellite loci and established DNA barcode for identifying cotton cultivars.

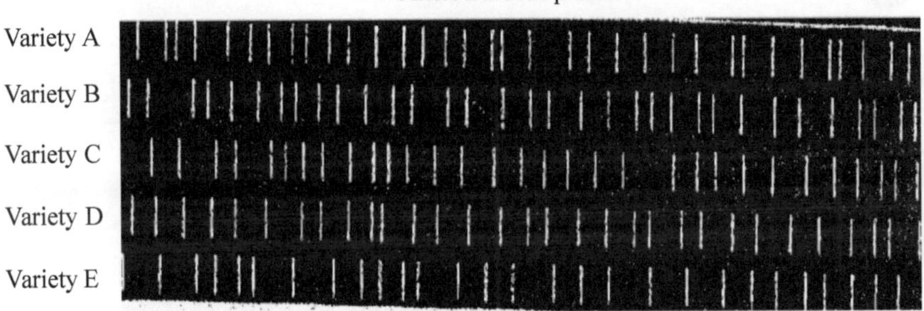

Fig. 17. Schematic representation of DNA Allele Barcode profile

DNA marker for New Variety Release: The ICAR has issued guidelines for testing of Crop varieties in the All India Coordinated Crop Improvement Projects before release of a new variety. ICAR has mandated that the variety proposed as a New Release should have been tested for DNA fingerprinting and to identify DNA markers specific for the newly proposed variety so as to maintain the identify (Tandon *et al.*, 2015).

DNA Fingerprinting Laboratory: DNA Finger Printing Laboratory was established in 2007-2008 in the Directorate of Seed Certification and Organic Certification, Department of Agriculture, Coimbatore-641003, Tamil Nadu location with the financial assistance of GOI under the Central Sector Scheme. This Laboratory has been notified as the State DNA Fingerprinting Laboratory under the Seed Act 1966 (G.O. Ms. No 81 dt 27. 6. 2014). This DNA fingerprinting Laboratory is the first of its kind among the public sector undertaking in India. This laboratory set up will help to enhance and augment the production of quality-assured seeds to the farming community. The DNA fingerprinting laboratory helps in the estimation of the genetic purity of a variety/ parent line within a short period of 3 to 4 days time and for the quick confirmation of the plant variety/ seed lot. This laboratory has been now recognized and accredited as the first in India by the International Seed Testing Association (ISTA). Seed certification for the export of seeds is also issued by this laboratory.

DNA Fingerprinting Protocol for Variety Purity Test

- Sample seeds collection
- Germinating the seeds
- DNA extraction and quality verification

- Primer addition
- Genomic DNA and DNA polymerase addition
- Amplification reaction

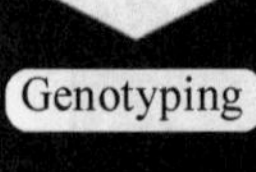

- Agarose gel electrophoresis
- Scoring number of alleles
- Interpretation of results

Sample seeds selection
↓
Seed sample germination, 5-7 days old seedlings for DNA extraction
↓
Mini-Prep. DNA extraction using Micro centrifuge tube
↓
DNA extraction and quality checking
↓
PCR amplication with addition of primers
↓
Genomic DNA and DNA polymerase addition
↓
PCR with annealing temperature 55ºC
↓
Gel. Electrophoresis
↓
Scoring number of alleles
↓
Result interpretation

Fig. 18. Molecular GOT for Purity Test

12

Nucleus and Breeder Seed Production of Field Crops

Cereals and Millets

1. Barley

(*Hordeum vulgare* L.) (2n=14)

Family : Poaceae

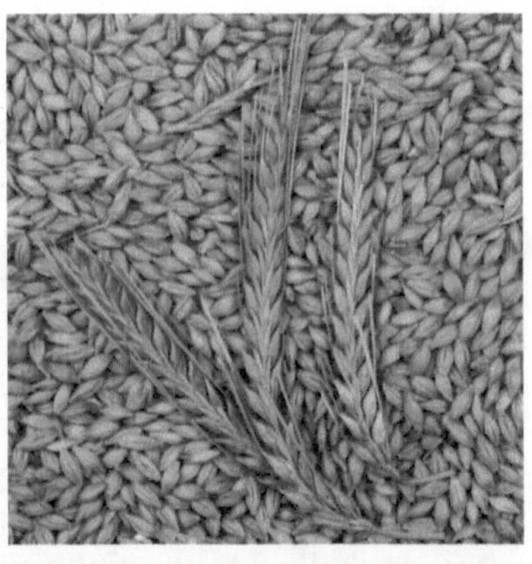

Leaves are alternate in distichous arrangement; lamina long and linear and leaf sheath covering the internodes; ligule membranous and auricle claw shaped. The inflorescence is a spike and florets are borne on rachis node. Each spikelet has three florets. In two row barely, lateral spikelets become sterile. The florets arise from the axils of bract / lemma and awn is borne at the tip of lemma. Each flower contains three anthers and one ovary bearing bifid feathery stigma. The characteristics of barley are detailed in Table 1. Pollination usually takes place in the closed flowers (cleistogamy) favouring self pollination. Little extent of cross pollination may occur to less than 5% (Raj Kumar *et al.* 2003).

Nucleus Seed Production: Source the seeds from the basic seed stock of the variety and raise the crop in isolation (5m.). Single plant per hill is maintained. Select 500 spikes or ear heads from the main tillers from single plants conforming to the varietal characteristics. Grains are separated ear wise and verified for the grain characteristics. Reject the spikes that are not conforming to the varietal description.

In the next season, ear to row progeny planting is made with wide spacing in rows of 4 m. long with isolation distance of 5 m. between varieties. Critical examination is made throughout the life cycle of the crop from juvenile to flowering to grain maturity stage looking for greater uniformity, stable expression of characteristics and true to type nature of rows. Rows with admixtures or deviant plants, are rejected and uprooted. The progeny rows showing deviant plants at flowering and the adjoining rows on either side are also uprooted to safeguard from contamination. Each ear head per plant in the progeny is harvested separately and seeds extracted progeny wise and examined for true to type. The seeds are then pooled together to constitute nucleus seed stage. In the same plot, 15 to 20 ears from main tiller in each progeny row are collected separately and individual spikes are verified for true to type nature and then well dried, cleaned and stored for next cycle of nucleus seed production. In the next season, pooled seeds from the ear rows are sown to raise nucleus seed stage II and the same protocol described in nucleus seed stage I is followed.

Nucleus Seed Stage II: To meet the increased demands of nucleus seed, NSS-II is followed further. If very high quantity of seeds are to be produced, NSS II preceding breeder seed production is helpful. Seeds of NSS-I are sown in progeny row plots (isolation of 5m.) utilising seeds of individual progeny rows. The progeny plots in NSS - II are examined critically for uniformity, genetic purity up to pre flowering and then for ear head and seed characteristics upto maturity stage. If any deviation in the variety characteristics is noticed, entire progeny row to discarded. In respect of the deviations noticed at flowering stage, the progeny rows exhibiting deviations and the adjacent rows are also uprooted to avoid accidental cross pollination with lines showing off-types. At maturity of the crop, individual progeny row plots are harvested separately and ear heads and grain samples are examined for colour, shape, uniformity and for the true to type nature. The seeds from uniform plots selected for true to type nature are bulked to constitute NSS-II for raising breeder seed stock.

Breeder seed production: The breeder seed plot is isolated by minimum distance of 5m. Nucleus seeds are sown in a clean, disease free, well prepared field not grown to barley in the last 2 to 3 years. No loose smut infected wheat, rye or triticale field should be within 5m. distance. From seed germination to pre flowering and there afterwards upto spike maturity, field examinations are made to examine all stages of plant development and for the descriptor characteristics. Rouging should be done and if deviants and rogues are found after the flowering stage, such plants should be removed / uprooted and simultaneously, plants around the rogue plants are also uprooted to avoid any chance cross pollinated seeds getting into the breeder seed. The breeder seed plot is monitored by the Monitoring Team. Seed testing with samples drawn from the breeder seed is done in the seed testing laboratory and the quality assured seed is then labelled with breeder seed tag. Grow out test is conducted with samples of the breeder seed as per the standard procedure to ascertain the genetic purity. Grain nutritive contents are estimated.

The Nucleus and Breeder Seed Production Steps for Barley are as Shown in Wheat Crop

Table 1. Characteristics of Barley

1.	Growth habit	Erect, semierect, prostrate
2.	Stem-basal pigmentation	Absent, Present
3.	Auriole flag leaf anthocyanin pigmentation	Absent, Present
4.	Upper node pigmentation	Absent, Present
5.	Flag leaf attitude	Erect, semierect, drooping
6.	Flag leaf waxiness of the sheath	Absent, Present
7.	Spike emergence(days)	Very early (<65), early (65 to 75), medium (75-86), late (86-96), very late (>96)
8.	Spike type	Two rowed, six rowed
9.	Lateral florets	Rudimentary, developed
10.	Spike waxiness	Absent, present
11.	Spike colour	Pale green, green, dark green
12.	Spike attitude	Erect, semi erect, drooping,
13.	Awn roughness	Smooth, rough
14.	Flag leaf length (cm.)	Short (<10), medium (10-14), long (>14)
15.	Flag leaf breadth (cm.)	Narrow (<1.0), medium (1 to 1.5), wide (>1.5)
16.	Awn tip pigmentation	Absent, Present
17.	Lemma pigmentation	Absent, nerve pigmented, present
18.	Spike basal sterility	Absent, Present
19.	Spike length (cm.)	Small (<7), medium (7-10), long (>10)
20.	Plant height (cm.)	Very short (<75), short (75), medium (85-90), tall (95-105), verytall (>105)
21.	Peduncle length (cm.)	Short (<22), medium (22-27), long (>27)
22.	Awn	Absent, Present
23.	Awn type	Hooded, awnletted, normal
24.	Awn length (cm.)	Short (<8), medium (8-11), long (>11)
25.	Spikelet clump	Lax, intermediate, dense
26.	Grain hullness	Naked, hulless, covered (hulled)
27.	Grain colour	White, yellow, purple, black
28.	Grain shape	Oval, oblong, elliptical
29.	Grain size 1000 grain wt. (g.)	Small (<30), medium (30-40), large (40-50), very large (>50)
30.	Grain surface	Smooth, wrinkled
31.	Rachilla hairs	Rudimentary, prominent
32.	Grain width	Narrow, intermediate, wide

Source: PPV & FRA (2007)

2. Maize

(*Zea mays.* L) (2n : 20)

Family : Poaceae

Maize is a monoecious crop with tassel positioned at the top of the stem and the cobs are borne in the axils of the lower leaves. Tassel or the male inflorescence is a terminal loose panicle with lateral branches and spikelets arranged in rows. Each spikelet consists of 2 glumes, lodicules and is staminate with three anthers. Pistillate inflorescence or cob is borne in leaf axils at the end of short axillary branch called shank. Husks cover the ear or cob which has a thick central axis on which the numerous paired spikelets are arranged in a series of longitudinal rows. Ovary is superior and has one ovule with basal placentation. The style is long, filiform and the stigma is bifid. The entire style is hairy and receptive upto 10 days from emergence. Maize is a cross pollinated species due to its monoecious and protandrous nature. The stigmatic surface of female flowers or the silk emerge from the leaf axils 2 to 5 days after the emergence of tassel thus rendering the maize crop cross pollinated in nature. The tassel sheds pollen for 5 to 8 days with major pollen shedding between 2 and 4 days. The fruit is a caryopsis and endospermous. The characteristics of maize are detailed in Table 1.

Nucleus seed production: Maintenance of Composites and OPVs. In OPV, 50 to 100 seeds from each representative cob are collected from randomly selected 1000 to 1500 true to type representative plants. Seeds are bulked and 1500 to 2000 seeds are planted and isolation distance of 1000m. is maintained. Single plant per hill is maintained. Monitoring is done on individual plants for uniformity and conformity for true to type. Off-types and deviating plants are uprooted. Very early and late flowering plants are discarded. Matured cobs and seeds are verified for true to type. Seeds from selected cobs are mixed. Nucleus seed is constituted.

Breeder seed production: Nucleus seeds are sown in an area of 2000 to 2500 sq.m. with isolation distance of 1000m. from other maize crops. Individual plant progeny rows are critically examined to remove weak plants and deviant types of plants. High degree of uniformity and true to type in plants and tasselling / silking is maintained. In advanced stage of seed chain, the metric / quantitative traits and ear characteristics should not show variation above ± 0.75s (s = std. deviation) and for qualitative traits acceptable range of variation can be ± 3%

(Singh *et al.,* 2003). Monitoring team visit is arrarged. Seeds from all selected ears may be bulked to constitute breeder seed (BSII). Second stage breeder seed is produced by growing the breeder seed in isolation of 1000m. distance. Nutritional and micronutrient contents are estimated in the grain. Maize Inbred seed production following half sib or ear to row is shown in Fig 1.

Nucleus Seed Production of Corn Inbred: Raise the inbred in an area of 2000 sq.m. retaining single plant per hill. Isolation distance of 1000m is provided. Select true to type vigorous and uniform plants for all the traits and for flowering (tassel and silk) and 2000 to 3000 uniform true to type plants are selected, self fertilised and cobs harvested. In the laboratory, cobs are examined for the uniformity and expression of shape and size, kernel rows, cob tip filling, grain shape and other trait features. After careful drying of individual cobs, 50 seeds in each cob are separated from all of the finally selected cobs.

Plant progenies or ear to row progenies are raised in the following season in isolation (1000m.) field plot. Progeny rows with deviant plants are uprooted before pollen shedding. In each progeny row, true to type and vigorous plants are selected and self fertilised. At maturity, self fertilised cobs are harvested and examined in the laboratory for trueness. In each selected cob, 50 seeds are separated and stored. In the next season, progeny rows or ear to row progenies are raised in isolation (1000m.) in two replications and each progeny row is critically studied for vigour, uniformity and true to type nature and flowering (tassel and silk). Progeny rows showing deviant plants are uprooted. Progeny rows performing above mean +2SE are identified and in each progeny 10 to 15 selected matured cobs are collected and examined in the laboratory and seeds extracted. In the selected cobs, 50 grains are bulked to make nucleus seed.

Breeder seed production : Breeder seed plots may be of size according to the demand of seed. Nucleus seeds of inbred are sown in rows in field with isolation of 1000 m. Single plant per hill is retained. Thorough monitoring is made for expression of characteristics specific to growth phases. All deviant plants are uprooted and roguing is frequently done. The plants at flowering stage are examined for uniformity, vigor and true to type with no variation. In advanced stages of seed chain, quantitative characters of the plants and ear should not show variation above ± 0.75 sd. and for qualitative characters variation should not exceed ± 3% (Singh *et al.,* 2003). Monitoring team visit is arrarged. The second stage of breeder seed is produced by growing breeder seed in plots provided with 1000m. isolation. Monitoring Team visit is arranged at flowering time or afterwards to adjudge the genetic purity. The breeder seed samples are drawn and tested for seed qualities. The quality assured breeder seed is labelled with Breeder Seed label. Seed samples from the processed breeder seed are tested in grow out test as per the prescribed steps to ascertain genetic purity.

Table 1. Descriptor of maize

S.No.	Characters	States	Stage of observation
1.	Anthocyanin colouration of sheath	Absent/weak/strong	Seedling
2.	Anthocyanin colouration of brace roots	Absent/present	Reproductive
3.	Width of leaf blade	Narrow/medium/broad	Vegetative
4.	Angle between main axis and lateral branches	Small/medium/large	Vegetative/reproductive
5.	Number of primary and lateral branches	Absent/few/many	Reproductive
6.	Plant height	Short/medium/long	Vegetative/reproductive
7.	Time of silk emergence	Early/medium/late	Silk emergence
8.	Anthocyanin colouration of base of glume	Absent/present	Reproductive
9.	Time of anthesis	Veryearly/early/medium/ late/very late	Reproductive
10.	Anthocyanin colouration of anthers	Absent/present	Reproductive
11.	Density of spikelets	Lax/medium/dense	Reproductive
12.	Anthocyanin colouration of silk	Absent/present	Silk emergence
13.	Length of peduncle	Short/medium/long	Reproductive
14.	Anthocyanin colouration of glumes of cob	White/light/purple/dark purple	Reproductive
15.	Ear diameter	Small/medium/large	Reproductive
16.	Ear shape	Conical/cylindrical	Reproductive
17.	Number of rows of grains	Few/medium/many	After harvest
18.	Type of grains	Flint/semi flint/semi dent/ dent pop/waxy/opaque/ opaquetinge/sweet	After harvest
19.	Row arrangement of grains	Straight/spiral /irregular	After harvest
20.	Grain shape	Shrunken/round/indented/ pointed	After harvest
21.	Grain size	Small/medium/bold	After harvest

Note: Description is based on Singh N.N. *et al.* (2003) and PPV & FRA

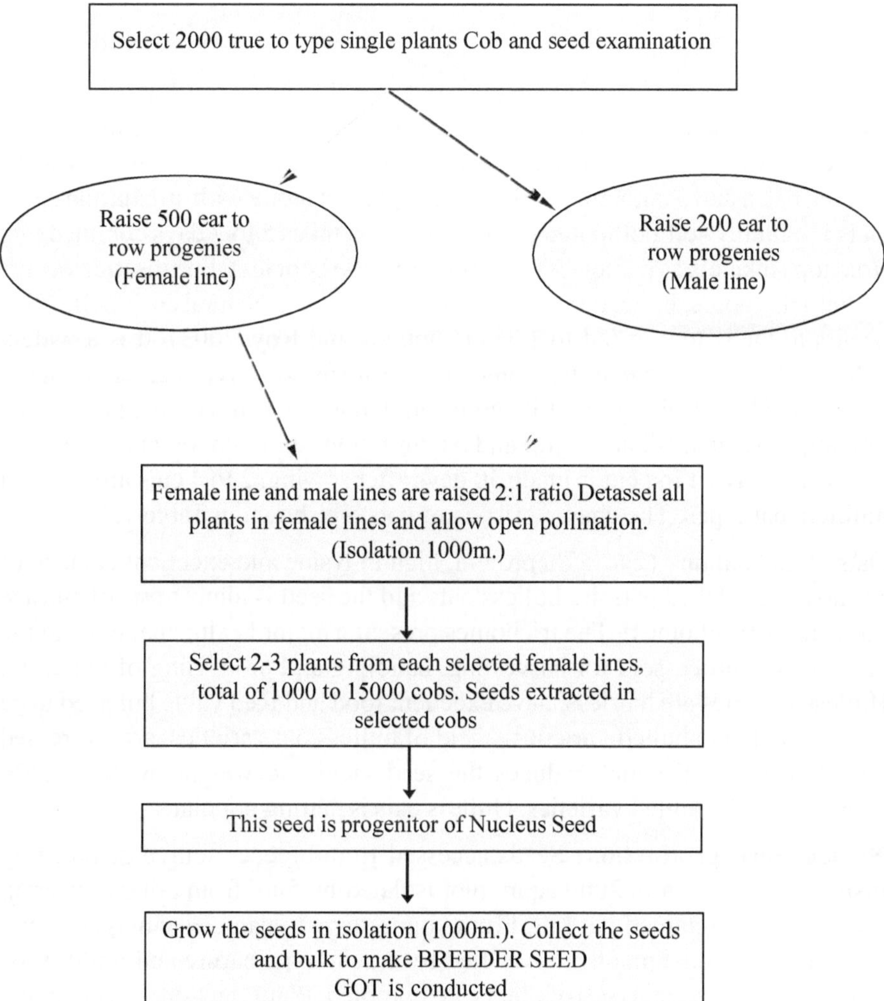

Fig. 1. Maize Seed Production in Maize Varieties

3. Oats

(*Avena sativa* L.) (2n=42)

Family : Poaceae

Oat is an annual grass. Most varieties are long day plants needing at least 12 hours of sunlight to induce flowering. There are winter and spring types of oats. Leaves alternately arranged, ligules present and auricle absent. Inflorescence is a compound panicle and branches in 5 to 7 whorls; spikelets have 2 to 3 florets connected by rachilla. Spikelets are surrounded by glumes. Floret consists of lemma, palea and 3 anthers. Single ovary is present along with stigma and style. Oats is a highly self pollinated crop. Pollination takes 5 to 7 days commencing from top spikelets as panicle is emerging from the boot leaf. Fertilisation occurs within 4 to 6 hours. Pollination occurs from 1 pm to 4 pm. Natural cross pollination occurs to the extent of 0.4 to 1.3% (Choubey and Roy, 2003). It is a widely cultivated forage crop in India. Some of them are multicut types green fodder. It is cultivated in tarai region of Uttaranchal, Uttar Pradesh, Himachal Pradesh, Punjab, Hariyana, Bihar, Gujrat and Andhra Pradesh. Fodder Oats matures in 135 to 150 days. First cut is made 50 days after sowing. 2 to 4 cuts are made in multicut oat types. The characteristics of oats are shown in Table 1.

Oats grain contains 12 to 15% protein, high in lysine and excellent balance of amino acids. Naked oats are hulless oats and the seed is almost free of surface borne hairs (trichomes). The trichomes present a major health challenge to the growers and processors for harvesting, handling and processing of the grain. Hulless varieties are hairless, have excellent food and feed value but need to be stored dried than hulled varieties. Yield of hulless oat varieties. are expressed after hull removal which reduces the seed yield and weight by 20 to 25% compared to the hulled varieties. Hulless oats is getting popular.

Nucleus seed production: Seeds accessed from breeder source or breeding Institution are sown in 2000 sq.m. plot isolated by 5 m. from other oats crop. Single plant is retained per hill. From young stage to panicles emergence and maturity, individual plants are verified for true to type nature and uniformity. Off-types and deviant types of plants are uprooted. Plants lacking in vigour and growth are also uprooted before flowering. Plants flowering early and late are removed so as to maintain flowering time uniformity. At the time of ear emergence and thereafter, ears are critically examined till maturity for uniformity and trueness of the spike, flower and pollen emergence. At maturity, 1000 spikes, one spike from the main tiller from each plant are collected and labelled. Panicles and seeds are examined in the laboratory for trueness and uniformity and variant types are discarded. Seeds separated from individual selected plants are labelled and stored (Fig 1).

In the next ideal growing season, in well prepared field 1000 single plant progeny rows are raised in two rows plots, 4 meter long, with the recommended spacing of 50 cm. x 10 cm. and adapting good agronomic production schedules. Only single plant per hill is retained. Border crop of the same oats variety is grown, two meter wide all around. Isolation distance of 5 meter is provided for the progeny row evaluation plot. Two rows plot of the concerned variety is grown after every 15 to 20 progeny rows for comparison. The progeny rows are examined up to maturity in growth phase specific schedule to assess the trueness to type and uniformity. Roguing is done in respect of off- types. Progeny rows showing deviating type of plants are uprooted. When deviant types are noticed after flowering, the adjoining progeny rows on either side are also uprooted including the plants in the surrounding rows that are likely to get pollen contamination. At full maturity, 1000 well matured spikes one each from the main tiller are harvested and dried. In the laboratory, the spikes are examined and then the seeds removed from the spikes are critically examined for trueness, selected and saved.

In the following cycle, 1000 progeny rows are raised in well prepared field plot (5m. isolation) and all nucleus seed production procedures as described earlier are followed. Only uniform and true to type progeny rows are retained till maturity. 1000 to 1500 single plants are selected and main tiller ear heads are harvested and seeds separated. Part of the seeds from individual plants are pooled together and nucleus seed is constituted. Rest of the seeds saved from each plant are utilised for next nucleus seed production cycle.

Breeder Seed Production: Nucleus Seeds are sown in well prepared fertile field not grown with oats in the precious two years. Single plant per hill is retained.All ideal crop production practices are provided. Isolation distance of 5m. maintained from other oats crop. Thorough roguing for off-types and deviating types of plants is made upto flowering. If deviating plats are noticed after the flowering, plants around such plants are uprooted and discarded to avoid pollen contamination. Early and late flowering plants are removed to maintain flowering uniformity. Visit of Monitoring Team is organised at mid maturity stage. At maturity true to type plant are harvested, earhead examined and seeds are bulked, dried and processed. Seed quality is assessed. Quality assured seed is labelled with breeder seed tag. Grow out test is conducted with samples of breeder seed. Nutritional and micronutrient qualities of grain and fodder are estimated.

Table 1. Descriptor of oats

S.No.	**Characters**	**States**	**Stage of observation**
1.	Growth habit	Erect/prostrate/semi-prostate	Vegetative
2.	Plant height	Verytall/tall/medium/short	Vegetative
3.	Hairiness of the uppermost node of the stem	Absent/present	Vegetative
4.	Leaf blade	Narrow/medium/wide	Vegetative
5.	Leaf sheath	Hairy/non-hairy	Vegetative
6.	Panicle emergence	Early/medium/late	Start of earemergence
7.	Panicle - orientation of branches	Unilateral/equilateral	Flowering
8.	Panicle - attitude of spikelets	Erect/pendulous	Flowering
9.	Glaucosity of the lemma of the primary grain	Absent/present	Maturity
10.	Glumes - length of glumes	Short/long/very long	Maturity
11.	Awns colour	Yellow/brown/black	Maturity
12.	Awns curvature	Straight/twisted	Maturity
13.	Awns nature	Weak/strong	Maturity
14.	Colour of the lemma of primary grain	White/black	Maturity
15.	Maturity	Early/medium/late	Maturity
16.	Basal hairs on primary grain	Abundant/sparse	Maturity/lab
17.	Basal fracture of primary grain	Straight/semi-oblique	Maturity/lab
18.	1000 seed wt. (g)	low<20, medium (20to 30), high>30	

Note : Description is based on Choubey and Roy (2003) and PPV & FRA (2018)

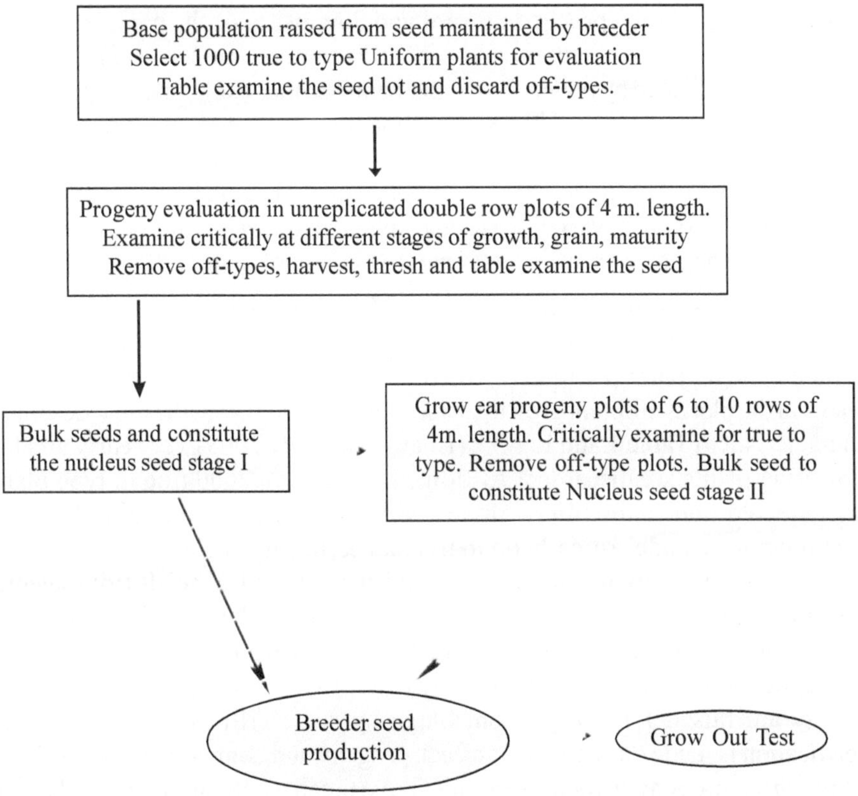

Base population raised from seed maintained by breeder
Select 1000 true to type Uniform plants for evaluation
Table examine the seed lot and discard off-types.

↓

Progeny evaluation in unreplicated double row plots of 4 m. length.
Examine critically at different stages of growth, grain, maturity
Remove off-types, harvest, thresh and table examine the seed

↓

Bulk seeds and constitute the nucleus seed stage I → Grow ear progeny plots of 6 to 10 rows of 4m. length. Critically examine for true to type. Remove off-type plots. Bulk seed to constitute Nucleus seed stage II

↓

Breeder seed production → Grow Out Test

Fig. 1. Nucleus and Breeder seed production in Oats

4. Pearl Millet

(*Pennisetum glaucum* L.) (2n:14)

Family : Poaceae

The stem is erect, leafy, mostly medium or high tillering. Inflorescence is a compound terminal spike. The central rachis bears fascicles on rachilla, each fascicle consisting of 1 to 2 spikelets encircled by involucre bristle whorls. The spike consists of 2 glumes and 2 florets, the upper one being bisexual and the lower one staminate. Ovary is monocarpellary with 2 styles and two bifid stigma that are brush like. Stamens are three with anthers that are pedicellate. Pearl millet shows typical protogynous nature. Styles emerge after 3 to 4 days of the panicle emergence. Styles from middle portion of the panicle emerges first and then proceed upward and downward of the panicle. Stigmas emerge faster than the anthers and remain receptive for 2 to 3 days. All the stigmas are pollinated by pollen from other panicles, thus excluding self pollination in the same inflorescence. Thus self pollination is avoided under natural open pollination condition. Only under self fertilised condition by bagging, self pollination is accomplished. The emergence of anthers usually occurs 3 to 4 days after the first stigma emergence. Variation is somewhat high between varieties / genotypes in respect of degree of protogynous nature.

Nucleus seed production in varieties (OPV): Grow sourced seeds of OPV in an area of 0.2 ha. in isolation of 1000 m. from other pearl millet crops and maintain single plant per hill. Detailed observation is made on the individual plants in the plot at different growth phases and compared with the OPV descriptor characteristics (Table 1 and Fig. 1). (Bhatnagar *et al.*, 2003) Very early and late flowering plants are discarded. At flowering, 1500 to 2000 true to type plants are selected and main tillers alone are harvested at maturity. Laboratory examination is made critically on individual earheads for shape, size and good seed set and on the grains for all the varietal characteristics. Seeds from selected ears are stored individually. In the next season, part of the seed from individual ears are sown in plant progeny plots in 2 replications and in 2 rows (4m. long) per replication. After every 10 progeny rows, two rows of the seeds from the variety are raised for comparison. Stage by stage critical examination and verification is made on each plant in each progeny and compared with the check. At maturity, progeny rows performing above mean +2 SE are identified and in each identified progeny row, 10 to 15 superior single plant selections are made and seeds stored for next cycle initiation. The remnant seeds of the selected plants used for raising progeny rows are bulked and grown in isolation for nucleus seed production and thereafter for breeder seed production.

Nucleus seed of B line: As a first step, 1000 to 1500 true to type and uniform plants are selected and self fertilised at the time of flowering from the B line source seed plots grown in isolation of 1000 m. from any other pearl millet crop. The harvested self fertilized ears and grains of individual B line plants are critically examined in the laboratory for conforming to the parent description for ear head and grain features. In the next season, part of the seeds from each of the self fertilised ear are sown in progeny row plots along with check progeny raised once in 15 progeny rows. Two rows plots are raised in two replications (4m. long) in each row. One thousand progeny rows are raised. Only single plant is maintained in each hill. Thorough examination of plants is made for descriptor characteristics in each of the growth phases. Progenies showing, deviating plants are uprooted along with adjoining row of plants. Fifteen plants in each of the selected progeny row of B line performing above mean +2SE are selected, and seeds stored for continuing the next cycle. Remnant seeds of the selected plants in the progeny rows are now bulked together and grown in the next season in plots with required isolation distance of 1000 m. for producing nucleus seed of the B line (Fig. 2).

Nucleus seed of A line: Breeder source seed of A line is sown in single rows in alternate to single rows of corresponding B line (1:1 row ratio) maintaining single plant per hill. The source of B line seeds is the nucleus seed of the same B line stock. Isolation distance of 1000m. is maintained from other pearlmillet crops. Careful observations are made on the different characteristics specific for each stage of development of the B line. At pre flowering stage, the plants from A line and B line that are true to type and uniform are selected and protected with selfing bags. At flowering stage, very early and late flowering plants are removed. Each of the earhead including the ones protected with selfing bags are examined in A line for pollen shedding and such plants showing pollen shedding are gently uprooted with all tillers intact and disposed. The next step is to make paired crosses in A x B combination. Sterile plants if found in B line are uprooted. The B line plants utilised for paired crossing is continued in self fertilised condition. At maturity, specific paired crosses of A x B are labelled as A1 x B1, A2 x B2, A3 x B3 ..... and An x Bn. Seeds set on individual A line plants (A and B paired cross plants) are collected and grain and seed features are examined and selected after the laboratory examination is made . The seeds from the selected ear heads of A and B pairs are separated, labelled and nucleus seed constituted.

Breeder Seed Production in A Line: In the following season, the seeds from each of the A x B paired cross are sown in 2 rows of progeny row of A x B along with one row of corresponding B line in plot isolated by 1000m. Around 600 to 700 progeny rows are sown in the season in two rows of A line (AXB) and one row of B line. Progeny plots are critically verified for true to type nature, uniformity and conforming to the descriptor characteristics. Progeny rows of AxB showing pollen shedders are discarded and all the plants of corresponding B rows and adjoining rows in A x B lines are also uprooted. In each of the selected (A1 x B1, A2 x B2, An x Bn) crosses with the respective corresponding B line rows ,10 to 15 plants in B are self fertilised and 10 to 15 A x B paired crosses are made with the corresponding plants in B. Seeds of A x B paired crosses made in each progeny row are collected separately along with the seeds of the corresponding B plants, labelled and stored. In the next season, part of the seed set in plants in A line are pooled together and part of the seeds in individual plants in B line are bulked and utilised for raising nucleus seed of A line by growing in plots with isolation distance of 1000 m. Seeds of A line are sown in 2 rows alternated with B line in one row. All the plants in the A line rows are closely monitored for pollen shedding / pollen fertility reversion. All plants showing pollen shedders are uprooted along with the other plants in the same progeny. The plants in the corresponding B line progeny are also uprooted. B line rows are harvested 3 to 4 days earlier to A line harvest and dried and seeds separated with highest precautions to avoid admixtures. The A line heads are harvested 2 to 3 days after B line is harvested and with all precautions the seeds are separated, labelled and stored. Part of the seeds from the A and B lines are bulked to constitute nucleus seed and the left over seeds are also utilised for producing nucleus seed in the next season. Seeds from paired crosses of A x B are maintained on individual A x B paired cross basis along with seeds of corresponding B plant of the same cross for utilisation to continue subsequent cycles of nucleus seed production.

Nucleus seed of R line: Source seeds of R line are raised in 0.4 ha area with single plant per hill. Isolation distance of 1000 m. is maintained. Based on the true to type expression, main tillers of 1500 to 2000 plants are self fertilised and seeds collected. In the next season, with self fertilised seeds from the selected plants, two rows progeny plots are raised in isolation of 1000 m. in two replications along with check for comparison. Examination of all the plants in each progeny is done for the true to type nature in characters corresponding to each of the growth phase till full seed set and maturity. Very early and late flowering plants in each progeny row are uprooted. Progenies performing above mean +2SE are identified. In each of the identified progeny, 10 to 15 ear heads of main tillers that are self fertilised are harvested. The ear heads are examined in the laboratory for panicle characters and good seed setting and for grain features. Seeds from selected ears are stored. In the next season, 500 to 700 progeny rows are grown with an isolation distance of 1000m with part of seeds collected from single R plants in the previous season. Individual plants are monitored for true to type nature till maturity. Seeds are harvested, examined and the nucleus seed is constituted.

The R line progeny rows are grown alternating with A line raised from the seeds from nucleus stock of A line. Thorough and very critical monitoring is made for plant expression in vegetative and reproductive phases to ascertain true to type expression in A and R lines. Main tillers of 10 to 15 plants in each progeny of R line are self fertilised and the secondary tillers in the R plant are utilised for making A x R paired crosses involving the main tillers of the A line that are already protected with selfing bags. In A lines, plants showing pollen shedders are discarded along with other plants in same progeny. The ears from self fertilised R lines are examined for ear shape, good seed setting and for grain characters and grains of the selected heads of R line are stored. The ears collected from A x R paired crosses are examined for shape and good seed set in the cross pollination and the crossed seeds of (A x R) paired crosses are stored crosswise (A1 x R1, A2 x R2, A3 x R3 and so on till (An x Rn).

In the next season, part of the seeds from selected plants in R line ears are sown in isolation plots in 2 rows progeny plots and examined thoroughly for true to type characters in the different development phases. Progeny rows showing deviant type plants are uprooted and the adjoining rows are also uprooted. In the same season at same time of sowing of R line progenies, the seeds of paired crosses of A x R are sown cross wise in a different isolated plot (1000m.) with labels. In every paired cross row, main tillers of 10 to 15 plants of A x R paired cross origin are also self fertilised. At maturity all the plants in all the A x R paired crosses are critically examined for uniformity , vigor and good grain/ seeds set. The progeny rows of A x R paired crosses and plants from the same self fertilised plants showing poor seed fertility restoration and poor/ less seed set are recorded and the corresponding R line progenies in R line testing plot are discarded. The A x R paired cross row growing indicates the pollen fertility restoration ability of the R line. Based on the simultaneous verification from A x R plots, progeny rows in R line plot are identified and self fertilised ears from the selected R line rows are harvested and seeds separated. In the next season, part of the seeds from the selected R line ears are pooled together and sown to raise R line nucleus seed production in isolation of 1000m distance. The remainder of the seeds from single plants of R line are stored for initiating next cycle of nucleus seed production. This procedure assures that all the individual plants in the nucleus seed stock of R line is endowed with pollen fertility restoration ability in A x R crosses.

Breeder Seed Production of R Line: Nucleus seeds are sown in a fertile clean plot isolated by 1000m. from other pearlmillet crops. Single plant per hill is retained by selecting vigorous seedling among the others germinated in the seed sown hills. All recommended crop production practices are followed at 30 to 35 days after sowing, Close examination is conducted and weak plants and deviating type of plants for stem and leaf characteristics and off types are uprooted. At flowering stage, very early flowering and late flowering plants are rejected. Examination is continuously made to reject deviating type of plants from the characteristics described. Pest and disease affected plants are removed. At maturity, plants with earhead and grain affected by diseases are uprooted. Visit of Monitoring Team is arranged at grain mid maturity stage. Individual healthy and matured plants are harvested and after verifying for the stem, leaf and ear head characteristics, the earheads are removed and placed in paper covers and dried. After table and laboratory examination for the shape of the ear, size, grain setting and for grain characters, seeds are separated as single plant collection. The seeds from the selected plants are bulked to constitute the Breeder Seed. Seed qualities are tested in the laboratory for quality assurance and packed and labelled with Breeder Seed Tag. Seed samples are drawn and GOT is conducted to ascertain genetic purity.

Table 1. Descriptor of pearl millet

S.No.	Characters	States	Stage of observation
1.	Plant height	Very short/short/medium/tall/very tall	Reproductive
2.	Plant anthocyanin	Absent/present	Seedling emergence
3.	Plant growth habit	Erect/intermediate	Vegetative/spike emergence
4.	Leaf sheath pubescence	Absent/present	Spike emergence
5.	Plant node pubescence	Absent/present	Spike emergence
6.	Plant time of spike emergence	Very early/early /medium /late/very late	Spike emergence
7.	Plant node pigmentation	Whitish/green/brown/red /purple	Reproductive
8.	Spike exsertion	Partial/complete	Reproductive
9.	Spike length	Very small/small/medium/long/very long	Reproductive
10.	Spike anthocyanin pigmentation of glume	Absent/present	Reproductive
11.	Spike bristle	Absent/present	Reproductive
12.	Spike bristle colour	Green/brown/red/purple	Reproductive
13.	Spike shape	Cylindrical/conical/spindle/candle/lanceolate/dumb- bell/club/ oblanceolate/globose	Reproductive
14.	Spike density	Very loose/loose/semi-compact/ compact/very compact	Harvest maturity/ lab
15.	Spike anther colour	Yellow/brown/purple	Anthesis
16.	Seed colour	White/cream/yellow/grey/deepgrey/ grey/brown/yellow brown/purple/ purplish black	Harvest maturity/ lab
17	Seed shape	Obovate/elliptical/hexagonal/globular	Harvest maturity/ lab
18	Seed size	Very low/small medium/bold/very bold	Harvest maturity/ lab

Note: Description is based on Bhatnagar *et al.* (2003) and PPV & FRA (2007)

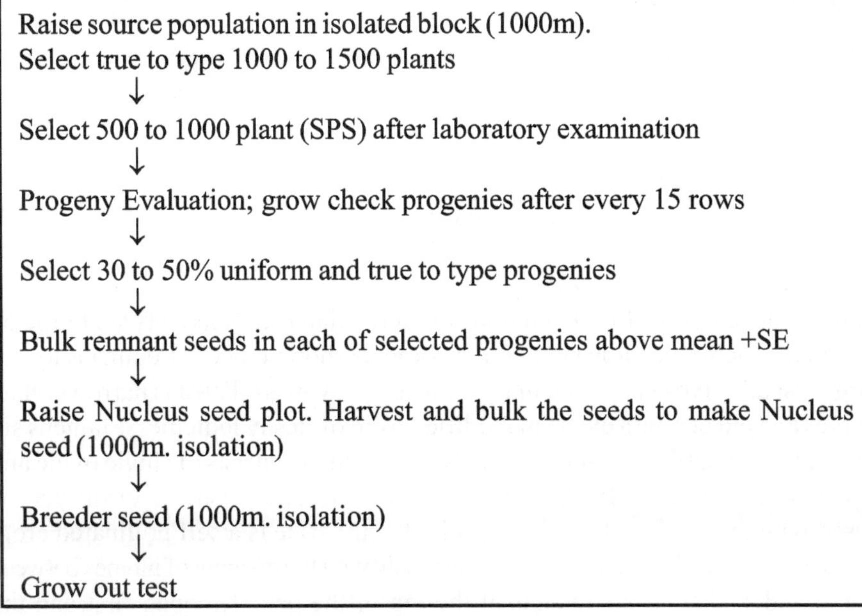
Raise source population in isolated block (1000m).
Select true to type 1000 to 1500 plants
↓
Select 500 to 1000 plant (SPS) after laboratory examination
↓
Progeny Evaluation; grow check progenies after every 15 rows
↓
Select 30 to 50% uniform and true to type progenies
↓
Bulk remnant seeds in each of selected progenies above mean +SE
↓
Raise Nucleus seed plot. Harvest and bulk the seeds to make Nucleus seed (1000m. isolation)
↓
Breeder seed (1000m. isolation)
↓
Grow out test

Fig. 1. Nucleus and Breeder Seed Production of OPV in Pearl Millet

1. Make 250 paired crosses of AXB (Isolation 1000m).
2. Paired crosses evaluation for uniformity true to type and maintenance of male sterility by growing A line (AxB) and B line in alternate rows.
3. Selection of uniform true to type AXB paired crosses.
4. Part of remnant seeds of selected paired crosses and constitute B line seeds.
5. Bulk the remnant seeds of A line (AXB) constitute Nucleus seed of A line and
6. Plant the nucleus seed stock of A line and corresponding B line in alternate rows and produce breeder seed stock A and B line in isolation of 1000m.
7. Conduct GOT.

Fig. 2. Nucleus and Breeder production in CMS (A) line and maintainer line in Pearl millet

5. Rice

Oryza sativa L. (2n:24)

Family : Poaceae

Stem erect, cylindrical and hollow at the internode, culm. Leaves alternate, leaf blades long and the leaf sheath encircles the whole or part of the internodes. Inflorescence is a loose panicle, erect to drooping. The base of the peduncle is enclosed in the sheath of the flag leaf. Spikelets are borne on secondary rachis, occur in singles and borne on short pedicels. Lemma is tough, large, usually five nerved and may be awned or awnless, Palea is narrower than nerves and the perianth exists in modified form of fleshy lodicules. Stamens six in number, versatile and filaments are slender. Pistil consists of single ovule and two plumose stigmas. In the fully developed seed, the pericarp/fruit coat is fused with the seed coat. Endosperm is starchy. Rice is a self pollinated crop. Pollination occurs just before or simultaneously with the opening of glumes between 8 am. and 11 am. The spikelets at the top of the panicles open first and the bottom spikelets later. The pollination takes place immediately after the spikelet opening. Nearly 4 to 6 days time is required for completion of flowering and pollination in a panicle. In fertile varieties, the glumes remain open for 30 to 60 minutes while in the male sterile lines, they remain open for 90 to 120 minutes. It is reported that the out crossing in hybrid rice seed production ranges from 20 to 40%. Minimizing pollen dispersal within the floret and maximizing residual pollen dispersal outside the floret (for pollen parent), prolonged and wider floret opening, period of receptivity and stigma exsertion (in seed parent) are considered important for out crossing and higher seed set in hybrid rice (Subba Rao *et al.*, 2003). The discriptors of rice are detailed in Table 1.

Nucleus seed production in Rice varieties: A base population from breeder source seed is grown in disease and pest free condition in 600 sq. m. area and from that 1500 to 2000 similar looking uniform true to type appearing plants for variety description identity are selected and serially numbered. The field is isolated by 5m. from other rice growing fields. At maturity the primary panicle from each selected plant is collected, dried and stored in separate envelopes in the laboratory. Individual panicles are closely examined for shape, colour, density, size and appearance. The seeds are separated from the selected panicles and are again verified for the shape, size, glume colour, hairiness, length and breadth. Grains are selected from individual plants and kernels separated and examined for kernel features. Only the selected single plant ear heads that confirm to the characteristics description of the variety concerned are selected and numbered finally (Fig. 1).

Seeds of selected single plants are sown in the next season in a well prepared single plant nursery of small size plots with utmost caution and care to avoid mixing of seeds. At the proper age of the single plant nursery appropriate to the duration of the variety, seedlings are pulled out and planted in a spacing of 30 cm. x 20 cm. with single seedling per hill. The age of the seedling is 20 to 25 days for the early (short) duration varieties, 25 to 30 days for mid duration varieties and 30 to 35 days for the long duration varieties (140 days and above). It is of utmost importance to plant the seedlings of proper age so as to establish uniform crop of synchronized development and ideal expression of characteristics and uniform flowering. Season coinciding with cooler and rain free months during seed filling is ideal.

The seedlings of single plants are planted in the two rows progeny plots of 4 meters long. In a well prepared ideal field, 500 progeny rows are planted with 60 cm. spacing between two rows plots and replicated twice. The required isolation distance of 5m. from other rice varieties is followed. The progenies are critically examined for true to type nature at each crop growth stage for uniformity in growth, tillering, flowering, panicle maturity and seed set. Progeny rows with off-types and deviating types plants are uprooted. Off-types occurring in progeny rows at flowering or post flowering are removed along with all other plants in that progeny row. Adjacent lines/progeny rows are also removed to ensure that there is no pollen contamination. All precautions are taken to avoid mixing of progeny rows and to tag the progenies, harvest and separate the seeds and dry them. Seeds from each progeny row is examined in the laboratory for uniformity and true to type nature of seed and kernel. Seeds from progeny rows showing yield above mean + 2SE and that conform to the varietal description alone are retained. Seeds from the selected progeny rows are pooled together to constitute the nucleus seed.

Breeder seed production: Nucleus seed is sown in a well prepared nursery in the favorable and proper season. Two meters border row is maintained around the nursery. At the recommended age, the seedlings are planted in the main field for breeder seed production with the isolation distance of 5 m. Single seedling per hill is maintained with the spacing of 30 cm. between rows and 20 cm. between plants. After every 10 rows, 60 cm. wide gap is left to facilitate field inspection and observation and to carry field operations. Continuous observation in every phase of the crop growth development is made to confirm the varietal characteristics and true to type nature at juvenile stage, active tillering stage, panicle emergence, pre flowering and post flowering and seed set. Offtypes and deviant plants are removed as and when they appear. Off-types and deviant plants noticed at time of flowering and post flowering are removed and plants around the off-type are also removed. Very early and late flowering plants are uprooted before their anthers dehisce so as to maintain the flowering duration within the specified period and to maintain flowering uniformity. Visit of Monitoring Team is arranged between the flowering and maturity period. Harvesting and seed processing are done with utmost care to avoid the seed mixing with other varieties. Seeds are dried, cleaned, processed and packed as per the specifications. Seed samples drawn from the breeder seed are tested for seed quality standards. Grow out test is conducted with seed sample from each lot. Breeder seeds are labelled with breeder seed label of Golden Yellow Tag. Nutritive and micronutrient contents are estimated in grain, kernel and straw.

Nucleus Seed Production of A, B and R parental lines (Three line system)

1. CMS line or A line (male serile)
2. Maintainer line or B line (male fertile)
3. Restorer line or R line (male fertile)

Source Nursery is raised as described below:

1. It is important to achieve synchrony in flowering in A, B and R lines so as to make AxB and AxR paired crosses in the Test Nursery. B line is sown 4 to 5 days after the A line is sown. Staggered sowing of R line is done depending on the difference in flowering period of A and R line.
2. 2000 plants in each of A line, B line and R line are planted. Single plant per hill is planted with the seedlings of proper age.
3. Every individual plants in A, B and R lines and A×B and A×R paired crosses are closely examined for deviations and off-types and they are uprooted. The off-types noticed at the time of flowering or afterwards are very carefully removed without disturbing them to prevent pollen dispersal. The plants around off-types are also removed to avoid contamination.
4. In each of A×B and A×R paired crosses, 300 to 350 crosses are made. While making the paired crosses, the same plant in A line is crossed with plant in B line by selecting 2 to 3 tillers per plant for each AxB and also in R line to make AxR paired crosses. The paired crosses of AxB is labelled as A1×B1, A2×B2 ..... An×Bn and A×R paired crosses are labelled as A1×R1, A2×R2, A3×R3.... An×Rn (Fig. 2).
5. Identification nursery is raised by sowing 15 to 20 seeds from each AxB crosses and all the seeds in AXR crosses and then transplanted in wide spaced manner. Paired crosses of A×B and A×R are observed for off-types and deviant plants and if identified such paired crosses are rejected. In A×B paired crosses, the plants should be highly male sterile and all plants therefore should maintain male sterility. Partial fertile A×B cross are identified

for rejection along with the seeds of A and B of that paired crosses. In A×R crosses, the hybrid A×R should be vigorous and fully pollen fertile. A×R crosses showing poor fertility restoration and vigour are identified and those A x R crosses showing poor vigour, off types and deviating plants are identified for rejection and uprooting along with the corresponding A line and R line.

6. **Multiplication Nursery for A line:** The remnant seeds in each A×B paired crosses showing complete male sterility (upon self fertilizing) are sown in a separate nursery, a month later. Seeds of B line of the corresponding A×B crosses are sown in an adjoining nursery three meters apart. The seedlings of A×B and B are transplanted in alternate rows in a separate block of progeny rows as described already. Isolation of 500 m. is maintained to protect the male sterile line (A×B) from pollen other than from B line. Very early flowering and late flowering plants in A×B and B are uprooted. Based on the observations in the identification nursery, such of those A×B and B lines showing off types and A×B crosses showing partial male sterility that are already identified are uprooted before flowering. The remaining plants in A (A×B) and B lines are allowed for cross pollination. At maturity, the B line is harvested 4 to 5 days earlier and threshed and seeds processed separately. A line panicles are harvested later and threshed separately. The seeds of A and B lines collected from the multiplication nursery grown in isolation are the source for nucleus seed production.

7. **Restorer line multiplication (R Line):** The R line panicle to progeny rows are raised in a plot with isolation distance of 500 m. Before flowering, the R line progenies are identified for rejection and uprooted on the basis of off type and deviating type of plants and those R lines showing poor or partial restoration of pollen fertility in A×R crosses are also rejected. Seeds of R line progenies with good pollen fertility restoration ability and vigour alone are collected and bulked as source seed for nucleus seed of R line.

8. **Nucleus seed production of A line:** Separate nursery is raised with 15 kg/ha. of A line seeds and 5 kg/ha. seed of B line. Seeds of B line are sown in an staggered manner. Half the quantity of seed in B line is sown three days after A line sowing and the remaining half of B line is sown 5 days after sowing of A line. The nucleus seed production plot is maintained with 500m. isolation distance. Spacing between A and B lines is 30 cm. and between two B lines is 30 cm. Plant to plant spacing is 15 cm. Row ratio of A:B is 6:2. Single seedling is planted per hill. The row direction is arranged perpendicular to the wind direction to facilitate higher out crossing. Off-types in A and B lines and pollen shedders in A line are carefully removed. Rouging is done in A and B line in all stages of crop growth. Application of GA3 (60 g/ha. mixed in 500 liters of water) to enhance panicle exsertion in

A line and seed set is made as per the recommendation. Supplementary pollination through rope pulling and shaking the plants are done to improve cross pollination. The B line is harvested first. After 3 to 5 days, the A line will be harvested and seed quality is tested and stored.

9. **Breeder seed production of A line:** Essentially the same procedure is followed for nursery raising to planting of A and B lines planted in row ratio of 8:2. The off types and deviant plants are removed in A and B lines. The pollen shedders in A line are removed before they shed pollen.Isolation distance of 500 m. is followed. Monitoring Team visit is arranged at flowering stage or after flowering time.
10. **Nucleus seed production of B line and R line:** Essentially the procedures are the same as being followed for the varieties (inbreds) in the nucleus seed production plots. Uniform looking progeny lines with true to type plants are selected and seeds from 1000 to 1500 superior single plants are saved for the next cycle. Isolation distance of 5m. is to be followed.
11. **Breeder seeds production of parent lines (B and R lines):** Breeder seeds of B and R lines are produced from the respective nucleus seed sources as per the method followed for the varieties/inbreds. Isolation distance of 5 m is provided. Off-types and deviating type of plants are removed. Monitoring Team visit is arranged at the time of flowering or after flowering but before full maturity of B and R lines respectively.
12. Separate Grow Out Test (GOT) is conducted as prescribed to ascertain the genetic purity for A, B and R lines respectively. Seed quality parameters are tested and quality assured seed is labelled and tagged with the Breeder Seed Tags. Nutrient qualities of B and R line seeds are estimated.

Temperature Sensitive Genetic Male sterility in Rice (TGMS)

This concept of two line approach emerged as an alternative to 3 line approach (Yuan 1987). The main features of TGMS lines are:

1. TGMS is controlled by recessive nuclear gens symbol.
2. Pollinator fertile lines carry dominant gene for fertility restoration,
3. Multiplication of female lines (TGMS) in locations / seasons where stable fertility inducing (Critical Fertility Phase – CFP) environment (temperature /photoperiod) condition prevails for a period of 30 days,
4. The sterility in the TGMS is induced when day time temperature and night time temperature is 32°C/24°C respectively in the critical sterility period (CSP). Fertility is restored when the temperature is 24°C/18°C (in day time and night time respectively) during critical Fertility phase (CFP).

5. Fertile lines express complete male sterility (reversion to sterility in the critical sterile phase) (CSP) when grown in locations/ seasons that induce male sterility (32°/24°C) (Fig. 3).
6. Reversibility from male fertile to male sterility and vice versa with changes in environmental factors such as temperature or day length during critical phases of plant growth (CSP and CFP)
7. Most sensitive phase for male phase sterility induction is from 5 days after panicle initiation to 10 days before flowering.
8. Seed multiplication of female line (TGMS) is done in suitable location with CFP (24°/18°C) by self pollination mode.
9. There is a wide choice / wide range of genotypes that can be used as male parent (R line).
10. There is no need for a separate maintainer line.
11. TGMS line is utilised as female parent (seed parent) in the sterility period. At the fertile phase TGMS line become seed fertile and can be seed multiplied.
12. Instability of male sterility expression in TGMS lines is noticed and such lines are rejected.
13. Transgressive segregation towards male fertility or sterility expression occurs in TGMS lines and when noticed such lines are rejected.
14. Low reversibility of fertility is due to large number of QTLs and their epistatic interaction.
15. Hybrid with TGMS line as female parent is fertile irrespective of day length and temperature conditions.
16. Negative effects of association with sterility inducing cytoplasm are absent in TGMS system.
17. In the TGMS line, pollen sterile expression coinciding with high critical temperature (>24°C) is not favoured for commercial exploitation since even a small shortfall of temperature may cause reversion to fertility. The TGMS line may lead to self seed set in hybrid seed production thereby resulting in low genetic purity (Arasakesary *et al.*, 2015). Two line hybrid rice breeding procedure is available in Virmani *et al.* (2003).

Seed production in TGMS line

Fertile Phase: Seeds of TGMS rice line that are produced in the previous fertile phase season with 24°C /18°C temperature in critical fertile phase(CFP) are sourced from the breeder stock seed of the Institution concerned.

In the following season coinciding with the fertile phase (24°C /18°C), seedlings are raised in a well prepared nursery which is at least one meter away from other paddy nurseries to avoid accidental mixing up of seed. Well developed healthy seedlings of 20 to 25 days age are transplanted in 400 sq.m. area field that is isolated by 10 m. distance from other rice crops. Single plant per hill is maintained and the spacing provided is 30 cm. x 30 cm. Continuous monitoring of individual plants is made at different stages of growth to verify the true to type, uniformity and stability nature. Off types and deviating plants are removed. At flowering phase, if deviating plants are noticed, they are uprooted along with the adjoining plants. Very early flowering and late flowering plants are removed to maintain flowering uniformity of the TGMS line.

At panicle emergence stage, 1000 uniform, vigorous and true to type plants are selected from middle of the field, labelled and the main panicles are self fertilized. At maturity, self fertilized panicles are harvested and examined in the laboratory for high seed set and trueness. Seeds are separated and examined for trueness and conformity for seed and kernel. Seed packets of selected self fertilized panicles are serially numbered and stored.

TGMS system in Rice

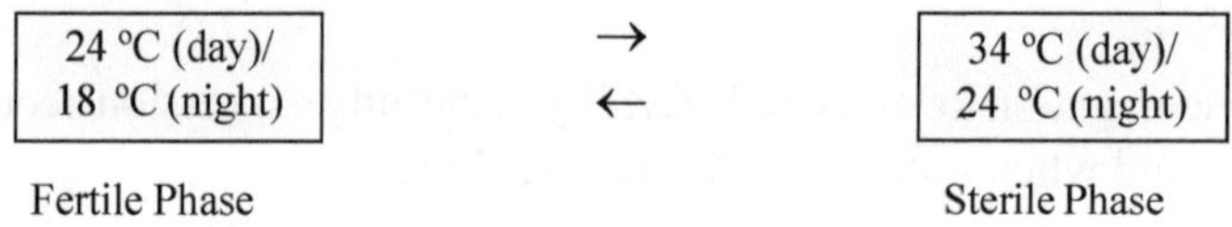

The same genotype exhibits differential response for fertility to sterility and vice versa, regulated by nuclear genes.

Sterile phase: In the next growing season, coinciding with the critical sterile phase, part of the seeds in each selected plant progeny rows are raised (isolation 500m.) to test the reversibility of the fertility to sterility by adjusting the sowing in such a way that the sensitive phase (CSP) of the crop coincides with 32°C / 24°C temperature respectively in day and night period. The sensitive period will coincide 5 days after flower initiation and 10 days before flowering. In this period of critical sterile phase/season, 500 progenies are raised in 3 rows progeny plot, each row 4m. long and with 30 cm. x 30 cm. spacing. Each individual plant in all the progenies is examined for true to type and conformity in the different phase of crop growth and development. Off-types are discarded. Deviant types of plants progeny are removed along with the other plants in the progeny. In each panicle progeny, 10 true to type plants are selected and the primary, secondary, tertiary and quaternary order panicles are protected with selfing bags. At maturity, the self fertilized plants in each progeny are harvested and the panicles of different order of the same plant are examined in the laboratory for expression of complete sterility (no seed set) in all the panicles. The stability in sterility manifestation across the panicles in each plant and across the plants of the same progeny and across the progenies are thus verified. If any panicle shows seed set on self fertilization, the entire progeny row is discarded. The progeny rows showing complete seed sterility confirming the TGMS nature alone are identified and selected.

Nucleus Seed Production of TGMS line: The remnant seeds saved from the individual plants grown in the previous season in the critical sterility phase (32°C/ 24°C) that gave rise to stable male sterile plants alone are now sown in separate single plant nurseries. Well developed healthy seedling of 20 to 25 days old are transplanted in progeny rows plot, each progeny in 3 rows and each row 4 m. long. Raising of nursery and planting are done in a favourable location for high fertility and seed set (critical fertile phase- 24°C/18°C) that will be very favourable for high fertility maintenance. The temperature of 24°C/18°C will prevail in the critical fertile phase facilitating panicle and seed development and maintenance of high seed fertility. Isolation distance of 50m. from other rice crops is maintained. Monitoring for trueness and uniformity, removal of offtypes and deviant plants are done as described earlier. Very early and late flowering plants are discarded. During flowering phase, if any deviant plants are noticed, progenies showing plants deviating from conformity are cut and removed. At maturity, progenies showing high seed set in all the plants in the primary, secondary, tertiary and quaternary tillers are selected. Harvesting and seed extraction are made progeny wise and seeds are tested for the seed qualities. Part of the seeds from each progeny is pooled to constitute the Nucleus seed. The remnant seeds of the progenies are saved progeny wise for initiating the next cycle of nucleus-breeder seed for TGMS line.

Breeder Seed Production: The ideal location of raising the breeder seed crop of TGMS in rice will be conducive for reversion of sterility to high fertility with ideal temperature of 24°C /18°C in day and night time respectively during the sensitive phase of panicle development. Seeds from nucleus seed source are sown in a well prepared nursery. Healthy vigorous seedlings of 20 to 25 days age are selected from the nursery and transplanted in the main field adopting a spacing of 30 cm. × 30 cm. Single seedling is maintained per hill. The field is isolated from other rice growing fields by 50 m. All good agronomic production practices are followed. Multistage examination of the plants for trueness, vigour, stability and free from pests and diseases is carried out. Off-types are removed. Plants showing deviation in characteristics prescribed for the TGMS line are uprooted. If the deviant plants are spotted after flowering, the deviant plants along with the surrounding plants are removed. Plants flowering very early and late are culled out to maintain well defined uniformity in flowering. Plants expressing poor vigour and less synchrony in tillering and flowering are uprooted.

At maturity, panicles are collected and examined in the laboratory for trueness of panicle shape and size and for the seed set and seed traits. Monitoring Team visit is organized at mid maturity of the TGMS breeder seed crop. Seed samples are drawn and tested for seed quality assurance and labelling with Breeder seed tag. GOT is conducted for verifying the genetic purity. Nutritional quality of rice are estimated. Detailed procedure is explained in Table 2.

Seed to Ratoon to Seed method: In this approach, seeds of TGMS line are raised following the procedures detailed for nucleus and breeder seed production. The seeds are raised in alternate generation in locations of sterility inducing environment that provides 32°C /24°C temperature respectively in day time and night time. The sterility expressing true to type individual plants are uprooted at maturity from the field without damage and multiplied by ratoon planting. In the next season, the ratoon stubbles are planted in location that provides 24°C/18°C temperature respectively in day time and night time during critical fertile phase of panicle development that is conducive for reversion to fertility from sterility. Panicles in the ratoon plants are self fertilized and harvested separately. After the laboratory examination of panicles and grains, they are selected and pooled together to constitute Nucleus seed of TGMS line. At each time of nucleus seed production, fertile plants are seed grown and sterility induced plants are selected in sterility inducing environment and stubble ratoon planted thereby reversal from fertility to sterility and vice versa are verified in the same plant. In this system of vegetative multiplying the rice by ratooning, the risks of crop health and pests and diseases carry over are likely to be present. Isolation distance of 50m. is provided for seed to ratoon and ratoon to seed method.

Photoperiod sensitive Genetic Male Sterility (PGMS) system in rice: The rice plants showed complete to high degree of pollen sterility when the line flowered before September 1 or 2 (day length 14 hrs). However, PGMS plants flowering after the date (day light less than 13.75 hrs) showed various degrees of spikelet fertility ranging from 10 to 40%. The sterility of the sterile stage was stable but the degree of fertility was not stable in fertile stage. Sterility in PGMS system is controlled by a single recessive gene and some modifiers (Virmani *et al.,* 1993).

Table 1. Descriptor of rice : *Oryza sativa* L.

S.No.	Characters	State
1.	Color of seedling base	Colorless, green, purple
2.	Basal leaf sheath color	Green, light purple, purple lines, uniform purple
3.	Leaf green color	Light, medium, dark
4.	Leaf anthocyanin coloration	Absent, present
5.	Leaf anthocyanin color distribution	On tips only/on margin only, in left side only/ uniform

6.	Leaf sheath anthocyanin coloration	Absent, present
7.	Leaf sheath intensity of anthocyanin coloration	Very weak, weak, medium, strong,very strong
8.	Leaf surface presence	Absent, weak, medium,
9.	Leaf auricle	Absent, present
10.	Leaf anthocyanin coloration of auricle	Colorless, light purple, purple
11.	Leaf collar	Absent, present
12.	leaf collar anthocyanin colors	Absent, present
13.	Leaf ligule	Absent, present
14.	Leaf-shape of ligule	Truncate, acute split
15.	Leaf ligule colour	White, light purple, purple
16.	Leaf: blade length (cm.)	Short (<30), medium (30-45), long (>45)
17.	Leaf: width of blade (cm.)	Narrow <1, medium 1-2, broad >2
18.	Culm attitude for floating rice only	Non procumbent, procumbent,
19.	Culm attitude	Erect, semi erect, open, spreading
20.	Time of heading 50% panicles(days)	Very early (<71), early 71-90, medium 91-110, late 110-130, very late >131
21.	Attitude of flag leaf	Erect, semi erect, horizontal,drooping
22.	Spikelet density	Absent, weak, medium, strong,very strong
23.	Male fertility	Absent, present
24.	Anthocyanin presentation of keel	Absent, very weak, weak, medium, strong, very strong
25.	Lemma anthocyanin coloration of area below apex	Absent, weak, medium, strong, very strong.
26.	Lemma anthocyanin coloration of apex	Absent, weak, medium, strong, very strong
27.	Spikelet color of stigma	White, light green, yellow, light purple, purple
28.	Stem thickness (cm.)	Thin (<0.40), medium (0.40–0.55), thick >0.55
29.	Stem length (cm.) excluding panicle	Very short <90, short (91-110), medium, (111-130), long (131-150), very long (>150)
30.	Stem- anthocyanin coloration of node	Absent, present
31.	Stem-anthocyanin intensity of nodes	Weak, medium, strong
32.	Stem-anthocyanin coloration of internode	Absent, present
33.	Panicle length of main axis (cm.)	Very short (<16), short (16-20), medium (21-25), long (26-30), very long (>30)
34.	Flag leaf attitude	Erect, semi erect, horizontal deflexed
35.	Panicle curvature of main axis	Straight, semi straight deflexeddropping
36.	Panicle number per plant	Few (<11), medium (11-20), large>20
37.	Spikelet color of lemma tip	White, yellowish, brown, red, purple, black
38.	Lemma palea color	Straw, gold, brown spot on straw, brown ferrous on straw, reddish to light purple spot, ferrous strawpurple, black
39.	Panicle awn	Absent, present
40.	Panicle color of awn	Yellowish white, yellowish brown, brown, reddish brown, light red, red, light purple, purple, black

41.	Panicle, length of longest awn	Very short, short, medium, long, very long
42.	Panicle distribution of awn	Tip only, upper half only, wholelength
43.	Panicle presence of secondary branching	Absent, present
44.	Panicle secondary branching	Weak, strong, clustered
45.	Panicle attitude of branches	Erect, semi erect, semi erect to spreading, spreading
46.	Panicle exsertion	Partly exserted/mostly exserted and well exserted
47.	Time of maturity (days)	Very early (<100) / early (101-120), medium (121-140), late (141-160), very late (>160).
48.	Leaf senescence	Early, medium, late
49.	Spikelet lemma color	Straw, gold, red, purple
50.	1000 grain weight (g.)	Very low (<15), low (15-20), medium (21-25), high (126-30), very high (>30)
51.	Grain length (mm.)	Very short (<6.0), short (6.1-8.5), medium (8.6-10.5), long (10.6-12.5), very long (>12.5)
52.	Grain width (mm.)	Very narrow (<2.0), narrow (2.1-2.5), medium (2.6-3.0), broad (3.4-3.5), very broad (>3.5)
53.	Grain phenol reaction of lemma	Absent, present
54.	Decorticated grain length	Short, medium, long, extra long
55.	Decorticated grain width (mm.)	Narrow (<2.0), medium (2.0-2.5), broad (>2.5)
56.	Decorticated grain shape	Short slender, short bold, medium slender, long bold long slender, extra long slender
57.	Decorticated grain color	White, light brown, dark brown, light red, red purple, dark purple
58.	Endosperm presence of amylose	Absent, purple
59.	Endosperm amylose content (%)	Very low (<18), low (10-19), medium (20-25), high (26-30), very high (>30)
60.	Vars with endosperm amylose absent only : porexsed grain exproxim of whitecore	Absent, very small, small, medium, large, fully chalky
61.	Gelatinisation temperature throug halkali spreading value	Low, medium, high medium, high
62.	Decorticated grain aroma	Absent, present

(*Source:* PPV & FRA, 2007)

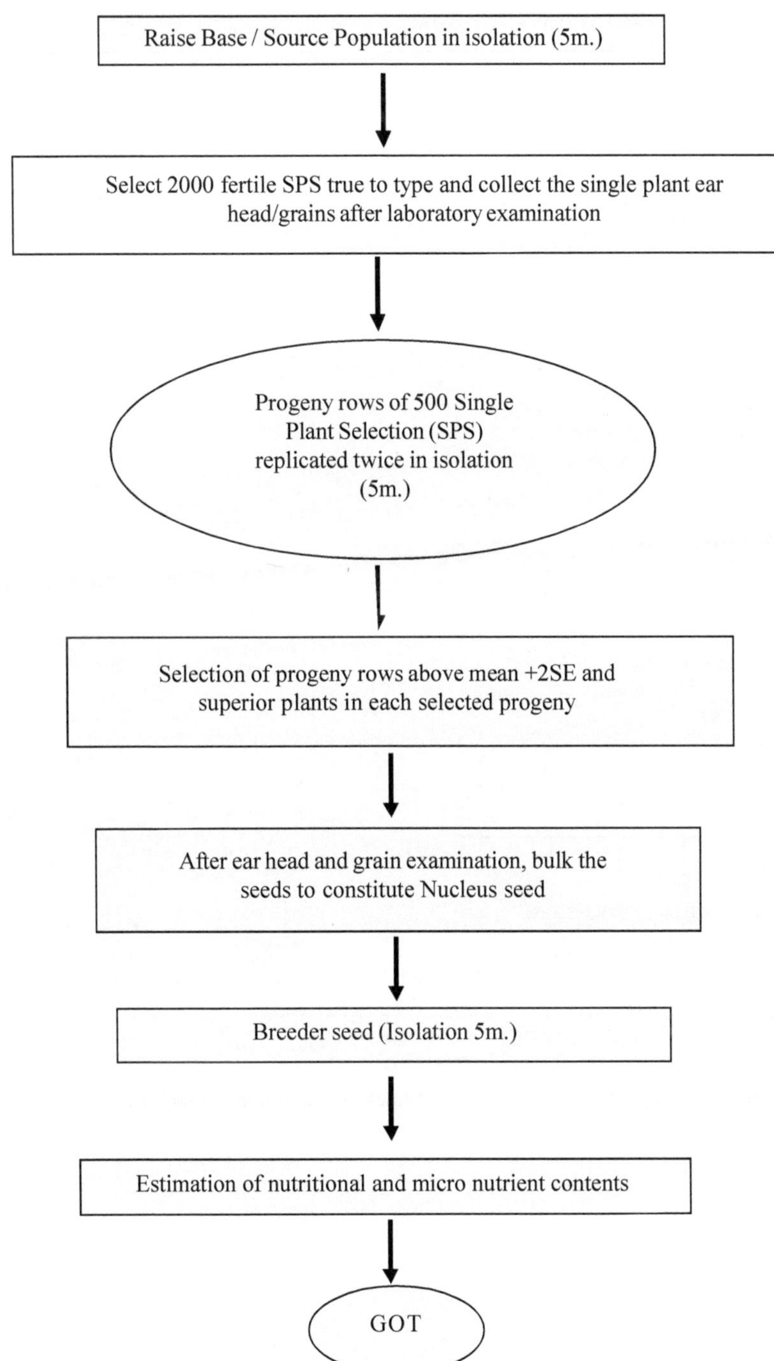

Fig. 1. Nucleus and Breeder Seed Production in Rice

SeasonI. Raising source nursery and test crossing:

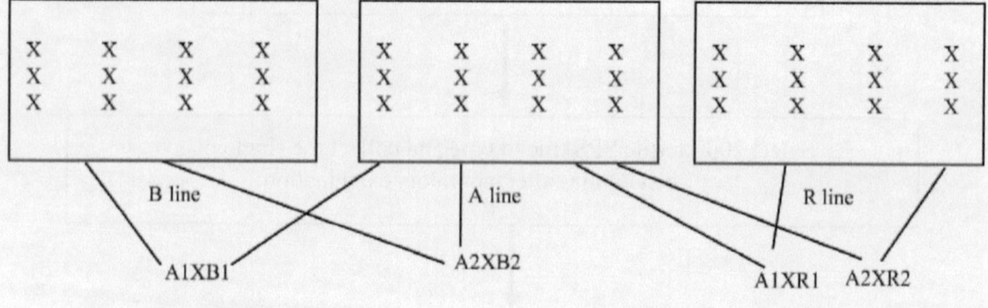

Season II. Raising identification Nursery (A)

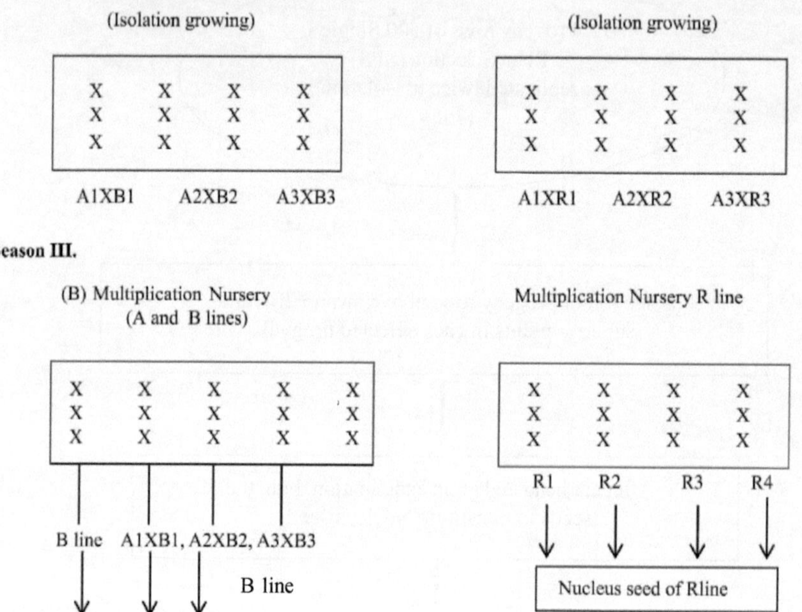

Nucleus seed of Line B → Nucleus seed of Line A

Fig. 2. Nucleus seed production of A, B and R lines in rice. (Isolation 800m.)

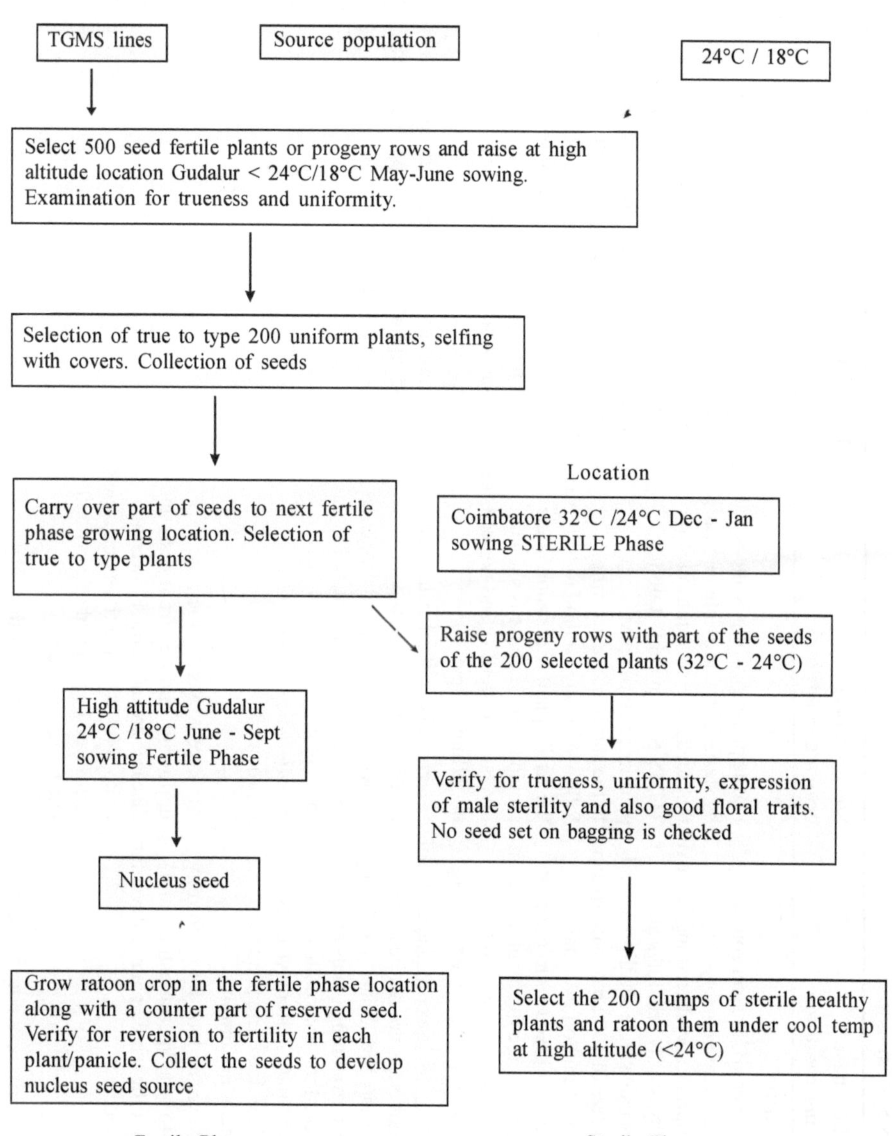

Fig. 3. Nucleus seed production in TGMS rice

Table 2. Seed production in TGMS System in Rice

Plant quality description	Fertile phase growing condition	Sterile phase growing condition	Fertile phase ratoon crop
Erect, uniformity, true to type, genetic purity, synchronous tillering, disease and pest free nature, erect fast growing, withstand weed competition, erect canopy, good panicle exsertion, high seed number and high seed set.	**NURSERY** Raise seeds of breeder source in well prepared nursery. At appropriate age of seedling, remove healthy, sturdy seedlings alone. Transplant single seedling per hill in an area of 400 Sq.m with spacing of 20x20 cm Follow usual production practices to establish healthy disease free field. Isolation distance of 50 m. is provided. Remove off-types and deviating	Fertile phase growing is alternated by sterile phase growing to ensure that seeds selected/multiplied for TGMS System exhibit pollen sterility ie sterile plants. In the sterile phase growing, following is the method to test that the fertile plant expression in the previous fertile phase is showing reversion to sterility and no seeds set on selfing. Steps:	Ratoon shoots of the sterile plants are carefully separated and planted in spacing of 30cm x30 cm and grown till flowering in a field isolated by 100 meters. Trueness and uniformity are verified at flowering and at panicle maturity. All the ratoon crop plants will bear grain set. Each main tiller panicle is also self fertilized to test full grain seed set and to verify the reversal of sterility to fertility in the fertile phase.
	plants up to flowering stage. When deviant types are noticed after flowering, the deviant plant along with surrounding plants are uprooted to avoid pollen contamination. Remove early and late flowering plants to maintain uniformity.	Select seeds of 200 fertile plants based on previous fertile phase growing. Now 200 plant progenies are raised in 4m long and in 2 rows plot. Isolation distance of 800 m. is provided.	Seeds are separated after the examination of individual earheads and then pooled together to constitute nucleus seed.
	At flowering look for uniformity, true to type in each panicle and uproot the whole plant if deviation is noticed.	All steps to verify trueness, uniformity and free from pests and diseases are taken up. Flowering uniformity within and between rows is ensured by removal of early and late flowering plants.	
	At maturity, examine matured	Deviant plants if noticed at	

panicles for deviations in panicle shape, size, grain colour besides clear exsertion of panicle in the primary, secondary, quaternary & pentanary tillers, Select/label 1000 plants. Harvest main panicle alone in each of the 1000 selected/ labelled single plants.	flowering and after flowering are uprooted as whole progenies.
In the laboratory examine panicle type, shape, size, grain shape, seed colour and for overall uniformity and true to type nature. Separate the grains from each panicle and save in packets plant wise, serially numbered.	At the time of flowering 10 true to type plants in each progeny row are selected. The main tiller panicle is first self fertilized and labelled. In all the selected plants for selfing the secondary, tertiary, quaternary and pentenary panicles are also self fertilized and labelled.
During the next fertile phase season/location, raise progeny rows with seedlings raised in single plant nursery beds avoiding mixing up.	All the self fertilized tillers in each plant are safely secured with a support from bamboo stick. The stick is tied with label indicating the progeny row and plant number.
The progeny row will be 4m long and in two row plots. Poorly established plants, unhealthy plants are uprooted. Step to ensure flowering uniformity by removing early and late flowering plants is completed. Individual plants in individual progeny rows and between progeny rows are very critically	At maturity, self fertilized panicle bags from each plant with all the selfed bags of the tillers are harvested. In the laboratory the selfed bags are serially arranged. Every bag is gently opened and critically examined for no seed set indicating that the plants grown in sterile phase is reverted to pollen sterility in all the self fertilized tillers in sequence first to fourth fillers.

examined for trueness (true to type), genetic uniformity and vigour and health.	This indicates stability for fertility reversion to pollen sterility. Next step is to uproot the plants from different progeny rows and carefully plant in paddy soil with marshy condition to encourage ratooning of the selected plants that express complete sterility.
Progeny rows showing deviating and off-type plants are discarded to ensure genetic uniformity and purity. At maturity, progeny rows showing uniformity for plant type, panicle shape, size, grain size and shape are identified and selected.	
In each of the selected progeny rows 10 to 15 main tiller panicles are individually collected. After laboratory examination of panicle and grain features, seeds of individual panicles are stored in individual packets.	
Parts of the seeds from each panicle packet are pooled together to constitute nucleus seed. The remaining seeds are retained for future use to develop nucleus seed.	

Two line hybrid rice seed production in Tamil Nadu

In Tamil Nadu, there is an exclusive centre for two line hybrid rice research under Tamil Nadu Agricultural university, Coimbatore. The centre was established during 1996 at Gudalur, Nilgiris District. The seed production procedure mentioned above is followed for evaluating TGMS line at Gudalur and Coimbatore. For exploiting the two line breeding system a stable TGMS Lines are needed and seed production to be standardised based on the line to be used and prevailing weather parameters in the particular location.

Manonmani *et al.* (2016) studied the stability of new TGMS lines for sterility and standardized the seed multiplication of TGMS lines at sterility and fertility favouring environments. The experiments were conducted at Paddy Breeding Station, Coimbatore, Farmers field at Sathiyamangalam (Sathy) and Hybrid Rice Evaluation Centre at Gudalur during the rabi and khariff seasons in 2013 & 2014 to assess the pollen fertility expression under different temperature regimes in newly developed temperature sensitive genic male sterile lines of TNAU. Weather parameters at Coimbatore and Gudalur during the past fifteen years was analysed for fixing the sowing season (Fig. 4&5). TGMS lines were evaluated under sterility inducing Environments viz., Coimbatore and Sathiyamangalam during the month of December 2013 and 2014 (Rabi 2013 and 2014). The same lines were stubble planted and evaluated for pollen sterility under pollen fertility inducing Environment during the month of July 2014 at high altitude (1500 MSL) with cool climate at Hybrid Rice Evaluation Centre, Gudalur. The new TGMS lines developed at TNAU viz., TNAU 45S, TNAU 60S, TNAU 95 S, TNAU 19S and TNAU 39S are evaluated for their stability of pollen sterility under different temperature regimes were given in the Table 3.

Table 3. Pollen sterility % of different TGMS lines in Rice

TGMS lines	Rabi 2013		Rabi 2014		Kharif 2013	Khariff 2014
	Coimbatore	Sathy	CBE	Sathy	Gudalur*	Gudalur*
TNAU 45S	100	100	100	100	5	4
TNAU 60S	100	100	100	100	3	5
TNAU 95S	100	100	100	100	6	5
TNAU 19S	100	100	100	100	7	9
TNAU 39S	100	100	100	100	4	6

*Fertility inducing environment

At sterility inducing Environments the lines showed 100% pollen sterility. These lines were seeded during December at Coimbatore and Sathiyamangalam to expose them to a sterility inducing temperature (>29°C /<23°C day night) during panicle initiation to flowering stage to test their sterility behavior so that their critical stage of flowering coincides with more than 29°C. Mean weather data for both locations was provided in the Fig. 4 & 5 and it showed that the temperature recorded in both the places exceeded > 25°C during the month of March, April and May. During the flowering stage all these lines showed 100% pollen sterility at both the locations for more than 60 days and was test verified for next year also.

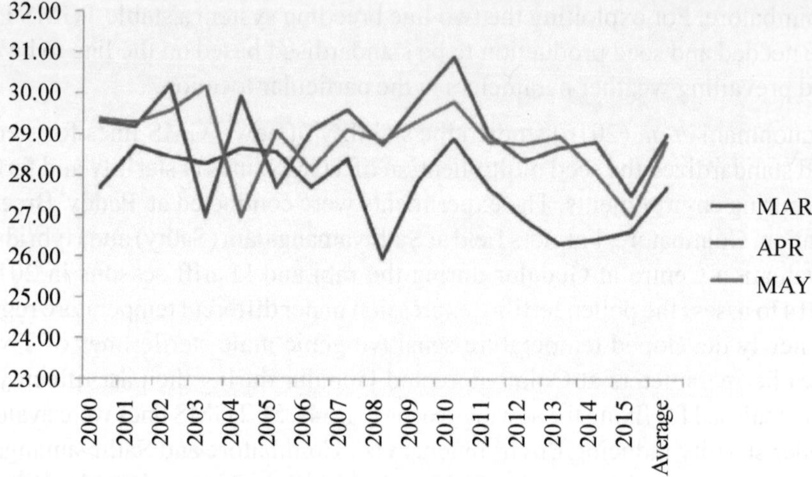

Fig. 4. Mean Weather data at Coimbatore Location (°C)

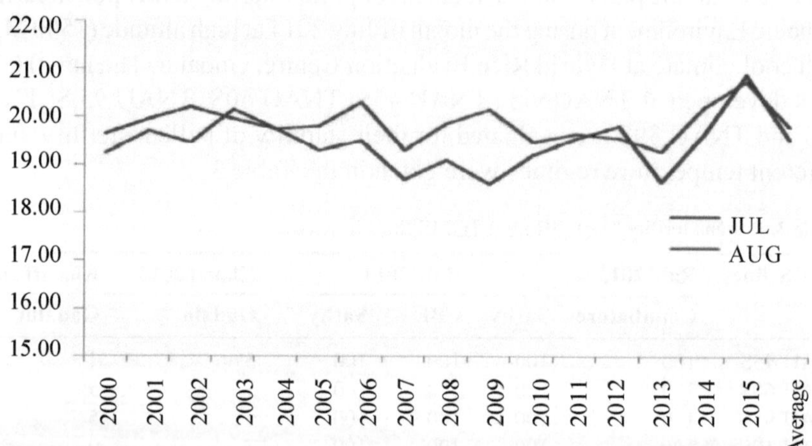

Fig. 5. Mean Weather data at Gudalur Location (°C)

The daily mean temperature of 24°C to 26°C was found to be the critical temperature for fertility alteration. The sterile stubbles of these lines were planted at HREC, Gudalur during May to induce fertility for their seed multiplication and were exposed at critical stages to fertility inducing temperature (24°C/18°C day/night). Maximum, minimum, and mean temperature significantly influenced the pollen and spikelet fertility in all five TGMS lines at high altitude. At Gudalur the temperature range during the month of July and August was less than 20°C. The appropriate sowing date of TGMS lines was fixed during June-July in such a way that the critical stages of panicle development would be exposed to the required temperature. The individual lines were maintained under isolation and genetically pure seeds were produced at Gudalur.

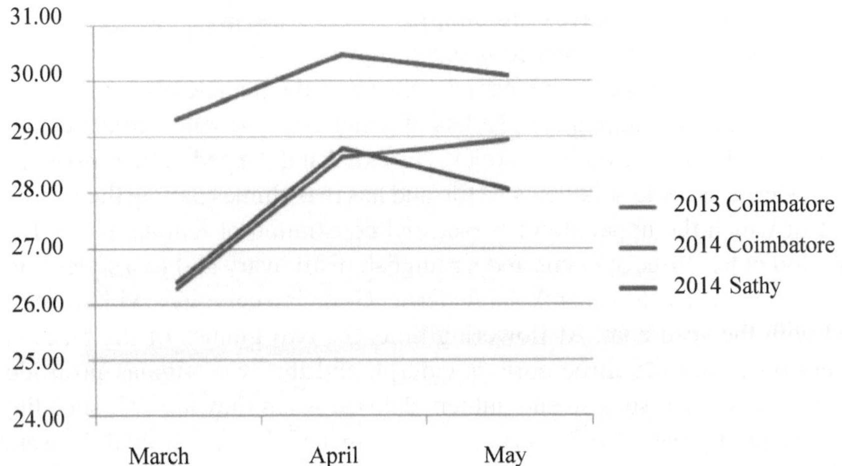

Fig. 6. Mean Weather data at Coimbatore and Sathyamangalam (°C)

The above TGMS lines with wider pollen sterility period under plains can be very well exploited for developing two line rice hybrids during the period of December to April at Coimbatore. The same lines can be easily seed multiplied at Gudalur during July to November.

Two line hybrid rice seed production

The TGMS seed produced from high altitude can be brought to locations like Coimbatore as proved above, where stable high temperature prevails during its sensitive stage for nearly 30 days. TGMS plants are planted in 6 rows and sandwiched with two rows of non TGMS good combiner lines on either sides. TGMS line must be randomly checked for complete pollen sterility during its sterile phase at high temperature. Supplementary pollination techniques as applied to three line system must be adopted such as GA3 spray, rope pulling, flag leaf clipping etc. to increase hybrid seed production. At the time of harvest care must be taken to harvest separately the non TGMS lines first. Seeds harvested from the TGMS line must be cleaned, packed, labeled, and distributed to farmers as two line hybrid seed.

Hence Coimbatore and Gudalur locations were identified respectively for the hybrid seed production and seed multiplication in TGMS in Tamil Nadu.

6. Sorghum

(*Sorghum bicolour* (L) Moench) (2n:20)

Family : Poaceae

Annual, herbaceous, erect and solid, single stalked or tillering; colour of the stem is green or with purple wash. Leaves are simple, alternate and long. Inflorescence is either compact or semicompact or lax panicle. Shape of the panicle also varies. The joints of the rachis bear paired spikelets of which one is sessile while the other spikelet is pedicellate. Sessile spikelet is bisexual and the pedicelled spikelet is male or sterile. Sessile spikelet is fertile and has two glumes having the flowers inside of which the upper floret is bisexual constituted of lemma, palea, two fleshy lodicules, three stamens and a roundish pistil ovary and two styles with feathery stigma. The lower spikelet is sterile. Grain is a caryopsis with pericarp fused with the seed coat. At flowering time, the two glumes of the bisexual flowers open and the three anthers emerge and the two stigmas protrude. Flowering occurs at sunrise and anthers dehisce when they are dry, shedding pollen on the stigmas. The flowering in sorghum panicle starts from the top and continues to lower whorls and over a period of 4 to 5 days flowering is completed. Pollen is available for nearly 2 weeks due to the panicles in the field population do not flower all at the same time. Sorghum is predominantly a self pollinated crop but cross pollination may occur to the extent of 2 to 10 per cent. (Kanna Babu and Seetharama 2003).

Nucleus seed production (OPV variety): Select from the middle of the field at least 1500 to 2000 similar looking true to type healthy plants grown with single plant per hill in 0.1 to 0.2 ha field sown with breeder source seeds of the OPV/ variety. Isolation distance of 400 m. is maintained. The uniform selected plants are self fertilised. The self fertilised ear heads are examined for good seed set, shape of ear head and seed features and selected. In the next season, 500 plant progenies of selected plants are raised in 2 row plots with 2 replications. The check row of the same OPV is raised once in 10 progeny plots. Isolation distance of 400m. is maintained. Thorough and close examination is made on each plant in the progeny plots for uniformity and true to type nature of the variety. The progenies showing deviating types of plants are uprooted and the adjoining progeny rows are also uprooted if the deviations in the plants are noticed at flowering stage. Progeny rows performing above mean +2SE are identified and selected. In each of the selected progeny plot, 10 to 15 healthy superior plants are self fertilised at maturity. All the ears of the self fertilised plants are harvested and examined in the laboratory for the ear head features, good seed setting and seed characteristics conforming to the original variety. Seeds from selected ears are labelled and carried forward (Fig. 1). Part of the self fertilised seeds in each plant are bulked together to constitute nucleus seed. The remainder of the seeds of each plant are utilised for growing progeny plots and initiating next nucleus seed production cycle.

Breeder Seed Production of (OPV) Varieties: Nucleus seeds are sown in a fertile clean plot isolated by 400m. from other sorghum crops. Single plant per hill is retained by selecting vigorous seedling among the others germinated in the seed sown hills. All recommended crop production practices are followed at 30 to 35 days after sowing, Close examination is conducted and weak plants and deviating type of plants for stem and leaf characteristics and off types are uprooted. At flowering stage, very early flowering and late flowering plants are rejected. Examination is continuously made to reject deviating type of plants from the characteristics described. Pest and disease affected plants are removed. At maturity, plants with earhead and grain affected by diseases are uprooted. Visit of Monitoring Team is arranged at grain mid maturity stage. Individual healthy and matured plants are harvested and after verifying for the stem, leaf and ear head characteristics, the earheads are removed and placed in paper covers and dried. After table and laboratory examination for the shape of the ear, size, grain setting and for grain characters, seeds are separated as single plant collection. The seeds from the selected plants are bulked to constitute the Breeder Seed. Seed qualities are tested in the laboratory for quality assurance and packed and labelled with Breeder Seed Tag. Seed samples are drawn and GOT is conducted to ascertain genetic purity. Seed samples are tested for nutritional and micronutrient contents.

Cytoplasmic Genic Male Sterility System

Nucleus seed production in A and B lines in sorghum: The nucleus seeds of A and B can be developed simultaneously. The seeds of A line and B line are obtained from the breeder source A and B lines (Fig. 2).

Rows of A line are raised alternated with B line rows (1:1) in plot with 800m.isolation. Single plant per hill is maintained. In the individual A and B progeny rows 1500 to 2000 plants are verified for the true to type character expression in the different growth phases and similar looking true to type plants are selected. At the time of flowering, 10 to 15 plants in each A and B rows are selected and protected with selfing bags. At anthesis initiation time the individual plants in A line protected with selfing covers are examined for pollen shedders and the plants in the A line showing pollen shedders are uprooted and rest of the A line plants are pollinated with pollen from plants of B line protected with selfing covers. Paired crosses, of AxB line are made. At least 1000 paired crosses of A×B are made and tagged with serial numbers such A1×B1, A2×B2 An×Bn. At maturity, earheads in A (A×B) and B are harvested and examined in the laboratory for ear shape, length, good seed set and grain characteristics. Panicles with deviations and poor grain set are rejected from A (A×B) along with corresponding B line of A×B paired crosses showing poor seed set. In the next season, in plots grown with isolation distance of 800m., progeny rows of A and B are raised in single rows retaining only single plant per hill. Critical monitoring of plants in A and B is made for true to type in the different growth phases and for phase specific characteristics and deviant plants are uprooted. Paired crosses of A×B are made between plants from adjacent rows. The paired crosses are serially numbered as A1×B1, A2×B2.... An×Bn and all the plants in B line involved in A×B crosses are self fertilised. Plants in A line showing pollen shedders are uprooted. At maturity, the ears of A (paired crosses of A×B) and B are examined thoroughly for ear and grain characteristics and deviant types are discarded. Seeds from selected ear heads of A and B are carried over for next season (Fig. 3).

In the next season, in isolation block (800m.), progeny row plots of A line (A×B) in two rows and corresponding B line in one row are raised alternatively. Monitoring of plants for true to type nature and for uniformity is done from juvenile to flowering and maturity stage and off types are uprooted. In each of the progeny plots, 15 plants in A line and 15 plants in B line are protected by selfing covers and A×B paired crosses are made besides self fertilising B plants. Pollen sterile plants, if any, in the B line are uprooted Laboratory examination is made on ears and grains of A and B for conforming to the parent characteristics. Part of the seeds from each earheads of A×B and B are bulked to constitute respectively of A line and B line nucleus seed.

Breeder Seed Production: Part of A seeds from each A×B cross and self fertilised B are sown in 800m. isolation in the next season, to raise progeny plots in 4:1 ratio of A and B rows for production of breeder seed. Pollination between A and B is promoted by 2 to 3 shakings of B plants during pollination period. Pollen shedding plants in A line rows are uprooted. At the end of the season, panicles of A and B with good set are collected and seeds separated and utilised for breeder seed production supply. Remnant seeds are carried forward for continuing the cycle once in 2 to 3 years.

Nucleus seed and Breeder Seed production in R line: R line plants are raised from source seeds in the field with 400 m. distance maintaining single plant per hill. 1000 to 1500 plants showing true to type uniformity and conforming to the R line characteristics are selected. The selected plants are self fertilised and ears showing high seeds *et al.* one are selected and examined in the laboratory for panicle and seed features. In the next season, three rows progeny plots in each progeny are raised and single plant per hill is maintained. Plants are examined for the growth phase specific characteristics in the different growth phases and the progeny row plots showing deviant types of plants are rejected and uprooted. In each progeny row, 10 plants are self fertilised and the self fertilised and matured ears are examined for good seed set, shape, size and grain characteristics and selected. Part of the seeds from each plant is bulked to constitute nucleus seed of R line.

In the next season, the self fertilised seeds of R plants are sown in progeny row plots in field with 400m. isolation. In each progeny plots 25 to 30 true to type plants are self fertilised. Simultaneously, the pollen collected from these selected self fertilised plants are crossed in A×R pair with A line plants grown at a corner portion of the plot. The cross pollinated panicles of A line are protected with paper covers. Deviant types and pollen shedders, if any, in the A line have been already uprooted. The paired crosses are serially numbered (A1×R1, A2×R2 ... and so on till An×Rn). In the following season, A x R paired crosses are raised in an observation plot and are assessed for heterotic vigour expression and complete pollen fertility restoration and good seed set in ears of A×R hybrid. Part of the seeds of selfed ears of R line plant that show high fertility restoration and hybrid vigour in A×R paired crossed are pooled together to constitute Breder Seed of R line. The remnant seeds are carried forward for repeating the nucleus seed production cycle followed by breeder seed production of R line once in 2 to 3 years.

Table 1. Descriptor of sorghum

S.No.	Characters	States	Stage of observation
1.	Plant height	Tall/medium/dwarf	Vegetative
2.	Plant colour	Pigmented (grey brown groups)/tan (yellowish)	Vegetative
3.	Stem juiciness	Juicy/corky	Vegetative
4.	Stem Leaf	Erect/semi-erect/droopy	Vegetative
5.	Leaf midrib colour	White/dullgreen/yellow/brown/purple	Vegetative
6.	Leaf colour	Dark green/green/light green	Vegetative
7.	Stem thickness	Thin/medium/thick	Vegetative
8.	Days to flowering	Early/medium/late	Reproductive
9.	Panicle type	Very lax/very loose with erect primary branches/very compactness and shape loose drooping primary branches/loose erect primary branches/loose drooping primary branches/semi-loose erect primary branches/semi-loose drooping primary branches/ semi compact elliptic /compact	Reproductive

		elliptic /compact oval /half broom corn/broom corn	
10.	Glume colour	White/straw/brown/light red/red/ yellow purple/black	Maturity
11.	Seed covering	25% grain covered/ 50% graincovered, 75% grain covered / grain fully covered/ glumes longer than grain	Maturity
12.	Awns	Awed /aweless	Maturity
13.	Days to maturity	Early/medium/late	Maturity
14.	Shattering	Low/medium/late	Maturity
15.	Seed colour	White/pale yellow/yellow/orange	Maturity
16.	Seed lusture	Lustrous/medium/non-lustrous	Maturity
17.	Seed sub coat	Present/absent	Maturity
18.	Seedplumpness	Dimple/plump	Maturity
19.	Seed form	Single/twin	Maturity
20.	Endosperm texture	Completely corneous/mostly corneous/ intermediate/starchy	Maturity
21.	Endosperm colour	White/yellow	Maturity
22.	Endosperm type	Normal/waxy/sugary	Maturity

Source: PPV & FRA (2007)

Grow source population in isolation of 400m.

↓

Select 500 true to type plants. Raise progeny rows in isolation (400m.) in 2 replications

↓

Evaluation of progeny rows and selecting true to type and superior performing progenies performing above mean +2SE. Laboratory examination of earheads and grains.

↓

Seeds separated from ears of superior plants in the selected progeny rows. Bulk the seeds and constitute Nucleus seed (Isolation 400m.)

↓

Breeder Seed (Isolation 400m.)

↓

GOT

Fig. 1. Nucleus and Breeder Seed Production in Sorghum

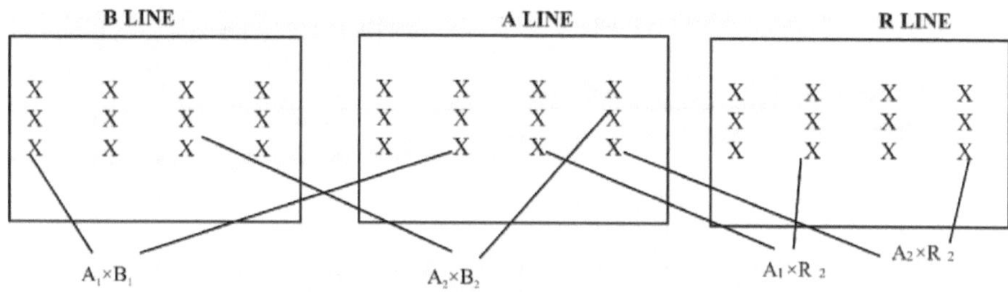

1. Raise identification nursery and make (1000 paired crosses (AXB and AXR)

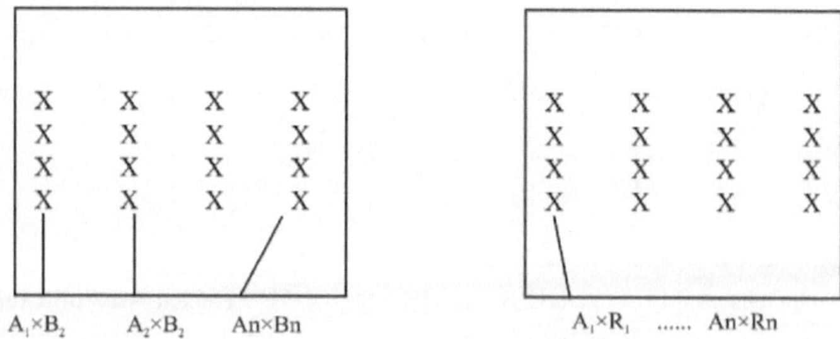

Fig. 2. Nucleus and Breeder Seed Production in A, B and R lines in Sorghum

Source: Population of A and B lines raised in isolation. Select 1000 true to type and uniform plants in each A and B. In R line source population, select 1000 true to type plants.

Raise progeny of A×B cross alternate with B line. Monitor for growth and true to type comparison (A×B) progeny and B line progeny. Select uniform and true to type progeny line in A×B cross and in B line progeny. Ear heads of A×B and B line are respectively harvested, tabulated and laboratory examination conducted for ear and grain and saved plant wise in A×B and B respectively . Seeds from individual ears/plant separated in A×B and B lines respectively and nucleus seed constituted. Similarly seed set in A×R are saved along with respective R line parent. A1XR1 to AnxRn are raised to verify pollen fertility restoration, high seed set and hybrid vigour. The selected R line plants are utilized to constitute R line nucleus seed.

Breeder Seed Production of A(A×B) line and B line is made in isolation (800m.) plot. Breeder Seed Production in R line is made in isolated plot field of 400m.

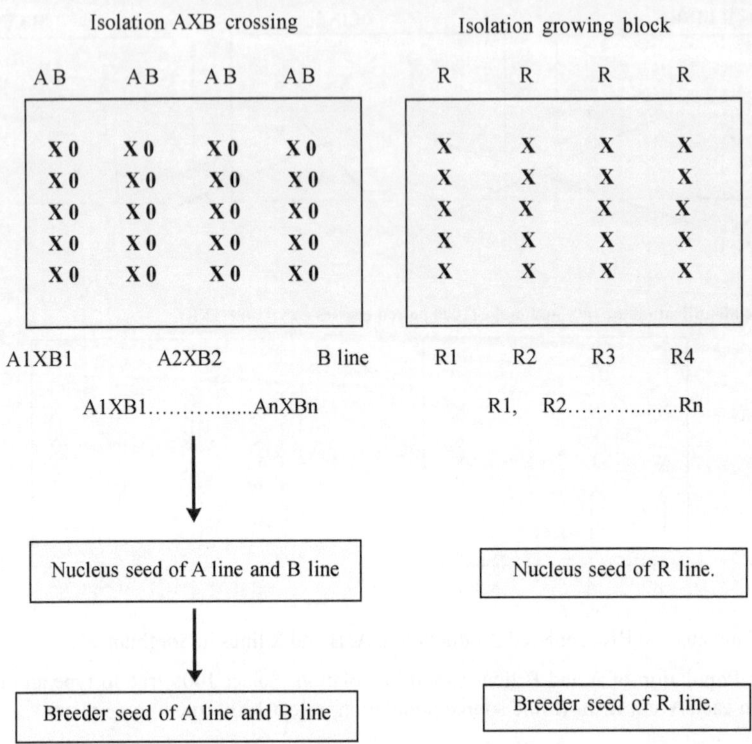

Fig. 3. Multiplication Nursery of selected A, B and R lines in Sorghum

7. Wheat

(*Triticum aestivum* L. em. Thell) (2n=42)

Family : Poaceae

In wheat, leaves are simple, alternate in distichous arrangement, long lamina, liguled, auricle membranous, prominent and claw shaped. Inflorescence is a terminal distichous spike with spikelets on rachis node. In each spikelet, 3 to 5 flowers (florets) are present. Lemma and palea are present in florets, lemma terminating with awn and palea is usually scaly. Three anthers and single ovary with bifurcated plumose stigma are present in each flower. Pollination occurs inside the closed flowers resulting in self pollination owing to the cleistogamous nature. Little out crossing to the extent of 5% may take place depending upon the floral morphology of some varieties like 2428 H.D and some Chinese varieties (Raj Kumar *et al.,* 2003). The characteristics of bread wheat are detailed in Table 1.

Nucleus Seed Production: Source the seeds from the basic seed stock of the variety and raise the crop in isolation (5m.). Single plant per hill is maintained. Select 500 spikes or ear heads from the main tillers from single plants conforming to the varietal characteristics. Grains are separated ear wise and verified for the grain characteristics. Reject the spikes that are not conforming to the varietal description.

In the next season, ear to row progeny planting is made with wide spacing in rows of 4 m. long with isolation distance of 5 m. between varieties. Critical examination is made throughout the life cycle of the crop from juvenile to flowering to grain maturity stage looking for greater uniformity, stable expression of characteristics and true to type nature of rows. Rows with admixtures or deviant plants, are rejected and uprooted. The progeny rows showing deviant plants at flowering and the adjoining rows on either side are also uprooted to safeguard from contamination. Each ear head per plant in the progeny is harvested separately and seeds extracted progeny wise and examined for true to type. The seeds are then pooled together to constitute nucleus seed stage. In the same plot, 15 to 20 ears from main tiller in each progeny row are collected separately and individual spikes are verified for true to type nature and then well dried, cleaned and stored for next cycle of nucleus seed production. In the next season, pooled seeds from the ear rows are sown to raise nucleus seed stage II and the same protocol described in nucleus seed stage I is followed.

Nucleus Seed Stage II: To meet the increased demands of nucleus seed, NSS-II is followed further. If very high quantity of seeds are to be produced, NSS II preceding breeder seed production is helpful. Seeds of NSS-I are sown in progeny row plots (isolation of 5m.) utilising seeds of individual progeny rows. The progeny plots in NSS-II are examined critically for uniformity, genetic purity up to pre flowering and then for ear head and seed characteristics upto maturity stage. If any deviation in the variety characteristics is noticed, entire progeny rows are discarded. In respect of the deviations noticed at flowering stage, the progeny rows exhibiting deviations and the adjacent rows are also uprooted to avoid accidental cross pollination with lines showing off-types. At maturity of the crop, individual progeny row plots are harvested separately and ear heads and grain samples are examined for colour, shape, uniformity and for the true to type nature. The seeds from uniform plots selected for true to type nature are bulked to constitute NSS-II for raising breeder seed stock.

Breeder seed production: The breeder seed plot is isolated by minimum distance of 5m. Nucleus seeds are sown in a clean, disease free, well prepared field not grown to wheat in the last 2 to 3 years. No loose smut infected wheat, rye or triticale field should be within 5m. distance. From seed germination to pre flowering and there afterwards upto spike maturity, field examinations are made to examine all stages of plant development and for the descriptor characteristics. Roguing should be done and if deviants and rogues are found after the flowering stage, such plants should be removed / uprooted and simultaneously, plants around the rogue plants are also uprooted to avoid any chance cross pollinated seeds getting into the breeder seed. The breeder seed plot is monitored by the Monitoring Team. Seed testing with samples drawn from the breeder seed is done in the seed testing laboratory and the quality assured seed is then labelled with breeder seed tag. Grow out test is conducted with samples of the breeder seed as per the standard procedure to ascertain the genetic purity. Grain nutritive contents are estimated.

Stepwise procedure/ protocols in Nucleus and Breeder Seed production in wheat is presented in Fig. 1.

Table 1. Characteristics in bread wheat (*Triticum* sp.)

S.No.	Characters	State
1.	Anthocyanin colouration	Absent, present
2.	Plant growth habit	Erect, semi erect, intermediate, semiprostate, prostate
3.	Foliage colour	Fade green, green, dark green
4.	Flag leaf anthocyanin colouration of auricle	Absent, medium, very strong
5.	Flag leaf hairs on auricle	Absent, medium, strong
6.	Plant flag leaf attitude	Erect, semi erect, drooping
7.	Ear-time of emergence (on 50% of ears)	Very early, early, medium, late, very late
8.	Flag leaf waxiness of sheath	Absent, weak, medium, strong, verystrong
9.	Flag leaf waxiness of blade	Absent, weak, medium, strong, verystrong
10.	Waxiness of Ear	Absent, weak, medium, strong, verystrong
11.	Culm waxiness	Absent, weak, medium, strong, verystrong
12.	Flag leaf length	Short, medium, long
13.	Flag leaf width	Narrow, medium, broad
14.	Panicle length excluding awn	Very short, short, medium, long, verylong
15.	Ear shape in profile	Tapering, parallel sided, club shaped, fusiform
16.	Ear density	Very lax, lax, medium, dense, very dense
17.	Ear length excluding awn	Very short, short, medium, long, verylong
18.	Awn or hairs presence	Both absent, awn present, hair present
19.	Scur tip	Short, long
20.	Awn length	Very short, short, medium, dark brown,black
21.	Awn colour	White, light brown, brown dark brown,black
22.	Awn attitude	Absent, medium, very strong
23.	Outer glume presence	Absent, medium, very strong
24.	Ear colour	White, light brown, dark brown
25.	Lower glume width	Absent, very narrow, narrow, medium,broad, very broad
26.	Lower glume shape	Sloping, round, straight, elevated, stronge levated
27.	Lower glume beak length	Very short, short, medium, long, verylong
28.	Lower glume-beak shape	Straight, moderately curved, strongly curved, geniculate
29.	Peduncle length	Short, medium, long
30.	Spike attitude at maturity	Straight, bent, crooked
31.	Grain variation	None, light, medium, dark, very dark
32.	Grain colour	White, amber, red
33.	Grain shape	Round ovate, oblong, elliptical
34.	Grain-germ width	Narrow, medium, wide
35.	Brush hair length	Absent, short, medium, wide
36.	Seed size (1000 grains weight)	Small, medium, large, very large
37.	Seam type	Written type, after nature type, springtype
38.	Grain hardness	Soft, semihard, hard

Source: PPV & FRA (2007)

First year (Ear-head collection)
Grow seed sample from original stock source in space planting.
Select 300 to 500 true to type uniform ear-heads.
Thresh selected ear-heads separately and examine in the laboratory
Reject ear-head produce of not true to type

↓

Second year (Nucleus seed stage – 1)
Grow ear-head progenies
Take critical observations at different growth stages of the crop for trueness
Rejection of progenies showing variation and non conformity
Keep proper pedigree record of progenies harvested
Bulk the seeds to make nucleus seeds.

↓

Third year (Nucleus seed stage – II)
Grow NSS-I harvest plots separately.
Observe each plot critically for any type of deviation and nonconformity at different growth stages. Discard plots showing variation
Thresh selected plots individually. and laboratory examine for grain characteristics
Bulk retained plots produce as nucleus seed stage II for raising breeder seed.

↓

Fourth year (Breeder Seed)
Sow breeder seed plots in rows.
Leave one row unsown after every eighth row.
Follow roguing norms in removing the off -type plants.
Arrange visit of Monitoring Team
Harvest the seed plots with utmost care to avoid any mechanical mixture.
Keep proper records of each lot separately.
Seed samples are tested in for different seed qualities. Quality assured seeds are then labeled with Breeder Seed Tag.
Grow Out Test is conduced to assess the genetic purity.
Nutritive contents are estimated in grain

Fig. 1. Nucleus and Breeder seed production in Wheat. Isolation distance recommended is 5m. from other wheat crops for nucleus and breeder seed.

(*Source:* Rajkumar *et al.*, 2003)

Small Millets

Nucleus Seed Production in Small Millets: The five small or minor millets namely. 1. Thenai - Foxtail Millet - *Setaria italica* 2. Kodo Millet - *Paspalum scrobiculatum* 3. Little millet - *Panicum sumatranse* 4. Proso Millet - *Panicum miliaceum* and 5. Barnyard Millet - *Echinochloa frumentacea* are highly self pollinated crops. The extent of cross pollination is less than one percent. Among them, kodo millet is cleisotogamous and self pollination is the rule (Seetharam *et al.,* 2003).

In the seed source nursery or in the standing crop of the small millet variety concerned, 1500 to 2000 uniform true to type plants are selected and the primary tiller panicles are self fertilised. The self fertilised panicles are examined in the laboratory for the panicle features such as shape, size and grain specific features of the variety. Seeds are separated from selected panicles and examined for true to type in the laboratory.

In the subsequent season, 2 rows progeny plots (4m. length) are raised in two replications in 5m. isolation from other milllets. Off types are removed. The progeny plots are examined for true to type in different stages of development. Progeny rows showing deviations in plant characteristics are uprooted in both replications. Progeny rows showing performance above mean +2SE are identified and selected. In each of the selected progenies, 15 to 20 true to type superior plants are selected. Part of seeds from the selected plants are raised in isolation (5m.) to progeny rows in the next season and seeds from the selected plants from typical true to type rows are harvested and bulked to constitute nucleus seed . The remaining seeds in each of the selected plants are saved for raising progeny rows and for selection and storing for next cycle of nucleus seed production.

Breeder Seed Production: Seeds from nucleus seed stock are sown in rows as per the production techniques of the respective crop. Isolation distance of 5m. is provided. The crop is thinned to retain single plant per hill. Continuous monitoring is done for adjudging true to type uniformity and vigour in seedling, juvenile, vegetative and floral traits. Plants showing variation/deviation are uprooted. If deviations are noticed in plants at flowering phase, the deviating plants and also plants around the deviating plants are also uprooted. Monitoring Team visit is arranged at the time of flowering or post flowering stage to assess the genetic purity. At crop maturity, panicles are harvested from individual plants. In the laboratory, panicles are closely examined for the characteristics conforming to the varietal description and selected (Fig. 1). Grains are separated from individual panicles and examined for true to type and uniformity for grain features and selected. Seeds from selected individual panicles are pooled together to constitute the breeder seed and then processed. Samples of the breeder seed are tested for the prescribed seed qualities and labelled with breeder seed label. Grow out test is arranged to ascertain genetic purity. Nutritional qualities and micronutrient contents in the grains are estimated.

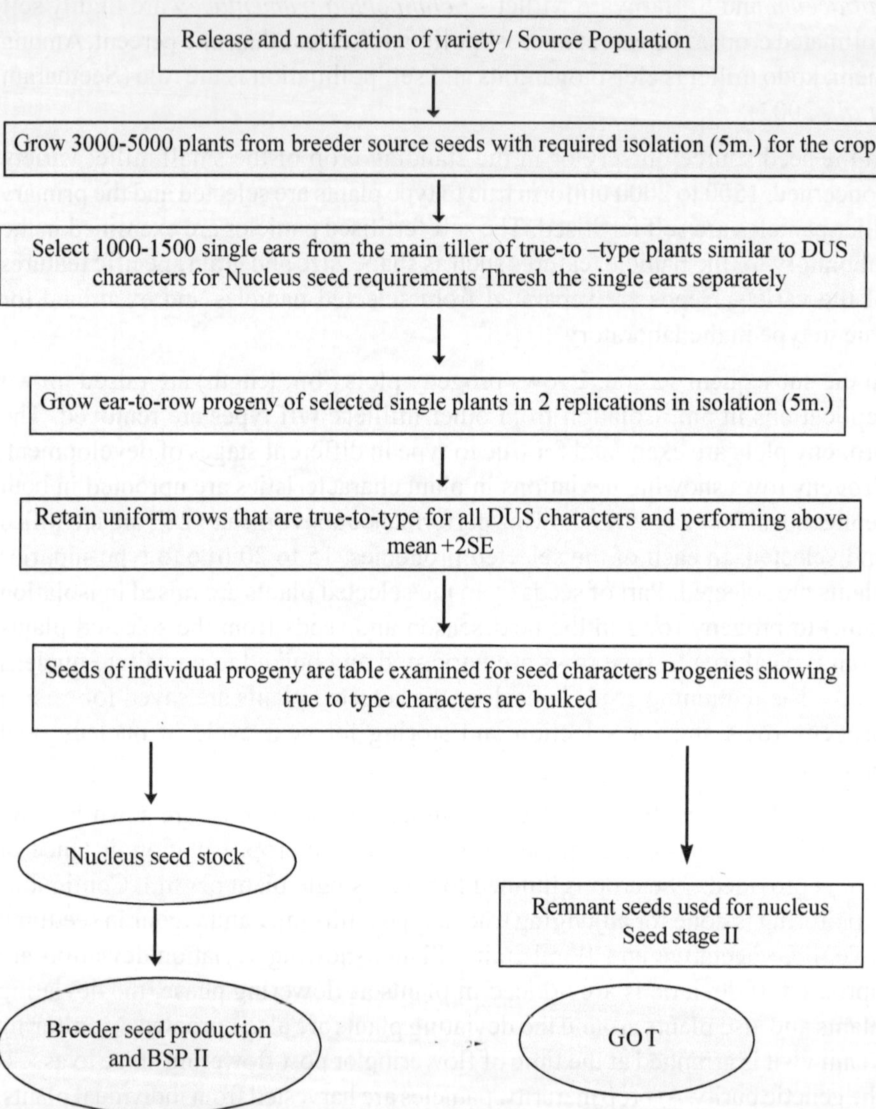

Fig. 1. Nucleus and Breeder Seed Production in Finger Millet, Fox Tail Millet, Kodo Millet, Proso Millets, little Millet and Barnyard Millet
(*Source* : Seetharam *et al.*, 2003)

Small Millets

8. Barnyard Millet

(*Echinochloa frumentacea* Roxb. (Linn)
(2n:36, 54)

Family : Poaceae

Annual erect and growing upto a height of 120 cm. Spikelets are two flowered, usually awnless and pedicels subtended by 2 glumes. Lower floret is sterile and upper floret is bisexual with lemma, palea and two lodicules. Stamens three with purple anthers. Ovary with two styles with plumose stigma. Grain is enclosed by shining lemma and palea. The characteristics are listed in Table 1.

Table 1. Descriptor of characteristics in *Echinochloa frumentacea* (Barnyard Millet)

S.No.	Characters	States	Stage of observation
1.	Growth	Erect/decumbent /prostrate	Vegetative
2.	Plant pigmentation	Pigment/ not pigmented	Flowering
3.	Degree of lodging	Slight / medium /extensive	Maturity
4.	Degree of culm branching	Low / medium / high (most nodes produce inflorescence	Maturity
5.	Plant height	Short / medium / tall	Maturity
6.	Number of basal tillers	Low / medium / high	Maturity
7.	Flag leaf size	Small / medium / large	Maturity
8.	Inflorescence colour	Green / light purple / dark purple	Maturity
9.	Inflorescence shape	Cylindrical /pyramided / globose to elliptic	Maturity
10.	Inflorescence compactness	Open / intermediate /compact	Maturity
11.	Spikelet arrangement	One side of the rachis /around rachis	Flowering
12.	Peduncle length	Short / medium / long	Flowering
13.	Days to 50% flowering	Early / mid / latc	Flowering
14.	Glume pigmentation	Present /absent	Dough stage
15.	Grain colour	Straw white / grey and straw white/brownish grey/grey/light grey	After threshing
16.	Grain shape	Concave/ oval	After threshing

Note: Description is based on PPV & FRA 2017.

9. Finger Millet (Ragi)

(*Eleusine corocana* (L) Gaertn) (2n : 36)

Family: Poaceae

It is an annual, high tillering, growing up to 40 to 100 cm. It is a self pollinated crop. Its inflorescence is terminal umbel with groups of digitately arranged spikes. Spikelets are arranged alternatively. Each spikelet contains 4 to 10 bisexual flowers. Terminal spikelets may be staminate or sterile. Each flower consist of lemma and palea, three stamens and single ovary having two styles with plumose stigma and two fleshy lodicules. Flowers open from the middle of the spike and progress towards up and down directions. Seeds are globose, 1 to 2 mm. dia. smooth or rugose. Grain colour varies such as orange red, reddish brown, dark brown nearly black and white coloured. The pericarp at maturity appears as papery structure enclosing the seed (Table 1). It is highly self pollinated. Natural cross pollination is less than 0.5%.

Nucleus seed production: Selection of representative plants of the ragi variety concerned is made from the breeder source seed raised crop and 1000 to 1500 true to type description plants are identified and main tillers are self fertilised. Isolation distance of 5m. is ensured. In the laboratory thorough examination is made of the selected single self fertilised ears for their shape, size and grain setting ability and the grain characteristics. Grains are separated from the selected ear heads.

In the following season, ear to progeny rows are raised in two row plots, each 4m. long, in two replications. Isolation distance of 5m. is followed. Observations for characteristics of the variety concerned are made in all the phases of crop growth. Off-types and deviating types of plants are culled out. Row plots showing deviations in plant type are rejected and uprooted along with the adjoining rows. In each of the identified progeny rows performing above mean +2SE, 10 to 15 typical and true to type superior plants are selected and the main tiller is self fertilised. From the seeds of self fertilised main tiller panicles, progeny rows are raised in the next season in 5m. isolation. In progeny row plots verifications are made in respect of all growth phases for uniformity and true to type nature of plants. Progenies showing deviant plants are uprooted. Within each progeny row, 10 to 15 plants are selected and seeds harvested from selected progeny rows are bulked together and utilised as source nucleus seed. The remainder seeds from the self fertilised ears are again utilised for raising ear row progeny row plots. This cycle is repeated once in 2 to 3 years to provide nucleus seed for raising breeder seed. (Seetharam *et al.*, 2003) (Fig. 1, Small millets).

Breeder seed production: Nursery is raised with nucleus seeds produced as described above. At the proper age of the seedlings (20 to 25 days), single seedling per hill is planted in a well prepared field protected by 5m. isolation from other ragi crops. At every phase of development upto maturity, individual plants are examined for true to type nature and crop uniformity for growth, tillering, flowering and maturity. Off-types and deviating plants are uprooted. At flowering, if deviant types are noticed they were uprooted along with the surrounding plants. Monitoring Team visit is arranged at the time of flowering of the breeder seed plot Ears are harvested on individual plant basis and examined in the laboratory for true to type nature of grain. Deviating type of ears are discarded. Seeds of selected individual ears are separated, examined in the laboratory and pooled together to constitute breeder seed. Seed testing is done to adjudge the seed qualities. Quality assured seeds are then labelled with breeder seed tag. Grow Out test is made to ascertain the genetic purity. Nutritive contents in grain and straw are estimated.

Table 1. Descriptor of Finger Millet

1.	Plant growth habit	Erect, decumbent, prostrate
2.	Plant pigmentation at leaf junction	Absent, present
3.	Leaf sheath pubescence	Absent, present
4.	Days to 50 % flowering	Early <50, medium (55-70), late >70
5.	Glume colour	White, light green, dark green, light purple, dark purple
6.	Stem colour	Absent, present
7.	Flag leaf blade length (cm.)	Short (<15), medium (15-30), long (30-45), verylong (>45)
8.	Flag leaf blade width (cm.)	Narrow (<1), medium (1-2), wide >2
9.	Peduncle length (cm.)	Very short (<10), short (10-20), medium (20-30), long (>30)
10	Ear shape	Fist type, compact, semi compact, open, droopy
11	Finger branching	Absent, present
12.	Finger, position of branching	In thumb finger, in all fingers
13.	Finger multiple row	Absent, present
14.	Earhead length (cm.)	Very short (<4), short (4-6), short to medium (6-8), medium (8-10), long (10-15), very long (>15)
15.	Finger length(cm.)	Short (<5), medium (5-7), long (>7)
16.	Finger width (cm.)	Narrow (<0.7), medium (0.7 to 1.0), wide >1.0
17.	Finger number on main ear	Low (<5), medium (5-8), high (>8)
18.	Number of productive tillers per plant	Low (<4), medium (4-6), high(>6)
19.	Plant height at maturity (cm.)	Very short (<40), short (40-60), medium (80-120) tall (120-160), very tall >160

20. Seed shattering	Absent, present
21. Seed covering by glume	Enclosed, intermediate exposed
22. Seed colour	White, light brown, copper brown, dark brown
23. Seed shape	Round, reniform, ovoid
24. Seed surface	Smooth, rough
25. Pericarp persistence after threshing	Non persistent, persistent
26. 1000 seed weight (g.)	Low (<2), medium (2-3), high (>3)

(*Source:* PPV& FRA, 2017)

10. Foxtail Millet

Setaria italica Beauv (2n:18)

Family : Poaceae

Annual, inflorescence is a spike like panicle. Rachis is ribbed and supports two flowered sessile spikelets (6 to 12 in number). Each spikelet is subtended by 1 to 3 bristles of varying colour and length in different varieties. Each spikelet has two elliptic glumes and two lodicules. Stamens are three with orange or white anthers. Ovary smooth with two long styles and plumose stigma. Foxtail millet is self pollinated. The grains are usually shining 2 mm. long and tightly enclosed by lemma and palea. Grain colour varies from cream to orange, yellow and brown (Table 1).

Table 1. Descriptor of foxtail millet (Setaria)

S.No.	Characters	States	Stages of observation
1.	Growth habit	Erect/decumbent/prostrate	Vegetative
2.	Plant pigmentation	Pigmented /notpigmented	Vegetative
3.	Degree of lodging	Slight /medium/extensive	Flowering
4.	Plant height	Low/medium/tall	Maturity
5.	Number of basal tillers	Low/medium/high	Maturity
6.	Leaf colour	Green/yellow/pigmented deep purple	Maturity
7.	Sheath pubescence	Glabrous/medium pubescent/ strongly pubescent	Flowering
8.	Leaf blade pubescence	Glabrous/medium pubescent/ strongly pubescent	Flowering
9.	Flag leaf size	Small /medium /large	Dough stage
10.	Inflorescence lobes	Absent/strong/long /large and thick	Dough stage
11.	Inflorescence bristles	Absent/strong/large and thick	Dough stage
12.	Inflorescence shape	Oblong /ovate/elliptic/obvate	At 50% flowering

13.	Inflorescence compactness	Loose/medium/compact/spongy	Maturity
14.	Apical sterility	Absent/present	Maturity
15.	Lobe compactness	Loose/medium/compact	Maturity
16.	Grain colour	Grey/brown/golden yellow/ straw white/cream	After threshing
17.	Grain shape	Oval/elliptical	After threshing
18.	Phenol colour reaction	Positive /negative	After threshing

Note: Description is based on PPV & FRA (2017)

11. Kodo Millet

(*Paspalum scrobiculatum* L.) (2n:40)

Family : Poaceae

Annual, grows upto 90 cm. height and tillering profusely. Spikes are 2 to 6 in number and 3 to 15 cm. long carrying sessile and short pedicellate spikelets that are arranged in two or more rows. Spikelets are flat, elliptic, awnless with 2 florets. Upper floret is round and lower is reduced to a valve. Kodo millet spikelets are highly cleistogamous and are completely self pollinated. Grain is enclosed in hard horny husks that are difficult to remove (Table 1).

Table 1. Descriptor of characteristics in *Paspalum scrobiculatum* (Kodo Millet)

S.No.	Characters	States	Stage of observation
1.	Growth habit	Erect/ decumbent /prostrate	Vegetative
2.	Degree of culm branching	Low/ medium / high	Maturity
3.	Plant height	Short / medium / tall	Maturity
4.	Number of basal tillers	Low / medium / high	Maturity
5.	Degree of lodging	Slight / medium / extensive	Flowering
6.	Internode pigmentation	Present / absent	Flowering
7.	Leaf sheath pigmentation	Absent / present	Flowering
8.	Sheath base pigmentation	Absent / present	Flowering
9.	Flag leaf size	Small / medium / large	Dough stage
10.	Leaf juncture pigmentation	Absent / present	Dough stage
11.	Nerves on glumes	Broad nerves (5 nerves/closed nerves (7 nerves)	Dough stage
12.	Panicle exsertion	Partial / complete	Dough stage
13.	Panicle appearance	Compact / semi-compact/open	Dough stage

14.	Length of peduncle	Short / medium / long	Maturity
15.	Length of inflorescence	Short/ medium / long	Maturity
16.	Spikelet arrangement on rachis	One side only / around the rachis	Dough stage
17.	Spike branching	Absent / present	Maturity
18.	Spikelet density	Sparse / intermediate /dense	Maturity
19.	Grain shattering	Non shattering / medium shattering / high shattering	Maturity
20.	Grain colour	Dark grey/ brown/light brown/ grey/ golden yellow/straw white/cream	Maturity
21.	Grain shape	Oval / elliptical	Maturity

Note: Description is based on PPV & FRA (2017)

12. Little Millet

(*Panicum sumatranse* Roth ex Roem.Schult) (2n:36)

Family : Poaceae

Annual, stem slender upto 90 cm. height; tillering varies from low to very high in different varieties. Panicle is made of spikelets mostly bearded on unequal pedicels. Flowers bisexual and open in basipetal pattern. Anthesis is rapid and brief and self pollination is the feature. Glumes open for a short time only. Seeds are small, 2 mm. long, oval shaped and colour varies from dark green to cream (Table 1).

Table 1. Description of characteristics *Panicum sumatranse* (little millet)

S.No.	Characters	States	Stage of observation
1.	Growth habit	Erect/erect geniculate decumbent/prostrate	Vegetative
2.	Plant pigmentation	Not pigmented/pigmented	Vegetative
3.	Pubescence of blades	Glabrous / medium pubescent / strongly pubescent	Vegetative
4.	Sheath pubescence	Glabrous / medium pubescent / strongly pubescent	Vegetative
5.	Ligule pubescence	Glabrous / medium pubescent / strongly pubescenct	Vegetative
6.	Lodging at maturity	Slight/ medium / extensive	Vegetative
7.	Shape of inflorescence	Diffused /arched /globose elliptic	Panicle maturity stage
8.	Compactness of inflorescence	Compact / intermediate /open	Panicle maturity stage
9.	Plant height	Short/ medium / tall	Maturity stage
10.	Days to flowering	Early / medium / late	Flowering stage
11.	No of basal tillers	Low / moderate /high /	Maturity

12.	Culm branches on the mainstem	Low / moderate /high /	Maturity
13.	Length of peduncle	Short / medium / long	Maturity
14.	Length of inflorescence	Short / medium / long	Maturity
15.	Inflorescence shape	Globose elliptic /arched /diffused	Maturity
16.	Spikelet /shape	Elliptical/ oval	After maturity
17.	Grain colour	Straw white, golden yellow light brown, brown, grey, dark grey.	After maturity

Note: Description is based on PPV & FRA (2017)

13. Proso Millet

(*Panicum miliaceum* L.) (2n:36, 72)

Family : Poaceae

Annual. Inflorescence is a drooping panicle with varying degree of compactness. Spikelet is ovate, pointed, 5 mm. long comprising two flowers, partially enclosed by glumes. Upper floret is grain bearing and lower floret is sterile. Fertile flower consists of three stamens, 2 lodicules and ovary with two long styles and feathery stigma. Flowers are self fertilized. Grain is globular and smooth and is cream, red, grey or yellow coloured. Dehusked grain is white or green in colour. The characteristics are given in Table 1.

Table 1. Description of characteristics in *Panicum miliaceum* (Proso Millet)

S.No.	Characters	States
1.	Plant growth habit	Erect/ decumbent /prostrate
	Basal tillers (no.)	Low (<5), medium (5-10), high (>10)
	Days to 50% flowering	Early (<35), medium (35-45), late (>45)
2.	Plant pigmentation at leafs heath	Absent, present
3.	Leafs heath pubescence	Glabrous, sparse, strong
4.	Ligule pubescence	Absent, present
5.	Leaf blade pubescence	Glabrous, sparse, strong
6.	Inflorescence shape	Arched, glabrous, elliptic
7.	Peduncle length(cm)	Very short (<10), short (10-20), medium (20-30), long (30-40), very long (>40).
8.	Flag leaf blade length (cm)	Short (<20), medium (20-35), long (>35)
9.	Flag leaf blade width (cm)	Narrow (<1.5), medium (1.5-2.5), wide (>35)
10.	Culm branching	Absent, present
	Panicle compactness	Compact, intermediate,
11.	Panicle length (cm)	Open very short (10), short (10-20), medium (20-30), long (30-40)
12.	Lodging	Absent, present
13.	Plant height (cm)	Dwarf (<60), semidwarf (60-90), tall (90-120), very tall (>120)
14.	Seed shattering	Absent, present
15.	Grain colour	Straw white, cream, golden yellow, gray, dark gray
16.	Grain shape	Elliptical oval
17.	1000 grain wt. (g.)	Low (<4), medium (4-6), high (>6)

Note: Description is based on PPV & FRA (2017)

Pulses

14. Black Gram (Urdbean)-*Vigna mungo* (L.) Hepper (2n:22, 24)

Green Gram (Mungbean)-*Vigna radiata* (L.) (Wilczek) (2n:22, 24) and

Family : Fabaceae

Vigna mungo and *Vigna radiata* are annual, herbaceous and are erect to semierect to trailing in habit. Stem much branched from base or middle. Leaves trifoliate, alternate, petiole pulvinate. Leaflets ovate, entire or acute. In blackgram, leaves are much hairy on both surfaces. Inflorescence axillary raceme with 5 to 8 flowers in blackgram. In green gram, flowers are 10 to 20 crowded on long pedicels. Flowers in both the species are yellow coloured and papilionaceous. Flowers pedicellate, bisexual, zygomorphic and hypogynous. Corolla polypetalous, papilionaceous, keel petals spirally coiled. Stamens 10 that are in the diadelphous (9+1) arrangement. Filaments are alternatively long and short. Ovary unilocular, axile placentation. In blackgram, ovules are 5 to 8, style terminal hairy. In greengram, ovules are 10 to 12, style hairy. Cotyledon white in colour in black gram and creamy colour in greengram. Hilum is concave in black gram while it is flat in green gram. Greengram differs from blackgram with 10 to 20 flowers in the axillary raceme. In green gram pods are longer than in blackgram and seeds per pod are 10 to 15 number, usually smaller in size. Greengram and Blackgram are self pollinated crops. Natural cross pollination less than 1.5% is recorded (Dixit and Tripathi, 2003).

Nucleus Seed Production in Mungbean and Urdbean: In a well prepared fertile. plot of 400 sq.m. basic seeds sourced from breeder or the institution concerned are sown in rows with a spacing of 10 cm. between plants and 45 cm. between rows. Isolation distance of 100m. is provided. Single plant per hill is maintained. The plants are closely and continuously examined for the true to type, uniformity, expression of diagnostic characteristics and growth. One thousand single plants are selected at the time of flowering, labelled and monitored till pod maturity stage. Very early and late flowering plants are uprooted to maintain uniformity in flowering and maturity. Pod characteristics are studied in all the selected plants. Off types and deviant plants are rejected as and when they appear. Selected true to type and tagged plants are individually harvested and their pods and seeds are examined in the laboratory. After confirming the pod and seed features for true to type, uniform maturity, size, shape and colour, the final selection of 1000 single plants is made and seeds separated, labelled and saved for next generation.

In the next favourable season, 400 to 450 single plant progenies are raised in 4 m. long two rows plots replicated twice. For every 10 rows, seeds of the source variety are sown in one row for comparison. The field is maintained with 100m. isolation distance. Good agronomic crop production practices are followed. Close and continuous monitoring is made from seedling, juvenile, vegetative, flowering and post flowering to pod development to pod maturity stages .Progenies showing off types and deviating plant types are discarded as and when they appear. When off types and deviating plants are noticed at flowering and post flowering stages, such progenies are uprooted along with the progeny on either side. At maturity, progeny rows conforming to the variety description are labelled and individual plants in them are verified for true to type and uniformity and plant type expression. The selected true to type progenies are harvested and yield recorded. Progenies that perform above mean + 2 SE are identified and retained for pod and seed separation. Pods and seeds are verified in the laboratory for trueness and progenies are finally selected. Seeds from single plant progenies are processed, dried and stored progeny wise. Samples of seeds from each progeny are bulked to constitute nucleus seed. The remainder of seeds are preserved for developing next cycle of nucleus seed. Seed samples of nucleus seed are tested for the nutritive content. Nucleus seed packets are labelled.

Breeder seed production: The nucleus seed is sown in a well prepared, well drained and high fertility status field provided with 100m. isolation distance. The seeds are sown in 2 row plots with wide spacing of 0.5 meter between plots to access for close examination and monitoring. At different stages of crop growth, individual plants are examined in the Breeder Seed Production plot for trueness, uniformity and stability. Early and late flowering plants are discarded. Off types are removed. Off types and deviant type of plants noticed at flowering and post flowering are removed along with the adjoining plants to avoid pollen contamination. At mid maturity stage, Monitoring Team visit to the Breeder seed Production plot is arranged. At maturity, uniformity in pod maturity and pod characters expression are carefully examined. Harvesting is done at proper maturity of pods with due precautions to avoid mechanical mixture. After drying and processing, seed sample is tested for seed quality parameters. Breeder Seed labelling or tagging is done after ascertaining all seed quality assurance aspects. GOT is conducted to confirm the genetic purity of the seed. Descriptor of characteristics is detailed in Table 1 and Table 2 respectively for Mungbean and Urdbean.

Table 1. Descriptor of Mungbean

S.No.	Characters	States	Stage of observation
1.	Growth habit	Erect/semi-erect/spreading	Vegetative/flowering
2.	Growth type	Determinate/semi-determinate/ indeterminate	Vegetative/flowering
3.	Color of base petiole	Green/green with purple splashes/purple	Cotyledons unfolded
4.	Leaflet type	Normal/lobed	Vegetative/flowering
5.	Leaf color	Light green/green darkgreen/purple	Vegetative/flowering
6.	Days to flowering	Early/medium/late	Start of flowering
7.	Raceme position	Above canopy/intermediate/no visible pods above canopy	Pods are green and premature
8.	Pubescence on pods	Absent/present	Pods are green and premature
9.	Pod color (pre-mature)	Green/green with pigmented sutures	Pods are green and premature
10.	Color of mature pods	Brown/black	Maturity
11.	Pod shape	Rectangular/elongated/drumshaped	Maturity
12.	Maturity	Early/medium/late	Maturity
13.	Seed color	Green/yellow/mottled	Maturity
14.	Seed shape	Oval/drum shaped	Maturity/after harvest
15.	Seed size	Small/medium/large	After harvest

Ref: P.C. Dixit and D.P. Tripathi: (2003) and PPV & FRA (2001)

Table 2. Descriptor of Urdbean

S. No.	Characters	States	Stage of observation
1.	Growth habit	Erect/semi-erect/spreading	Vegetative/flowering
2.	Growth type	Indeterminate/semi-determinate/ determinate	Vegetative/flowering
3.	Shape of terminal leaf	Ovate/lanceolate/cuneate	Vegetative/flowering
4.	Leaf color	Light green/green/darkgreen	Vegetative/flowering
5.	Days to flowering	Early/medium/late	Start of flowering
6.	Length of peduncle	Short/medium/long	Pods are green and premature
7.	Leaf petiole color	Green/green with purple splashes/purple	Vegetative/flowering
8.	Pubescence on pods	Absent/present	Pod are green & premature
9.	Color of mature pods	Brown/black	Maturity
10.	Maturity	Early/medium/late	Full maturity
11.	Seed shape	Globose/oval drum shaped	Maturity
12.	Seed color	Green/greenish brown/black/mottled	Maturity
13.	Seed size	Small/medium/large	Maturity

Note: Description is based on Dixit and Tripathi (2003) and PPV & FRA. (2001)

15. Broad Bean (Field Bean-Faba Bean)

(*Vicia faba* L.) (2n:12)

Family : Fabaceae

Annual, erect, winter and spring season adaptation. Stem is square shaped. Branches are 1 to 2 in spring season varieties and 4 to 6 per plant in winter season varieties. Leaves alternate, two leaflets in bottom leaves and 6 to 8 leaflets in top leaves. Flower axillary, 2 to many, papilionaceous. The flowers are 1.0 to 2.5 cm. long, calyx campanulate. Petals are five with the standard white petal, two wing petals white with a true back black spot and white keel petals. Stamens 10 in 9+1 diadelophous arrangement. Ovary superior, monolocular, ovules arranged alternately on opposite placenta, 3 to 4 ovules / ovary. Seed colour ranges from yellow, green, brown to black. Anthesis occurs in the early morning. Anthers dehisce before flower opening. Honey bee and Insects visit frequently and aid in tripping through rupture of stigmatic surface. Natural out crossing is estimated from 2% to 84% with mean of 32% and frequently from 32% to 40% .Broad bean is a partially cross pollinated crop. Autofertility in fertilization without insect trip is very frequent in broad bean by excluding insects during pollination (Hari Har Ram, 2005). The descriptors are presented in Table 1.

Nucleus seed production: Seeds obtained from the institution / breeder source are raised in 1000 sq,m. plot that was not cultivated with Broad bean crop in the last 2 years. Isolation distance of 100 m. is maintained. Single plant per hill is maintained. Weak and unhealthy plants and off-types are removed from the seedling establishment stage to pod set and pod maturity stage. At pod maturity, heathy and uniform looking 1000 true to type seed fertile plants conforming to the variety description are selected from the middle of the plot. They are labelled and harvested. Seeds from the selected 1000 single plants are packed separately and examined in the laboratory. Non conforming seed samples are discarded.

In the following season in two row plots, each 4m. long, plants are raised in two replications in 100m. isolation plot. Individual progeny rows plot of 500 selected plants are raised. In each block, 50 progenies are raised. Observations are made in all the plants in all the progeny rows for the prescribed traits at different growth and development phases of stem, foliage, flower, pod set and pod maturity. Trueness and uniformity in flowering, high extent of pod set ,uniform pod maturity and seed development are critically examined. Progeny rows with plants failing to show conformity criterion are uprooted. Based on the pod and seed yield, progenies showing performance above mean + 2 SE are identified. From all over the selected progenies, 500 elite superior true to type plants are identified and harvested and their pods and seeds are tested and verified in the laboratory. The seeds of selected progenies are carried forward. In the next generation, 300 progeny rows are raised in progeny plots. Off-types and progeny rows showing plants with non conformity are uprooted. At maturity, elite and uniform progeny rows are selected and pods from them are collected separately. In the laboratory, pods and seeds are examined and progeny rows are selected. Seeds samples from each of the selected plants are pooled together to constitute the nucleus seed.

Breeder seed production: Breeder seed is produced in well prepared clean field that is not grown with broad bean crop in the past 2 years.The field is isolated by 100 m. from other broad bean crops. Border rows are raised with the same variety around the plot. Single plant per hill is maintained. Continuous examination is made of every plant from germination to maturity phase. Plants are selected based on the true to type, uniform looking and stability in expression. Non conforming plants are removed at different crop growth phases. Monitoring Team visit is arranged at midpod maturity stage. Pods are harvested at maturity. In the laboratory, matured pods and seeds separated from them are examined and selected. The seeds are pooled together to constitute Breeder seed. Seeds are dried, processed and tested for seed qualities. Quality assured seed is labelled with Breeder seed label and samples drawn are tested in GOT for genetic purity. Nutritional quality of fleshy pods and seed is estimated.

Table 1. Description of Broad Bean

1.	Foliage intensity of greencolour	Low, medium, dark
2.	Foliage greyish of green colour	Absent, present
3.	Time of flowering	Very early, early , medium, late
4.	Wing-melanin spot	Absent, present
5.	Wing-colour of melanin spot	Yellow, brown, black
6.	Flower length	Short, medium, long
7.	Flower standard width	Narrow, medium, broad
8.	Leaflet length	Short, medium, long
9.	Leaflet breadth	Narrow, medium, long
10.	Plant growth type	Determinate, indeterminate
11.	Plant height	Short, medium, long
12.	No. of nodes on stem	Few, medium, many
13.	Pod length	Short, medium, long
14.	Pod width	Narrow, medium, broad
15.	Pod-green colour intensity	Light, medium, dark
16.	Seed shape	Circular, non circular
17.	Seed testa colour	Light yellow, green, green black
18.	Hilum pigmentation black colour	Absent, present
19.	100 seed weight	Very low, low, medium, high, very high

16. Chickpea

(*Cicer arietinum* (L.) (2n:16)

Family : Fabaceae

Temperate or subtropical in adaptation. Annual, herbaceous, with spreading branches and glandular hairs present all over plant parts. Leaves smaller, dark green and purple coloured flowers. Flowers are cleistogamous, papilionaceous type . Anthers in 9+1 diadelphous placement. Pods are mostly double seeded and distinctly larger in Kabuli type varieties and smaller in Desi types. Seeds are brown coloured in Desi type, wrinkled and small in size. Kabuli type varieties are generally tall, with pale green leaves and white flower and setting white large seeds that are generally smooth. Chickpea is a predominantly self pollinated legume. The extent of out crossing is recorded upto 5% (Dua and Sewak, 2003). The white bud stage is very critical and at this stage anther filaments elongate and keep the anthers close to stigma at the time of dehiscence. Flower type is cleiostogamous and full bloom of flowers occurs 24 h. after the pollen is shed.

Nucleus seed production in Chickpea: Source or base population of the variety is raised in 800 sq.m. area with an isolation of 100 m. distance. Single plant per hill is maintained. One thousand uniform looking single plants are selected and tagged. The tagged plants are repeatedly examined for the characteristics and true to type and uniformity. After maturity, the harvested pods and seeds of single plants are examined in the laboratory for characteristics conforming to the variety. Seeds of 500 selected plants are stored in separate packets. The characteristics of chickpea are presented in Table 1.

In the next favorable season, 200 plant progenies are sown in 2 row plots along with the check variety for comparison. The progeny rows are raised in block of 50 progenies. Isolation distance maintained is 100m. The progeny rows are examined at different stages of growth and development from the germination, juvenile to flowering and pod maturity. Progeny rows showing off types or deviating plants are rejected and uprooted. At flowering and post flowering stage, if any progeny is showing off-type or deviant type of plant, that progeny row along with the adjoining rows are uprooted. The pods of selected progenies are harvested and labelled. The pods and seeds separated from the individual progeny rows are examined for the uniformity and true to type characteristics and seeds of selected progenies are packed separately in seed packets. Seeds from these selected progenies are bulked together and nucleus seed is constituted. Seed quality testing is conducted and the nucleus seed packet is labelled. The remnant seeds of single individual plants selected on the basis of progeny testing are stored separately with labels for next nucleus cycle.

Breeder seed production: Nucleus seeds are sown in rows with spacing of 40 cm. x 10 cm. in well prepared fertile field that is isolated by 100 m. distance from other fields grown with chickpea. The seed rate followed is 50 kg/ha. for small seeds, 75 kg/ha. for medium sized seeds and 100kg. per ha. for bold seeded varieties. Progeny rows are provided with 0.5 meter space between beds of 5 rows for easy monitoring. Regular and frequent examinations are made on all the plants in the Breeder seed plot to verify for the uniformity, trueness, stability and disease free nature. Roguing operations to remove off types and deviating types of plants are performed at different stages of the chickpea crop. Monitoring Team visit is arranged to inspect the breeder seed plot. At proper maturity, individual plants are harvested and selected after the examination of pod characteristics. Precautions are taken to avoid physical or accidental mixing of pods and seeds. Seed quality parameters are tested and the quality assured seed is labelled with the Breeder seed label. GOT is conducted with seed sample drawn. Nutritive quality of the chickpea seeds is tested. Detailed procedure is given in Fig. 1.

Table 1. Descriptor of chickpea

S. No.	Characters	States	Stage of observation
1.	Plant type	Erect/semi erect/semispreading/prostrate	Vegetative
2.	Plant pigmentation	No anthocyanin (stems and leaves pale green)/no anthocyanin (stem and leaves green)/ low anthocyanin (stems and leaves partly light/ purple), high anthocyanin (stem and leaves predominantly purple)/dark purple	Vegetative
3.	Plant hairiness	No hairs (including glandular ones) on stems, leaves & pods/lightly pubescent/pubescent/densely pubescent	Vegetative & reproductive
4.	Leaf type	Normal (uni - imparipinnate)/simple (leaf lamina not differentiated into leaflet and rachis)/multi-pinnate (leaf lamina differentiate more than once	Vegetative
5.	Days to flowering	Early/medium/late	Flowering
6.	Flower color	White/pink/blue	Flowering
7.	Number of flowers and pods/peduncle	Single/twin/multiple	Reproductive
8.	Days to maturity	Early/ medium /late	Maturity
9.	Pod size	Short/ medium/ long	Maturity
10.	Number of seeds per pod	Single/double/triple	Maturity
11.	Seed shape	Pea shaped / owl's head/angular	Maturity
12.	Testa texture	Rough/ smooth/ tuberculate	Maturity
13.	Seed color	Beige/brown beige/green/yellow/brown/darkbrown/pink	Maturity
14.	Seed size (wt. of 100 seeds)	Very small (<20g), small (20-25g), medium (26-35g), large (36-45g), very large (>45g)	Maturity
15.	Seed Type	Desi, kabuli	

Note: Description is based on Dua and Sewak (2003); PVP & FRA.

Grow basic seed of variety in 200 sq.m. area with 100m. isolation Select 500 to 1000 single plants with characters identical to variety. Select 500 to 1000 plants for Progeny evaluation

↓

Laboratory evaluation of selected plants for seed characters like size, shape, colour. Reject deviating types. Select, keep seed of each plant separately Grow plant progeny rows in 2 replications with proper plant and row spacing (100m)

↓

Discard all progenies that are not uniform and not conforming to varietal character. Discard all off type plant, pest and disease affected plants. True-to-type single progenies harvested and threshed separately Identify progeny rows showing above mean +2SE

↓

Seeds of individual plants laboratory examined for seed characteristics The selected true-to-type progenies are bulked and stored with 10% moisture content

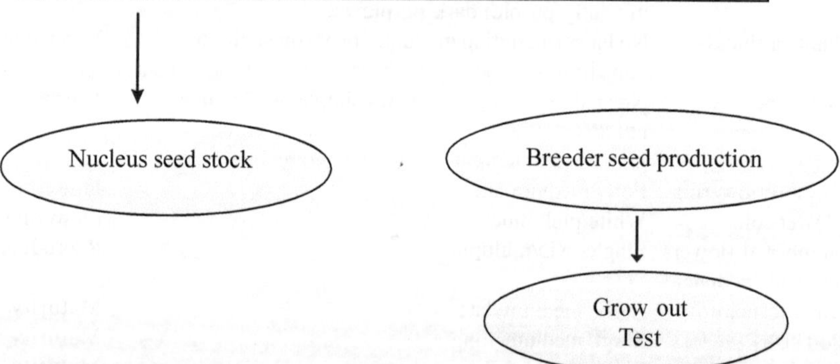

Fig. 1. Nucleus and Breeder Seed Production in Chickpea

17. Cowpea

Vigna unguiculata Walp. ssp *unguiculata* (2n:22)

Vigna unguiculata Walp. *sesquipedalis* (L.) verdec (2n:22)

Vigna *unguiculata* L. Walp (2n :22)

Family : Fabaceae

Vegetable and pulse grain Vigna types are annual, herbaceous, with somewhat hairy and ridged culm. Modern cowpea varieties are determinate to semideterminate with low degree of tendril or vine growing tips. Root system is somewhat sturdy and nodulating. Leaves are alternate, foliaceous, dark green to green. Petiole long, pulvinus. Leaflets three, varying in size and shape. Seedling with epigeal germination; cotyledons oblong or sickle-shaped, thick; first two leaves simple and opposite, subsequent leaves alternate, 3-foliate. Inflorescence axillary racemes, bearing flowers in condensed fascicles. Flowers are white to dull white to yellowish coloured. Flowers bisexual, zygomorphic, papilionaceous; pedicel 1 to 3 mm. long, with spatulate, deciduous bracteoles; calyx campanulate, lobes narrowly triangular, corolla showy, pink to purple, sometimes white or yellowish, standard petal very broadly obovate, hood-shaped, 2.5 cm. long, wings obovate, 2 cm. long, keel boat-shaped, 2 cm. long; stamens 10, 9 fused and 1 free in diadelophous (9+1) arrangement. Anther filaments are long and short. Anthers are uniform size and shape.; ovary superior, 1.5 cm. long, laterally compressed, style upturned, with fine hairs in upper part, stigma obliquely globular. Fruit a linear-cylindrical pod, 8 to 30 cm. long, straight or slightly curved, with a short beak, glabrous or slightly pubescent, pale brown when ripe, 8 to 30-seeded. Seeds oblong to almost globose, often laterally compressed, 0.5 to 1 cm. long, black, brown, pink or white; hilum oblong, covered with a white tissue with a blackish rim-like aril. Cultivated vigna types find use such as grain pulses, fleshy, inflated vegetable pods, green leaf vegetable and fodder and feed. It is an arid legume and drought tolerant crop. The vegetable type cowpeas are almost very low fibre content in pods at the time of harvest of immature pods. Cowpea is a highly self pollinated crop. Cross pollination by means of bees visit is reported upto 15%. The different characteristics of cowpea are furnished in Table 1 for vegetable, grain and fodder types.

Nucleus seed production: There are improved cultivars in cowpea types such as vegetable cowpea and pulses grain cowpea. Seeds originating from breeder sources are sown in 400 sq.m. plots isolated by 100 meters from other types of cowpea crops. Systematic observations are made on germination, seedling vigour, health, uniformity in foliage and plant canopy up to flowering phase. Off-types are uprooted. At flowering , pod set and pod development stages, plants showing deviation for plant type, foliage, flower size, colour, pod size, pod colour and seededness are uprooted as early as possible. Very early and late flowering plants are removed to achieve flowering uniformity. Based on critical observations, 1000 single uniform and true to type plants with good pod bearing and conforming to the variety description are tagged. In case of vegetable types, at peak immature pod development, plants with immature pods that are fleshy, soft and very low or no fibre content are identified and tagged. At maturity, the labelled plants are cut close to the ground and examined in the field and 400 elite superior plants with high pod setting, free from insect and disease infections are selected and are again examined in the laboratory for number of pods, seeds per pod and seed features such as colour, shape and size and hilum expression. Out of the 400 plants, 200 elite plants conforming to the varietal description and that are uniform, true to the type and stable are identified, seeds separated and saved in individual packets.

In the following generation, in the favourable season, 200 progeny rows are raised in well prepared field in 2 row plots in 2 replications following the spacing of 45 cm. x 45 cm. Isolation distance of 100 m. is maintained. The land selected was not previously grown with cowpea crop in the past 2 seasons. Check variety is grown in 2 row plots once in 10 progenies. The seeds are inoculated with the rhizobum. Good agronomic practices and plant protection measures are adopted. Examining the single plants in each progeny row at different stages of growth from germination is done as per the protocols and procedures described in earlier generations. Off- types and progenies showing deviating types of plants for any one of the traits are uprooted. Evaluation of immature pods for suitability for fresh vegetable market is made. At maturity, data of progeny rows are recorded for overall productivity, pod and seed number. Based on the analysis, progenies performing above mean + 2 SE are identified. Seeds samples drawn from identified elite plants from each selected progeny are examined in the laboratory and pooled together to constitute nucleus seed. The remnant seeds are utilized for developing nucleus seed in the next cycle.

Breeder seed production: Nucleus seed is sown in a well prepared isolated field by 100 m. from other cowpea crops. The field selected is clean and not grown with cowpea in the past 2 years. Border rows with the same variety are raised all around the breeder seed production plot. Single plant per hill is maintained. Ideal growing conditions are maintained. Removal of off-types and non-conforming plants is stringently done. Only disease free healthy vigorous plants that are true to type are maintained. Early flowering and late flowering plants are discarded. Visit of monitoring team is organized. In case of vegetable type cowpea variety, samples of fleshy immature pods of appropriate size are evaluated for the vegetable market suitability. Nutritive quality of fleshy pods is studied. Individual plants are harvested that have uniformity for plant type and pods. The pooled seeds from all the selected plants are dried and processed. Seed samples are drawn and tested for seed qualities. Quality assured seeds are labelled with Breeder Seed Tag. GOT is conducted for assessing genetic purity. If more quantity of Breeder seed is required, BS II stage multiplication may be taken up. It is necessary to repeat nucleus seed - breeder seed production cycle once in 3 years. Nutritive qualities of immature pods, tender green leaves and seeds are estimated.

Table 1. Descriptor of cowpea

1.	Characterization of Traits	
	Botanical type (specify)	
	Growth habit	Erect (bushy), semi-erect, semi-spreading, spreading
	Growth pattern	Determinate, semi-determinate, average vigour, poor
	Basic plant vigour	Highly vigorous, vigorous, average vigour, poor
	Raceme position	Mostly above canopy, upto upper 1/3rd of canopy, throughout the canopy
	Pod attachment of peduncle	Pendent, 30°-60° from erect, erect
	Utility type	Vegetable type Dual-1 (vegetable + seed) type Seed type (grain pulse type) Dual-2 (seed + fodder) type Fodder type Dual-3 (fodder + vegetable) type
2.	Flowering characteristics	
	Days to first flowering	
	Duration of flowering	Short, medium, long
	Number of flushings	
3.	Maturity Traits	
	Days to first pod picking (green stage)	
	Days to first pod maturity	
	Days to 50% maturity	Uniform maturity, overlapping maturity (over 2 consecutive flushings), highly variable maturity ranges
	Days to complete maturity	
	Senescence	
4.	Plant Canopy Characteristics	
	Plant height (cm.)	
	Number of primary branches	
	Leafiness	
	Leaf size- large, medium, small	
	Leaf density-dense, medium, sparse	
	Leaf shape-normal, moderately dissected, highly dissected	
5.	Number of clusters per plant	
6.	Pod Characteristics	
	Number of pods per cluster	
	Pod length (cm)	
	Pod shape	Straight, slighty curved, highly curved
	Pod fibrousness (at green picking stage)	Very soft, average, highly fibrous

7.	Seed Characteristics	
	Number of seeds per pod	
	Seed size	Bold, medium, small
	Seed shape	Round, oval, rhomboid, crowedge,others (specify)
	Seed colour description (specify)	
	100-seed weight (g.)	
8.	Shattering	Non shattering, slightly shattering, medium shattering, shattering, highly shattering
9.	Yield Traits	
	Yielding ability aspect (yield potential) (visual score) : based on number of seeds per pod/pods per cluster/ clusters per plant etc.	Good, above average, average, below average, poor
	Seed yield per plant (g.)	
	Seed yield per plot (kg/ha)	
	Seed production potential (visual score) : based on matured pods/plant, plant canopy structure	
	Vegetable pods production (visual score)	Potential, high, medium, low, duration of picking, extended overlong period, uniform single picking
10.	Disease/Pest Status	0-9 Scale (specify disease/pest), 0 immune, 3 low susceptibility, medium susceptibility, highly susceptible
11.	Quality Traits	Protein content, Cooking ability (time to cooking, minutes)

Note: Descriptors based on Gupta *et al.* (1995); PPV-FRA (2020). Hari Har Ram (2005)

Descriptors for different types of cowpea namely vegetable cowpea, grain type cowpea and fodder cowpea are presented in Tables 2, 3 and 4.

Table 2. Descriptor for Vegetable Cowpea

1.	Days to 50% flowering	Early (<40), medium (40-50), late (>50)
2.	Climbing tendency	None, slight, pronounced
3.	Plant type	Determinate, indeterminate
4.	Plant grown habit	Erect (bush), semi erect, horizontal, pole type
5.	Twining habit	Viny, non viny
6.	Pod colour	Green, darkgreen, purple
7.	Pod surface	Smooth, rough
8.	Pod pubescence	Absent, present
9.	Pod length (cm)	Short (<20), medium (20-30), long (30 to-)
10.	Pod thickness (cm)	Thin (<0.5), medium (0.5 to1.0), thick (>1.0)
11.	Pod stringiness	Absent, present
12.	Pod straight nature	Straight, curved
13.	Pod maturity days	Early (<50), medium (50), late (>50)

14.	Pod cross section through seed	Cordate, circular, 8 shaped, oval
15.	Number of seeds per pod	Few (<10), medium (10-15), high (>15)
16.	Seed shape	Elliptical, kidney, rhomboid
17.	Seed length (cm)	Short (<1), long(>1)
18.	Seed colour	White, brown, brickned, black
19.	Flower colour	Yellow, white, violet

(S.No. 1 to 5 and 19; 35 to 60 DAS. 6 to 15 at 45-80 DAS; 16 to 18 at 70 to 110 DAS)

Table 3. *Vigna unguigulata sub* sp. sesquipedalis- Cowpea Grain type

1.	Stem colour	Green, light purple, purple
2.	Stem pubescence	Absent, present
3.	Days to 50% flowering	Early (<40), medium (40 to 60), late (>60)
4.	Plant growth habit	Erect, semierect, spreading
5.	Leaf colour	Green, dark green
6.	Immature pod colour	Green, dark green
7.	Pod anthocyanin colour	Absent, present
8.	Days to maturity	Early <70, medium 70 to 80, late >80
9.	No. of seeds per pod	Low (<14), medium (14 to 18), high (>18)
10.	Seed eye colour	Tan, brown, red, black
11.	Seed length (mm)	Shoot (<5), medium (5 to 10), long (>10)
12.	Seed shape	Kidney, elliptical, rhomboid
13.	Seed Test weight (100 seeds) (g)	Small (<7), medium (7-10), high (>10)
14.	Pedunde length (cm)	Short (<60), medium (60 to 80), long (>80)

Table 4. Fodder Cowpea

1.	Plant height (cm)	Dwarf (<50), tall (50-60), extratall (>60)
2.	No. of primary branches/plant	Low (<5), medium (5 to 8), high (>8)
3.	Days to 50% flowering	Early (<45), medium (45 to 55), late (>55)
4.	Plant growth habit	Erect, semierect, spreading
5.	Terminal leaflet length of penultimate main leaf (cm)	Short (2.5), medium (5 to 8) long (>8)

18. Field Pea/Garden Pea

(*Pisum sativum* L.) (2n:14)

Family : Fabaceae

Peas is a highly valued fresh vegetable crop. Plant is an annual herb, with angular stem, winter season adaptation and requires staking. Leaves glaucous and alternate, distichous and rachis terminates into simple or branched tendrils. Inflorescence is axillary raceme with upto 3 flowers. Flower is papilionaceous, calyx is campanulate and corolla colour is white, pink or purple. Stamens 10, diadelphous (9+1), ovary superior , ovules upto 13 in number that are alternately attached to the two placenta. Stigma becomes receptive to pollen from several days prior to anthesis. Pea is a highly self pollinated crop.White flowered horticultural or garden pea is also known as Sweet Pea. Pods are straight or slightly curved. Seeds are smooth or wrinkled. Matured pod is brown in colour. In early varieties, first harvest can be made 50 days after sowing and in the late varieties, first harvest can be made 65 to 70 days after sowing. Single stem that is erect and non lodging, , with high yield, green pod with more number of seeds, good taste, non shattering of matured pods and resistance to diseases and pests are the desired characters.

Nucleus seed production: Planting season is February - March and October-November. Seeds of the designated variety from authentic sources are sown in 45 cm. x 10 cm. spacing in an area or 400 sq. m. in a well prepared plot that is isolated by 100 m. from other pea crops. Weak and unhealthy plants are removed. Proper staking is provided to each plant. Type of tendril is also studied in all the plants. True to type and uniform looking pea plants alone are retained and staking done. At flowering phase, flower colour and size and at pod developing and pod maturity stages, pod size and shape are verified. Plants showing deviations from true to type and description are uprooted. Border rows are excluded for selecting the single plants. One thousand plants that are true type and uniform are selected from the center of the field. Pods from individual plants are packed separately and dried. In the laboratory, pods and seeds separated from pods are closely examined for trueness and similarity. Non conforming seeds of individual plants are discarded.

In the next favourable season, seeds are sown in 2 rows progeny plots in 45 cm. x 10 cm. spacing and each row four meters long .Between two progeny plots 60 cm. spacing is provided. The progeny rows plots are raised in 2 replications. In each block 50 progenies are raised. Isolation distance of 100m is maintained. All the plants are supported with staking. Plants in the individual progeny rows are studied from the germination to maturity stage for true to type of the crop traits specific to the growth phases. Progeny rows showing deviating type of plants are uprooted. Pods at fresh vegetable market stages collected from individual progenies are examined for more number of green pea seeds per pod, taste of seed and size and consumer acceptance. The performance of progeny rows in respect of flower prolificity, fresh green pod yield and seed yield are evaluated at harvest and progenies performing above mean + 2SE are identified. Pods from 200 elite superior plants from the identified and selected progeny rows that are uniform, and prolific in pods set and uniform seed development are harvested. Pods and seeds of the 200 selected individual plants are examined in the laboratory for true to type, uniformity and stability and forwarded to the next season.

In the subsequent season, the seeds of the selected individual plants are raised in progeny rows in well prepared clean field not grown with pea crop in the last

2 years. The plot is isolated from other pea crops by 100m. All plants in the progeny rows are examined at different stages of crop development from seed germination, seedling emergence and field establishment. Progeny rows showing off types and plants that are deviating from the description of characters of the original variety are uprooted. Pods suitable for fresh vegetable market stage are harvested and examined from each progeny. This is an important step. Progenies fulfilling the fresh vegetable market norms and acceptability alone are selected and 50 matured plants in each of the selected progeny rows are harvested and the seeds pooled to constitute the nucleus seed. The remnant seeds are utilized for next cycle of nucleus breeder seed production chain. The characteristics in pea are listed in Table 1.

Breeder seed production: The nucleus seed is sown in the next favourable season in a well prepared clean field that is isolated by 100 meters from other pea crops. In this selected field, pea crop has not been grown in the last 2 years. Good agronomic and vegetable production practices and staking of plants are adapted and spacing of 40 cm. between rows and 10 cm. on the rows is provided. Staking of the individual plants is done. Border rows (8 to 10 rows) are raised around the plot. The individual plants are continuously observed for field establishment, health and trueness in stem, foliage and tendrils, flowers and pods and qualities of green pods for their suitability to fresh vegetable market for green peas. Plants that are in nonconformity for the trait expression are uprooted.Monitoring Team visit is arranged to judge the breeder seed crop. Matured plants that are free from foliar diseases are harvested and verified. The seeds of selected plants are examined in the laboratory for true to type and selected and pooled together for pod size, shape, seed filling and maturity and conforming to the description. The pooled seeds are dried, processed and packed. Sample seeds are drawn and tested for seed qualities. Quality assured seeds are labelled with Breeder seed tags. Grow out test is conducted with seeds samples drawn from Breeder seed to ascertain the genetic purity. In case of more quantity of breeder seed is required, BS I and BS II stages may be taken up. Nutritional quality of green peas and dried grain peas is estimated.

Table 1. Descriptor of peas (*Pisum sativum*)

S.No.	Description	Characters
1.	Plant anthocyanin	Present, absent
2.	Stem fasciation	Present, absent
3.	Stem length	Short, medium, long, very long
4.	Stem – number of nodes up to and including first fertile node	Very few, few, medium, many,very many
5.	Foliage color	Yellowish green, green, blue green

6.	Leaf leaflets	Absent, present
7.	Leaflet length	Short, medium, long
8.	Leaflet dentation	Absent, weak, medium, strong,very strong
9.	Stipule length	Short, medium, long
10.	Stipule flecking	Present, absent
11.	Stipule flecking density	Sparse, medium, dense, verydense
12.	Petiole length	Short, medium, long
13.	Time of flowering	Very early, early, medium, late very late
14.	Only with varieties with stem fasciation/ absent maximum number of lowers per node	One, two, three, four or more
15.	Standard flower color	White, white with cream, cream
16.	Shape of base of standard petal	Raised, moderately raised, levelarched, strongly arched
17.	Peduncle length of spur	Long, medium, short
18.	Pod length	Very short, short, medium, longvery long
19.	Pod width	Narrow, medium, broad, verybroad
20.	Pod parchment	Absent, partial, entire
21.	Pod curvature	Weak, medium, strong, verystrong
22.	Pod color	Yellow, green, blue green, purple
23.	No. of ovules/pod	Few, medium, many
24.	Seed shape	Ellipsoid, cylindrical, rhomboid, irregular
25.	Seed cotyledon color	Green, yellow orange
26.	Seed hilum color	Same color as testa, darker than testa
27.	Starch type in seed	Simple, compound
28.	Seed weight	Very low, low, medium, high

Note: Description based on UPOV (1998)

19. French Bean

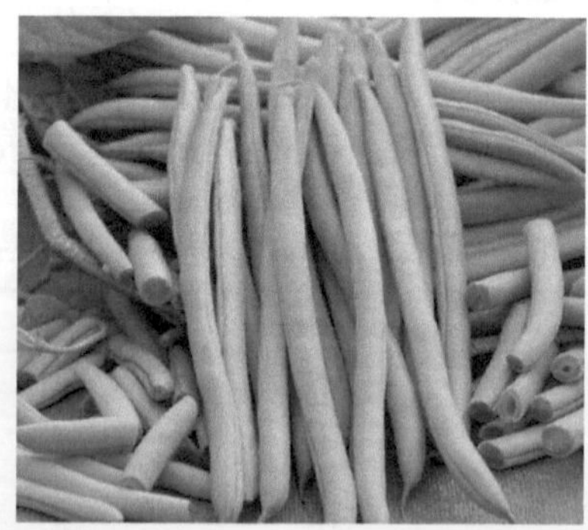

(*Phaseolus vulgaris* L) (2n : 22) (String Bean, Snap Bean, Common Bean, Kidney Bean and Rajmah Bean)

Family : Fabaceae

French bean is largely cultivated for fresh green pods as vegetable. Snap bean is used for dry bean production. There are three types of French Bean - Bush types with short internodes, Semipole type with longer internodes than those in bush type and Pole type with internodes longer than that of semipole type. India produces nearly 50% of world's production of dry bean (Rajmah) and snap bean. India registers top in the top ten countries for dry bean production and fourth ranking in respect of fresh pod vegetable bean. French bean is a day neutral cool season vegetable. The descriptor for french bean is given in Table 1.

French bean is an annual, erect, determinate, herbaceous, bushy legume with flat pods in concentrated fruiting. Vegetable grade green pods show very low fibre content. Determinate French beans have low number of nodes and short internodes as compared to indeterminate forms. Stem is somewhat thick, leaves trifoliate, alternate and leaflets are large sized. Varieties show wide differences for size and shape of pods and mature seeds. Flat or round pods are common in determinate and early types. Primary branches are high in number. Inflorescence is axillary or terminal, in false racemes or solitary flowers in the inflorescence axis. Calyx five, campanulate, corolla five, white in colour to pale, purple and red purple. Anthers 10 arranged in 9+1 diadelphous manner. Ovary superior, monolocular, ovules 10 to 20. Pods are flat or round and space between seeds inside the pods is almost nil. Pods 10 to 25 cm. long with prominent pod peak, mostly straight. Seeds are somewhat compressed and kidney shaped and varying in seed coat colour from white, pale to dark or blackish. First harvest is made 45 to 50 days after sowing and harvests completed in stages in 7 to 12 days after flowering. It is a self pollinated crop. Isolation distance recommended (for nucleus seed and breeder seed production) is 100m. from other French bean crops.

Nucleus seed production: Seeds accessed from the Institution or Breeders concerned are raised in 400 sq.m. plot previously not grown to French bean. Single plant is maintained per hill. Isolation distance of 100 m. from other broad bean crops is followed. Vigorous, healthy and uniform looking plants alone are selected. At flowering, uniform and true to type plants in flower colour, developing pods, marketable immature bean and pods are retained. Off types and progenies having deviating types of plants from conformity are uprooted. At pod maturity, pods and seeds are examined for true to type in the pods shape, size and 1000 plants are selected from middle of the plot and harvested. Seeds are separated plantwise and stored in separate paper covers. The seeds of single plants are studied in the laboratory for conformity of trueness and seed features and 500 individual plants are selected.

In the next generation, seeds of 300 progeny rows are sown with recommended cultivation practices for French bean in two row progeny plots in 2 replications and each row 4 m in length. Single plant is maintained per hill. Isolation distance of 100m. is maintained. Progeny rows are assessed for fast seed germination, healthy seedlings, good field establishment and fast growth and uniformity in development in every crop growth phase and flowering uniformity in French bean. Off types and progeny rows showing non conformity for any of the prescribed traits are uprooted. Characteristics of stem, foliage, flower size and colour and size of pod and seed and uniform pod development at different harvest stages, quality of immature pods for vegetable market and consumer acceptance, no or low fibre content in pods are assessed and progenies showing conformity and true to type nature alone are retained. Fresh pods are harvested and recorded from 2 replications. Progenies showing performance above mean ± 2SE are selected. Matured pods are harvested from the selected progenies. In the laboratory, seeds are examined for uniformity, true to type and stability and non conforming progenies are discarded. Seed samples from each of the selected progenies are then pooled together to constitute nucleus seed. Remainder of the seeds are saved for further utilization of nucleus seed / stock seed development. Nutritional qualities of fresh pods are evaluated.

Breeder seed production: Breeder seed production is taken up in a well prepared field not grown with French bean in the last 2 years. Seeds are inoculated with rhizobium bacterium. Isolation of 100 m. is followed. Good agronomic and vegetable production practices are adapted. Single plant per hill is maintained. Vigorous and healthy plants alone are retained. Off types and non conforming plants are uprooted. Prerequisites such as flowering uniformity, good pod set and development and quality and standards of the marketable fresh pods and no fibre or very low fibre content in pods respectively in first, second and third picking are carefully verified and plants are selected. At midflowering stage of the Breeder seed crop, visit of Monitoring Team is arranged. It is advisable to assess the nutritional and vegetable qualities of the composite fresh green pod samples. Pods are harvested from individual plants and seeds extracted and studied in the laboratory. Seeds that show conformity are pooled together, dried and processed. Seed samples are tested for seed qualities. GOT is conducted for ascertaining genetic purity. The quality assured seed is labelled with the Breeder seed Tags. BSP I and BSP II stages of Breeder seed production are taken up if the breeder seed requirement is high.

In respect of dry beans (Rajmah types), single plant selection from the Breeder stock seeds, progeny rows evaluation and selection and constitution of nucleus seed are made after critical evaluation for plant type, size and shape of matured pod, colour, size and shape of dry seeds and cooking quality of dry beans.Breeder seed production is done following the protocols for nucleus and breeder seed production of vegetable French bean in all its aspects with added emphasis on the dry pods and dry seeds characteristics and nutritional quality.

Table 1. Descriptor of French Bean (*Phaseolus vulgaris* L.) and (Rajmah)

S.No.	Characters	State
1.	Time of flowering (days)	Early (50<), medium (50-75), late (76-150), very late (>150)
2.	Stem of terminal leaflet of firstflowering node	Absent, present
3.	Size of terminal leaflet of firstflowering node	Small, medium, large
4.	Plant growth type	Erect, semi erect, spreading
5.	Plant twining habit	Viny, non viny
6.	Plant habit	Determinate, indeterminate
7.	Leaf green color	Light, dark
8.	Central leaflet shape	Cordate, ovate, rhombohedrick, hastate
9.	Flower color-standard petal	White, yellow, pink, violet
10.	Flower outside of standard petal	Striped, non striped
11.	Pod curvature	Absent, medium, strong
12.	Pod shape cross section	Cordate, circular, eight shaped, oval
13.	Pod shape	Concave, convex
14.	Pod shape, distal part	Acute, acute to truncate, truncate
15.	Pod color	Pale green, green, purple
16.	Pod stringiness	Absent, present
17.	Pod pigmentation on pod shell	Absent, present
18.	Plant height (cm.)	Short (<40), medium (45-75), tall (>75)
19.	Seed shape	Circular, circular to elliptic, elliptic, kidney shaped
20.	Seed size: 1000 seed weight (g.)	Small (<250), medium (251-300), large (351-400), very large (>400)
21.	Seed testa color	White, brown, red dark red and black
22.	Seed testa variegation	Absent, present

Source: PPV & FRA, 2007; Dixit and Tripathy, 2003

20. Guar (Cluster Bean)

(*Cyamopsis tetragonoloba* L.) (2n:14)

Family : Fabaceae

India produces 80% of World's production of guar seeds and meets 40% of the World's guar gum supply. Guar gum (galactomann) is derived from the endosperm of guar seeds. Gum constitutes 29 to 30% of endosperm. Guar seed growing area in India is 42 lac. ha. and the seed productions is 24 lac MT.

Cluster bean is an annual, herbaceous, 1.0 to 1.2 m. tall, bushy, with 1 to 4 basal branches. Stem is erect, angular and hollow. Leaves are alternate, dark green to green, trifoliated leaflets ovate and flowers 10 to 15 in axillary racemes. Flower colour is white to purple. Pods are borne in clusters and are many (10 to 15 per cluster). Pods are compressed, flat, double ridged on dorsal side, tip pointed, seeds 5 to 12 per pod, small, oval, 5 mm. long. Matured seeds whitish or grey. It is a drought tolerant multipurpose arid leguminous crop grown for green manure, forage, vegetable and guar gum for industrial purposes. Guargum (galactomann) is well known product. Tender pods in clusters can be harvested for fresh vegetable consumption 40 to 45 days after sowing and harvests continue upto 85 to 90 days. Dry pods are harvested for seed extraction in guar gum varieties when pods turn yellow and drying. Guar seed type varieties and fresh market vegetable type varieties are distinct in stature and pod features. Guar is a self pollinated crop. The descriptor for guar is presented in Table 1. The genetic variability and diversity in clusterbean are reported by Kgakudi *et al.* (2020) and the performance of different clusterbean for yield and quality traits published by Bakang *et al.* (2020).

Nucleus seed production: Seeds are accessed from stock seed maintained by the breeder and are sown in a well prepared 200 sq.m. area. Only one plant is maintained per hill. Isolation distance of 100m. from other guar fields is maintained. Observation is made from seedlings stage onwards. Weak unhealthy and disease affected plants are uprooted. At the commencement of flowering, every plant is examined from basal leaf axils upwards for stature, branching and flowering uniformity. Very early and late flowering plants are removed. At midmaturity of pods in the basal 1 to 2 clusters, uniformity, vigour of plants and true to type and conformity of stem, foliage size and colour and cluster size (number of flowers or pods per cluster), immature pod colour and size, basal branching (1 to 2 in number) and non lodging nature are studied. Off-types and non conforming or deviating plants are uprooted. Plants showing high variation for cluster size on the mainstem are discarded. Superior plants, 300 in number are finally selected and labelled. Matured dry pods are harvested intact in 2 to 3 clusters, labelled plantwise and saved. In the laboratory, the intact cluster is assessed for the synchrony in maturity and uniformity of matured pods, pod shape and seed size and colour.

In the next generation, seeds of 100 selected plants are sown in 2 replications in 2 rows progeny plot, 4 m. long and 60 cm. between 2 rows plot. The check variety for comparison is raised in 2 row plots in every ten single plant progeny rows. The plot is isolated by 100m. from other guar crops. Systematic and continuous monitoring is done to examine individual plants in each progeny row in every stage of growth and development. Off-type and progeny rows with deviating or non conforming plants for any of the prescribed characteristics are uprooted. Non lodging nature, uniformity in flowering, flower cluster and pod development features are critically assessed. Based on the true to type conformity, the progeny rows, with performance above the mean +2SE are identified. From among the identified superior plants, 200 plants are selected and fresh vegetable market suitability of immature pods are examined for fleshiness, less fibre nature, slow seed maturity and consumer appeal. Two to three clusters with dried pods intact are harvested from each selected plant. In the laboratory, pod clusters from individual single plants are assessed for uniformity and synchronised development and maturity. Seeds extracted from the respective plants are examined for conformity of seed size, shape and colour besides the suitability of immature fleshy pods for the fresh vegetable market. Seeds from selected 100 elite plants are bulked to constitute the nucleus seed.

Breeder seed production: The nucleus seeds are sown in well prepared field not grown with the cluster bean in the past 2 to 3 seasons. The field is isolated by 100 m. from other cluster bean crops. All good agronomical and plant protection practices against leaf hopper, pod borer, leaf spot and powdery mildew are adapted. Off-types and deviating plants, if any, are discarded Monitoring Team visit to the BSP plot is arranged. Two to three clusters (pods) alone are harvested from each plant. Market acceptability of tender pod judged. In the laboratory, critical examination is made for the trueness of immature and mature pods and seed characteristics. Seeds are separated and bulked to make Breeder seed. Samples of seeds are tested for different seed qualities and quality assured seeds are labelled with Breeder Seed tags. GOT is conducted as per standard protocols. Nutritional qualities of the fresh market stage pods is estimated.

In the breeder seed production method followed for the Guar varieties for seeds for gum production, single plant selections in nucleus and breeder seed production stages are critically examined and identified for the maximum extent of pod number, seed number and size of seeds per pod and per se plant yield of guar seeds. Besides, the guar gum content (galactomann guar gum) and quality in the selected progenies in nucleus and breeder seed generations are estimated and the prescribed seed quality standard is maintained.

Table 1. Descriptor of guar (cluster bean)

S.No.	Description	Characters
1.	Seedling vigour	Vigorous, less vigorous
2.	Plant habit	Erect, semi secret, prostrate
3.	Plant height	Tall, medium, short
4	Branching	Absent, 1 to 2, more than 2
5.	Stem thickness	Thin, medium, thick flat
6.	Foliar size	Large, medium, small
7.	Foliar color	Dark green, green, light green
8.	Leaf lamina attitude	Erect, flat
9.	Size of flower cluster	Small, medium, large
10.	Fruit attitude	Erect, radiating
11.	Fruit size	Small, medium, large
12.	Fruit breadth	Narrow thin, medium broad, broad and thick
13.	Immature fruit surface	Smooth, protrusions present
14.	Maturity of flowers in cluster	Less uniform, uniform
15.	Fruit tip	Long pointed, medium long pointed, short and pointed
16.	Seed size	Small, medium, large
17.	Guar seed type	Small, medium, large

21. Horse Gram

Macrotyloma uniflorum Lam. Verd. (2n:24)

Syn. : *Delichos uniflorum*

Family : Fabaceae

Horsegram is an annual grain legume. It is a drought tolerant pulse grain crop and highly adapted to soil moisture deficit and harsh and fragile production environments. The crop residues such as stem, leaf and husk are utilized as livestock feed. Horsegram varieties are mainly suited to rabi season growing. Horsegram shows high degree of varietal and land races diversities and variation especially for maturity duration, growth habits, canopy spread from bushy to semi bushy stature, trailing to suberect habit, hairiness, pod shape and size and seed shape and colour. The horsegram pod is usually sickle shaped and seeds are flattened and compressed.

Horsegram stem is mostly viny, slender and intervining. Leaves are trifoliate, leaflets compound and alternate. Inflorescence axillary racemes. Flowers are white to violet coloured and are hermaphrodite, axillary; calyx five and corolla five, papilionaceous. Stamens ten in diadelphous condition (9+1). Ovary superior and ovules 5 to 7 in number. Pod short curved enclosing seeds with light brown, red, brownish grey and black mottled. Seed coat is shining. It is a self pollinated crop. Characters of Horsegram are furnished in Table1 (Kanaka Durga *et al.*, 2015; Rani *et al.*, 2013).

Nucleus seed production: The seeds are accessed from institution / breeder source. Seeds are examined in the laboratory for size, shape and colour and uniform flat nature. Samples conforming to the colour and size of seed are selected and the selected true to type source seeds are sown in a wide space in a plot of 400 sq. m. and maintaining isolation of 100 m. distance from other horsegram fields. At the juvenile stage, weak unhealthy and disease affected plants are removed. At flowering stage, plants showing differences in canopy size, habit, foliar density and hairiness are discarded. The remaining plants from flowering to pod development stage are critically examined for trueness. Deviating plants and off-types are removed. At maturity, 500 to 600 single plants of uniform size and conforming to the variety description are harvested and pods from each plant are packed separately and dried intact in packets. In the laboratory, numbers of pods, seeds per pod and seed colour, seed shining and size and total seed weight are recorded. Single plant seeds are selected based on the superiority of the selected and harvested plant. Seeds of 200 single plants are finally selected.

In the following rabi season, 200 single plant progenies are raised in two replications in 2 row plots, 4 m length and 30 cm. spacing between rows. Each progeny row plot is 45 cm. on either side from the next progeny row plot. For every 10 progeny rows plot, one check plot is raised in 2 rows for comparison. Isolation distance provided is 100 m. Single plant per hill is maintained. Observations and monitoring are critically made on the crop establishment, vigour, uniformity in canopy size, growth rate, branching and foliar traits. Progenies showing off types and deviant plants are uprooted in the preflowering stage. Early flowering and late flowering plants are discarded to maintain uniformity in flowering and maturity. At flowering stage, data on days to 50% flowering, flower colour and profuseness of flowering are recorded. At pod setting and pod maturity phases extent of biomass and pod setting are recorded. Progeny rows, showing true to type and performance above mean +2SE are selected. Around 50 to 60 superior plants in selected progenies are gently uprooted, harvested and stored in separate container bags. Plant type, seeds / pods, extent of pod setting and pod colour, shape and size of pods are examined. Seeds from each progeny are verified for true to type and genetic uniformity for the seed coat shining and colour and for seed flatness. Portion of seeds from selected progenies are pooled to constitute nucleus seed. Progeny basis remnant seeds from each progeny are saved for developing nucleus seed to make the next nucleus to breeder seed cycle. The nucleus seeds are tested for seed quality.

Breeder seed production: Nucleus seeds are sown in rabi season well prepared field and that was not grown with horsegram for 2 or more seasons and also isolated by 100m. from other horse gram growing fields. Border rows with the same variety are grown around the field. All good agronomical practices are followed in seed production. Frequent and systematic monitoring of the breeder seed crop from germination to harvest is made for stage specific traits and features as described in nucleus seed development. Roguing the disease affected plants, off- types and deviant plants are done. In essence, every plant is examined for the true to type conformity. Uniformity of the crop is studied. Monitoring Team visit is arranged /organized to assess the breeder seed crop at mid maturity stage. After harvesting the individual plants, the matured pods and seeds are examined in the laboratory for genetic uniformity, trueness and stability. Seeds are tested for seed quality. The quality assured seeds are then packed and Breeder seed label attached. Grow Out Test is conducted to confirm the genetic purity. Nutrient content in seeds is estimated.

For kharif season adopted horsegram which are only a few, nucleus and breeder seed production is organized in the Kharif season specific zone.

Table 1. Characters of Horsegram

1.	Growth type	Erect, spreading
2.	Leaf size	Small, medium, large
3.	Leaf pubescence	Light, medium, dense
4.	Leaf colour	Light green, green, dark green
5.	Colour of leaf petiole	Green, green with light purple tinge, green with dark purple tinge
6.	Leaf shape	Ovate, ovate to linear
7.	Stem colour	Green, purple
8.	Stem pubescence	Sparse, medium, dense
9.	Vein colour	Green
10.	Flower colour	Green, yellow
11.	Twine colour	Green, green with purple tinge, green purple
12.	Twining habit	Twining
13.	Immature pod colour	Green
14.	Matured pod colour	Straw, straw with purple straw with dark purple streaks
15.	Pod pubescence	Sparse, medium, dense
16.	Pod curvature	Slight, medium curved, sickle shape
17.	Seed colour	Straw, straw with green tinge light straw, straw with black tinge, straw with green tinge, straw with orange tinge, black

22. Indian Bean Dolichos Lab Lab

(*Lab Lab purpureus* (L.) (2n : 22, 24)

Lab Lab var typicus - Garden bean

Lab Lab var lignosus - Field bean

Family : Fabaceae

Varieties of *Lab lab purpureus* varies from annual to perennial form. Perennial types are season specific, trailing on supports. Immature pods and seeds of perennial types are used as fresh vegetables. The lab lab purpureus var typicus is an annual, bushy, tending to trail with branch apices forming tendrils. Annual types are non photosensitive, determinate and bushy in habit and are cultivated in all the seasons for the tender immature fleshy pods, consumed as fresh vegetable. The descriptors are given in Table 1 and 2.

Leaves are greenish to dark greenish, alternate, pulvinate, pinnately trifoliate. Leaflets have smooth margin and acute to obtuse shape. Inflorescence axillary racemes. Flowers borne on long peduncle 15 to 25 cm. long. Flowers arise from axiliary buds of the nodes on inflorescence. At each node, 3 to 5 flowers arise in congested manner. Flowers are zygomorphic, calyx five, gamosepalous, two lipped, upper two and lower three, petals five, papilionaceous in arrangement. Stamens 10 in diadelphous manner and didynamous (9+1). Anther filaments are long and short and are alternately placed. Fruit is a fleshy pod and on drying the pods dehisce. Seeds are ovate to oblong, compressed to different extent. Seed colour ranges from white to brown to purple and dark brown. Hilum is somewhat linear and prominent with whitish axil.

Modern varieties of Dolichos garden beans are non photosensitive and are bushy and compact plant type, 3 to 4 erect branches and long peduncles and infloresecences, each inflorescence supporting 10 to 14 flowers. Around 10 to 14 racemes per plant are produced. At each node on the infloresecence, 2 to 4 flower nodes arise. Anthesis and pollination occur from 8.30 am. to 10.30 am. Around 30% of flowers alone set pods of varying length from 7 cm. to 8 cm. in broad flat types and from 12 cm. to 14 cm. in narrow tubular pod types. The fleshiness and tenderness of the immature pods show considerable variation. Seeds are from 7 to as many as 12 to 13 per pod. The immature fleshy part of the pod & seeds are the fresh vegetable parts. Fleshy pods vary from flat thick broad to medium thick to hollow tubular. Pod shape may be somewhat curved with the pod tip very acute. Pod margins are fibrous. In modern varieties, tendril forming tendency is almost absent. In case, tendril formation is noticed, such tendrils are pinched off so as to arrest the trailing tendency and to encourage flowering.

Nucleus seed production: Seeds of bushy types of lab lab typicus variety (garden bean) are sourced from institution / breeder or from the advanced variety evaluation of stage of prerelease. Seeds are sown in 100 sq.m area plot with an isolation distance of 100m. from other dolichos crop. Individual plants are raised in 60 cm. x 60 cm. spacing and are monitored from the seedling stage. Weak, unthrifty and unhealthy plants are uprooted. Detendrilling is done whenever tendrils are noticed. At mid flowering stage, plants showing low number of infloresces (less than 9 to 10) indicative of low productivity are culled out. At harvestable stage of immature fruits, around 500 uniform and highly productive plants as indicated by high number of panicles and more pod set are identified and labelled. True to type nature, shape, size and fleshiness of immature marketable size pods and slow seed maturity, flavor, smell, free from pests and disease are scored in the selected plants. Synchronous development and uniformity of harvestable pods in each inflorescence is critically assessed. Based on the overall plant type, foliar and reproductive traits, 300 plants are finally selected and the uniformity in pigmentation, colour, shining and attractiveness of the pods are again ascertained. At maturity, dry pods from selected single plants are collected and stored in separate packets with record of number of flower stalks and number of pods. In the laboratory, individual pods and seeds are studied for trueness, seed size, colour and uniformity.

In the next season, 250 selected single plant progenies are raised in 2 row plots, each row 4 m. long, in two replications adapting a spacing of 45 cm. x 45 cm.

For every 10 progeny rows, one two row plot of the original variety is raised for comparison. Ideal field growing conditions are provided for raising the vegetable crop. Isolation distance of 100m. is provided. Border rows of 2m. width are grown all round the plot with the same dolichos variety. All observations and procedures as described in the previous generation are followed. Off-types and progenies showing deviating plants are uprooted. Uniformity in flowering is maintained by discarding the rows that flower very early and very late. Every progeny row is adjuged for trueness ,uniformity, growth and vigour. Colour of flower, pod colour, fleshiness of marketable immature pods and extent of pod set are especially looked into in each progeny. Two to three harvests of immature marketable pods are made in each of the progenies in the 2 replications. Data on number of pod bearing peduncles, number of fleshly fruits and total harvests were recorded. Progenies showing values above mean + 2SE are selected and the market acceptability of fleshy immature vegetable pods is assessed. At pod maturity, pods from selected plants in the identified progenies are harvested and saved in separate covers with labelling of the progeny. Pedigreed seeds of 200 such superior plants are thus saved and part of seeds of the selected progenies are pooled together to constitute the Nucleus seed. The remnant seeds are stored for the next cycle nucleus seed production. Nutritive contents are estimated in the seeds and in the fleshy pods.

Breeder Seed Production: Breeder seed is produced from the nucleus seed source adapting the production practices for dolichos/garden bean and with isolation distance of 100 meters. Off types and plants showing deviations from trueness are uprooted before flowering and after flowering. Monitoring Team evaluation of the Breeder seed production plot is done at midflowering or midmaturity stage of the crop. The fleshy pod samples are adjudged for pod quality, fresh vegetable market suitability and the consumer preferences and plants are selected. Pods and seeds harvested at maturity from selected individual plants are examined in the laboratory. Seed quality and nutritional quality testing are done. The quality assured seeds are labelled with Breeder Seed Label. GOT is conducted. If the seed quantity required is high, BSI and BSII stages of Breeder Seed Production could be organized.

Dolichos lab lab for grain /fodder purpose

Field lablab : *Lablab purpureus var lignosus*

Garden lablab : *Lablab purpureus var typicus*

Dolichos is grown for grain / seed or vegetable purpuses. Also utilized for forage and livestock fodder. It is autogamous annual herb. Outcrossing is neglibile. Grain and vegetable types are erect busy with low tendency for tendril formation. Flowers are white or purple. Traditional vegetable types are pandal or climber on bowers supported.

Table 1. Descriptor of *lab lab purpureus* var typicus

S.No.	Description	Characters
1.	Cotyledon leaf	Small, medium, broad
2.	Plant habit	Erect, semi-erect, spreading, trailing
3.	Tendril	Shy, medium, profuse
4.	Leaf lobe size	Small, medium, large
5.	Leaf-trifoliate leaflet margin	Smooth, wavy
6.	Leaflet tip	Acute, obtuse
7.	Stem-green pigmentation	Green, light green, red
8.	Panicle (raceme) stalk	Short, medium, long
9.	Fruit pod size	Narrow, medium, broad
10.	Fruit shape (l.s)	Tubular, semi-tubular, flat
11.	Fruit color	Whitish green, green, dark green
12.	Fruit shape at blossom end	Acute, obtuse, round
13.	Fruit/pod coat	Fleshy, medium fleshy, fibrous
14.	Seed number per pod	<3, 3 to 4, >4
15.	Immature seed shape	Elongate, round flat, globular round
16.	Seed color of matured pod	Whitish, green, greenish white, dullbrown
17.	Bean odour	Mild, medium, strong
18.	Seed arrangement in the pod	Parallel to the suture, vertical to the suture

Table 2. Characteristic of *Lablab purpureus* L. cultivars

1.	Emerging cotyledon colour
2.	Hypocotyl pigmentation
3.	Stem pigmentation
4.	Leaf anthocyanin
5.	Leaf colour- Palegreen, green, dark green, purple
6.	Leaf length
7.	Leaflet breadth
8.	Number of primary branches
9.	Number of secondary branches
10.	Days to flowering
11.	Days to maturity
12.	Number of pods/ plant
13.	Length of pod
14.	Breadth of pod
15.	Pod weight (g.)
16.	Weight of 100 seeds
17.	Number of seeds per pod
18.	Seed yield / plant
19.	Straw yield / plant
20.	Harvest index %
21.	Plant height
22.	Plant duration
23.	Plant habit
24.	Tendrils
25.	Flower colour
26.	Pod-flat, long flat, broad
27.	Vegetable type or grain type

(*Source* : Keerthi and Byre Gowda *et al.*, 2015)

23. Lentil (Masur)

(*Lens culinaris* Medik.) (2n:24)

Synonym : *Lens esculenta* Moench (1794), *Vicia lens* (L.) Coss. and Germ. (1845).

Family : Fabaceae

Lens culinaris (cultivated lentil) is an annual bushy herb, slender, almost erect or suberect, much-branched, softly hairy. Stem is slender, angular, upto 75 cm. tall. Leaf compound , alternate, pinnate, usually ending in a tendril; leaflets 4 to 7 pairs, alternate or opposite. Flowers are small, pale blue, purple, white or pink, in axillary 1 to 4 flowered racemes. Flowers are borne on a single peduncle and a single plant can produce upto 10 to 150 peduncles each being 2.5 to 5 cm. long (Muehlbauer *et al.*, 2005). Flowering proceeds acropetally. Flowers consist of 5 sepals, 5 petals comprising of one standard, two wings and two keels, 10 stamens in diadelphous (9+1), 9 stamens fused to form a staminal column. Ovary is short with 1 or 2 ovules, style curved and hairy on its inner surface. Pods are short, flattened, with curved beak. Pods contain mostly 2 seeds. Seeds are often light brown in colour and lens shaped., Weight of 100 seeds range from 2 to 8 g. Cotyledons will be red, orange, yellow, or green, bleaching to yellow, often showing through the testa, influencing its apparent color. Two types, namely; macrosperma, found mainly in the Mediterranean region and the New World (seed size ranging from 6 to 9 mm. in diameter and yellow cotyledons with little or no pigmentation), and microsperma (2 to 6 mm. diameter with red orange or yellow cotyledons) found in the Indian subcontinent. Germination is hypogeal and this keeps the developing seedlings below ground level which reduces the effects of freezing and other desiccating environmental conditions (Muehlbauer *et al.*, 2005). Lentil is self-pollinated crop (Ahlawat, 2012). The descriptor for lentil is given in Table 1.

Nucleus seed production: Basic seed sourced from breeder is grown in an area of 400 sq.m. adopting a spacing of 25 cm between lines and 5 cm. spacing between plants. The base population is grown with utmost good agronomic care for better expression of traits and for selecting true to the type single plants. Standard agronomic practices must be followed to raise the basic population. Lentil may be sown around mid-October. Individual plants are verified at different stages of growth till flowering. Off-types and deviant plants if any are discarded before flowering. Uniform and true to type plants, 1000 to 1200 in number are selected before flowering are tagged. Plants showing variation in flowering are removed along with the surrounding plants. Selected tagged plants are harvested separately. Pods and seeds collected from the individual plants are examined in the laboratory for trueness and uniformity and seeds of those plants not confirming to the pod and seed characteristics of the variety are rejected.

In the next cropping season, the individual plant progenies are grown in rows in two replications following recommended row spacing in a well prepared, fertile and disease free field that is not grown with lentil crop in the preceding 2 to 3 years. Isolation distance of 100 m. around the plot is provided from other lentil fields. Frequent and regular verification of the individual plants for the characteristics of the variety concerned from germination to flowering, pod development and maturity stage is done. Off-types and deviating type of plants are removed before flowering are removed. Progeny rows showing deviating plants at flowering are uprooted. Progeny rows showing uniformity, true to type nature maintaining varietal purity and stability of expression alone are labelled. The performance of the true to type progeny rows are evaluated for productivity. The true to type single plant progenies showing performance above mean + 2 SE are identified and the seeds from them are saved. The saved seeds of seed lot of the identified individual progeny are examined in the laboratory with reference to seed size, shape and colour. Any progeny showing deviation from the varietal seed characteristics is rejected.

The seeds of the selected true to type progenies are raised in clean field the subsequent cropping season adopting recommended spacing and production practices. Isolation distance of 100 m. is provided. Individual plants in each progeny are continuously monitored for the conformity of the characteristics of stem, branches, leaves, leaflets, tendrils, flowers and matured pods. Off-types are removed and the progenies showing deviating type of plants are discarded along with progeny rows on either side. At, maturity, superior plants in each disease free progenies are tagged and harvestd. Pods from them are removed and separately saved. In the laboratory pods and seeds separated are verified for trueness and uniformity and selected. Part of the seeds from each selected plants are bulked to make the nucleus seed. The remnant seeds are preserved for the next cycle of nucleus seed production. The seed quality of nucleus seed are tested. Nucleus seed labelling is done.

Breeder seed production: In the next ideal season, the nucleus seed is planted in a disease free, well prepared and fertile plot having recommended isolation distance of 100 m. from other field having same crop. The recommended spacing of 25 cm. between lines and 1 to 2 cm. spacing between plants. is adopted. Single plant per hill is maintained. The breeder seed production field is not grown with the same crop in preceding 2 to 3 seasons. Planting is done with the required seed rate leaving sufficient space after each bed for easy monitoring of the field. The crop is maintained free from diseases and insect pests from the establishment to the harvest stage. Individual plants are verified for trueness, uniformity and stability of the characteristics expressed by the stem, branches, leaf, flowers, and pods. Off-types and deviating plants noticed upto flowering stage are uprooted. Deviating plants occurring in flowering and post flowering stages are removed along with the plants around them. At midmaturity of seeds, visit of Monitoring team is arranged to adjudge the field and genetic purity of the crop. At maturity, individual plants are harvested and examined in the laboratory for true to type in pod and seed characteristics. Seeds from the selected plants are bulked to constitute breeder seed. Samples of breeder seed are tested for seed quality parameters and quality assured seeds are labelled with Breeder Seed Label. Grow out test is conducted with seed samples drawn from breeder seed to confirm the genetic purity of the seed lot. Laboratory estimation of nutritive quality of lentil seed is estimated.

Table 1. Descriptor of *Lens culinaris* (Masur)

S.No.	Characters	State
1.	Foliage intensity of green color	Light, medium, dark
2.	Stem anthocyanin coloration	Absent, present
3.	Leaf pubescens	Absent, present
4.	Time of flowering (days)	Early, medium, late (<60), (60-80), (>80)
5.	Leaflet size length	Small, medium, large
6.	Plant growth habit	Erect, semierect, horizontal
7.	Flower color of standard	White, pink, blue, violet, others
8.	Plant height (cm.)	Short (<40), medium (40-60), high (>60)
9.	Pod anthocyanin color	Absent, present
10.	Seed size (100 seed weight) (g.)	Small (<2.0), medium (2 to 2.5), large (2.6 to 3.0)
11.	Seed testa color	Grey, green, pink, brown, black
12.	Seed testa mottling	Absent, present
13.	Cotyledon color	Yellow, olive green, orange

(*Source:* PPV & FRA, 2007)

24. Lima Bean

Phaseolus lunatus L (2n:22)

Family : Fabaceae

This is a herbaceous annual, erect, and bushy. Not commonly grown in India. Plants grow to 0.9 to 1.0 meter. Leaves are trifoliate and alternate. Inflorescence is axillary and terminal flowering raceme. Flower is hermaphrodite, papilionaceous, calyx five, campanulate, corolla five, stamens 10 in diadelphous arrangement and ovary superior. Fruits are 5 to 12 cm. long and 2 to 4 seeded, dehiscent pods. Seeds are small and plumpy and are in different colours. Seeds are cooked as vegetable. It is a self pollinated crop.

The protocols for nucleus seed and breeder seed production are essentially the same being prescribed for French bean and other self pollinated beans.

25. Pigeon Pea (Red Gram)

(*Cajanus cajan* L. Mill sp.) (2n:22)

Family : Fabacea

Pigeonpea is a short to long duration maturity crop. Stem erect, woody, showing green colour to purple streaks. Leaves trifoliately compound, narrow lanceolate leaflets, inflorescence axillary or terminal racemes on long peduncle; flowers are papilionaceous, calyx 5, gamosepalous, corolla 5, polypetalous with standard, two wings and two keel petals, stamens 10 (9+1), diadelphous, ovary superior, bicarpellary and unilocular. Fruit is a pod, variable in shape, size, colour and pubescence. Seed colour ranges from cream to black. Cotyledons deep yellow to light yellow in colour.

Floral structure of pigeonpea was initially adapted for self pollination which changed over time to partial out breeding. The duration of flowering varies according to the climate and environment. Maximum anthesis occurs between 10 and 12 hrs. The flowers that open in the evening remain open in the night and close before noon on the next day. The stigma receptivity starts 68 hours before anthesis and continued for 20 hours after anthesis. In 90% of the flowers, pollination is over before the flower opening. (Vishwa Dhar *et al.,* 2003). The characteristics of pigeonpea are presented in Table 1. Singh and Singh (2016) have published an account on heterosis and inbreeding depression in pegeon pea.

Nucleus seed production: Raise the breeder source seeds of the variety in an area of 2000 sq.m. retaining only one plant per hill. The field is well protected by distance isolation of 100 m. from other pigeonpea crops. Careful examination is made at preflowering and off types and all the atypical plants were uprooted.

At flowering, rouging is done to remove off-types and plants showing deviations. Representative plants 1000 to 1500 in number, showing uniformity and true to type for phenotype and flowering nature and conforming to the description of the variety concerned are selected and labelled. At maturity, pods are collected and examined for pod setting and pod filling, pod colour, number of seeds per pod and other characteristics. Seeds are separated from individual plants and examined in detail matching with the varietal description. Seeds from the laboratory selected plants alone are saved and labelled (Fig1). Different system of male sterility utilised for hybrid seed production are illustrated in Fig. 1a.

Breeder seed production: In the next season, in 4 meter long, 2 rows progeny plots, seeds from each selected individual plants are sown in progeny rows in two replications (100m isolation). Critical examination is made on every plant from germination to pre-flowering stage. Progeny rows showing off-types and variant plants are uprooted. At flowering time, close monitoring is made to identify progeny plots showing deviant plants and the progeny rows as such and the adjoining rows are uprooted to prevent vicinism. Early and late flowering plants are removed from all the progeny rows. At maturity, progeny rows performing above mean + 2SE are selected and 10 to 15 superior plants in each selected progeny are labelled. Plantwise pods and seeds are examined in the laboratory and selected. Part of the seeds from each of the single plants are bulked to constitute Nucleus Seed. In the following favourable season, nucleus seed is sown in a plot not grown to redgram in the past 2 to 3 seasons. Isolation distance of 100m. is provided. Single plant per hill is maintained. All precautions and monitoring procedure and thorough roguing and removing deviant plants from sowing to harvest are made. True to type and uniformity are adhered to. Monitoring team visit in arranged. The seeds produced from the breeder seed production tested for nutritive content.

Genetic Male Sterile line (GMS)

Nucleus seed production of GMS line: The Genetic Male Sterile plants(GMS) are female fertile and male sterile. The source seeds are obtained from the breeder or from the breeding organization and when the plants are raised, the plants will be in equal proportion of male fertile and male sterile plant genotypes.

Sourced seeds of GMS line (female parent) are raised in an area of 2000 sq.m that is isolated by 800m. distance. Plants are examined individually for true to type from germination stage onwards till flowering initiation. Off-type and deviating plants are uprooted. When the flowering commences, carefully select and label the male sterile plants that are similar to the fertile plants for all the characteristics and conforming to the descriptor. The anthers in the genetic male sterile line plants will be pale white and shrivelled after flower opening and are male sterile. Paired crosses between male sterile and male fertile plants are made by hand pollination under caged condition to avoid any chance pollination with alien pollen and to maintain genetic male sterile plants in genetically pure condition. At least, 500 pair wise crosses are made under caged condition between genetic male sterile plants and male fertile plants with total similarities.

Seeds are harvested separately from each paired cross and progeny of individual paired crosses are raised in clean field with 1600m. isolation, the original genetic male sterile line is raised once in every 10 rows maintaining only single plants per hill. The paired cross progenies are now critically evaluated. Deviant types of plant, if any, are uprooted. Uniform individual progenies that are similar to the original genetic male sterile line raised for comparison are covered with nylon net cages and are maintained. The progenies will segregate in 1:1 ratio for male fertility and male sterility. In the progenies protected by nylon cages, the genetic male sterile plant is pollinated with fertile plant counterpart by hand pollination and labelled. The pods set in the male sterile plants are collected on individual paired cross basis and critically examined for pod set, pod and seed filling and all seed related characteristics. Seeds set in the male sterile plants from the different paired crosses are now selected and based on purity and true to type nature seeds are bulked to constitute the nucleus seed.

Breeder seed production of GMS line (female parent): The strategy is to identify true to type GMS plants to develop breeder seed. This is achieved by raising breeder seed production plot in isolation (1600m.) and identifying male sterile but female fertile plant alone and tag them and sib mate with male fertile plants (Fig. 2). Nucleus seed set in the male sterile plants is sown in plot with isolation distance of 800 m. All recommended good production protocol is followed. Off-types and deviating type of plants are roughed out. Very early and late flowering plants are removed to maintain uniformity in flowering. At the time of flowering, flowers in each and every plant are examined to identify the male sterile or male fertile nature. Pollen production is not observed in male sterile plants while abundant pollen production by well developed fertile anther in the male fertile plants is noticed. The anthers are translucent white with sterile pollen in male sterile plants and this is judged well by collecting 2 to 5 flowers and opening and visualising and examining for shrivelled anthers, anther colour and sterile pollen. Fertile plants have yellow anthers. The male sterile plants are identified and tagged till 90% of the male sterile plants are tagged. The rest of the GMS plants are removed. Pods set on the true to type male sterile plants are harvested. The natural crossing with pollen from fertile sibs facilitate seed set in the GMS plants. Pods are harvested first from the tagged male sterile plants to avoid mixing of seeds from fertile plants. Fertile plants are also cut by this time and removed. Visit of Monitoring Team is arranged in mid flowering and early pod development period.

Seeds set on individual GMS plants are examined for similarity with the original GMS line. The labelled seed stock is bulked to constitute breeder seed. In the laboratory, seed testing is done to evaluate the different seed quality parameters. The quality assured breeder seed is then labelled with breeder seed tag. GOT is conducted with seed sampled drawn from breeder seed. The seed samples from the breeder seed lot is evaluated for nutritional qualities.

Table 1. Descriptor of pigeonpea

S.No.	Characters	States	Stage of observation
1.	Plant type	Compact/semi-spreading	Vegetative
2.	Growth habit	Determinate/indeterminate/ semi-determinate	Flowering
3.	Pigmentation on stem	Green/purple	Vegetative
4.	Leaf shape	Oblong/obovate	Flowering
5.	Days to flowering	Early/medium/late	Start of flowering
6.	Color of the base of petal	Light yellow/yellow/orange yellow/purple /red	Flowering
7.	Pattern of streaks on petal	Absent/sparse/medium/dense	Flowering
8.	Pod color	Green/green with brown streaks/green with purple streak/purple/ dark purple	Premature pods
9.	Pod pubescence	Absent/present	Full podding
10.	Seed color	Cream/brown/dark brown/grey/purple	Maturity
11.	Seed shape	Oval (egg shape) / globular (pea shape)/angular)/elongate	Maturity
12.	Seed size	Small/medium/large/extra large	Ripe seeds

Source: Viswa Dhar *et al.* (2003) and PPV & FRA (2007)

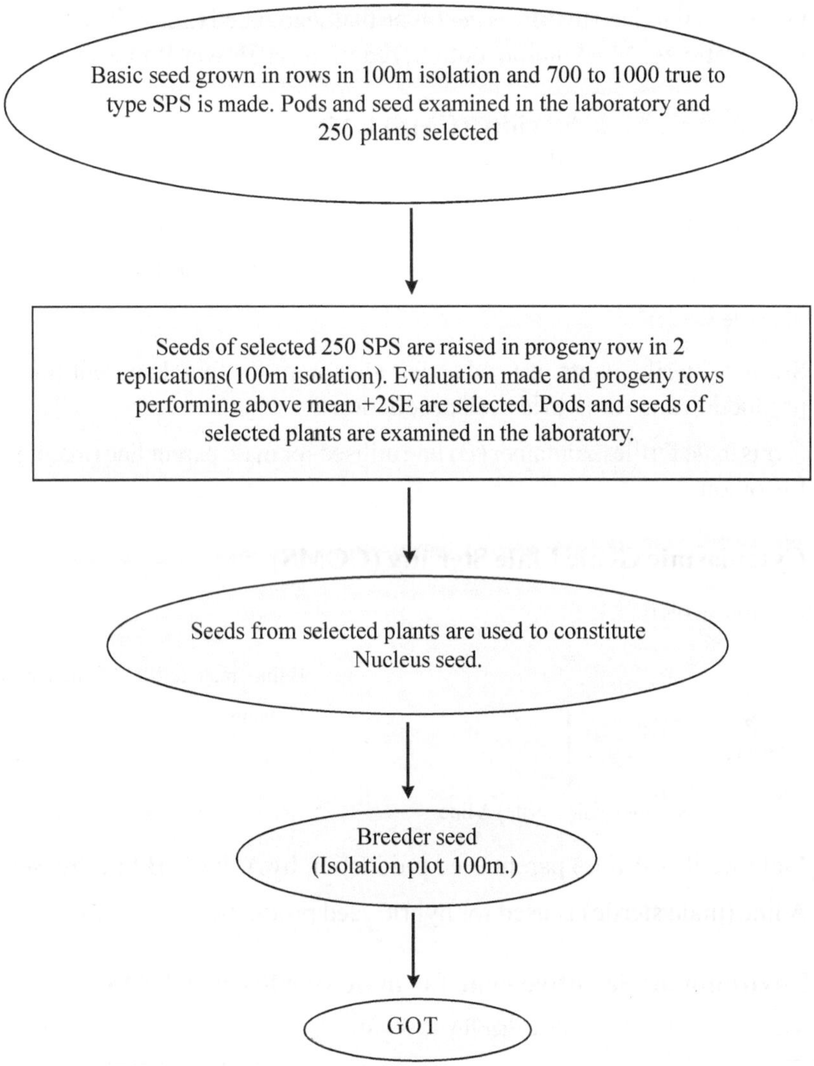

Basic seed grown in rows in 100m isolation and 700 to 1000 true to type SPS is made. Pods and seed examined in the laboratory and 250 plants selected
Seeds of selected 250 SPS are raised in progeny row in 2 replications(100m isolation). Evaluation made and progeny rows performing above mean +2SE are selected. Pods and seeds of selected plants are examined in the laboratory.
Seeds from selected plants are used to constitute Nucleus seed.
Breeder seed (Isolation plot 100m.)
GOT

Fig. 1. Nucleus and Breeder Seed Production in Pigeon Pea (varieties and male parents)

I. Genetic Male Sterility (GMS)

X

Msms (Sterile) ↓ Msms (Fertile)

msms: Msms
Sterile, fertile
1:1

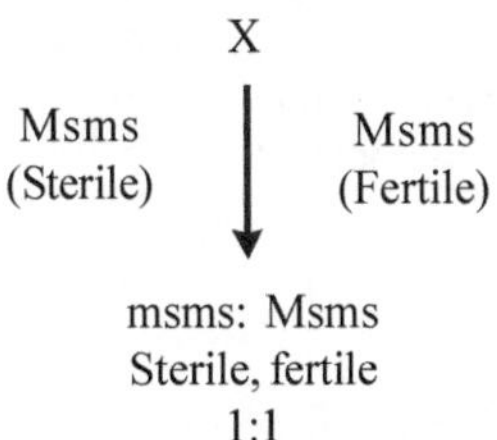

1. Utilise msms as female parent line for hybrid seed procuction

2. GMS is maintained by Msms genotype that segregated to sterility and fertility in 1:1 ratio eg. Muskmelon, cotton, redgram, safflower Rapeseed mustard

II. Cytoplasmic Male Sterility (CMS) CMS

1. Srr male sterile plants utilised as female parent for female parent line seed production and for hybrid seed production
2. N rr is male fertile maintainer (B) line utilised for male parent line (maintaines) Eg. onion.

III. Cytoplasmic Genic Male Sterility (CGMS)

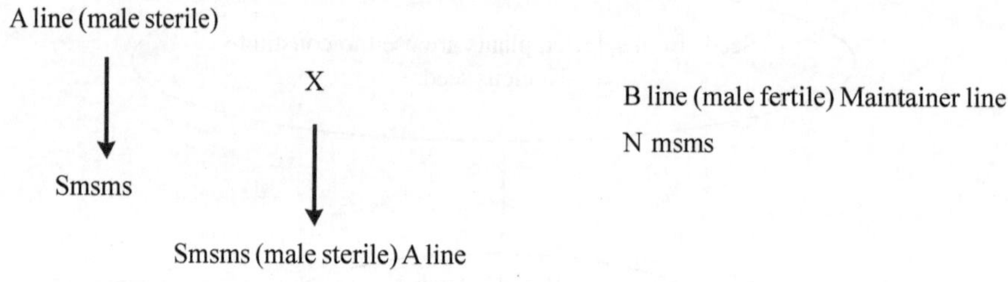

1. Male sterile (A line) parent seed production by A line < B line crosses.
2. A line (male sterile) is used for hybrid seed production.

IV. Environment Sensitive gene for male sterility or EGMS

Temperature sensitive Genic Male Sterility (TGMS)

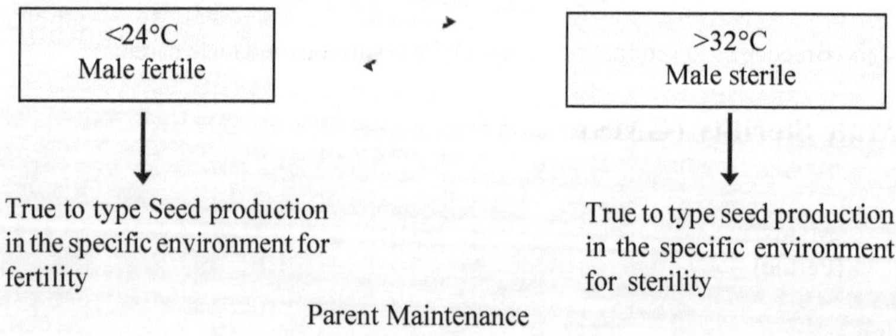

Examples: Rice, Pistillate plants in castor, now proposed in pigeon pea.

Fig.1a. Male Sterility System and Seed Production

Raise 500 plants from seeds of GMS female parent with isolation of 1600m.

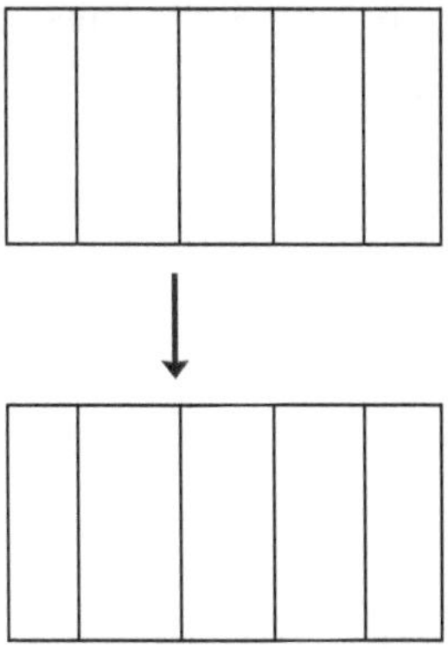

Make paired crosses of GMSx male fertilie sib plant (GMSx fertile counterpart) (Inside the net cages)

↓

Paired crosses

Segrgating in 1:1 proportion for GMS and male fertile nature

Make paired crosses with true to type GMS and male fertile sib couter part. Hard pollination inside nylon net cage.

Fig. 2. Nucleus and Breeder seed production in GMS (Female parent) line in pigeon pea.

Seeds set in GMS plants are harvested. Laboratory examination is done, selected and bulked the seed to constitute nucleus seed. This type of hand pollination within nylon net cages can be repeated to the extent needed year by year or ones in 2 to 3 years.

In Nucleus seed production in isolated plots (800m.) is allowed for natural cross pollination of GMS plants with pollen fertile sibs. Collect the seeds set in GMS plants.

Breeder Seed Production in isolated plots (800m.) with natural cross pollination. Collect the seeds set in the GMS plants.

Nucleus and Breeder Seed Production in A, B and R lines in pigeonpea/ Redgram.

Sourcing: Source the seeds of A, B and R lines from the breeder or the Institution concerned.

Growing the Source Material: Raise the A and B line in 1:1 row ratio in isolation of 1600m from other redgram varieties/ crop with recommended spacing of 45 cm x 20 cm. Critically examine the individual plants in A and B lines for conforming to the prescribed diagnostic characteristics, plant height, branching, foliage, flower, pod and seed characteristics. Deviating plants and off-types are removed. Uniform looking plants only are retained. Single plant per hill is maintained. Very early and late flowering plants are uprooted.

The A line plants are examined for the presence of pollen shedders and in B line, the sterile plants and if found they are uprooted. Plants of A and corresponding B line are isogenic and look similar but for the pollen fertility in B line and pollen sterility in A line. Anthers in the A line will be shriveled with sterile pollen content and the anthers in the B line are plumpy and filled with pollen grains.

Select true to type healthy plants in A and B. Make 300 paired crosses (A×B) with selected plants in A line and B line. Self fertilise the corresponding plants in B line in each cross of A×B. Save the seeds of individual plants in A (A×B) and B cross wise and serially number A1B1, A2B2, …AnBn

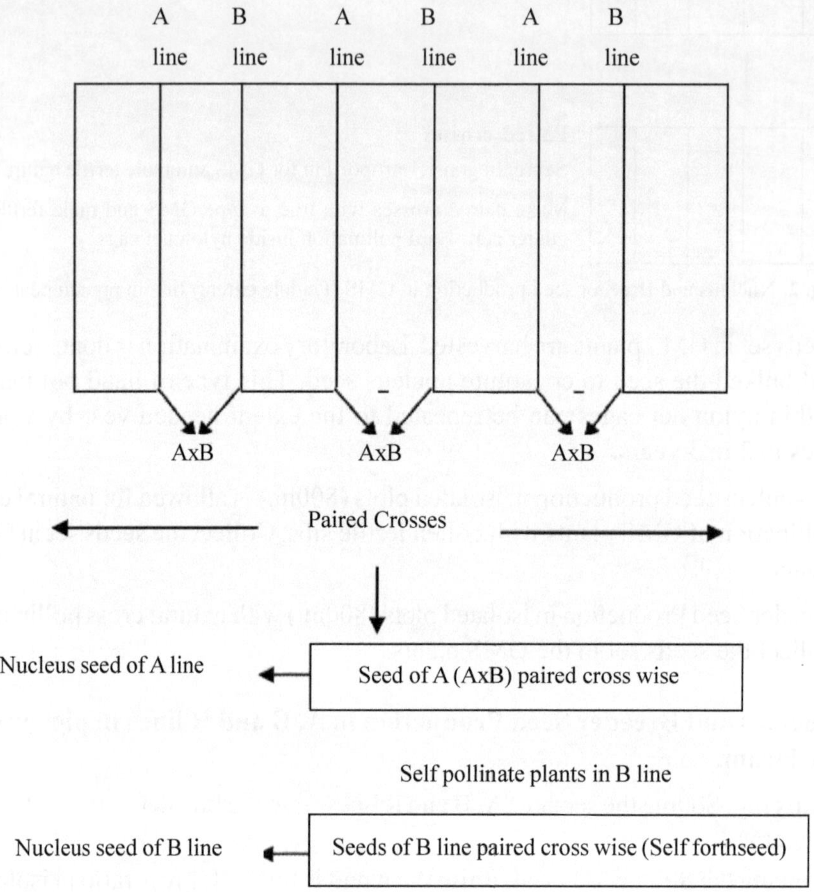

Fig. 3. Nucleus Seed Production in A and B lines

Nucleus Seed Production in A and B line: In the following season, raise A seeds of A line (A × B) and B line in 1:1 row ratio in the pairwise manner in the field plot with 1600m. isolation from other redgram crops. All plants in A and B are examined in different growth stages for true to type and the conformity with the characteristics prescribed. Deviating types of plants and off types in A and B line are uprooted. Plants in A line in all the paired crosses are examined for the presence of pollen shedders and if found the line along with B line of the A×B paired crosses are discarded. Paired progenies of A×B are evaluated for synchrony in flowering in A and B uniformity and vigour and the seed set in A line plants in each paired cross. Paired crosses performing better than check A×B lines are selected at maturity. Seeds of the self fertilised B plants in the selected paired crosses of A ×B are harvested 7 days earlier to the harvest of A line and labelled. Seeds of A line and B line are examined in the laboratory for seed characteristics and selected plant wise (A×B). Seeds from each of A line plants are bulked separately. Also the seeds set in plants of B line are bulked together. Part of the seeds in each A line and B line is bulked to constitute the Nucleus seed. The remnant seeds of A and B line are utilized for the next Nucleus – Breeder cycle of seed production (Fig. 3).

Cages made of Netlons/Nets cages can be made use of to produce A×B crosses in limited numbers. Large quantity of A×B crosses can be also carried inside large size net houses.

Breeder Seed Production in A and B line: Seeds of A line and B lines are sown in 3:1 (A:B) rows in a fertile well prepared field isolated by 1600m. from other redgram fields. All round the plot, 5 rows of redgram of B line is sown to provide pollen supply adequately. The crop is grown adapting the recommended crop production management schedules. Individual plants in A (A×B) lines are examined for uniformity in plant type, vegetative and floral parts and free from diseases and pest, off-types and deviating types of plants. Plants are monitored for conformity to the diagnostic characteristics. Pollen shedders, if noticed in A line, the plants are uprooted. Early and late flowering plants are removed. Visit of Monitoring Team is arranged at mid maturity of the A and B lines. At maturity roguing for sterile and late pod setting plants in B and A line is carried out. Pods of B line are harvested separately 7 days earlier to that of pods in A line. They are separately dried and seeds separated. Laboratory examination is made of individual plants for pod and seed characteristics and selected. Seeds from plants in A line and from B line are saved separately, dried and packed. Samples of seeds are tested for seed qualities. Quality assured seeds are then labelled as Breeder Seed of A line and B line respectively. Nutritional and other qualities are estimated with the seed samples (Fig. 4).

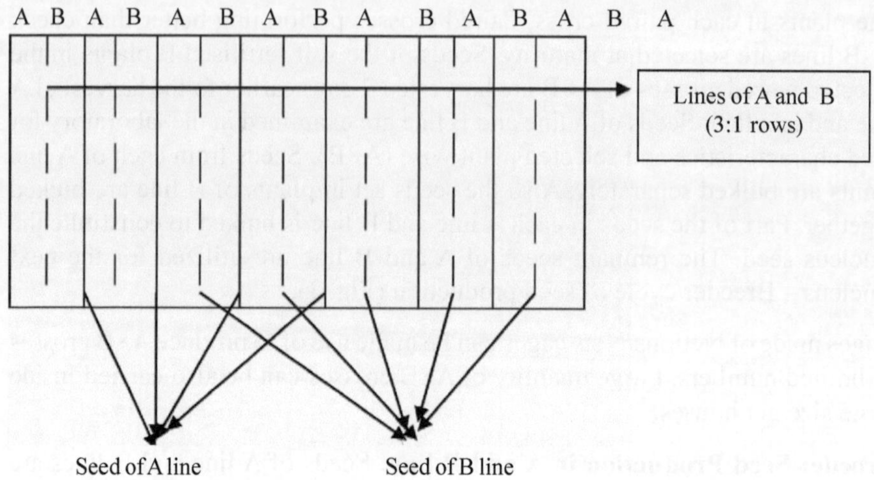

Fig. 4. Breeder seed production

Nucleus and Breeder seed production in R line (Restorer line): Raise large number of plants in R line in isolation of 100m. in fertile, disease free field. Examine for true to type and uniformity for prescribed characters at different stages of plant growth to maturity stage. Deviating types of plants and off type plants, disease and pest affected plants are uprooted and discarded.

Self fertilise 500 plants conforming to the characteristics of R line at the time of flowering. At maturity, collect the pods and seeds and after laboratory examination, pods and seeds of selected 250 plants are saved plantwise.

Nucleus seed Production: In the following season, raise 250 progeny rows of R line in 2 row plots in 2 replications. Retain portion of self fertilised seeds in each plant as remnant seeds. Evaluate the progeny rows for true to type, uniformity, vigour and disease and pest free nature. At maturity, the progeny rows are evaluated for economic characters, high pod number and high seed set, number of pods and seeds per pod; Select the progeny rows performing above mean ± 2 SE for yield and within each such progenies select 250 to 300 superior plants. Save pods or seeds plantwise and examine in the laboratory for true to type. Select seeds of 200 plants.

In the next season, progeny rows of selected 200 plants in R line are raised in an isolated (100m.) well prepared disease free fertile soil field. The progeny rows are examined for vigour, field establishment of seed, fast growth and high degree of expression of yield related traits and maximum expression of flower traits and anther development. Plants not satisfying the conditions are uprooted. Simultaneously, part of nucleus seeds of A line of redgram is sown in another isolated plot and all the plants are examined for free from pollen shedders. At flowering selected plants in A line are crossed with selected plants of R line performing above mean +2 SE by aided hand pollination. The plants in R line used for A×R crosses are self fertilised. Seeds set in A×R crosses are saved and labelled as A1×R1, A2×R2, _____, A200×R200. Seeds set in A×R crosses and seeds of self fertilised R line plants are saved separately (Fig. 5).

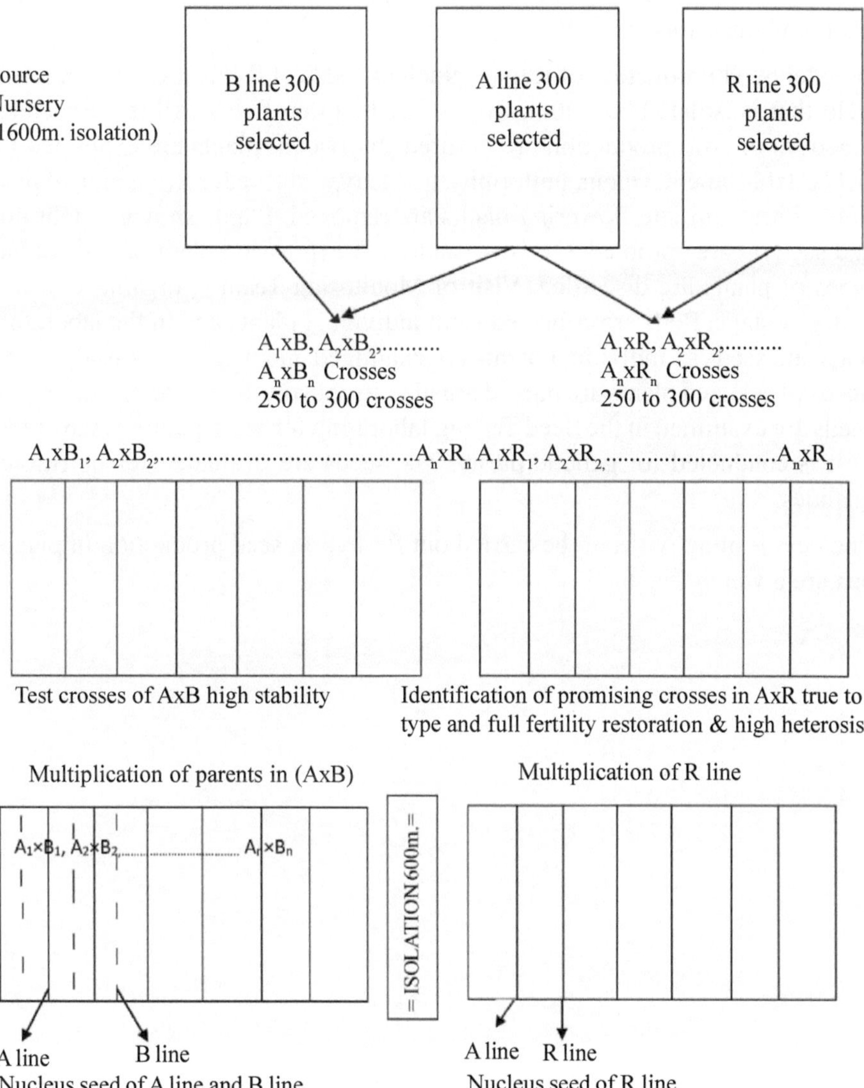

Fig. 5. Nucleus seed Production of A, B and R line

In the next season, seeds of A×R crosses are raised in a well prepared fertile field and examined for heterosis and enhanced expression of plant characteristics, panicle and flower production and pod sets. Hybrids (A×R) progenies showing high heterotic expression and high pollen fertility restoration and high pod and seed yield are recorded. The higher yielding A×R crosses are identified. The self fertilised seeds of the plants of R line corresponding to the selected highly heterotic A×R crosses are then bulked together to constitute Nucleus seed of R line originating from the R line with proven pollen fertility restoration ability and high combining ability.

Breeder Seed Production of R line : Nucleus seeds of R line are sown in fertile field that is isolated by 100m. from other redgram fields. All recommended practices of crop production are ensured. Individual plants are examined for field establishment, vigour, uniformity, true to type and better expression of plant traits. Early and late flowering plants are removed. Plants showing poor pod and seed set are uprooted. Off types and weak expression plants and deviating types of plants are discarded. Visit of Monitoring Team is organized at mid maturity stage. Pods are collected from individual plants and in the laboratory pods and seeds of individual plants are examined and verified for true to type. Seeds of selected plants are bulked together to constitute Breeder Seed. Sample seeds are examined in the Seed Testing laborating for seed qualities. Grow Out Test is conducted for genetic purity. The seeds are evaluated for nutritional qualities.

The details of activities to be carried out for hybrid seed production in pigeon pea are given in Fig. 6.

A line B line	R line
S (msms) × N (msms)	N (MsMs) Maintain R line by raising 300 progeny rows. Selecting for true to type high productivity. R line progeny selection above mean +2SE
Make 300 paired (A×B) crosses	Make A×R cross - 300 crosses
1. Select A line plants that are true to type and high pollen sterility	Evaluate 300 paired crosses of A×R
2. Remove off-types and pollen shedders in A	Select high performing heterotic, fully fertile superior A×R crosses. Identify superior hybrid. Self fertilise R plants in the corresponding R lines of the superior A×R crosses to identify and select high pollen fertility restoring and high sca parent for hybrid pigeonpea.
3. Sterile plants in B line discarded Make A×B	
Select true to type plants in A and B	Develop nucleus and breeder seed by isolation growing of R line plants with proven fertility restoration and high sca /heterosis.
Parent line A×B maintenance in isolation (1600m.)	
Nucleus seed: A and B line in 1:1 row	
Breeder Seed: A and B line in 3:1 row in breeder seed production	

Fig. 6. Hybrid Seed Production in Pigeon Pea utilizing A, B and R line parents

26. Soybean

(*Glycine max* (L.) Merrill) (2n:40)

Family : Fabaceae

Herbaceous annual, temperate & subtropical adaptation. Habit erect, rarely climbing, presence of pubescence in all plant parts. Leaves are alternate; leaflets three, ovate to lanceolate. Inflorescence axillary racemes. Flower small upto 15 per raceme. Flower colour white or violet purple. Papilionaceous flower, stamens 10, monadelphous, ovary with short style. Ovules few, 2 to 5. Fruit is a dehiscent pod with globular seeds that are dull white, cream or black. There is a very high percentage of self fertilization due to the presence of cleiostogamy. The natural outcrossing is generally less than one percent. (Hussain *et al.,* 2003). The descriptors are presented in table1.

Nucleus seed stage I: A population of the specified variety of soybean from breeder source seeds is established in an area of 400 sq.m. with 10m. isolation. From this population, a minimum of 1000 true to type and uniform looking plants are selected and labelled and compared to the morphological description and identity. The selected single plants are labelled and features of pods and seeds are examined in the laboratory and then seeds are separated from selected plants, labelled and stored at 25°C and 50 to 60% RH. Examination of seeds should be very critical even to deduct small variations.

In the next season, well drained and fertile field is selected and well prepared for growing soybean crop. Seeds of single plants are sown in single progeny rows of 4 m. length in two replications along with the seeds of the variety once in 10 rows for comparison. The field is isolated by 10m. from other soybean crops. In each block, 50 progenies are raised. The progeny rows are critically studied for uniformity and true to type nature. When off-types and deviant plant types are deducted, the entire progeny row showing them is uprooted. The progeny rows showing off-types and deviant plant types at flowering and post flowering are removed along with removal of the progeny rows on either side. The progeny rows that are remaining uniform,stable and true to type are serially numbered and each row harvested separately. Progenies performing above mean + 2SE are selected. The pods and seeds of each selected progeny are carefully examined in the laboratory to observe deviant types of plants and such progenies are rejected. The cleaned seeds from different progenies are bulked to form nucleus seed. The remnant seeds of the single plants that are finally selected on progeny performance basis are stored as a future source seed stock for raising plant progenies to constitute nucleus seed.

Nucleus Seed stage II (NSS II): The bulk of the individual plant progenies is utilized for breeder seed production,when the breeder seed requirement is high. Nucleus seed stage II production is organized by sowing bulked seeds of the selected single plant progenies. The seed production in nucleus seed stage II is done exactly following the entire protocol as being adopted similar to the Nucleus seed stage I.

Breeder seed production: Isolation distance required is 10 m. from other soybean crops. Seed source is either from Nucleus seed stage I or II. In exceptional circumstances, breeder stage I can be also used to produce Breeder seed stage II by following the procedure of selection for purity, uniformity and true to type of individual plants. Off-types and plants showing deviations, disease and pest affected plants are removed. Monitoring Team visit is arranged to inspect the breeder seed crop. Laboratory selection of plants for pod and seed characteristics conforming to the cultivar characteristics is followed after the pods are harvested on single plant basis. The seed separation, drying and processing are done as per the recommended practices for the seed crop of soybean.The breeder seed is tested for the seed quality parameters and seed quality is thus assured. Breeder seed is labelled with the breeder seed label. GOT is conducted following the prescribed protocol for the soybean crop. Protein and oil content are estimated. The characteristics of soybean are detailed in Table 1.

Table 1. Descriptor of soybean

S.No.	Characters	States	Stage of observation
1.	Hypocotyl anthocyanin coloration	Absent/present	Seedling
2.	Plant growth habit	Erect/semi-erect/semi-erect to horizontal	Vegetative
3.	Shape of lateral leaflets	Lanceolate/triangular/pointed/ovate/rounded ovate	Vegetative
4.	Size of lateral leaflets	Small/medium/large	Vegetative
5.	Time of flowering	Early/medium/late	Start of flowering
6.	Plant growth type	Determinate/semi-determinate/indeterminate	Reproductive
7.	Flower color	White/violet	Start of flowering
8.	Presence of hairs on pod	Absent/present	Reproductive
9.	Pod color of hairs	Grey/tawny	Reproductive
10.	Plant height	Short/medium/tall	Vegetative/reproductive
11.	Seed size	Small/medium/large	Harvest maturity
12.	Seed shape	Spherical/sphericalflattened/elongated/elongated flattened	Harvest maturity
13.	Colour of test a excluding hilum	Yellow/yellow green/green brown/black	Harvest maturity
14.	Seed lusture	Shiny/intermediate/dull	Harvest maturity
15.	Seed hilum color	Grey/yellow/brown/imperfect black/black	Harvest maturity

Note: Description is based on Hussain *et al.* (2003); PPV & FRA (2007)m

Oilseeds

27. Castor

(*Ricinus communis* L) (2n:20)

Family : Euphorbiaceae

Castor is a cross pollinated crop. Stem is pithy, glabrous with prominent nodes. Bloom is present on stem and leaves with varying intensity in different varieties. Leaves are alternate and palmately lobed. Inflorescence is monoecious terminal panicle with female flowers in the upper part and male flowers at the base of the panicle. In a few varieties, the male and female flowers are interspersed or mixed together. In male flowers, perianth is splitting to 3 to 5 segments. Stamens are many borne on branched filaments. Anthers are usually round and yellowish. In female flowers, ovary is superior, tricarpellary, style is short and bifid or trifid and branched. Fruit is a schizocarp dehiscing into three one seeded units. Seed testa is hard, smooth and coloured. The embryo is tiny, cotyledons two and papery and endosperm is large. Castor is wind pollinated. Pollen dispersal depends on the wind velocity. Besides, insect pollination also occurs.. A minimum isolation distance of 3000 m. for female parents and 1600 m. for varieties is recommended to produce genetically pure seed. (Vishnuvardhan and Hanumantha Rao, 2003). The descriptors are given in table1.

Castor varieties and male lines: Source seeds of variety/ male line are sown in an area 1000 sq.m. in field with 1600 m. isolation from other castor crops . At pre-flowering stage 500 to -600 true to type plants are selected. Deviant plants in respect of vigour, internode length, stem pigmentation, shape and colour of leaf, bloom (nil, or double or triple) and days to first flowering and off types are removed. Very early and late flowering plants are culled out. In the selected 200 plants, primary, secondary and tertiary spikes are self fertilized. Self fertilized spikes are harvested and examined in the laboratory for capsules and seeds characteristics. Deviant types and off types are removed. In the next season, about 500 progeny row plots are raised in two replications with self fertilized seeds in primary, secondary and tertiary spikes of selected plants, maintaining 20 to 25 plants in each progeny row. The progenies are examined upto pre-flowering stage for all the vegetative characteristics. The off-types are rogued at primary spike formation stage and 250 uniform looking true to type progenies are selected and labelled. In each of the progenies, 5 to 10 single plants are selected and 2 to 3 spikes other than the primary spike are self fertilized. At maturity, the primary, secondary and tertiary spikes are examined for all details and harvested and capsule number, spike density, seed yield and oil content are recorded. All the spikes are carefully labelled plant wise.

Castor–Pistillate line: Castor is sexually polymorphic and the sex expression and the sex reversal is highly influenced by the environment conditions. Pistillate line is a recessive mutant with interspersed staminate flowers (ISF) providing pollen for self fertilization. There is no role for maintainer line in this system. Inbred pistillate lines in castor are available. Castor VP1 pistillate line (ff) is often used as female parent to develop heterotic hybrids in castor. VP1 line possesses genes for production of ISF usually at 32°C and above. Usually hybrid castor seed production utilising VP1 is taken up in kharif season which is cool enough to encourage stable female /pistillate nature expression upto 4th and 5th order of pistillate spikes.

In the summer season sown female (pistillate) lines, staminate flowers (ISF) arise in the internodes of the panicles when the temperature rises beyond 32°C (32°C & above) and provides pollen for pollination of the pistillate (ff) flowers to set seeds. The seeds, thus set, will be true breeding for pistillate phenotype only. The pistillate parent seed production is taken up in summer season by sowing the seeds in Jan-Feb with isolation distance of 2000 m. from other castor crops. As the plant develops, and reaches the reproductive phase, the temperature rises up to 32°C or more and this condition is favourable for inducing staminate flowers (ISF). This behavior is taken advantage to produce seeds of pistillate lines without pollinator or maintainer line. In cool seasons, the pistillate nature is stable.

Nucleus seed production in Pistillate line: Seeds are obtained from the breeder source stock and 2000 plants of pistillate line are raised in summer season in plot with isolation of 2000 m. from other castor crops. The spikes selected in true to type pistillate plants are self fertilized and seeds collected on single plant basis. In the next favourable season, (Jan sowing) 500 progeny rows plots with 3 to 4 rows per plot are raised in isolation of 2000 m. The characteristics in vegetative phase for vigour, stem colour, blooming, leaf colour and shape are critically examined in individual plants. The deviants and off types are rogued out at preflowering, plants expressing male flowers are pulled out before flower opening. At flowering stage, 400 true to type plants are selected (3 to 5 plants in each progeny) for uniformity in flowering and for true to type characteristics based on foliar features, node number and female sex expression (complete pistillate nature) in the primary rachis. In the selected plants, primary spike, secondary, tertiary and quaternary spikes are individually self fertilised. At maturity stage, the primary spikes that are self fertilised are examined for capsule set. If capsule set is noticed in the primary spike that particular plant is rejected. Only from such of the selected plants in which no capsule set is noticed in the self fertilized primary spikes, the self fertilised seeds from the secondary, tertiary and quaternary spikes are collected and examined for the extent of capsule set. The capsule and seeds are then thoroughly examined for true to type of the prescribed characteristics and uniformity. The seed packets are then numbered progeny wise and spike wise and carried forward.

In the following summer season, the same procedure as in the previous summer season is followed. Around 100 to 125 progeny row plots are raised in 2 rows plots, in blocks replicated twice. In each block, 50 progenies are sown with isolation of 2000 m. from other castor crops. In each progeny plot, 15 plants are labelled based on the primary spikes bearing pistillate flowers alone and all other prescribed characteristics. Rest of the plants are removed. Meticulous verification is done for occurrence of any revertants in the late formed spikes such as secondary, tertiary and quarternary and all such plants with late revertants are removed in each and every progeny plots. Meanwhile, spikes of secondary, tertiary and quarternary are self fertilised and at maturity, self fertilised seeds are collected from selected plants that have not shown revertants upto pentanary spikes. The self fertilised spikes are examined in the laboratory for conformity of capsules and seed characters.

Seeds of selected spikes bulked from the pistillate plants are constituted as nucleus seed for producing breeder seed. Seeds from the selected self fertilised spikes are grown as progeny rows in the next summer season for continuing the nucleus seed production cycle.

Breeder seed production in Pistillate line: Nucleus seed is sown in summer season in breeder seed production plot that is isolated by 2000 m. from other castor growing fields. Good agronomic practices are followed to raise healthy and disease free crop. At different stages of crop growth, the characteristics of the pistillate line for juvenile, stem and foliage pigmentation, vigour, leaf features and other vegetative traits are examined .The deviating type of plants and off types are discarded. Plants expressing male flowers are uprooted. Early and late flowering plants are removed to ensure uniformity in flowering. At flowering, the primary panilcle emergence and its true to type nature are verified. The individual plants showing complete pistillate expression in the primary panicles alone are retained and rest of the plants are removed. Very careful observations are made to notice any sex revertants in the primary and late formed spikes and such revertant plants are uprooted. In the selected true to type plants, the primary and the spikes of the secondary, tertiary, quarternary and pentenary order are self fertilized. At maturity, the self fertilised primary panicle of the single plants are verified for no seed set and such plants showing no seed set only are retained for further verification for not showing revertants upto pentenary spikes. The self fertilised seeds are collected from the selected plants that have shown stability for pistillate nature. The self fertilised spikes are examined in the laboratory for conformity of capsules and seed characters in the spikes of secondary, tertiary and quaternary spikes. Visit of Monitoring Team is arranged at the midmaturity stage. Seed quality parameters testing and GOT are conducted.

Breeder seed labelling is done. Oil content is estimated in the breeder seed. Analysis on agronomic traits is made and mean and SE are computed and plants showing value above mean +2SE are selected. The harvested seeds of self fertilised spikes in each selected plant progeny are bulked and 100 to 120 progenies which exceed general mean by 2SE. Around 50 seeds in each of the selected plants with exceeding general mean by 2SE are utilised as source basic seeds for further maintenance and bulk the remaining selfed seeds from selected progenies to constitute breeder seed to take up breeder seed production (Fig. 1-3).

Table 1. Descriptor of castor

S. No.	Characters	States	Stage of observation
1.	Anthocyanin pigmentation of young emerging leaves	Absent/present	Emerging fresh leaves
2.	Waxy bloom on ventral side of leaf	Absent/present	Emerging fresh leaves
3.	Waxy bloom on dorsal side of leaf	Absent/present	Emerging fresh leaves
4.	Waxy bloom on stem	Absent/present	Vegetative/flowering
5.	Stem color	Green/pink/mahogany/ red/dark red	Vegetative/flowering
6.	Leaf shape	Flat/shallow cup/deep cup	Vegetative/flowering
7.	Number of leaf lobes	Few/many	Vegetative/flowering
8.	Leaf hairiness	Absent/very weak/ medium /strong	Vegetative/flowering
9.	Branching	Convergent/divergent	Vegetative/flowering
10.	Branching pattern	Top/basal	Vegetative/flowering
11.	Days to flowering	Early/medium/late	Start of flowering
12.	Color of stigma	Pink/red/dark red	Start of flowering
13.	Type of flowers on primary raceme	Monoecious/inters persed/pistillate	Start of flowering
14.	Capsule color	Sulphur white/green/ purple/mahogany/red/black	Dough grain
15.	Capsule spineness	Non-spiny/semi-spiny/spiny	Dough grain
16.	Capsule length	Small/medium/long	Dough grain
17.	Spike compactness	Loose/medium/compact/ very compact	Dough grain
18.	Capsule dehiscence	Non-dehiscent/partially dehiscent/dehiscent	Harvest maturity
19.	Seed color	White/pink/dark chocolate/brown /red/dark/ red black	Harvest maturity/lab
20.	Seed mottling	Less conspicuous/ conspicuous	Harvest maturity/lab

Note: Description is based on Viswanathan and Hanumanta Rao (2003) and PPV & FRA

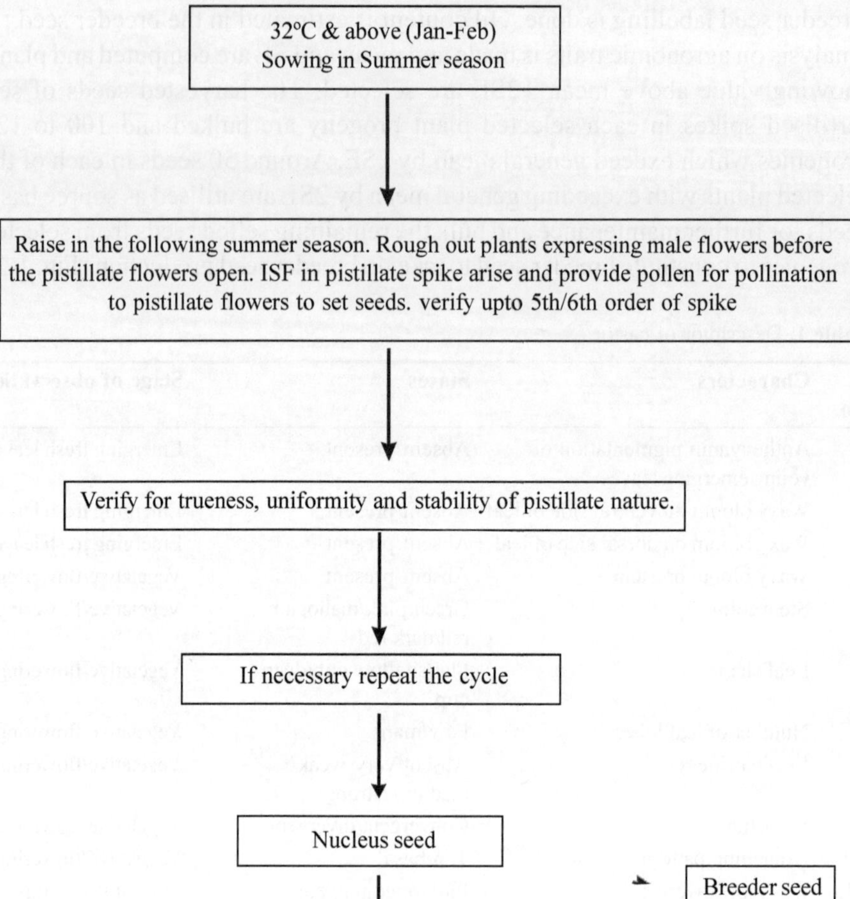

Fig. 1. Pistillate line seed production in Castor

Nucleus seed production

>32°C pistillate plants in castor. Pollinated by ISF flower

ISF staminate flowers produced by pistillate plants. Maintenance line is not required (>32°C)

(>32°C)

Raise pistillate plants in isolation of 2000m. At preflowering rough out monoecious plants expressing male flowers. Retain pistillate plants and select 150 to 200 uniform, true to type healthy pistillate plants not showing reversion to monoecious nature (2). Stability for pistillate nature with ISF for pollen supply, upto 5th order of branching.

(>32°C)

Raise 150 plants progenies in summer season (>32°C)cull out monoecious plants at pre flowering stage. True to type uniform plant progenies are identified that do not show reversion to monoecious nature. 100 uniform pistillate plants with ISF are selected from the progenies. Seeds from individual plants are examined plant wise, spike wise and progeny wise

Seeds from selected pistillate plants are bulked to constitute nucleus seed

(>32°C)

Breeder Seed Production (as in nucleus seed)

GOT

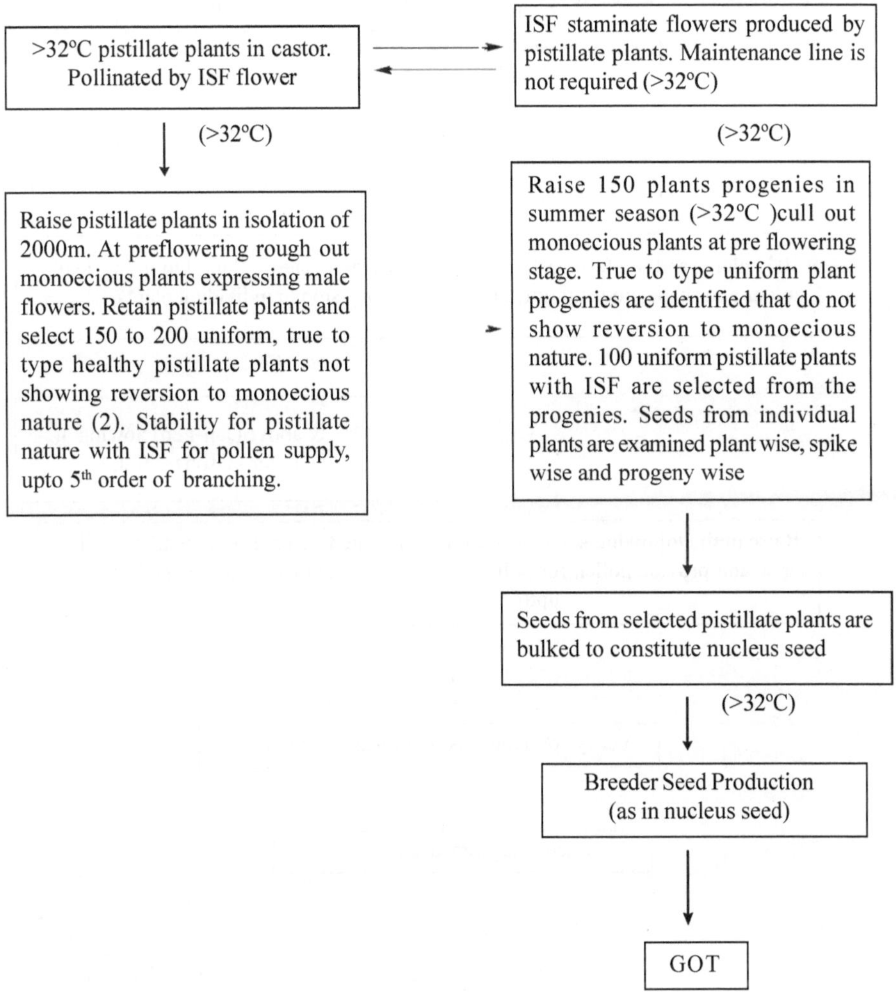

Fig. 2. Castor pistillate line Nucleus seed production

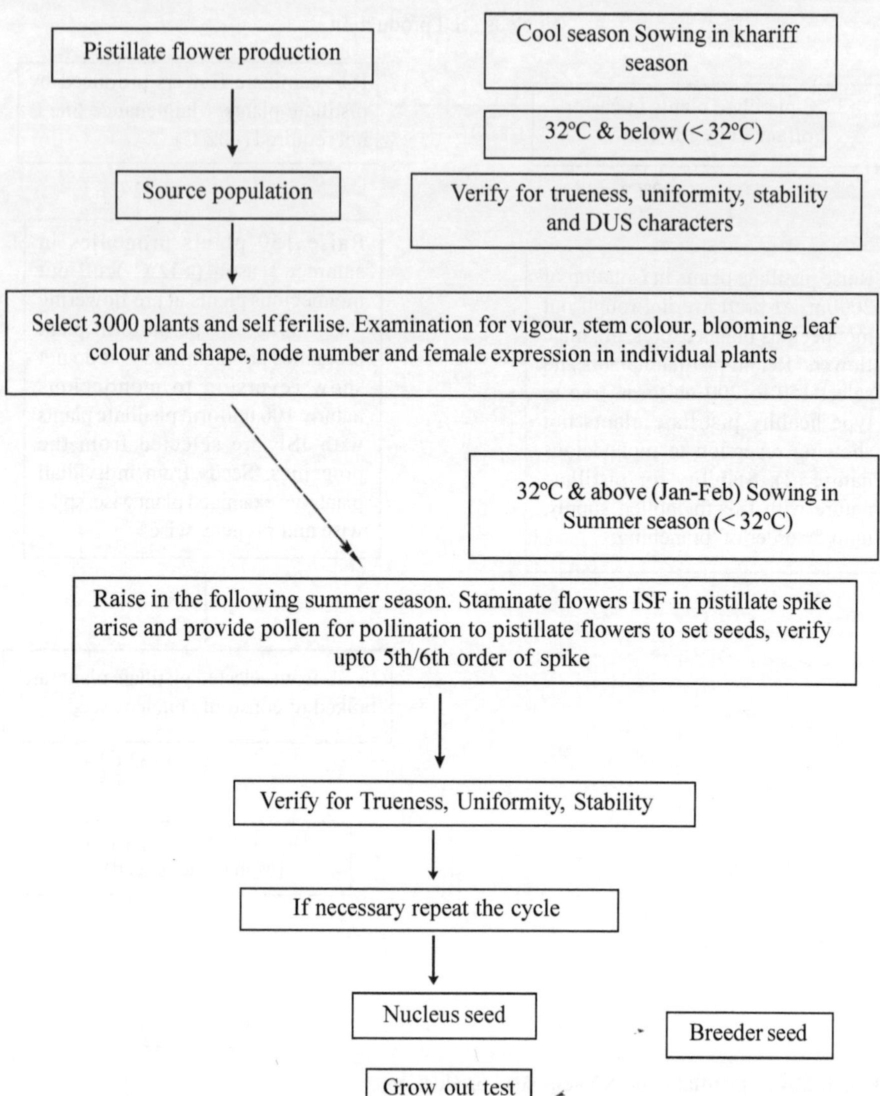

Fig. 3. Pistillate line seed production in castor

28. Linseed

(*Linum usitatissimum* L.) (2n:30)

Family : Linaceae

In tropical countries Linseed is grown as an oilseed crop. It is grown for its fibre in temperate countries. Linseed is an annual, erect herb with slender stem. Leaves are linear / lanceolate, alternate and small in size. Flowers are arranged in terminal or axillary racemes. Colour of flower is blue, white or lilac. Flowers 2 to 3 cm. long, pedicellate, pedicels 1 to 1.5 cm long, hermaphrodite and hypogynous. Sepals five, petals 5, stamens 5 and a compound pistil of 5 carpels in a circular arrangement. The calyx consists of 5 connate sepals and the corolla is polypetalous and deciduous. Stamens 5, anthers 2 celled and longitudinally dehiscing. Gynoecium superior, five united carpels and 5 styles. Ovary is 10 celled with false septum in each capsule. Fruit is an indehiscent capsule, ten loculed, each with one seed, that is oval, pointed, smooth and shining yellow to brown. Pollination occurs before the flower opening. Natural cross pollination is recorded up to 0.5 to 0.9%. The flowering period ranges from 25 to 40 days. Linseed is a self pollinated crop and pollination takes place in matured buds before flower opens (Kherki *et al.*, 2003). The characteristics of linseed are presented in table 1.

Table 1. Descriptor of linseed

S.No.	Characters	States	Stage of observation
1.	Growth habit	Erect/semi-erect/bushy	Vegetative/reproductive
2.	Plant height	Dwarf/medium/tall	Vegetative/reproductive
3.	Flower size	Small/medium/large	Flowering
4.	Petal color of corolla	White/tinge blue/lilacblue/light violet blue/red - violet	Flowering
5.	Petal aestivation (arrangementof petals)	Twisted/semi-twisted/valuate	Flowering
6.	Petal venation color (fullydeveloped flower)	Colorless/light-violet/violet/blue	Flowering
7.	Stamen filament color (immediately after flower opening)	Colorless/violet/blue	Flowering
8.	Anther color (immediately after flower opening)	Cream/grey/violet/blue	Flowering
9.	Capsule size (fully developed capsule)	Small/medium/bold	Harvest maturity
10.	Capsule dehiscence	Dehiscent/semi-dehiscent non-dehiscent	Harvest maturity
11.	Capsule tip shape	Pointed/blunt	Harvest maturity
12.	Seed color	Fawn/yellow/lightbrown/dark brown	Harvest maturity/lab
13.	Seed length	Small/medium/large	Harvest maturity/lab
14.	Seed size	Small/medium/bold	Harvest maturity/lab

Note: Description is based on Kherki *et al.* (2003); PPV & FRA.

Nucleus seed production: Source seeds are sown in an area of 400 sq.m. of loamy fertile and productive field that are not sown to linseed in the past two years. Source seeds may originate from the Breeder or from prereleased variety. Isolation distance of 10 m. is provided. Off types and deviant type of plants are uprooted as and when they appear. Single plants numbering 1000 that are uniform and conforming to the varietal description and characteristics are selected and labelled. At maturity, 500 true to type plants are selected and harvested and their capsules stored in individual packets and numbered plant wise. After threshing, seeds are separated, cleaned and examined for size, shape, colour / appearance. Off types or deviating types are rejected. The seeds of selected plants are dried and stored.

In the next season, usually in October-November, 200 progeny rows are sown in two rows plots with 20 cm. spacing row to row in 2 replications. The row length is 4 m. Once in every 2 beds 40 cm. space is provided for inspection. Ten border rows of linseed variety (source seeds) are sown. Isolation distance of 10m. from other linseed crops is provided. Examining the progeny rows continuously from juvenile to vegetative to preflowering stage is done and the progeny rows showing off types and deviant plants are uprooted. Progeny rows with off types and deviant types occurring at flowering and post flowering stage are uprooted along with the adjoining rows. At full maturity, progeny rows performing above mean +2 SE are identified and five hundred single plants are selected in the identified rows and the capsules stored for future use after drying. In the next season, 200 progeny rows are raised and examined for true to type, uniformity, health and vigour. Progeny rows showing off types and deviant plants are discarded at pre and post flowering stages. At maturity 200 true to type single plants are selected for uniformity, seeds separated and examined in the laboratory for uniformity in shape, colour and seed filling in capsules and also for seed features. All progenies showing deviations are discarded. Seeds of uniform progenies are bulked to constitute Nucleus seed. Seeds are checked for seed quality oil content and labelled.

Breeder seed production: Nucleus seeds of the variety is sown with spacing of 30 cm. line to line, 60 cm. from block to block. Ten rows blocks are raised. The isolation distance of 10m. is adopted.Single plant per hill is maintained. Three rows of border rows are grown all around. Inspection for genetic purity, true to type nature and uniformity is made frequently at different stages of crop growth from juvenile to flowering and maturity. Off types and deviant plants in comparison with the diagnostic characteristics are uprooted. The off types and deviant type plants noticed during flowering stage are uprooted along with the adjoining plants. Monitoring Team visit is organized at flowering or at midmaturity stage to evaluate the field stand and purity. Harvesting and seed separation and processing are done with utmost care and seed sample is tested for seed quality parameters. The seed bags of suitable sizes are labelled with Breeder Seed Tags. Seed oil content is estimated. GOT is conducted with the seed samples as per the procedure.

29. Mustard

(*Brassica nigra* L.) (2n=16)

Family : Brassicaceae

Oilseed Brassicas: Herbaceous annual, leaves alternate, simple, often pinnately lobed. The lower leaves are termed as stem leaves and upper leaves as floral leaves. Inflorescence is a long raceme with few bracts. Flowers with four erect sepals and four petals that are with narrow basal claw. Stamens six –two outer and shorter and four inner taller than outer ones. Ovary superior, two joined carpels with single short style and capitate stigma; ovules many in parietal placentation. Fruit is a siliqua, longitudinally dehiscing. Seed colour is dark brown, black, light brown or yellow coloured. The descriptor of Rapeseed mustard is shown in Fig. 1.

Rapeseed-mustard group of crops are (Yadava and Singh, 2003):

Self compatible (self pollinated) species:

1. Yellow sarson (*Brassica napa* var. yellow sarson)
2. Gobhi sarson (*B. napus*)
3. Ethiopian mustard (*B. carinata*) and
4. Rapeseed: *B. juncea* (Indian mustard).

Self incompatible (cross pollinated species) include:

1. Toria (*B. napus* var. toria)
2. Brown sarson (*B. napa* var. brown sarson) and
3. Taramira (*Eruca sativa*).

Rapeseed (*Brassica juncea*)-Indian Mustard is largely grown crop and the extent of cross pollination is from 5 to 15 percent. Isolation distance of 1600 m is to be provided for self incompatible (cross pollinated) species.(*B. napa* var. toria, *B. napa* var. brown sarson and *E. sativa*) and 200m. for the self pollinated (self compatible) species including *B. juncea* (Indian mustard), *B. napa* (yellow sarson) *B. napus* (ghobi sarson) and *B. carinata* (karan rai).

Nucleus seed production: A base population is raised from source seeds of breeder or Institution or previous nucleus seed crop with 200m. isolation. The crop is examined for uniformity and 1000 true to type plants are selected for the nucleus seed production in self pollinated species. In respect of cross pollinated or self incompatible crop, 2500 to 3000 true to type plants are selected.

Five border rows of the same variety are planted around the plot. The selected true to type plants are again verified for characteristics conforming to the variety and labelled, harvested and threshed plant wise separately. The seeds are closely examined in the laboratory for seed shape, colour and size and then selected. Off-types and deviant type of plants are discarded and seeds from true to type plants alone are forwarded for raising nucleus seed plot. Seed oil content is estimated in a composite sample.

In the next season, seeds from selected plants are sown in 1000 progeny rows in isolation plots (200m). Each progeny is replicated 2 times and in each replication 2 rows plot, each row 4 m. in length is grown. Each plant in progeny plots is examined for true to type during different stages of growth till flowering and pod set. The progeny rows showing off-type/deviant types are uprooted. If any of the progeny rows show off types at flowering time, that progeny row and progeny rows on either side of it are also uprooted so as to prevent any chance of cross pollination or contamination. Very early and late flowering progenies are discarded. At maturity, 1000 single plants in self pollinated Brassica and 2500 plants in cross pollinated / self incompatble Brassica are selected and seeds are examined for conformity with the original variety. Progeny rows performing above mean +2SE are identified for forwarding with selected plants within each identified progeny row.

Seeds from selected single plants are stored for raising next cycle of nucleus seed production. Part of the seeds from the selected progenies are bulked together and nucleus seed is constituted. Seed oil content in nucleus seed is estimated (Fig. 1).

Breeder seed production: Seeds from the nucleus seed production plots are raised in wide space in proper favourable season. Ten rows of the same variety are raised all around a border rows. Isolation distance of 200m.is provided. All good agronomic practices are followed. Only one plant per hill is retained. Roguing to remove off type and deviating type of plants is done with utmost care at juvenile, active growth upto pre-flowering, at the time of flower initiation and pod seed development stages. Off type plants noticed at pre-flowering are removed while off- types occurring at flowering stage are removed along with adjacent plants to prevent pollen contamination and crossing. The uniformity in flowering is well maintained in the plot by removing very early flowering and late flowering plants and 50% flowering within the specified period. Monitoring Team visit is arranged to adjudge the crop purity. After harvest, all the plants are examined for the uniformity in pod (siliqua) and seed characteristics and true to type nature. The breeder seed thus produced are dried, cleaned and processed. The seed quality parameters are tested and the quality declared. Quality assured seeds are labelled with breeder seed lebel. Grow Out Test is arranged in the specified manner and the genetic purity of the breeder seed is ascertained. Oil content in seed is estimated.

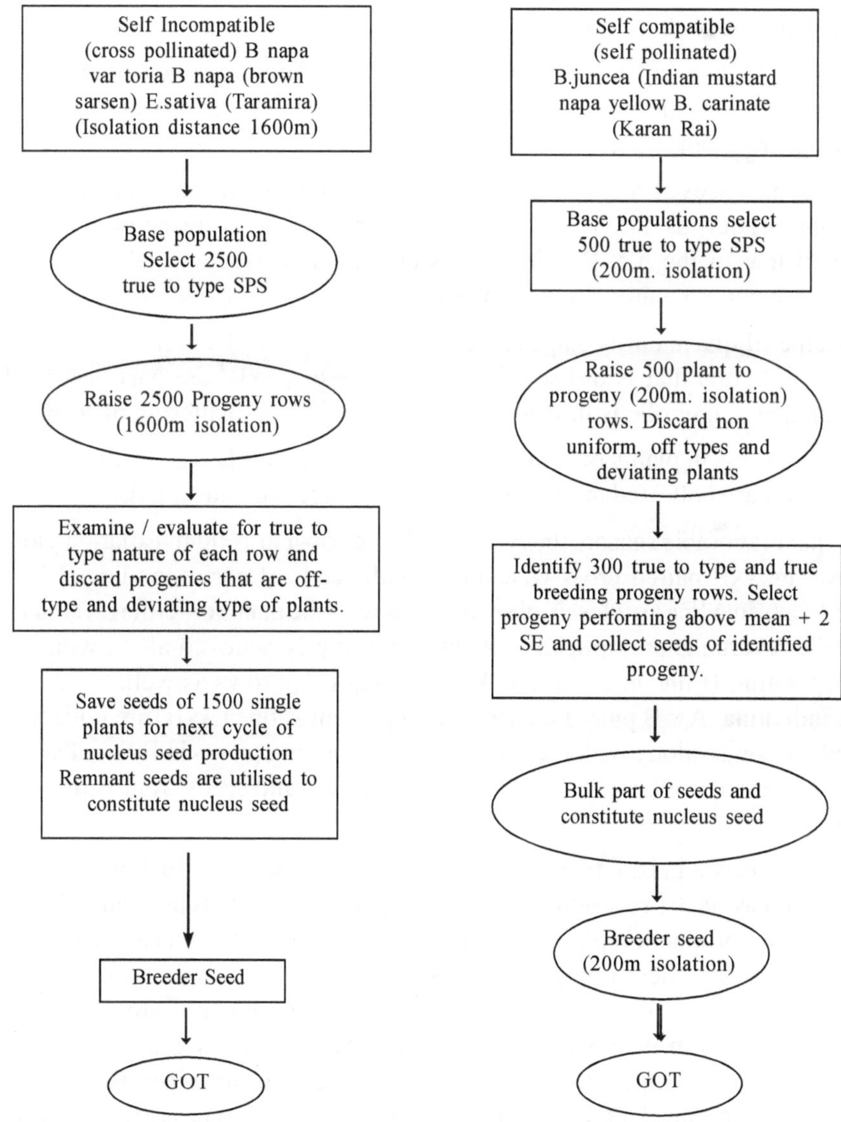

Fig. 1. Nucleus and Breeder Seed Production of OPV in Repe-seed-Mustard

CMS (A) line, B line and R line Nucleus seed production

The following steps are adapted for nucleus seed production of A line and B line in the self pollinated Brassica crop (*B. juncea*).

1. Seeds of A and B line from basic seed sources are raised in single rows with space planting in row ratio of 1:1 of A and B respectively. Isolation distance of 1600m. is provided.
2. Thorough roguing and removal of off-types and deviant plants in A and B line from juvenile to vegetative flowering and seed formation stage in B line are carried out.
3. Paired crosses of A X B lines, 500 to 600 in number, are made and labelled appropriately as A1xB1, A2xB2, A3xB3 AnxBn. The corresponding plants in B line of the A x B paired cross are self fertilised and labelled with corresponding paired cross numbers.
4. At the siliqua maturity stage ie. before its dehiscence, the individual siliqua in the B line that is utilised for paired crosses (AxB) are harvested and labelled. Other plants in B line are harvested before shattering of seeds.
5. Siliqua sets in plants of A line utilised in A1xB1, A2xB2 ... AnxBn paired crosses are collected at maturity, plantwise, labelled and stored.

In the next favorable season, the A line seeds are sown in 1600m. isolation plots in rows in AxB paired cross wise along with seeds of corresponding B line plant. Plants in A line are closely observed for any fertile plants or fertile revertants. If fertile revertants are noticed in A line, that line is removed along with the corresponding B line and also the A line in adjoining rows as well. Seeds set from individual A x B paired crosses that represent A line (AxB) are collected paired crosswise along with the seeds of corresponding plants in B line. Part of (paired crosses) seeds of A line are bulked together and seeds from B line are also pooled.

In the next season in isolation plots for A line nucleus seed production, A and B lines are raised in 3:1 row ratio. Isolation distance of 1600m. is maintained. The nucleus seed production plot is grown with 5 to 7 rows of B line all around to ensure adequate pollen supply. Roguing is done to remove any fertile revertants in A lines. If any sterile plants are noticed in B line, such plants are discarded. Roguing is done in A line and B line for the deviant plants and off types in characteristics to maintain true to type nature and genetic purity. Seeds collected from superior A line plants and B line plants that are true to type are respectively bulked to make nucleus seed of A and B line.

Breeder Seed Production in A and B lines: The seeds harvested from A line and B line are further utilised in the breeder seed production of A line with the same procedures followed in nucleus seed production. The male sterility maintaining ability of B line is ascertained. In every stage of A line production by growing A : B in row ratio of 1:1 or 3:1 manner, strict attention to harvest the B line just at time of A line seed maturity so that shattering and mixing of seeds from B line to A line is avoided. Seeds set in A line plants are collected and processed. Seeds set in B line is collected separately for Breeder Seed. Seed quality is tested and labelled with breeder seed tag. Grow out test is conducted with samples drawn from seeds of A line (AxB). Visit of Monitoring Team is arranged at midmaturity of pods in A line (A×B).

Breeder Seed Production in R line: In respect of R line for producing nucleus and breeder seed, the practices being followed in respect of varieties are adhered to. Required isolation distance is 200m. Grow out test is conducted with seed samples of R line.

In addition, steps to ascertain the fertility restoration of R line in A x R crosses are fulfilled as described herein. In an isolation plot with 1600m. plants of A and R lines are raised in 1: 1 row ratio and 500 to 600 paired crosses of A x R are made. R line is harvested earlier than A line. Seeds of R line are collected from individual plants utilised in specific A x R paired crosses (A1xR1, A2xR2 ...... AnxRn). In the next season, hybrid seeds of A x R crosses (paired cross wise) are sown and AxR plants are observed for heterotic vigour and pollen fertility restoration and good seed set in the hybrids. R lines grown in separate but isolated plot (200m.) are selected based on the pollen fertility restoration and vigour in the corresponding combination of AxR hybrid. Seeds from R line plants showing fertility restoration in A x R cross are pooled and is utilised in R line multiplication in isolation of 200m. The seeds of the R line thus obtained is utilised in nucleus and breeder seed cycle following the procedure of nucleus and breeder seed production in varieties.

Table 1. Descriptor of rapeseed-mustard

S.No.	Characters	States	Stage of observation
1.	Leaf shape	Serrated/non serrated	Vegetative
2.	Leaf type	Sessile/petiolate	Vegetative
3.	Leaf color	Light green/medium green/dark green/purple green/purple	Vegetative
4.	Leaf hairiness	Present/absent	Vegetative
5.	Calyx color	Green/light green	Flowering
6.	Corolla color	Dark yellow/yellow/ cream yellow/white	Flowering
7.	Petal shape	Narrow/broad	Flowering
8.	Plant type	Erect/semi-erect/spreading	Vegetative/reproductive
9.	Plant height	Dwarf/medium/tall	Vegetative/reproductive
10.	Main shoot length	Short/medium/long	Vegetative/reproductive
11.	Siliqua Arrangement	Appressed/semi-appressed/spread	Maturity
12.	Siliqua surface	Smooth/intermediate/constricted	Maturity

13.	Siliqua beak	Short stout/long slender	Maturity
14.	Seeds silique	Less/average/more	Maturity
15.	Pod locule	Unilocular/bilocular/trilocular/tetra locular	Maturity
16.	Seed color	Yellow/dull gray/reddish brown/brown/black	Maturity/ post
17.	Seed size	Small/medium/bold	Post harvest
18.	Oil content	Low/medium/high very high	Post harvest

Note: Description is based on Yadava and Singh (2003); PPV & FRA (2007)

30. Niger

(*Guizotia abyssinica.* L) (2n: 30)

Family : Asteraceae

Herbaceous annual, erect and smooth stemmed. Leaves sessile with serrated margin. Inflorescence is a head. Bracts are in 2 rows of involucre bracts. Ray florets are female flowers and produce seeds. Disc florets are hermaphrodite arranged in 3 whorls. Florets are tubular with five lobes, surrounding the stamens and style. Style is bifid. The florets open around 8 am. and the anthers liberate the sticky pollen. The styles elongate with the stigmatic lobes well above the anthers so that there is no contact. Ray florets open first followed by disc florets. Niger is an entirely cross pollinated crop. Bees and other insects frequent the niger. The cross pollination aspect in niger flower is reinforced by a sporophytic self incompatibility system. The mechanism is sporophytic, protandrous and self incompatibility. Mode of pollination is solely by entomophilous agents. Very high inbreeding depression associated with poor seed set is observed with advancement of inbred generations. With the initiation of flowering, selfing or sibmating is done by enclosing 2 or more adjacent representative plants of a line inside a fine muslin cloth or mosquito net. Shaking of bagged plants ensures self pollination. The descriptors are given in Table 1.

Table 1. Descriptor of Niger

S.No.	Characters	States	Stage of observation
1.	Leaf color	Pale green/light green/green/darkgreen	Vegetative
2.	Leaf width	Very narrow/narrow/medium/broad/verybroad	Vegetative
3.	Leaf margin	Entire/serrate/dentate	Vegetative
4.	Stem color	Green/purplish green/purple/deep purple	Vegetative
5.	Days to flowering	Early/medium/late	Flowering
6.	Branching habit	Erect/semi-erect/semi-spreading/spreading	Flowering
7.	Flower color	Pale yellow/yellow/orange	Flowering
8.	Capitulum size	Small/medium/large	Maturity
9.	Maturity of heads	Synchronous/asynchronous	Maturity
10.	Seed color	Grey/brown/dark brown/black	Maturity
11.	Seed shape	Cylindrical/curved/sickle	Maturity

12.	Days to maturity	Early/medium/late	Maturity
13.	Seed size	Low/medium/high	Post harvest
14.	Seeds per capitulum	Low/medium/high	Post harvest
15.	Oil content(%)	Low/medium/high	Post harvest

Source: Duhoon S.S. (2003)

Nucleus seed production: Seeds from the source population are sown in an area of 800 sq.m. plot provided with 1600m distance isolation. Zn S04 is applied (15 kg/ha) to the crop. Based on the uniformity and true to type nature of characteristics that conform to the original variety description, 1000 to 1200 representative plants are selected and labelled. The single plants are then harvested. The heads are examined in the laboratory for size, shape, seed set and seed filling. The seeds are checked for the colour, size and shape. Seeds from the selected 250 individual plants are stored in separate packets and labelled. Niger has orthodox seed storage.

In the next season, at the time of sowing, part of the seeds in each selected plant is advanced for raising single plant progenies and the remnant seeds stored for constituting the nucleus seed in future cycle. The sowing is done in a favorable season for seed production in plots isolated by 1600m. distance from other niger crops. Single plant progeny rows, 200 to 250 in number, are raised in two row plots in 2 replications. Parent row is raised once in 10 progeny plots. Depending on the number of progeny rows, total number of blocks may be decided, usually in each block 50 progenies are grown in two replications. The progeny rows are critically examined for the field establishment, vigour, foliar expression, free from diseases, uniformity in flowering and maturity and seed set. Off types are removed along with the plants in the same progeny row. Progeny rows showing off types at flowering and post flowering are also removed along with the adjoining progeny rows to prevent cross pollination and vicinism. At the maturity, selected true to type progeny rows in conformity with the variety description are harvested progeny wise. The characters of the heads and seeds are examined in the laboratory for true to type nature, genetic purity and uniformity. The progenies are then examined for yield equal to or higher than the mean +2 SE and such progenies are finally selected. The oil content of the selected progenies is estimated .The seeds of the plants selected in the progeny evaluation are then bulked to constitute nucleus seed. This completes one cycle of nucleus seed production The oil content of the nucleus seed and seed quality parameters are tested. Nucleus seed labelling is done in nucleus seed packets (Fig. 1).

Breeder seed production: The breeder seed production is taken up in a well drained, fertile loamy soil in a very favourable season and with isolation of 1600 m. distance. Zinc sulphate application at 15 kg/ha is made to the field. The breeder seed crop is continuously examined for genetic purity, true to type and uniformity in flowering. Off types and deviant plants are removed at different stages of crop development. Off type noticed at flowering or post flowering stage are removed along with the plants around to prevent cross pollination. Very early and late flowering plants are discarded. At maturity stage or prior to that, the visit of Monitoring Team is organized. The harvested heads single plantwise are verified for true to type and dried and seeds are separated, threshed and processed. Seed sample is drawn and tested for oil content and seed quality parameters. Breeder seed packets are tagged with Breeder seed labels. GOT is raised to adjudge the genetic purity. Being a highly cross pollinated crop, the cycle of nucleus and breeder seed production should be repeated very critically at least once in 2 years to maintain the genetic purity and to arrest the genetic deterioration. Seed oil and protein contents are estimated.

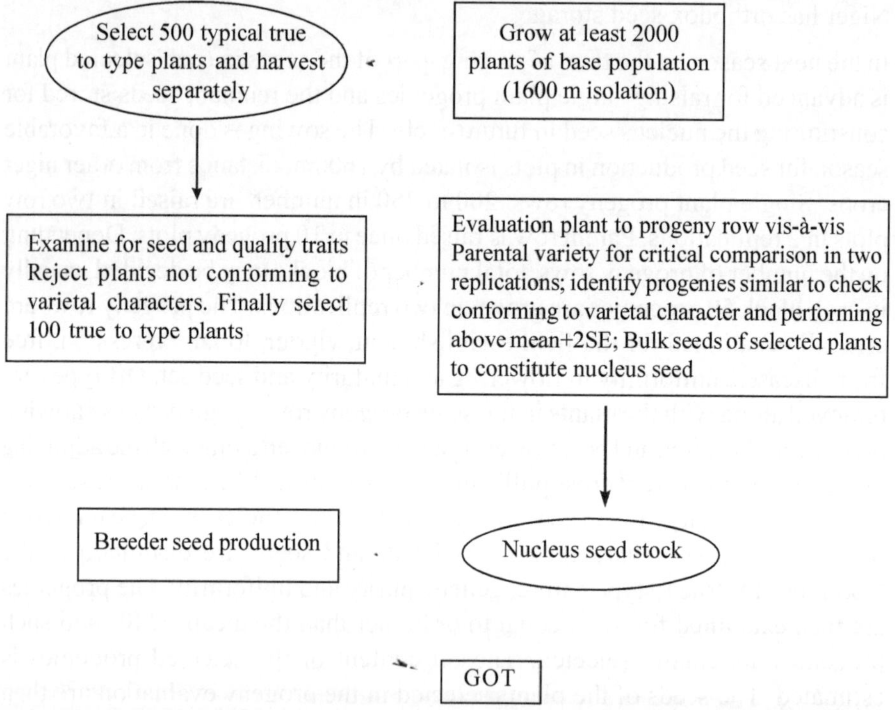

Fig. 1. Nucleus and Breeder seed production in Niger (1600m - isolation)

31. Peanut/Groundnut

(*Arachis hypogea* L.) (2n:40)

Family : Fabaceae

Peanut /Groundnut is an annual herb with habits like bunch, semierect and spreading. Leaves paripinnately compound, 2 pairs of leaflets and adnate stipules. Flowers are sessile, borne in leaf axils and solitary to 3 to 4 in number. Petal yellow coloured, calyx tube long, ending in 5 lobes, Corolla five papilionaceous, keels are united. Stamens monadelphous, eight fertile anthers and two staminodes. Eight anthers are dimorphic and didynamous with 4 longer lobes and 4 with round anther lobes. Ovary monocarpellary, superior, style is slender, stigma terminal and hairy. Ovary unilocular with 1 to 3 ovules in marginal placentation. Fruit is an indehiscent pod, 1 to 3 seeded. Testa is commonly rose or infrequently from dark to red coloured and occasionally variegated. The peg is positively geotropic. Flowers open at sunrise in some genotypes and anthers dehisce in 7 to 8 hours before flower opening. In others, anthers do not dehisce even at flower opening. Stigma becomes receptive about 24 hours prior to anthesis and receptivity persists for 12 hours after anthesis. Groundnut is predominantly self pollinated crop and about 2% is the natural outcrossing (Chunni Lal *et al.*, 2008). The characteristics in groundnut forms are given in Table 1&2 and Fig. 1.

Nucleus Seed Production: In plot of, 2000 sq.m. seeds of groundnut variety sourced from the institution are sown in lines in wide spacing. Critical observations are made in juvenile and adult phases for vegetative, floral, peg and pod characteristics. At harvest, 2000 to 2500 plants are selected based on the verification of detailed dry pod and seed characteristics and pod development uniformity. Further selection is made in the single plants based on the peg, pod and kernel characteristics. The pods are separated, dried and stored as selections. This constitutes the source population. In the next season, 200 single plant selections are sown in progeny rows with proper isolation distance (5 meters). Border rows, 3 m. wide are raised all around with the same source seed. Frequent and continuous field monitoring is done from germination to juvenile to vegetative stage for foliage colour and branching characters, flowering duration and pattern and uniformity and pod characters among the progenies and between the progenies. Progeny rows showing off types are removed at different stages of observation. Within each progeny row, twenty uniform plants are selected at harvest time and dried well. Pod and seed characters are critically examined and selected pods are stored as single plant produce for continuing next cycle of nucleus seed production (Fig. 2).

In the next season, 200 single plant progenies are raised in 2 row plots, 4m length and in two replications. In each block, 50 progenies are raised. Critical examination is made on individual progeny rows for trueness, uniformity, vigor and free from diseases. At flowering stage, conformity for floral traits description of the concerned variety is checked. At maturity, harvest of pods is done by gently lifting the plants from soil and the true to type nature of pods set and all features/ characteristics of the dried pods verified. The performance of each progeny row is recorded. Progenies with performance equal to or higher than mean +2SE are identified. Dried pods and kernel traits are studied for conformity. Pods of selected progenies are well dried and seeds separated in the individual selected plants. Shelling percentage in selected plants and oil content of kernel in nucleus seeds are estimated. Part of seeds from individual plants are pooled together to constitute nucleus stage. Seeds obtained from the nucleus seed stock are multiplied in 2 stages of nucleus seed multiplication stage I and which in turn is utilised for nucleus seed multiplication stage II.

Breeder seed production: Breeder Seed Production is practised adopting all precautions in selecting fertile and right kind of field not grown with groundnut in the previous 2 to 3 seasons. The main produce is pods borne on pegs below the soil. All efforts are made to avoid volunteers and rogue plants from the previous season. After raising the breeder seed crop with isolation distance of 5m. critical and continuous examination is made at different crop growth phases to identify deviant plants and rogues, if any, and to remove them. Monitoring Team visit is organized to assess the genetic purity and stand of the crop in the breeder seed plot. After harvest and drying the pods with vines intact ,the individual plants are examined in the threshing floor for uniformity and trueness of pods. Deviating plants in respect of pod and seed characteristics are discarded. The seeds from selected plants are dried and pooled and processed. The seeds are packed and samples tested for seed quality parameters. Oil content in seed is estimated. Quality assured seed is then labelled with the beeder seed tag. Grow Out Tests are conducted with the breeder seed samples following the standard procedure to ascertain the genetic purity of seeds. Breeder Seed Stage II multiplication can be taken up, if need be in view of low seed multiplication ratio.

The multiplication ratio in groundnut varies from 1:5 to 1:10 and extra care is to be taken for good agronomic production practices to ensure quality seed production in the required quantity of nucleus seeds requirement. NSP II stage is taken up to meet the nucleus seed requirement. Oil content and protein content in the seeds are estimated/recorded.

Table 1. Descriptor of groundnut

S.No.	Characters	States	Stage of observation
1.	Growth habit	Erect/semi-erect/prostrate	Flowering to pod formation
2.	Flowering pattern	Alternate/sequential/irregular with flower on main axis/irregular without flower on main axis	Flowering to pod formation
3.	Leaflet color	Light green/green/dark green	Flowering to pod formation
4.	Leaflet size	Small/medium/large	Flowering to pod formation
5.	Leaflet shape	Cuneate/obcuneate/elliptic/oblong/ elliptic/narrow-elliptic/wide-elliptic/suborbicular/ovate/obovate/ oblong/oblong- lanceolate/lanceolate/ linear-lanceolate	Flowering to pod formation
6.	Leaf tip	Obtuse/acute/mucronate	Flowering to pod formation
7.	Stem pigmentation	Absent/present	Flowering to pod formation
8.	Type of inflorescence	Simple/multiple/elongated	Flowering to pod formation
9.	Standard petal color	White/lemon yellow/yellow/ orange/orange/dark orange/ brick red	Flowering
10.	Peg pigmentation	Absent/present	Pod formation
11.	Time of maturity	Early/medium/late	Harvesting
12.	Pod beak	Absent/slight/moderate/ prominent/very prominent	Harvesting
13.	Pod reticulation	Absent/slight/moderate/ prominent/very deep	Harvesting
14.	Pod constriction	Absent/slight/moderate/deep/very deep	Harvesting
15.	Seed color	One color/variegated	Harvesting
16.	Kernel shape	Spheroid /cylindrical/ fusiform	
17.	Fresh seed dormancy	Absent/ present	
18.	Weight of 100 kernels (g)	Low (<36), medium (36-50), high (51 to 65)	

Note: Description is based on Chunilal *et al.* (2003); PPV & FRA.

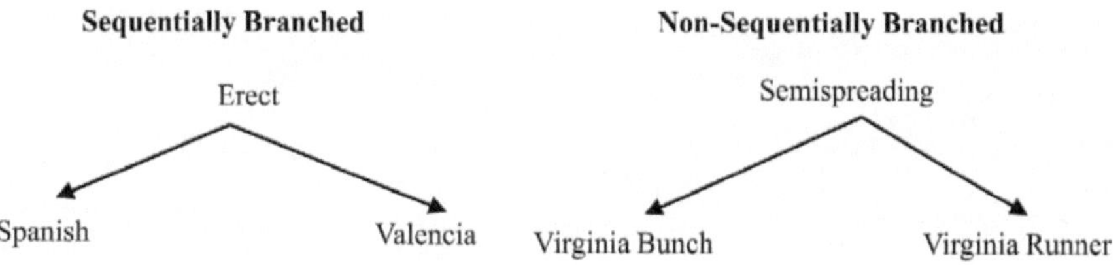

Fig. 1. Different Botanical Groups of Groundnut

Table 2. Groundnut - Morphological differences among the three botanical varieties of cultivated Arachis hypogaea L

S.No.	Spanish (Var. vulgaris)	Valencia Var. fastigata	Virginia var. hypogaea
1.	Bunch (**erect**)	Bunch (**erect**)	Semi spreading and spreading
2.	Sequentially branched 7 to 8 side branches and short internodes vegetative branches on primaries, occasional or irregularly placed	Sequentially branched, 4 to 5 tall branches, vegetative branches on primaries usually absent	Profuse branches, branches absent on main stem, alternate with 2 vegetative and 2 reproductive nodes on primary branches
3.	Stem. Green tinge pubescent	Purple	Green tinge, weak pubescence, haulms as fodder
4.	Medium maturing	Early maturing	Late maturing
5.	Leaves- small or medium light green, dark green, elliptic, pointed tip	Medium or large round tip and base	Small leaves, dark green, inverted egg shaped, medium foliage
6.	Uniform ripening of pods, pod setting compact	Pods around main stem	Bold pods
7.	Pods small 2 seeded	Medium 2-4 seeded	Large 2 seeded
8.	Testa colour, tan red, white or purple	Tan, fleshy red, white, yellow purple, variegated.	Tan, fleshy red, white, yello wish, Purple, variegated.
9.	Shell thin	Thick	Thin to thick
10.	Flowers on the axils of main stem	Flowers on main stem	No flowers on main axils, alternate 2 vegetative nodes with 2 reproductive
11.	Fresh seed dormancy usually absent	Usually absent	Present upto 30 to 60 days
12.	Peg colour-green, pigmented	Green	Pigmented

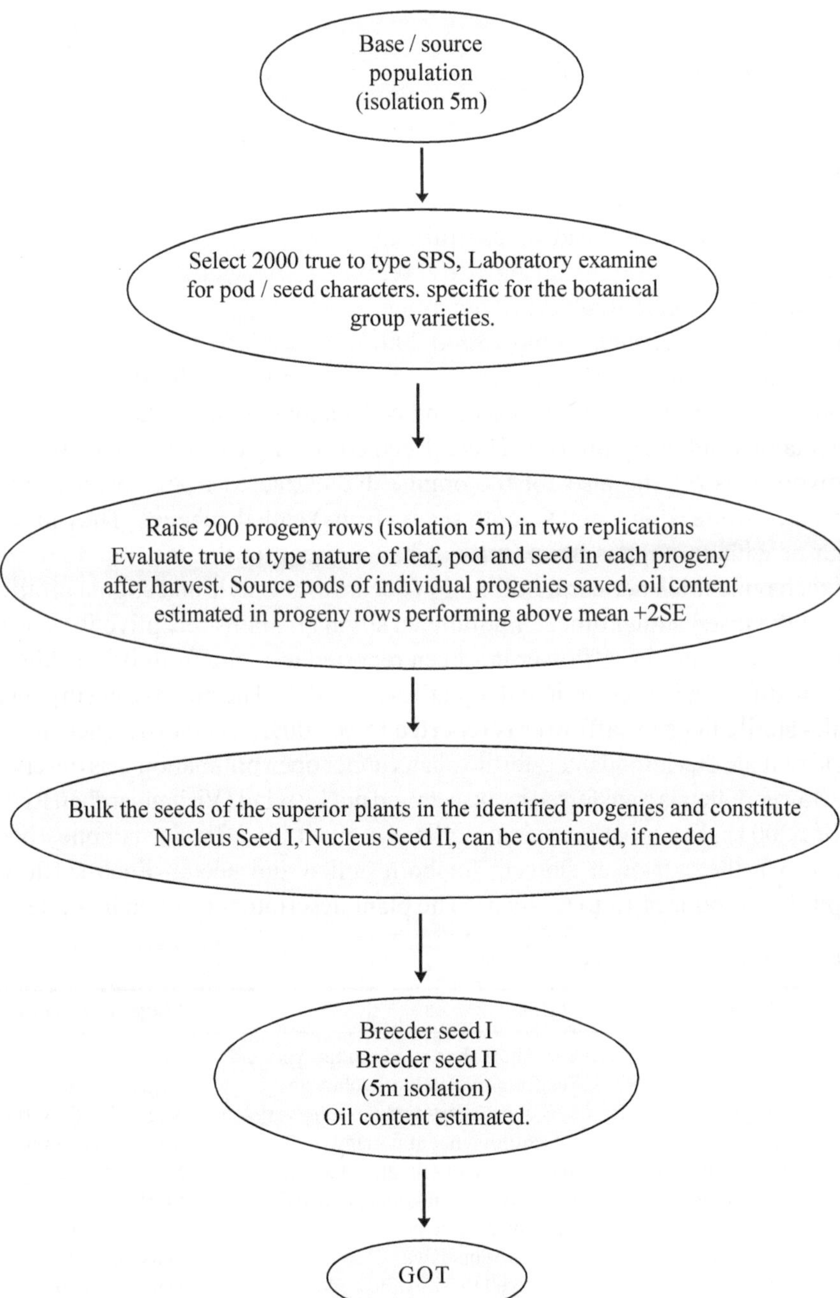

Fig. 2. Nucleus and Breeder Seed Production in Groundnut

32. Safflower

(*Carthamus tinctorius* L.) (2n:24)

Family : Asteraceae

Plant, annual herb, glabrous or pubescent. Leaves alternate, entire, margin serrated or not. Inflorescence is head or capitulum, homogamous, outer bracts are ovate, spinous or not, inner involucre of bracts are oblong. Florets are many, upto 150 to 200 florets, tubular, light yellow to orange red in colour, hermaphrodite, bracteoles modified to soft hairs. Calyx absent and corolla long tubular, stamens five and epipetalous and ovary inferior. This oilseed crop is a cross pollinated crop. It is grown for its oil and also for the orange dye extracted from flower head. It produces white, shiny, and smooth seeds (fruits) called achenes. They may be with or without pappus hairs (tufts of hair present on the seed) and are four sided, having thick pericarp. Pollination occurs as the style and stigma protrude out of the fused anther tube. Unpollinated stigma remains receptive for several days. Outcrossing in safflower has been reported to range from 0 to as high as 59% in different genotypes in India (Patil *et al.*, 1987). The outcrossing in genetic male- sterile lines in safflower is reported to be 100%, as no difference in seed yield of male-sterile and male-fertile plants under open pollination was observed. Safflower pollen is transferred by insects and not by wind (Vishnuvardhan Reddy *et al.*, 2003). The most prevalent pollinator agent in safflower is honey bees. Bees visit the safflower flowers for both pollen and nectar. Each safflower capitulum produces 15 to 60 seeds. The plant descriptors is given in Table 1.

Tabel 1. Descriptor of safflower

S.No.	Characters	States	Stage of observation
1.	Plant height	Very short/short/medium/tall/very tall	Vegetative/flowering
2.	Leaf shape	Fusiform/ovate/elliptic/obovate	Vegetative/flowering
3.	Length of leaf blade	Very short/short/medium/long/very long	Vegetative/flowering
4.	Spines on leaves	Absent/few/medium/many	Vegetative/flowering
5.	Time of flowering	Very early/early/medium late/late	Start of flowering
6.	Petal color	White/yellow/pink/orange/mixed	Start of flowering
7.	Capitulum diameter	Small/medium/large	Dough grain
8.	Seed size	Small/medium/large	Harvest maturity
9.	Seed color	White/yellowish white/yellowish brown/brown	Harvest maturity
10.	Seed oil content	Low/ medium/high	After harvest
11.	Seed percentage of oleicacid	Low/medium/high	After harvest

Note: Description is based on Vishnuvarthan *et al.* (2003); PPV & FRA

Nucleus seed production: Source seeds of the variety from the breeder are planted in September - October months with isolation distance of 800 m. from other safflower crops. Single plant per hill is maintained. The crop is carefully monitored for any off type or deviant plants for the prescribed diagnostic characteristics in growth phases. Off types occurring during flowering period are removed along with plants around so as to totally avoid cross pollination and vicinism. Very early and late flowering plants are discarded. True to type well developed healthy 600 to 700 plants are selected and 50% of heads in each selected plant are self fertilized and the remaining heads are allowed for open pollination. At maturity, seeds are collected from each plant separately from self pollinated and open pollinated heads and are examined for the extent of seed set. Seeds are separated and seed characteristics are compared with the variety description. Seeds from self fertilized and open pollinated heads from each plant are extracted separately. Plants with self fertilized seeds showing weight more than the general mean are retained and selected for raising progeny rows in the next season.

In the next favourable season, progeny rows are raised in 2 rows plot, 4m. long and replicated twice. Isolation distance of 800m. is maintained. After every ten progeny row plots raised from self fertilized seeds of selected plants, one check row sown with seeds from open pollinated heads is raised for comparison and evaluation of the progeny rows from self fertilized seeds for true to type nature of different diagnostic characteristics of the variety concerned. At every development stage of the crop, progenies showing off types are uprooted. If deviant type of plants are noticed during flowering stage, the progeny row showing the deviant type of plants and adjoining rows are removed. Plants/ progeny flowering for extreme earliness or lateness are removed. At maturity, uniform looking true to type progeny rows from self fertilised seeds are selected and retained for harvest while unselected rows are discarded. Seed yield and oil content of the selected progenies are estimated and compared with the check. The progenies possessing higher yield and oil content above mean ± 2SE are selected. Seeds from the selected progenies are bulked to constitute the nucleus seed.

Breeder seed production: Isolation distance of 800 m. is maintained in respect of the breeder seed production plot. Seeds from the nucleus seed lot are wide space dibbled in proper season in well prepared field not grown with safflower crop in the past 2 to 3 years. Only one plant per hill is retained. All recommended good agronomic production practices are adapted to facilitate better expression of the crop. Periodical examination of the individual plants is made at different stages of crop growth to identify and remove off types and deviant type of plants upto pre flowering stage. Off types and deviant type plants found at flowering phase are removed along with the plants around them to avoid out crossing and vicinison. Very early and late flowering plants are removed to maintain flowering uniformity. Healthy, true to type and uniform true to type plants alone are retained and matured heads are harvested, seed separated and dried. At flowering stage of the crop, Monitoring Team visit is arranged. The seed quality is tested in seed testing laboratory and the quality assured breeder seed is labelled with Breeder Seed label. Seed oil content is estimated/recorded. GOT is conducted from samples drawn from breeder seed to assess the genetic purity.

Nucleus seed production in GMS lines (Female parents) in safflower: The Nucleus seed production field of GMS line is prepared with isolation distance of 800 m. from other safflower crop fields. Sourced seeds from previous nucleus seed or breeder seed stock of GMS line are sown with wide spacing and finally, only single plant per hill is retained. Careful selection is made for pairs of GMS and normal fertile plants in the same rows at the time of flowering and all GMS plants are carefully labelled. Uniformity for flowering is maintained by removing early and late flowering plants. Utmost care is bestowed to identify and removing off-types or deviants at different stages of plant growth. When off-types / deviant plants are noticed at flowering phase, the off-type and deviant plant and plants around them are also removed. Hand pollination is made between GMS x normal pollen fertile sibs within the same progeny row under nylon cage protected conditions and the hand pollinated GMS plants are labelled. At least 100 paired crosses are made under nylon cage protection and at maturity crossed seeds set on individual GMS plants are collected separately paired cross wise. Crossed seeds are examined in the laboratory for genetic purity and true to type nature.

In the next season, plant progeny rows from seeds set in GMS plants of individual paired crosses are raised in clean and fertile field along with the original GMS line for comparison. Isolation distance of 1600m. is maintained. Plants in each progeny row are examined for genetic uniformity and true to type nature. The off- types and deviant plants are removed when noticed at different stages of crop growth. The progeny rows are also verified at flowering for segregation in 1:1 ratio for male sterile and pollen fertile plants. Uniform looking progeny rows segregating 1:1 for male sterile and pollen fertile plants are covered / protected with nylon net cages separately at flower initiation stage. GMS plants are again verified and labelled. Hand pollination is made between male sterile and male fertile sibs of the same progeny row and labelled.

Seeds of GMS x male fertile plants harvested from individual progeny rows are examined for the characteristics, true to type nature and for seed yield and oil content. Seeds collected from labelled GMS plants from similar looking progenies rows are now bulked together to constitute nucleus seed of GMS safflower. In each selected progeny rows, seeds set on 10 to 15 heads of GMS plants are examined for the characteristics and genetic purity and stored for next cycle of nucleus seed production (Fig. 1).

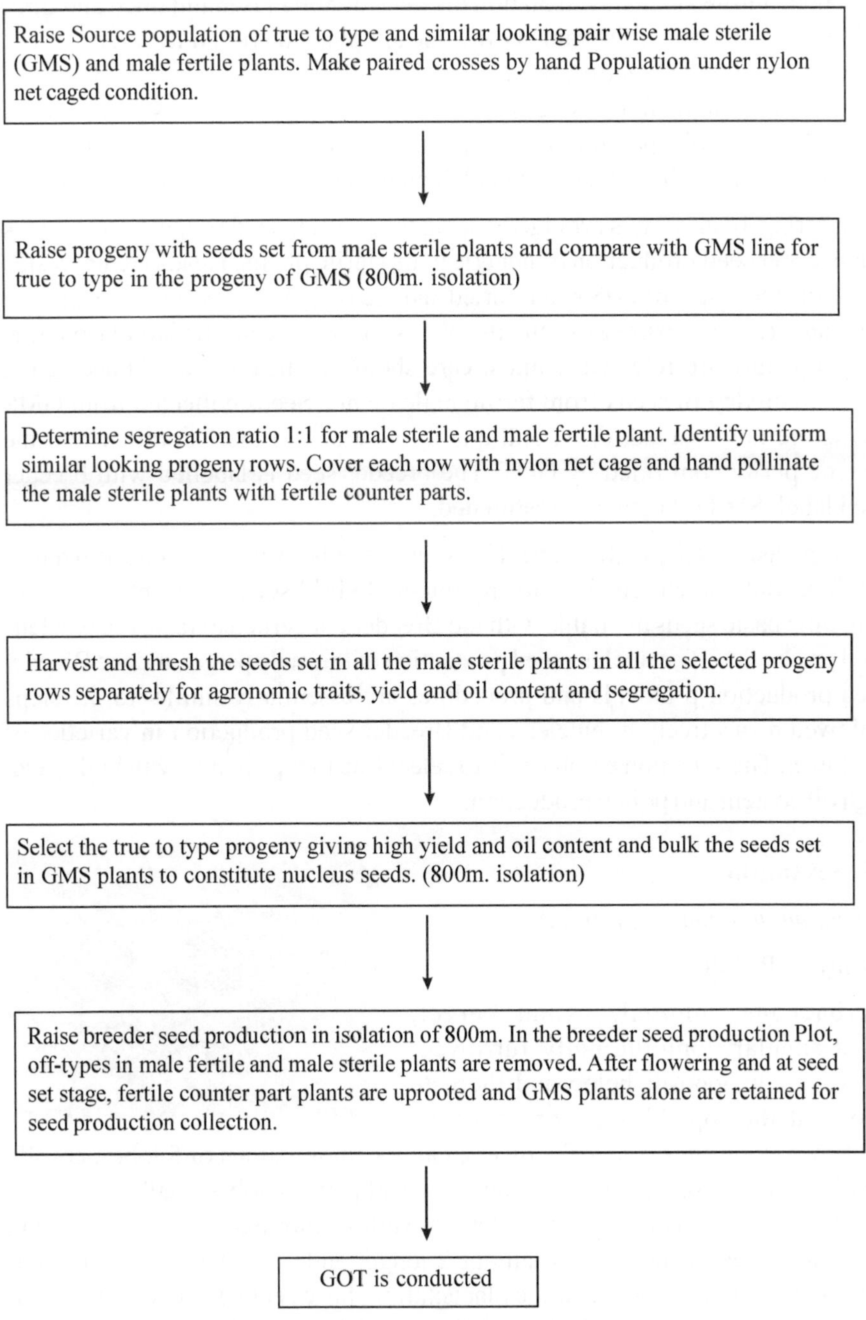

Fig. 1. Nucleus and Breeder seed Production in GMS line in Safflower

Breeder seed production of GMS line (Female parent): The breeder seed production plot is raised with nucleus seeds collected from GMS plants pollinated with male fertile sib plants. Isolation distance of 800 m is maintained and three meters border rows of the same GMS line are planted around the breeder seed production plot. The entire plot is examined on plant basis for off types and deviant plants at different stages of development up to maturity and they are removed. The off types and deviant plants that appear at flowering phase are uprooted along with the plants around them to avoid pollen contamination.

Monitoring Team visit is arranged during the breeder seed production to check the Breeder seed production of genetic male sterile plants. Heads are harvested first from the labelled GMS plants, dried and seeds separated. Seeds are examined for characteristics expression and the plants showing seeds deviating from true to type nature are rejected. Utmost care should be taken to avoid accidental physical mixing of seeds from fertile male plants. Seeds collected from GMS plants are now bulked, dried and quality assessed in the laboratory and the genetic purity confirmed by GOT. The breeder seed is labelled with breeder seed label. Seed oil content is estimated.

Male parent seed production: The source seeds of male parent utilized in GMS X Male parent crosses (for producing hybrid seeds) are obtained from authentic basic seeds available with the Breeder concerned to raise the Nucleus seed production. The nucleus seed production of the male parent and the Breeder seed production protocols and procedures are essentially similar to the steps followed respectively in Nucleus and Breeder seed production in varieties of safflower. The selection emphasis is to select true to type plants with high yield, high oil content and pollen production.

33. Sesamum

(*Sesamum indicum* L.) (2n:26)

Family : Pedaliacea

Herbaceous annual, stem erect, quadrangular, longitudinally furrowed. Leaves are in opposite below and alternate above at the top. Flowers are axillary, solitary or in groups of 2 to 3 flowers; calyx gamosepalous of 5 lobes, corolla tubular, five, gamopetalous, bell shaped; upper lip of two lobes smaller and lower lobe is bigger and made of 3 lobes. Flowers white to purple coloured. Androecium and stamens epipetalous, didynamous 2 longer and 2 shorter; ovary superior, bicarpellary, ovules many in axile placentation, false septa present. Extra floral nectary gland is present at the base of the peduncle. Fruit oblong to ovoid. Fruit loculicidal dehiscing. Locules 4. Seeds are compressed; white to brown or black. Flowers open between 5 and 7 a.m. The stigma is irritable and the 2 lobes separate when touched and the stigmas are covered with pollen by 5 a.m. (Duhoon, 2003). A greater extent of self pollination occurs before the sunrise. However, owing to insect pollination, a little pressure of pollinating agents and weather conditions may increase the natural cross pollination extent. Natural cross pollination occurs from 4 to 5%. Sesamum is considered as Queen of Oilseeds. The descriptors are given in Table 1.

Nucleus seed production : Source seeds of sesamum variety are raised in plot with isolation of 100m. distance and closely observed for uniformity, true to type, branching pattern and flowering among the plants and for the characteristics expressed in each of the development phases. Off types, deviant plants and disease (phyllody) affected plants are uprooted. During active flowering stage, 2000-2500 uniform and true to type plants are selected, self fertilised and labelled. At maturity, selected single plants are harvested and their pods and seeds are verified for true to type nature and uniformity. Laboratory examination is made for pod and seed characteristics and seed oil content in composite samples.

In the following season, single plant progeny plots are raised in two rows in 2 replications in 2 rows plot with 4m. long rows. Five to six border rows of the same variety of sesamum are raised all around the field. Isolation distance adapted is 100m. The progeny plots and all the plants in each of the progeny are examined in multistage of growth and development for uniformity and expression matching with the prescribed characteristics. Early and late flowering progeny rows are discarded. The deviant plants are removed at vegetative, branching, preflowering and capsule stage. When deviant or off type plants are detected at flowering stage, the entire progeny row is removed along with the adjoining rows on either side. Last round of examination of the selected progeny row plots is made to ascertain the true to type nature and uniformity for vegetative and reproductive characteristics. Finally progeny plots, that are above mean +2 SE for performance are selected. The selected / identified progeny plots are individually harvested and examined for capsule and seed characteristics and yield. Oil content is estimated. Poor yielding and low oil content progenies are rejected. The estimation of oil content concomitant with yield is repeated for selection in each cycle of nucleus seed production. Single plant progenies conforming to the characteristics and yielding higher than mean + 2SE are bulked together to constitute the nucleus seed. This process of single plant selection in authentic source population, single plant progeny evaluation and identifying higher yielding true to type progeny lines and estimation of seed oil content need to be repeated year after year in cycles (Fig. 1).

Breeder seed production: Seeds from the nucleus seed stock are sown in the field not grown with sesamum in the past 2 years. The crop is raised in the very favourable season with isolation distance of 100 meters. All good agronomic practices for good crop growth for seed crop and application of $ZnSO_4$ @10 kg/ha for favourable expression of characteristics are practised. Careful examination is made on individual plants at every growth phase to ascertain true to type nature and uniformity. All off-types and deviant plants are removed upto pre-flowering phase. In the flowering phase, if off-types and deviant types are noticed, they are uprooted gently avoiding pollen dispersal and plants all around are also uprooted and due care is taken to ensure genetic purity. Very early flowering and late flowering plants are discarded. Monitoring Team visit is arranged. After the harvest, capsule and seed characteristics are checked in the laboratory for confirming ~~to~~ the true to type nature. Quality assured seeds are packed and tagged with Breeder seed label. Grow Out Test is conducted as per the stipulated procedure to ascertain the genetic purity. Oil content in the seed is estimated.

Table 1. Descriptor in sesame

S.No.	Characters	States	Stage of observation
1.	Seedling vigour	Low/medium/high	Seedling
2.	Branching pattern	Basal branching/top branching	Vegetative
3.	Branching habit	No branching/low branching/medium branching/profuse	Vegetative branching
4.	Stem hairiness	Absent/sparse/dense	Green capsule
5.	Flower color	White/light purple/dark purple/violet	Flowering
6.	Days to flowering	Early/medium/late	Flowering
7.	Corolla hairiness	Absent/sparse/dense	Flowering
8.	Capsule hairiness	Absent/sparse/dense	Green capsule
9.	Capsule shape	Tapered/narrow oblong/broad oblong/square	Green capsule
10.	Capsules per leaf axil	One/more than one	Maturity
11.	Seed coat color	White/light brown/dark brown/grey/black	Maturity
12.	Capsule length	Short/medium/long	Maturity
13.	Days to maturity	Early/medium/late	Maturity
14.	Seed size	Small/medium/bold	Post harvest
15.	Oil content(%)	Low/medium/high	Post harvest
16.	Reaction to diseases phyllody (% plant infected), macrophomina stem/root rot (% mortality), phytophthora blight (%mortality), cercospora leaf spot (%leaf area covered)	I/R/MR/MS/S/HS; I/R/MR/MS; I/R/MR/MS/S/HS; I/R/MR/MS/S/HS;	Flowering Maturity All stages Vegetative
17.	Reaction to pests Antigastra gall fly (% damage)	R/MR/MS/S/HS; R/MR/MS/S/HS;	All stages Capsule

Source: Duhoon. SS (2003); PPV & FRA (2003)

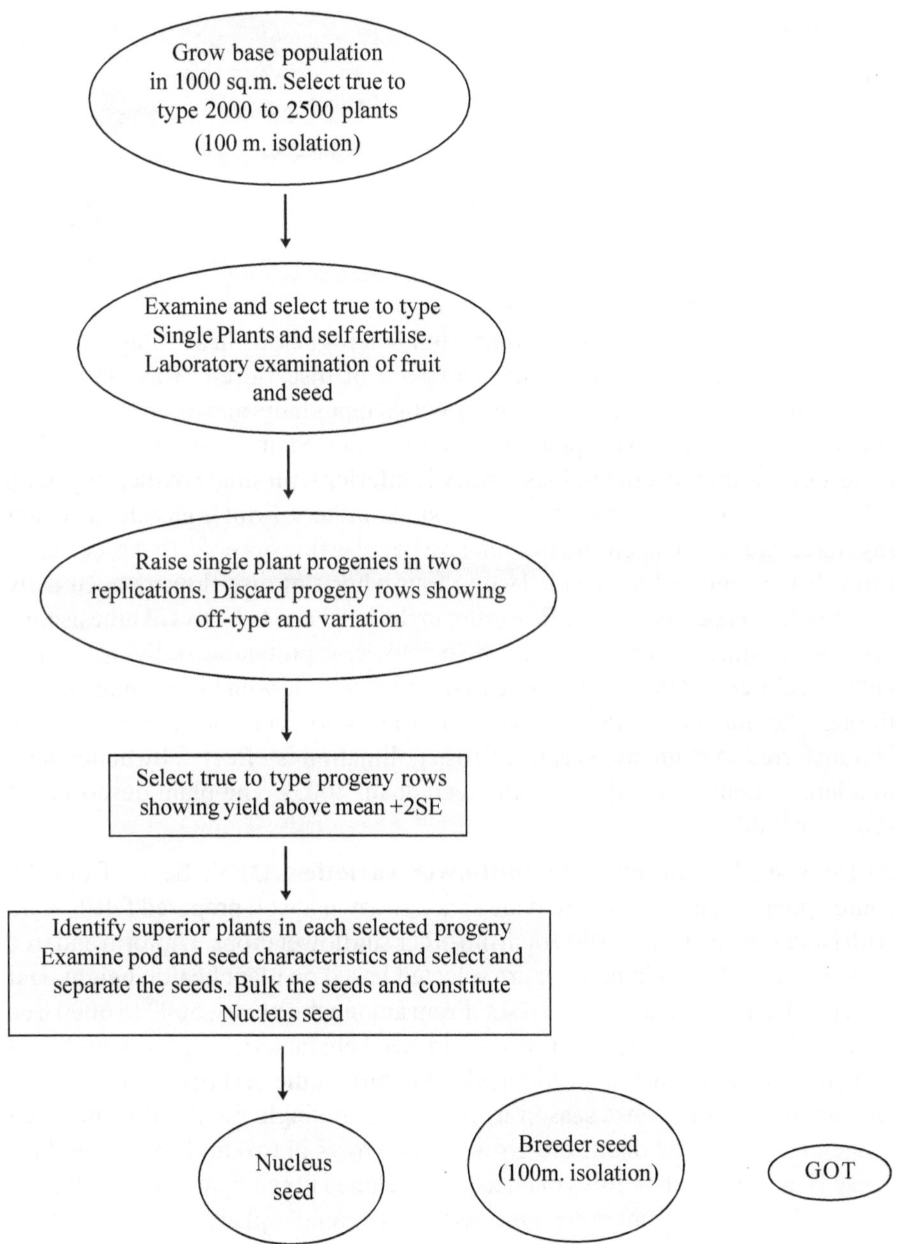

Fig. 1. Nucleus and Breeder Seed Production in Sesamum

34. Sunflower

(*Helianthus annuus* L.) (2n:34)

Family : Asteraceae

Sunflower varieties vary from dwarf to tall types and unbranched to branched types. Stem hairy, sturdy and pithy. Leaves are simple, large, alternate, angularly serrated in margins and dark green to green in colour. Inflorescence is terminal and axillary and heterogamous head. Outer rows in the head are ray florets and inner rows are of disc florets. Ray florets are zygomorphic, ligulate, yellow in colour with pappus hairs and often sterile. Disc floret is bisexual, actinomorphic and gamosepalous. Stamens are five in number, syngenesious united, epipetalous. Ovary is inferior with single ovule, style with bifid stigma that passes through syngenesious anthers. Fruit is an achene. Outer ray florets are first to open and the outer row of disc flowers open first proceeding towards the center of the head. Two to four whorls of disc flowers open daily and disc flowers complete the flowering in the head in 5 to 8 days. Anthesis time is in the morning between 6 and 8 am. Sunflower is protandrous. Syngenesious anthers dehisce and the pollen is released while the style with bifid stigma pierces through the anther tube . Pollination and fertilisation occur when the spiny pollen is transferred to stigmatic surface. Cross pollination is effected by honey bees to a large extent. (Giriraj and Vishnuvaradhan, 2003). The plant descriptor is shown in Table 1.

Nucleus seed production in sunflower varieties /OPV: Seeds from the source population / variety are wide space sown in a well prepared fertile field with isolation distance of 1000 m. from other sunflower crops. Uniform and true to type plants, 10000 in number are selected based on plant health, height, leaf size, head diameter and high seed set .From among the plants, 5000 to 6000 true to type plants are selected and seed yield, seed characteristics, test weight, oil content and hull content are estimated. One third of the best plants are selected and advanced to the next season as progenies of single plants. Five hundred progenies of selected plants are grown in two rows in two replications and for every 10 progenies, two rows of check are included for comparison. The field is isolated by 1000m. from other sunflower crops. Single plant is maintained per hill. The check entry is the original variety. Remnant seeds of each progeny plot are stored separately. If off types and deviant plants are noticed, the entire progeny row is uprooted. At different stages of crop growth, each progeny plot is observed for true to type, stability and uniformity in stand, flowering, flower heads size and maturity and good filling of heads and seed set. Seed yield is recorded and based on the performance, 250 to 300 high performaning progenies above the mean +2 SE are selected. Oil content of the seeds of the selected progenies is estimated. Final selection of 200 progenies is made based on the progenies performance with high seed yield and high oil content. The remnant seeds of the finally selected 200 progenies are bulked to constitute the nucleus seed (Fig. 1).

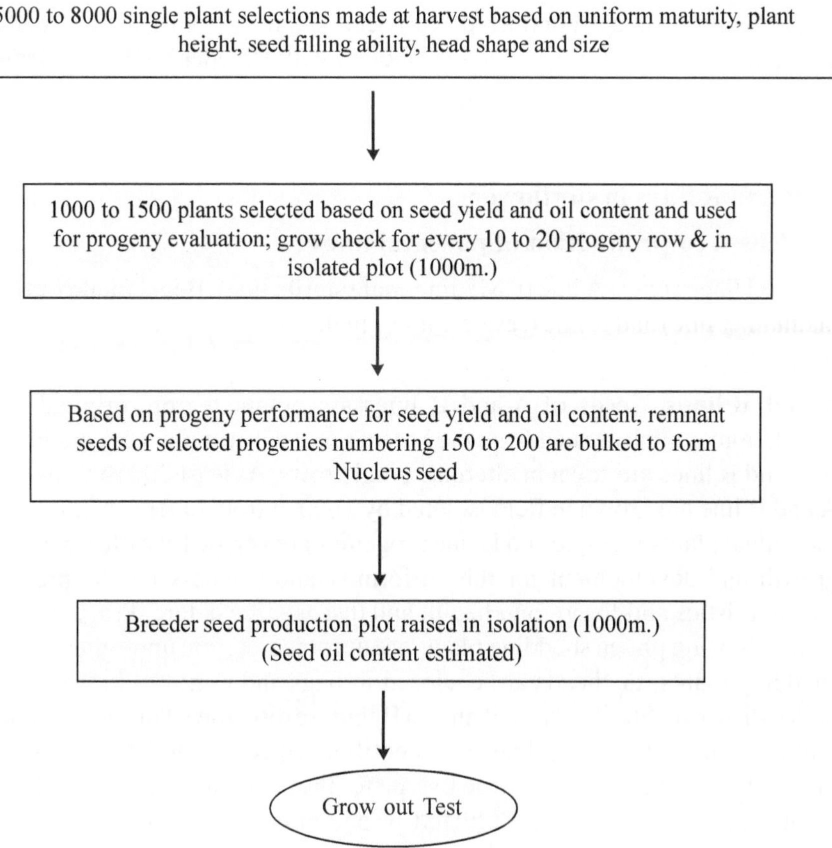

Fig. 1. Varietal maintenance in Sunflower (Pustovit scheme)

Breeder seed production: Nucleus seed is sown in well prepared fertile and clean field not grown to sunflower in the past 2 to 3 years. Isolation distance of 1000 m. from other sunflower crops is provided. Single plant is maintained per hill. Careful verification of trueness of each plant is made from the early juvenile to flowering and from flowering to seed set and maturity. Off types and deviant types of plants are removed and if they are noticed at flowering or later, they are removed along with the adjoining plants as well. During flowering, hand pollination between different plant heads is made to ensure adequate cross pollination. Monitoring Team visit is arranged at flowering or at mid maturity stage of the crop to judge the uniformity and true to type nature. At maturity, plants showing uniformity and true to type nature and high seed filling in heads, ideal head shape, seed yield and disease resistance are selected. Seeds from the selected plants are bulked to constitute breeder seed. In the laboratory seeds are examined for trueness. Seed oil content is estimated from seed samples. Seed quality tests are conducted with seed samples from Breeder seeds lot. The quality assured Breeder seed is labelled with Breeder seed tags. Grow out test is conducted with seed samples from breeder seed to verify the genetic purity.

A, B and R lines in sunflower

Nucleus seed production of parental lines

Parental lines are viz A line (CMS line-male sterile line), B line (male fertile and maintainer line) and R line (restorer line – male fertile and male fertility restorer line).

A and B lines: Seeds of A and B lines are obtained from original source population or nucleus seed from the breeder or the institution concerned. Seeds of A and B lines are sown in alternate single rows. At least 200 rows in each of A and B line are grown in field isolated by 1600 m from other sunflower crops. Individual plants in A line and B line progenies are verified at different stages of growth and development for the uniformity and trueness for the prescribed characteristics and for vigour, health and disease free nature. Progeny rows in A line showing pollen shedding plants are uprooted. Before uprooting the pollen shedding plants, the heads are enclosed in bags and removed in tact to avoid pollen dispersal. Sterile plants, if any, in B line are discarded. Uniform and similar looking plants of A and B lines are identified, selected and labelled. Selected plants should be uniform and true to type for the characteristics. The selected B plants are self fertilized and utilized in paired crosses of A x B. The paired crosses are made by aided hand pollination of A line plant with pollen collected from B line plant of the paired cross. Five hundred paired crosses are made. The A x B paired crosses are labelled cross wise, A1xB1, A2xB2 …… AnxBn. At maturity, individual plants in B lines are harvested first and seeds separated and labelled corresponding to each paired cross. The A line is harvested later and seeds are separated according to A1xB1, A2xB2 …… AnxBn and labelled as A1, A2, A3… An respectively and labelled (Fig. 2&3).

The seeds of respective A X B paired crosses are raised in one row as A line (female) alternated by seeds from the corresponding plant in B line of the paired cross. Isolation distance of 1600m. is maintained. A and B lines are observed for vigour, growth, true to type and conformity with all diagnostic characteristics.

The lines (AxB and B) showing off types or deviations are discarded. At flowering, if the A line plants are observed for pollen shedders, the entire A line along with corresponding B line are rejected. The heads of pollen shedding plants are enclosed in bags, gently removed and disposed outside the field. Very early and late flowering lines are discarded and 100 best progenies of A x B pairs (A line and B line) performing above mean + 2 SE are selected. Laboratory examination is carried out on the sunflower heads and selected. Seeds are separated and verified for the trueness. Seed oil content is estimated. Seeds from individual progeny lines are stored separately. Part of the seeds from each pogeny are bulked to constitute the nucleus seed of A and B respectively. The remnant seeds of the best selected progenies (A and B lines) are now bulked into A and B separately for continuing the nucleus seed cycle.

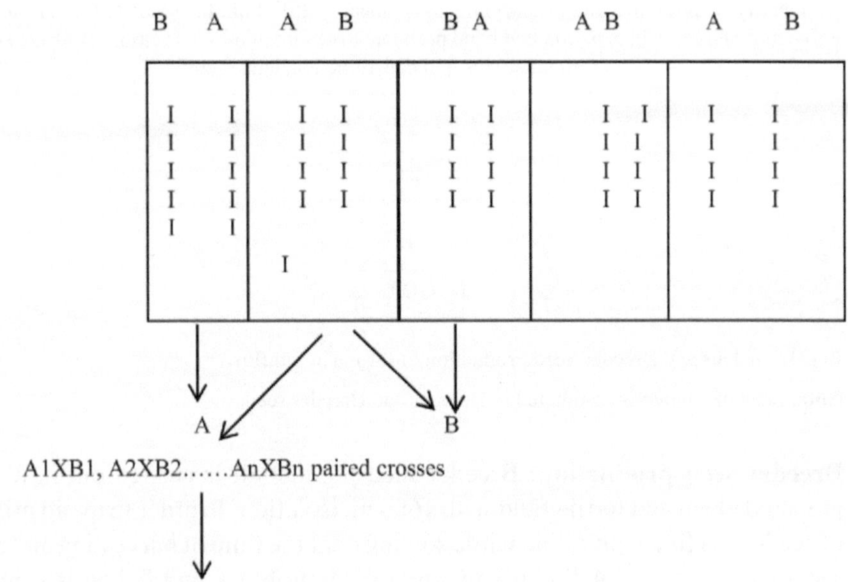

Plant A1XB1,A2XB2 A_nXB_n(A) and B line of individual plant in alternate rows.

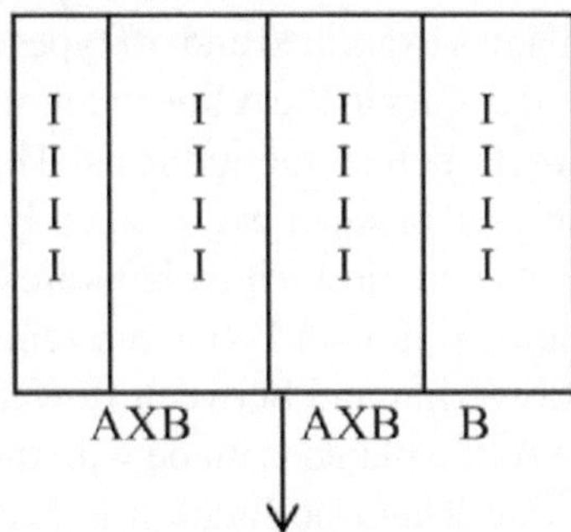

Fig. 2. Nucleus and Breeder Seed Production in sunflower A and B lines: (1600m. isolation)

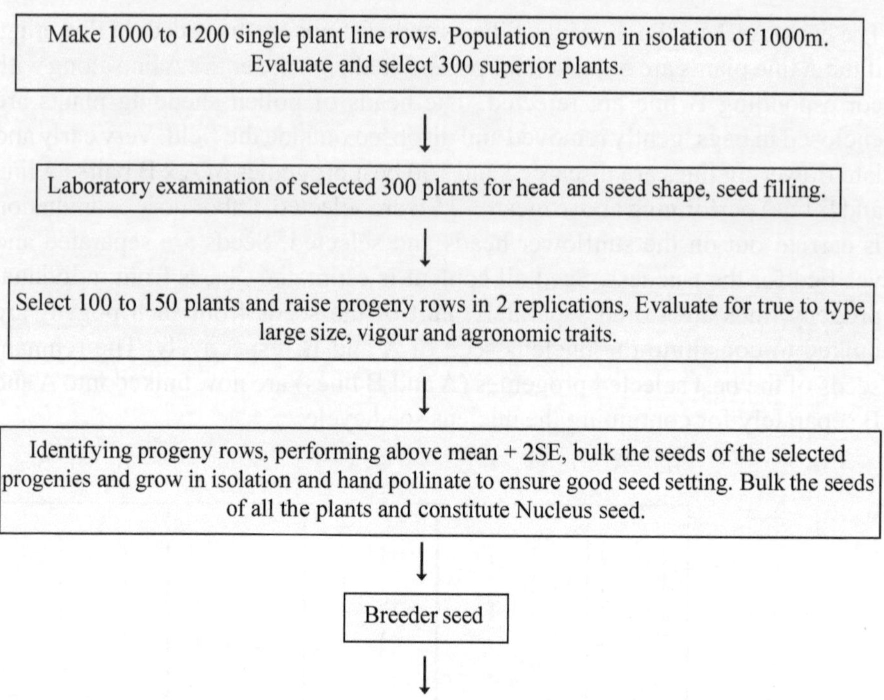

Fig. 3. Nucleus and Breeder Seed Production in R Line in Sunflower

Note: Seed oil content in estimated in Nucleus and Breeder seed.

Breeder seed production: Breeder seed production is carried out in a well prepared clean and fertile field with 1600 m. isolation. In order to avoid mixing of seeds of B line with A line while sowing or at the time of harvesting and seed separation, seeds of A line is sown in a separate block and B line is sown in another separate block (in 3:1 ratio area) in the same field. Between the A and B line plots, 2 m. space is left to avoid mechanical mixing of A and B plants. In A and B lines every individual plant is examined and off types and deviant plants are removed. Pollen shedders that occurs in the A line are immediately removed after unfolding of outer ray florets before the anthesis. The heads of A line showing pollen shedding are enclosed in paper bag and gently removed with out allowing pollen dispersal. Early and late flowering plants are discarded in A and B line blocks. Pollen sterile plants, if any, in B line are removed. Monitoring Team visit is arranged during flowering and between flowering and maturity. Hand pollination of male sterile A line plants is made with the pollen collected regularly from the heads of B line. Hand pollination is performed by gently dusting and transferring the pollen of B line to the heads of A line. All adequate steps are taken to maintain 100 per cent genetic purity in A line and B line. At maturity, B line is harvested first, dried and threshed and stored in separate containers. The heads of A line is harvested later, threshed and seeds dried. During these operations care is taken to avoid the mechanical mixing with seeds of B line. Seeds of A line are properly labelled and stored separately. Seed oil content is estimated. The seed quality parameters are tested and the quality assured Breeder seed of A line are labelled with the breeder seed label. GOT is conducted and genetic purity is verified.

Breeder Seed Production of B line: The seeds from the nucleus seed of B line that has the ability to maintain male sterility nature in A line are sown in plots isolated by 1000 m. from other sunflower crops. Continued examination at all stages of crop developmen is made for the true to type and uniformity nature.In case if male sterile plants occur in B line plants such sterile plants are uprooted. Thorough rouging of off types and deviant type of plants before and after flowering is carried out. Monitoring Team visit is arranged. Heads that show conformity with the descriptor alone are selected and seeds separated. Oil content in seed samples is estimated. Seed testing is done to ascertain the seed quality standards.The quality assured seed of B line is labelled with the Breeder seed label. GOT is conducted to ascertain the genetic purity.

Nucleus Seed Production of R line: Seeds of R line from source population /line are sown in an area of 500 sq. m. in isolation of 1000 m. distance from other sunflower crops. Uniform looking and true to type plants conforming to the R line characteristics are selected at pre-flowering stage and 1000 plants are self fertilized. Hand pollination is made to improve seed set. At maturity self fertilized individual heads are collected. Based on the field study and laboratory examination of the self fertilized heads for shape, size and seed set and for seed characteristics, 300 plants are finally selected after the laboratory examination of the seeds. Oil content estimation of seeds is done.

In the next season, 250 to 300 plant progenies are grown in 2 rows plot in two replications. Portion of selfed seeds is retained as remnant seeds. The progeny rows are critically examined for presence of offtypes and deviant plant types. Progeny rows showing such plants are rejected and uprooted. If offtypes and deviant type plants occur after flowering, the progenies are rejected along the adjoining rows on either side and rejection of progenies is made in both replications. Very early flowering and late flowering progeny rows are discarded. In an isolated plot (1000m), A line plants raised with seeds from nucleus / breeder stock of A line are sown. At the flowering time, plants are labelled and crosses made with pollen collected from R lines that are grown along with the A lines. Crossed seeds are collected and saved as A×R hybrid. The A×R hybrids seeds are sown A×R crosswise in the next generation and evaluated for vigour, health, pollen fertility restoration and high seed setting and head filliing. The heterotic vigour of the A×R cross are noted and poor A×R are rejected with the R. The selected R progeny rows in both the replications are harvested and seeds of the progeny rows showing performance above mean + 2SE are selected and seeds separated. Part of the seeds from each selected progeny are bulked to constitute nucleus seed.

Breeder Seed in R: Seeds from the R lines that showed fertility restoration, heterotic vigour in A×R crosses are sown in 1000m. isolation and Breeder seed is produced. Monitoring Team visit is organized at mid maturity. Matured heads of true to type plants are harvested and seeds collected after laboratory examination.

Table 1. Descriptor of sunflower

S.No.	Characters	States	Stage of observation
1.	Leaf size	Very small/small/medium/large/very large	Vegetative
2.	Leaf color	Light green/medium green/dark green	Vegetative
3.	Leaf blistering	Absent/medium/strong	Vegetative
4.	Fineness of serration	Fine/medium/coarse	Vegetative
5.	Flowering	Early/medium/coarse	Vegetative
6.	Ray flower color	Ivory/pale yellow/yellow/orange/ purple/red brown/multi colored	Flowering
7.	Head diameter	Small/medium/large	Harvest maturity
8.	Head shape	Concave/flat/convex	Harvest maturity
9.	Plant height	Very short/short/medium/tall/very tall	Vegetative/ reproductive
10.	Seed length	Short/medium/long	After harvest
11.	Seed base color	White/grey/brown/black	After harvest
12.	Stripes on seed	Absent/present	After harvest

Note: Description is based on Giriraj and Vishnu Varadhan (2003); PPV & FRA

Select true to type A and B lines and grow them in alternate rows. Make AXB paired crosses.

Select 100 lines true type and uniform looking lines of (AXB) and true to type B rows and selected paired crosses of AXB and B are evaluated for trueness and uniformity and heads are selected. Part of the seeds from individual A×B plants are bulked to make nucleus seed. The rest of the A×B cross seeds saved for nucleus seed production in the next cycle.

Breeder Seed Production following the steps in nucleus seed production (in isolation 1600m)

Fibre Crops

35.Cotton

Silk cotton

(Gossypium sp) (2n:52)
Family : Malvaceae
Gossypuim hirsutum (2n=52)
G. barbadense (2n=52)
G. arboreum (2n:26)
G. herbaceum (2n:26)

Herbaceous, annual, primary axis erect and dimorphically branched as monopodia (vegetative) and sympodia (fruiting) branches. Monopodia do not directly bear flowers. Sympodia are present in varying number. Sympodia or fruiting branches arise from the axils of subtending leaf on monopodia. Leaves are alternately arranged, large to small, palmate to lobed to different degree and prominently petiolate. Hairiness on leaf shows wide variation from smooth to highly and densely hairy. Flowers are simple, solitary, axillary or extra axillary. In the flowers, three free serrated, persistent and green bracteoles are present, collectively termed as epicalyx. Each bracteole exhibits a nectar gland. Calyx is green, five, fused at the flower base and cup shaped. Petals 5, polypetalous, free. At the inner side of the petal base, petal spots of different size and colour intensity are present in some species while absent in certain other species. Staminal column enclosing the stylar column displays anthers firmly attached to the monadelphous staminal column. Anthers are reniform. Length of anther column and anther number show considerable variation. Ovary superior, 5 loculed, the slender style passing through the staminal column terminates with 3 to 5 stigmatic lobes. Placentation is axile. Each locule contains 5 to 8 or more ovules. Capsule fruit is loculicidally dehiscent. Seeds are mostly linted or hairy, seldom naked. Hairs are long linted or short fuzzy. At maturity, capsules dehisce & are termed as squares. Lint is of varying length. Cotton is considered as an often cross pollinated crop. The detailed crop characteristics are presented in Table 1.

Table 1. Descriptor of cotton

S.No.	Characters	States	Stage of observation
1.	Hypocotyl pigmentation	Absent/present	Seedling
2.	Stem pigmentation	Absent/present	Vegetative/reproductive
3.	Stem hairiness	Absent/medium/strong	Vegetative/reproductive
4.	Plant height	Dwarf/medium/tall	Vegetative/reproductive
5.	Leaf shape	Palmate (normal/digitate okra/ semi-digitate lanceolate)	Vegetative/flowering

6.	Leaf size	Small/medium/large	Vegetative/flowering
7.	Leaf color	Light green/green/light red/dark red	Vegetative/flowering
8.	Leaf pubescence	Absent/medium/strong	Vegetative/flowering
9.	Leaf nectaries	Absent/present	Vegetative/flowering
10.	Leaf petiole pigmentation	Absent/present	Vegetative/flowering
11.	Bract type	Normal/frego	Vegetative/flowering
12.	Sepal pigmentation	Absent/present	Flowering
13.	Petal color	White/cream/yellow/pink/ red/bicolor	Flowering
14.	Petal spotting	Absent/present	Flowering
15.	Position of stigma	Embedded/ exerted	Flowering
16.	Anther color	White/cream/yellow/purple	Flowering
17.	Boll size	Small/medium/large	First boll bursting
18.	Boll shape (longitudinal section)	Rounded/elliptic/ovate	First boll bursting
19.	Boll tip prominence	Blunt/pointed	First boll bursting
20.	Boll surface	Smooth/pitted	First boll bursting
21.	Boll opening	Close/semi-open/open	First picking
22.	Fibre length	Very short/short/medium/ long/extralong	First picking
23.	Fuzz color	White/grey/brown/green	Harvest maturity
24.	Fibre color	White/off white/brown/green	Harvest maturity
25.	Fibre strength	Weak/medium/strong	Harvest maturity
26.	Ginning percent	Low/(<31), medium (31-35)/ high(36-40)/very high (>40)	After ginning
27.	Density of fuzz	Naked/semi-fuzzy/fuzzy	After ginning

Nucleus seed production: Base population of variety is grown in 1000 sq. m. field with isolation distance of 100 m. from other cotton growing fields with the seeds that are sourced from the concerned breeder or from the maintenances breeding source stock. The field should be fertile, well prepared and all good agronomic production practices are followed. Weak plants and disease and pest affected plants are uprooted as and when they appear upto preflowering stage. Off types and deviant type of plants for stem pigmentation, leaf colour, leaf lobing and hairiness are uprooted. Uniformity to first flowering is observed and very early or late flowering plants are removed. After flowering stage, continuous observations are made on each and every plant for stem and leaf characteristics and for flower (petal) colour, anther colour, pollen colour and stigma exertion. Plants deviating for any of the characters are rogued out. Developing bolls are examined for size, shape and pittedness and plants showing deviations, if any, are uprooted. All plants in the source field are monitored and only those plants that are uniform, true to type and conforming to the description of characteristics are retained and from them 250 to 300 plants are selected and self fertilized and such self fertilized plants are verified at active boll development phase / boll bursting phase for uniformity for boll size, shape and locule number and good boll bursting. Self fertilised and selected superior plants are harvested individually and labelled. The fibre properties are evaluated to determine 2.5% span length, micronnaire, uniformity ratio, bundle strength and compared to the prescribed quality standards for that variety. Single plants that match with the varietal specification of fibre properties and which lie within mean ± SD are selected (Manickam *et al.,* 2003), (Fig. 1).

Self fertilized seeds from the selected plants in the previous season are now raised in two row progeny plots replicated twice in a compact area with 100 m. isolation distance. All recommended agronomic production practices for cotton are followed. The progeny plants are carefully examined for true to type on individual plant basis for juvenile, vegetative, flower, boll development and boll bursting stage as described earlier. If any deviant plant is identified for any of the characteristics, the progeny row as a whole is rejected. Fifteen true to type and healthy plants are selected in each identified progeny and self fertilized. Fibre quality data are recorded in the selected progenies wise and statistically analysed. The progenies that fall within the mean ± CD at 5% are selected.(Manickam *et al.,* 2003). The selected progenies are tested in microspining test . Part of the selfed seeds from the selected progenies are bulked to constitute the nucleus seed. The remnant selfed seeds of the selected progenies are stored and forwarded to the next cycle of nucleus seed production. In case, the entire quantity of selfed seeds is required for production of large quantity of Breeder seed, the next stage of seeds from the progeny bulk seed may be taken as Nucleus seed stage II (Fig. 1).

Breeder seed production: The nucleus seed is sown in an ideal location suitable for quality seed production by maintaining isolation distance of 100 m. Single plant per hill is maintained. All good agronomical and seed production practices are adapted. Ten rows of border crop with the same variety are raised on all sides of the plant. Examination of all the individual plants from juvenile to maturity stage is thoroughly made and off types and deviant plants for vegetative, floral and boll characters are uprooted. Visit of Monitoring Team is arranged to the field and to assess the crop purity and uniformity. Fiber quality evaluation and microspinning tests are conducted with adequate samples of lint and only after satisfying the variety specifications, seeds of individual plants are extracted from kapas. Seeds from the selected plants are bulked to constitute the breeder seed. Samples of the breeder seeds are tested for seed qualities and quality assured seeds are labelled with Breeder seed label. Grow out test is conducted with the samples drawn from the breeder seed lot. Cotton fibre qualities are tested.

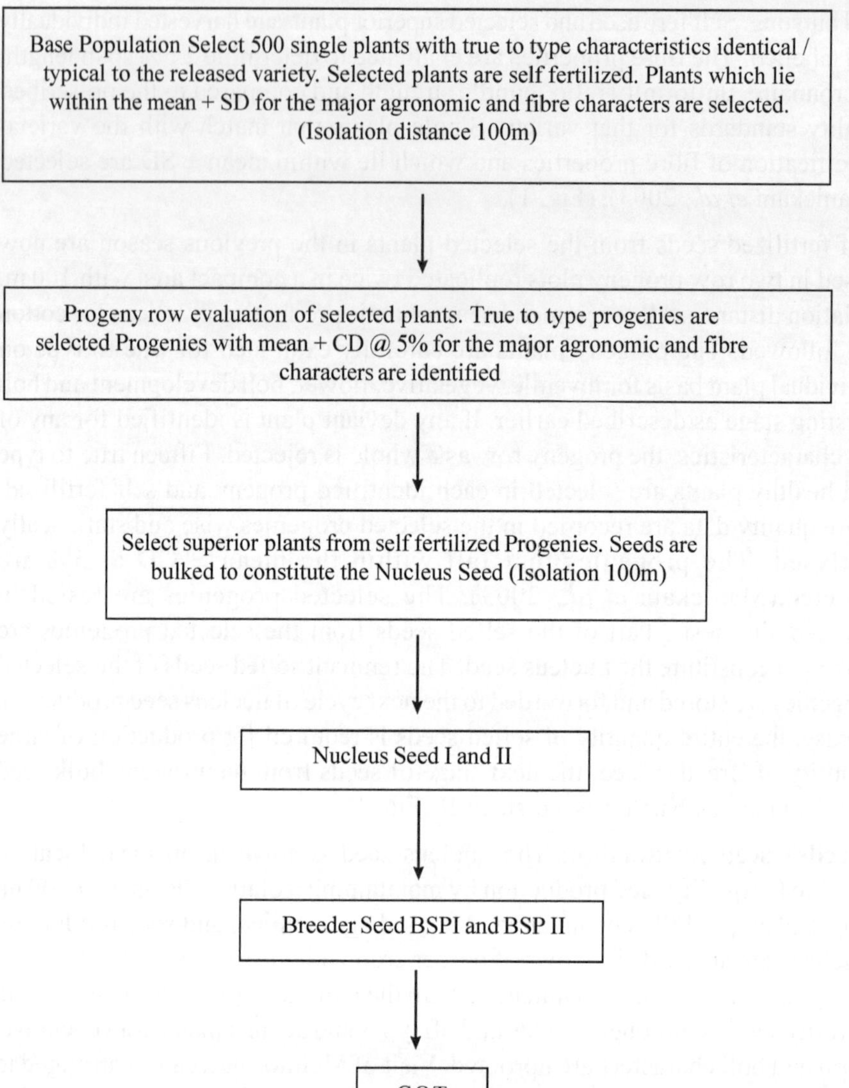

Fig. 1. Nucleus and Breeder seed production in Cotton:

Parental lines of commercial cotton hybrids

Nucleus and Breeder seed production: The nucleus and breeder seed production of male and female parental lines are taken up in the same way as being done in case of nucleus and breeder seed production in a variety. The required isolation distance of 100 m. is maintained in the seed production plots between male and female parents and also from other cotton crops being grown.

All meticulous efforts in verifying for the different characteristics, rouging off types and deviant plant types and maintaining true to type and uniform plants are fulfilled. At final stage, 500 paired crosses are made between the selected plants of male and female parent lines grown in separate isolated plots and enough crossed seeds are obtained. Seeds from the corresponding male and female parent plants are saved.

In the following season, the crossed seeds are raised along with male and female parents. The hybrids are assessed for its heterotic vigour, boll set, yield ability and fiber quality traits. Hybrids between such of those lines showing highest SCA and yield are identified and seeds of the corresponding parent lines are utilized for futher multiplication in nucleus and breeder seed production plots as the case may be. In order to retain the high yield potential and fiber qualities of the parental lines and to realise highest heterosis and SCA in the resulting hybrids, this procedure is repeated in every season.

'A' line (CMS A line), B line (maintainer line) and R line (restorer line)

Nucleus seed production: 'A' line (CMS A line): The CMS male sterile A and male fertile B lines are isogenic but alloplasmic. The A line is maintained by pollinating the A line with the corresponding B line. Nucleus Seed production of A line is done in fields isolated by 100 m. distance from other fields being grown with cotton. The B line seeds are sown in alternate rows to A line . All steps are taken to verify the characteristic of CMS A line and B line with the parent description, true to type nature and thorough rouging is done to maintain purity. At flowering stage, all the individual plants in A line are observed for pollen shedders and such pollen shedder plants are uprooted. In B line, pollen sterile plants ,if any, are uprooted. Plants in B line are selected and self fertilized and the same plants of B line are utilized for making A x B paired crosses. The A x B paired crosses are serialy numbered as A1 x B1, A2 x B2, A3 x B3.... An x Bn, and 200 to 250 paired crosses are made and crossed seeds are obtained along with the self fertilized seeds of corresponding plants in B line.

In the next season, portion of the A x B crossed seed and selfed seeds of B line are sown in alternate rows of A and B respectively. The A and B lines are observed for the prescribed characteristics in pairwise cross origin. Plants of A line are also examined for male sterility expression. A lines showing pollen shedders and the corresponding B line of that paired cross are uprooted. Uniform looking true to type A and B lines conforming to the parental lines characteristics are identified and selected. The lint quality of the B line is tested before finalizing selection of A x B paired crosses. The remnant crossed seeds of the selected paired crosses of A x B that are free from off types, deviant plants and pollen shedders in A line and conforming to the parent standards in lint quality are selected and bulked to constitute nucleus seed of A line for producing breeder seed of A line (Fig. 2&3).

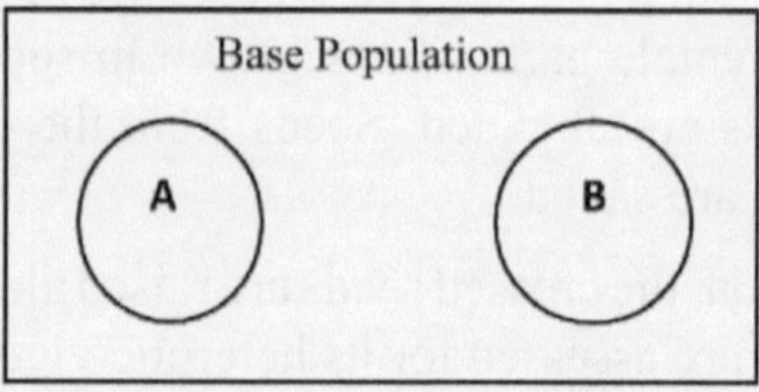

Select 100 to 150 true to type superior and similar looking A and B plants and make AXB paired crosses.

Raise progeny rows of A (A×B) and B

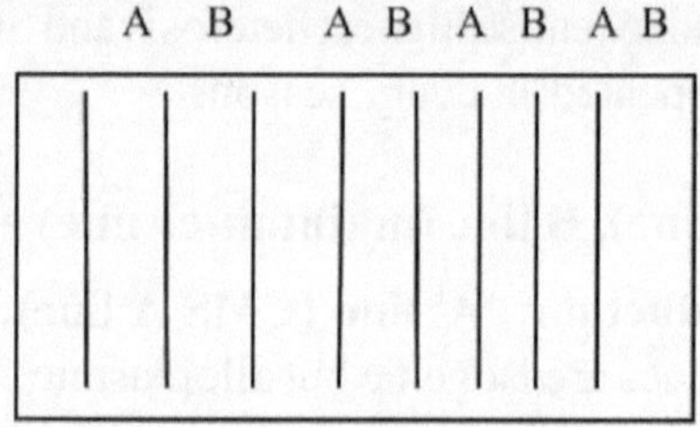

Identify uniform true to type progeny rows in A (A×B) and B respectively and perform selected paired crosses of similar looking A and B plants:

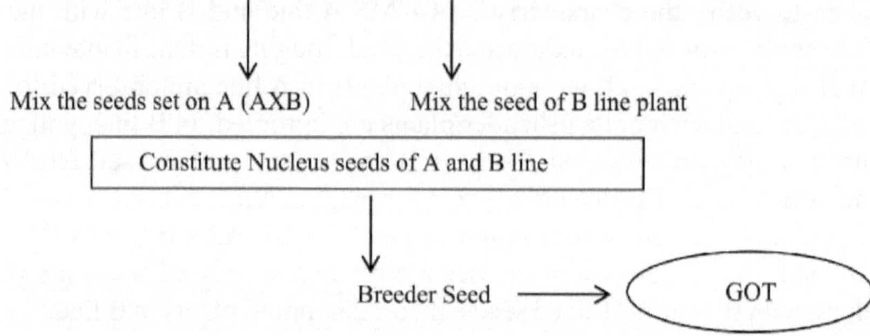

Fig. 2. Nucleus and Breeder Seed Production in CMS (A) and maintenance (B) lines in Cotton

Breeder seed production in A line: In the breeder seed production of A line, sib mating of A line and B line is followed by raising A and B lines in 1 to 1 row ratio in isolated plots with 100 m. distance. The plants in A and B line are verified for trueness and uniformity. Deviant type of plants are removed at preflowering stage. Deviant plants and offtypes noticed during flowering phases are uprooted along with the adjoining rows to prevent cross pollination. Plants in A line are also examined for male sterility expression. A lines showing pollen shedders are discarded.Plants in B line showing pollen sterility expression are uprooted. Male sterile A line is cross pollinated with pollen from isogenic B line plants with due precautions. Supplementary aided pollination can be done between A and B lines. Monitoring Team visit is arranged at the boll bursting stage.The plants in B line are cut and discarded well before the harvest of bolls in the A line to avoid mechanical mixing of bolls .At maturity bolls set in the plants in A line are harvested and examined and selected for seed separation.Seed samples from A line are tested in the laboratory for seed qualities and the quality assured seed is labelled with breeder seed label. Grow out test is conducted with the seed samples from the breeder seed of A line.

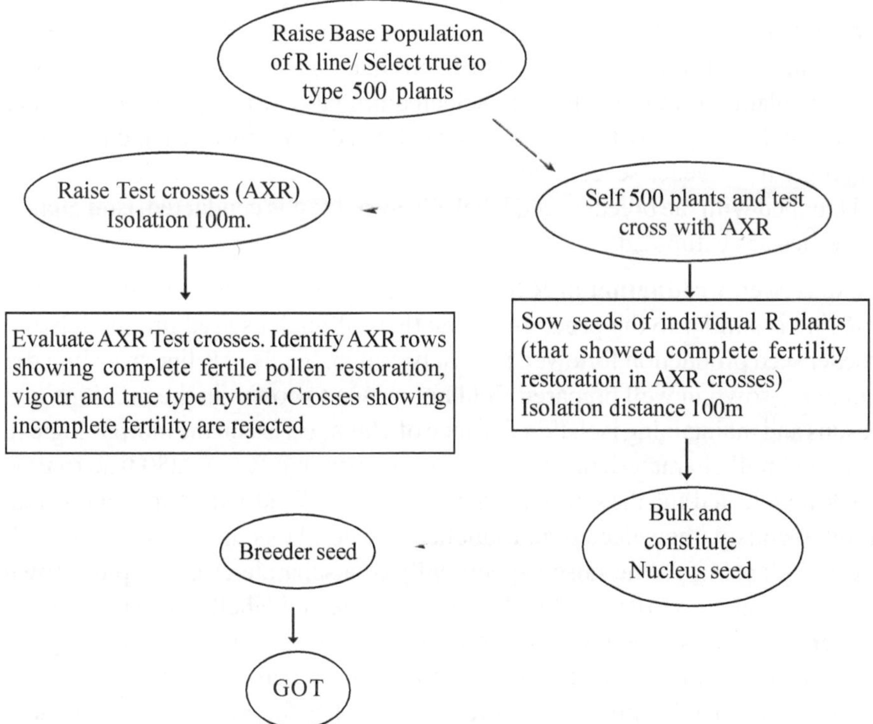

Fig. 3. Nucleus and Breeder Seed Production in R line in Cotton

Nucleus seed production in B line: In respect of production of nucleus seed of B line which is the isogenic maintainer line of the pollen sterility trait in A line, selection is made for 200 to 300 individual plants of B line that are already tested for the ability to maintain male sterility in A x B paired crosses. Isolation distance of 100m. is maintained from other cotton crops. Selected plants are verified for conforming to the B line true to type characteristics and self fertilised. Progeny rows of B line are raised and evaluated for vigour, growth rate and true to type for foliar and floral traits and kapas and seed features. Fiber properties are tested for the individual B plants and plants are finally selected and seeds extracted. Seeds bulked from the selected plants from the superior progeny rows constitute nucleus seed of B line.

Breeder seed production in B line: In the following season, nucleus seeds of the B line are sown in well prepared field plot with isolation distance of 100 m. All good agronomic and seed production protocols are followed. Off types and deviating plant types are discarded. Plants that are uniform, true to type, stability in expression of characteristics are selected at flowering. Early and late flowering plants from the days to first flowering recorded in the corresponding A line are removed. Deviating plants noticed in the post flowering phase are removed along with the adjoining plants around to avoid pollen contamination and vicinism. Monitoring Team visit is arranged at boll, maturing stage. The superior plants from the B line are selected at maturity and matured bolls are harvested. Fiber quality tests are performed. Seeds are separated and bulked to constitute the breeder seed. Seed samples are tested for seed quality assurance and labelled with the breeder seed label. Grow out test is conducted. Lint quality properties are estimated.

Nucleus Seed Production of R line: Production of nucleus and breeder seeds in R line is similar to the production practices of nucleus seed production and breeder seed production followed in variety cotton. Seeds of R line from breeder source are sown in well prepared field not grown to cotton in the previous two seasons and maintaining isolation distance of 100m. Based on the morphological, floral and boll characteristics matching to the original R line, 250 true to type plants are selected and the selected plants are self fertilized and serially numbered. Simultaneously, the selected and labelled R line plants are test crossed with selected plants of A line raised specifically in a separate crossing plot grown with A line and test crosses of AxB are also made and labelled serially. Part of self fertlised seeds of R line are raised in next season as R line plant progeny rows along with the corresponding test cross (A x R) grown at 100m. isolation distance. Thorough comparison for R line progenies for conformity to trueness with R line characteristics is made. R line progenies showing off types and deviating plants are rejected. The test cross progenies of A x R are observed for high heterosis and heterotic vigour, complete pollen fertility restoration and high boll set. Such of those R line progenies not capable of complete restoration of pollen fertility are rejected and uprooted. The remnant selfed seeds of the selected R line progenies conforming to the original R line and proven ability of fertility restoration and heterosis in test crosses (A x R) alone are identified and bulked to constitute the nucleus seed of R line.

Breeder seed production of R line: The breeder seed production of R line is done in isolation plot (100m.). Border rows are raised on all sides with the same

R line seed. The procedures described earlier in respect of rouging deviant plants and off types and selecting plants that conform to the description of R line in respect of morphological, flower and boll characteristics are followed. Lint qualities are tested in the selected plants of R line. Monitoring Team visit is arranged at the boll bursting stage of R line. The seeds of R line is tested for seed qualities and quality assured seeds are labelled with breeder seed label. Grow out test is conducted with breeder seed samples of R line to assess the genetic purity (Fig. 3).

GMS Line:(Genetic Male Sterile line)

Nucleus and Breeder Seed Production of GMS line in cotton: Genetic male sterile plants (GMS) are maintained by sibmating between the GMS plants and the corresponding male fertile sib plants. The sib mated progenies segregate in 1:1 ratio for male sterile and male fertile nature. The male fertile plants are culled out and seeds set on the male sterile plants alone are utilized as female parent. Isolation distance of 100 m. is maintained in the seed production plot.

Five hundred fertile plants from the base population that is heterozygous for male fertility and similar to the GMS female parent line are selected. The selected 500 plants will show segregation in the progenies in equal proportion for male fertility and male sterility. Sib mating is done between the male sterile (GMS) and male fertile plants within the segregating progenies. In the next year, the portion of seeds collected from GMS plants of each sibmated crosses are sown in progeny rows in 100m isolation plot. The progenies are evaluated for morphological characteristics and fibre quality. Segregation for male sterility and male fertility in 1:1 ratio is also checked in each progeny. Progenies that conform to the description of GMS line and segregating for male sterility and male fertility in 1:1 ratio alone are selected and the rest of the progenies rejected. Boll and fibre characters of the fertile plants are tested. The remnant seeds set in the male sterile plants of the selected progeny rows are bulked to constitute Nucleus seed for breeder seed production (Fig. 4).

Breeder seed production of GMS parent: Seeds obtained by sibmating the GMS plants with the male fertile counterparts are sown in isolated field plot with 100 m. isolation distance. Single plant per hill is retained All good agronomic and seed production practices are followed.The individual plants are verified for true to type nature and uniformity and off types and deviant type of plants are uprooted as and when they occur. Plants are examined for the male sterile nature and the male sterile plants are tagged. The total similarity of the fertile counterpart plants with the GMS plants is verified. Hand pollination of the sterile plants with the pollen from fertile plants is carried out to the extent necessary. Field visit of the Monitoring Team is arranged. At maturity, the male fertile plants are removed by uprooting and discarded. Matured bolls set in the tagged GMS plants alone are harvested separately and seeds extracted, dried and processed. Lint fibre qualities are also tested in the bulked samples of male plants drawn from the kapas samples. Seed samples are tested for seed qualities and the quality assured. seeds are labelled with Breeder seed label. Grow out test is conducted with seed samples from the breeder seed to judge for the genetic purity of the GMS parent line. Complete expression of morphological characters essential for the comparison is obtained from 800 plants in the GOT to confirm the genetic purity.

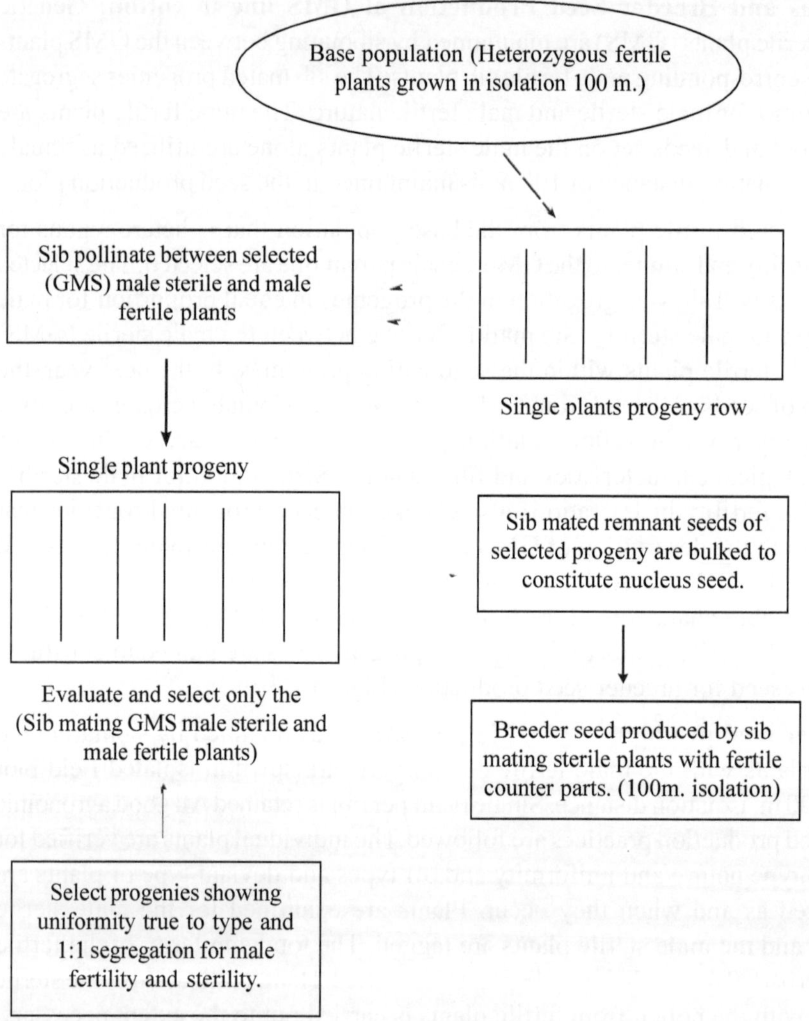

Fig. 4. Nucleus and Breeder seed Production in GMS parent in Cotton

36. Jute

Corchorus capsularis L. (2n:14) and *Corchorus olitorius* L. (2n:14)

Family : Tiliaceae

India is the largest producer of Jute which is a sustainable versatile fibre for raw jute bundles production. The tender leaves are the leafy vegetable sold for 2 months in the market out of 4 months crop cycle of jute. Tender wood stems are utilized in the paper manufacturing. Nine hundred kg of fibre is obtained from one acre. Jute is easily blended with natural and synthetic fibres, Jute is a carbon neutral crop. Jute is a versatile crop.

C. olitorius: Herbaceus annual. Stem greenish or reddish. *C. olitorius* is more branching than *C. capsularis*. Leaves are glabrous. Flowers axillary, stamens many, 30 to 60; ovary elongated, 5 to 6 carpelled. Fruit elongated capsule with ridges lengthwise on capsule. Seeds are blackish to bluish green in colour.

C. capsularis: Herbaceus annual, stem slender, cylindrical with sparsely produced branches. Leaves simple, alternate, ovate to oblong. Flowers small sized and in axillary cymes of flowers. Stamens 20 to 30, superior ovary, carpels syncarpus and placenta axile. Fruit is a capsule, five loculed, smaller and brownish in colour.

In both the species, flowering is completed in a period of 4 to 6 weeks. Flowers open 2 to 3 at a time. Flowering is prolonged for about 10 weeks in *C. capsularis*. Flower opening commences from 8.30 am. to 10.30 am. and closes at 3.30 pm. In *C. capsularis*, anthers dehisce one hour before flower opening rendering them into cleistogamous system. In *C. olitorius* anther dehiscence commences one hour before sunrise and is completed just before flower opening Both the species of Corchorus are predominantly self pollinated. The extent of out crossing is 1.04% to 16.4% in *C. capsularis* and 3.30 to 17.70% in *C. olitorius* (Kumar *et al.,* 2003).

Nucleus seed production: Quality seeds of the variety maintained by the breeder are sourced and sown with wide spacing. The crop is raised in a well drained fertile field. Single plant per hill is retained. Isolation of 100 m. from other jute crop is ensured. After careful examination of the health of plants, 1000 individual plants that are uniform looking, true to type and conforming to the varietal characteristics and flowering are selected and labelled. At maturity, single plants are harvested separately and examined in the laboratory for the capsules and seed features and the plants are selected based on the uniformity and true to type nature of seeds are carried forward as plantwise seeds.

In the next season, that is ideal for the jute crop, 500 plant progeny are raised in 2 rows plots; 4m long plots with a spacing of 40cm. x 30cm. in 2 replications in isolation of 100m. In each block 50 progenies are sown along with 2 rows of the variety for compossion. Five rows of seeds of the same variety are also raised in all the borders. Based on the requirement of nucleus seeds, number of blocks is decided. All good agronomic practices are provided to raise a healthy crop showing good expression. Single plant per hill is maintained. Careful observations are made on individual plants in all the progenies from juvenile to preflowering, flowering and capsule and seed maturity stages. Progenies showing off types and deviant plant types are uprooted to maintain genetic purity. When off types and deviant type plants occur at flowering and post flowering stages, such progenies and also progenies on either side are uprooted to prevent cross pollination and contamination. At maturity, single plant progenies are harvested separately and the capsules and seeds are critically examined for any deviations and progeny rows showing deviating plants are rejected. Plants with uniform true to type progeny rows showing mean values equal to or more than mean +2SE are identified. Seeds are selected from superior plants in each identified progeny and seed packets labelled. Required number of individual progenies are separately stored progeny wise for next cycle of nucleus seed production. Seeds of the remaining progenies are bulked together to constitute nucleus seed (Fig. 1).

Breeder seed production: Nucleus seed is sown in a well prepared fertile field with 100 m. isolation distance. In well spaced rows, one plant is retained per hill. Good agronomic practices are followed to establish an ideal and healthy plant expression. Continuous plant to plant examination is made at different stages of growth from seedling to maturity and the growth stage specific characteristics are compared with the variety. All off types are uprooted at every growth stage. Off types and deviant type of plants appearing at flowering or post flowering are uprooted along with all adjoining plants. Monitoring Team visit is organized at post flowering stage. After harvest, utmost care is taken to avoid mechanical mixtures. Seeds are separated, processed and packed as breeder seed. The breeder seed sample is tested for breeder seed quality standards. Quality assured seed is labelled with breeder seed label. Grow out test with seed samples from breeder seed is conducted to test the genetic purity. Important characteristics in Jute are presented in Table 1.

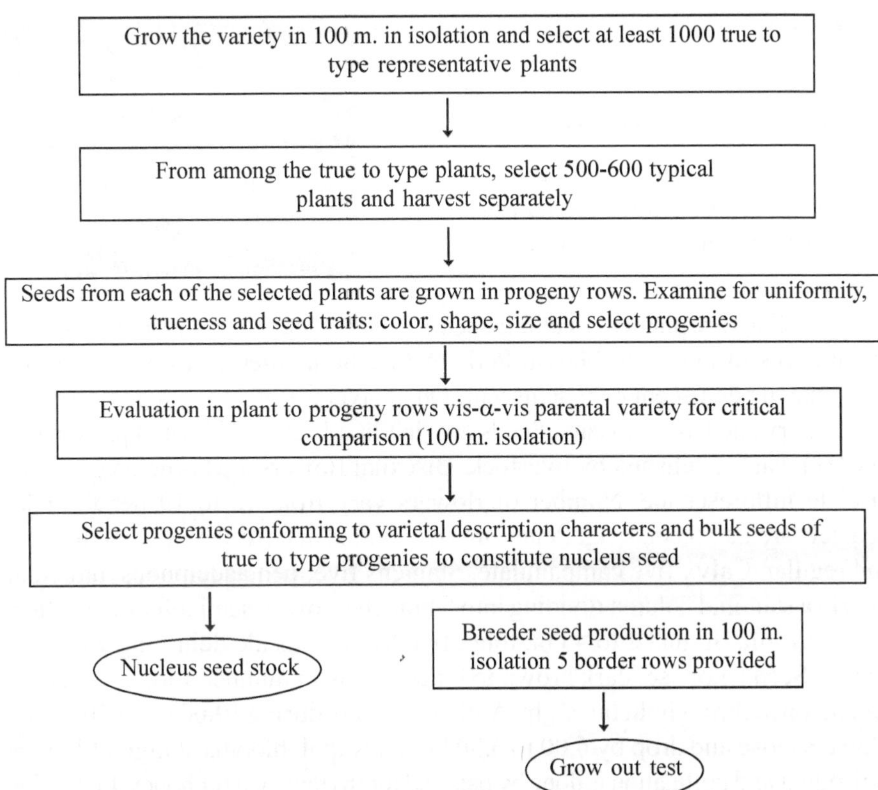

Fig. 1. Nucleus and Breeder seed production in Jute

Table 1. Important diagnostic characters in jute

S.No.	Character	States
1.	Early plant vigour	Low, medium, huge
2.	Leaf colour	Green, red
3.	Leaf vein colour	Green, red
4.	Stem colour	Green, red
5.	Stipule	Exstipulate, stipulate
6.	Stipule colour	Green, red, tip red
7.	Branching habit	Non branching, or very weak, weak, intermediate, strong, verystrong
8.	Leaf angle	Erect, semierect horizontal, drooping
9.	Leaf shape	Ovate, lanceolate, elliptical, cordate
10.	Pigmentation of flower bud	Green, red
11.	Days to flowering	Early, medium, late
12.	Fruit pigmentation	Green, red, others
13.	Days to maturity	Early, medium, large
14.	Pod dehiscence	Absent, present
15.	Seed size	Small, medium, bold

(*Source:* Kumar *et al.* 2003; PVP & FRA)

37. Kapok

Ceiba pentandra L. (Gaertn.) (2n = 72 to 88)

Family : Bombacaceae (APG) Malvaceae

Kapok is a highly drought tolerant multipurpose tree valued for its fibre, wood and green foliage as livestock feed.

Kapok commences flowering in 3 to 4 years after seedlings are planted. Leaf fall occurs in December-February and flowering commences in February-March. Pods mature in another 60 to 65 days time. Pods may be dehiscent or indehiscent. On an average 300 to 400 pods per tree can be harvested after 5 years. Seeds are rich in oil (20 to 25%) and protein (20 to 24%) that are relished by livestock. Bisexual flowers are borne on axiallary fascicle inflorescence. Number of flowers vary from 10 to 19 per fascicle. Pedicels are 2 to 5 cm. long. Petals are five, united at base showy, pentamerous and regular. Calyx five campanulate. Stamens five, pentaadelphous, united at base in a staminal column dividing into 5 branches; ovary semi inferior, 5 celled, style filiform. Stigma 2 to 4 cm. long, five lobed.Capsule dehiscing with five valves. Seeds globose, dark brown to black, many in number. Flowers (petals) remain open throughout the night. Anthesis occurs during 19.00 to 19.30 hours. Flowers close and drop by 6.00 to 6.30 hours. Kapok blooms at night. Flowers emit odour and pollination is done by bats and moths that visit for honey. Flowering phenology is suitable for nocturnal pollination. Self fertility or self compatability is reported in Ceiba. It is estimated that 46% fruit set is obtained by self fertilization. Insect (moth) pollination results in 56% fruit set while open pollination mainly by visit of bats leads to 80% fruit set (Nathan *et al.,* 2005).

Kapok trees can be propagated by seeds, seedlings and planting of orthologous branch cuttings and coppice shoots. Modern kapok commercial plantatins are in need of quality assured planting material developed from elite, superior high yielding trees, that are bushy or semicompact bushy and lacking spines on plant trunk and branches. Kapok could be planted in high density planting, yielding higher number of indehiscent oblong / obovate pods / capsules with white quality floss.

Nucleus Seed Production: Survey is made in high productive plantations for seed sourcing. High performing elite plus and healthy bushy kapok trees of 6 years age and more, bearing regularly higher number of good sized indehiscent pods are identified and selected. Laboratory examination is made on pods, seeds and floss qualities and saved as individual tree seeds. Flowers from the selected elite trees that will open next day are selected. Only one flower per inflorescence is retained and self fertilized by covering with butter paper bags in the forenoon or in the afternoon time and labelled. Self fertilized seeds from elite plus individual trees are sown in well prepared raised bed nurseries in progeny rows. Well developed 40 vigorous, true to type uniform and healthy seedlings, per progeny are lifted from the nursery after a month growth and single seedlings are planted in polybags and looked after for 6 to 8 months. From the polybag seedlings, well developed heathy twenty true to type seedlings are selected in each progeny and palnted in well prepared pits at 5 m. x 5 m. spacing with 20 plants in each row for each plant progeny. Minimum of 20 to 25 plant progenies are thus raised. The land needed will be 5000 sq.m. accomodating 200 plants. The plants need 7 m. x 7 m. spacing for full span of 10 years or more. It is expected that the plants will reach maximum pod bearing potential from 5th year onwards and hence to accommodate more number of plants, a spacing of 5 m. x 5 m. can be adapted. Isolation distance of 800 m. from other kapok fields is provided.

Alternatively, it is suggested that vegetative propagation of orthologous branch cuttings from elite superior trees of 5 years or more regular pod bearing stand may be followed for establishing nucleus and breeder stock seed production plots in kapok. The steps are:

1. Select the elite plus trees conforming to the semicompact or semibushy habit and highly productive at the age of 5 to 6 years. The trees should be regular bearers of indehiscence pods. The size of pods and floss colour should be of high expression. Select cuttings from orthotropic branches.
2. Seven to ten cuttings of uniform size and thickness from orthotropic branches are prepared from each selected plus tree identified by surveying the highly productive tree plantations. Identify at least 20 elite trees and from each tree 10 to 12 cuttings from orthotropic branches are prepared.
3. The cuttings are planted in suitable size polybags for rootings. IBA 1000 ppm dip for ten seconds be applied to promote good rooting.
4. Basic seed orchard: In a compact field, 10 true to type rooted cuttings (4 to 5 months old) from each selected elite tree are planted in well prepared pits of 30 cm. x 30 cm. x 30 cm. size and secured with good staking support. Spacing of 5 m. x 5 m. is provided between plants and isolation distance of 800m. is maintained from other kapok plants. Matured pods, floss and seeds of the selected trees from each branch cuttings are examined in the laboratory for conformity of the traits and true to type nature of each tree.
5. Critical performance observations are made from 2nd year or first time flowering onwards to establish true to type nature for branch, foliage, flower, pods, floss and seed features. Off types / non conforming and less vigorous and early and late flowering plants are removed.
6. Periodical agronomic and field management practices and plant protection measures are applied and the plantation is well protected.

Seed collection: Fully matured and indehiscent pods of typical shape and size are harvested from each tree and labelled. After drying, the pods are broken to open to remove the floss and to separate the seeds. Laboratory examination is made of the white colour floss and its fineness and other fiber and market qualities and % of floss to pod weight. The seeds are examined for trueness, uniformity and selected seed samples are graded and tested for the seeds qualities. The samples of the quality assured seeds from each tree is saved separately to develop nucleus seed stock. The remnant seeds of each tree are bulked together as stock seed.

Breeder Seed: The plants raised from the nucleus stock in 800 m distance isolation will form the Breeder seed stock plantation and to produce breeder seeds. The breeder seed is tested for the quality assurance and labelled.

38. Mesta

Hibiscus cannabinus L. and *Hibiscus sabdariffa* L.

Family : Malvaceae

Mesta is a herbaceous annual crop. Stems may be smooth or with prickles. Leaves lobed variously and alternate. Flowers axillary, solitary and epicalyx some what stiff. Corolla longer, spreading and stamens many and numerous and present in a staminal column. Style is branched stigma. Fruit is a capsule with pointed tip. Seeds glabrous and is rich in semidrying oil. Bulk of the fibre is present in the lower portion of the stem. Mesta fibre is an important commercial fibre in India.

In both the species namely, *Hibiscus cannabinus* and *H. sabdariffa*, flower blooming extends over a period of 4 to 5 weeks. Flowers open before dawn in *H. cannabinus* and remains open for 8 to 10 hrs and close between 3 and 3.30 pm. Anthesis is completed before the flower opening. Anthers are very close to the stigmatic surface enabling self pollination only.

In *H. sabdariffa*, flowers open one hour later than *H. cannabinus* and remain open for a lesser time and closes shortly afternoon. In this species, anthesis is completed before flower opening. Stigma never comes out of the staminal column ensuring self pollination as a rule under Indian conditions. The natural cross pollination in *H. sabdariffa* is as low as 0.23% and in *H. cannabinus*, the natural cross pollination ranges from 9.94 to 15.96%. (Kumar *et al.* 2003).

Nucleus seed production: The seed crop is grown with seeds from source population / variety in a fertile soil that enables the mesta crop to have a good and healthy expression of various characteristics. Minimum of 50000 plants are grown. The crop is grown with isolation distance of 100 m. At different stages of crop growth, up to maturity, off types and deviant plants occurring in the crop are uprooted. Finally 800 to 1000 uniform healthy and true to type plants for the characteristics are finally selected and labelled. After harvest, capsules and seeds of individual plants are examined in the laboratory for expression of characteristics conforming to the source variety. Seeds from selected plants are carried forward to the next season to raise 2 rows of single plants progeny plots. Due to large number of progenies, 50 progeny plots along with one plot for parent variety may be grown in one block. Number of blocks are planned as per the demand for the seed. Border rows (5 to 10 rows) are grown all around. Early and late flowering progeny rows are discarded. Careful and continued examination of the progeny plots is made and off types are removed. If the off types and deviant type of plants are found at or after flowering stage, that progeny row is uprooted along with the progeny row on either side. At maturity, single progeny plots are harvested separately and capsules and seeds are examined for the different characters. The single plant progenies which confirms the uniformity and true to type nature alone are selected and saved.

In the next season, the seeds from the selected progenies are grown in progeny row plots in two replications with 50 progeny plots in a block along with the check variety. All good agronomic practices are followed. Isolation distance of 100m. is maintained. Individual progenies are carefully checked and deviating plant types are uprooted and the entire progeny row is removed. When off types and deviant plant types are detected at flowering stage, the progeny showing the off type and deviant type plants and the progeny plots on either side are also uprooted to avoid cross pollinated by off type plants. Final evaluation of progeny rows is made at maturity. Based on the performance, progeny rows performing above mean +2SE are identified and capsules from superior plants are collected and the capsules and seeds are examined for the uniformity and then bulked to constitute nucleus seed. The seeds are tested for seed quality standards.

Breeder seed production: In a fertile field with fine tilth, a good breeder seed production plot is raised with 5 border rows all around in a very favorable season. Isolation distance of 100 m. is maintained. One plant is maintained per hill. The plants are examined critically from establishment to different growth and development stages till maturity. At each stage, characteristics specified in the variety description are compared with the individual plants to verify the trueness and uniformity. Off types and deviant plants are removed as and when they appear. When off types and deviant plants are noticed at flowering and later stages, they are removed along with the plants around them. Monitoring Team visit is arranged. At maturity, plants in the border row are not included in the harvested capsules. Capsules and seeds collected from main breeder seed plot alone are considered. Seed testing is done to assess minimum seed quality standards. Grow Out Test is then conducted as prescribed. After ascertaining the genetic purity, seeds are packed and Breeder Seed Tag is fixed to the quality assured breeder seed. The Nucleus-Breeder cycle is followed once in 2 to 3 years period. Important characteristics in mesta are presented in Table 1.

Table 1. Important characters in Mesta

S.No.	Character	States
1.	Early plant vigour	Low, medium, huge
2.	Leaf colour	Green, red
3.	Leaf vein colour	Green, red
4.	Stem colour	Green, red
5.	Stipule	Exstipulate, stipulate
6.	Stipule colour	Green, red, tip red
7.	Branching habit	Non branching, or very weak, weak, intermediate, strong, verystong
8.	Leaf angle	Strong Erect, semierect horizontal, drooping
9.	Leaf pubescence	Absent, present
10.	Pigmentation of flowerbud	Green, red
11.	Days to flowering	Early, medium, late
12.	Flower colour	Yellow, purple, pink
13.	Plant height	Dwarf, medium, tall
14.	Fruit pigmentation	Green, red, others
15.	Fruit pubescence	Smooth, hainy bristled, others
16.	Days to maturity	Early, medium, large
17.	Pod dehiscence	Absent, present
18.	Seed size	Small, medium, bold
19.	Seed shape	Reniform, sub-reniform, angular
20.	Seed coat colour	Brown, grey

(*Source:* Kumar *et al.,* 2003; PVP & FRA)

39. Tobacco

(*Nicotiana tabacum* L.) (2n:48)

Family : Solanaceae

In India over 85% of the tobacco area is cultivated with *N. tabacum* L and the rest of the area is with *N. rustica* L. The leaves are broad, altenately arranged and are harvested and cured for commercial purposes. The inflorescence is terminal, large with many flowers. Tobacco flowers are with short pedicels and consist of 5 tubular united cylindrical corolla,5 petals and 5 stamens. Fruit is an elliptic capsule with many small sized seeds. Tobacco crop is self pollinated. Cross pollination is recorded from 1 to 5% mainly due to insect pollination (Deo Singh *et al.*, 2003). The characteristics in tobacco are listed in Table 1.

Table 1. Descriptor of characteristics in tobacco

S.No.	Characters	States	Stage of observation
1.	Plant habit	Closed/semi open/open	Vegetative
2.	Plant type	Cylindrical/conical/bushy	Vegetative
3.	Plant height	Short/medium/tall	Vegetative
4.	Stem color	Green/creamy white	Seedling/vegetative
5.	Stem shape	Round/angular	Vegetative
6.	Leaf petiole	Sessile/petiolate-long to short/broad/narrow	Vegetative
7.	Auricle	Absent/present- narrow/wide/entire/winged	Vegetative
8.	Lamina	Erect/semi-erect/drooping	Vegetative
9.	Leaf color	Light/medium/dark green	Vegetative
10.	Lamina shape	Lanceolate/boat/elliptic, elongated/ cordate/ovate	Vegetative
11.	Lamina thickness	Thin / medium/broad	Vegetative
12.	Lamina width	Narrow/medium/broad	Vegetative
13.	Lamina surface	Glabrous or hairy smooth or puckered, dry or gummy	Vegetative
14.	Lamina margin	Entire/undulated/serrate	Vegetative
15.	Midrib	Thick/medium/thin	Vegetative
16.	Veins	Prominent or thin	Vegetative
17.	Panicle	Compact/open/loose	Reproductive
18.	Panicle length	Short/medium/long	Reproductive
19.	Corolla color	Creamy white/pink, red/yellow	Reproductive
20.	Limb shape	Narrow/wide	Reproductive
21.	Capsule size	Small/medium/large	Reproductive
22.	Capsule shape	Round/elliptic/conical	Reproductive
23.	Seed color	Light to dark brown, creamy white	Reproductive
24.	Seed size	0.4 to 0.6 mm long	Reproductive

Source: Description ley Deo Singh *et al.* (2003): PPV & F.R.A

Nucleus seed production: One thousand true to type single plants are selected from the evaluation plot isolated by 10m. and raised from the seeds of original source crop (nucleus/breeder seed crop) based on morphological identity, uniformity and genetic purity. A single capsule may yield from 2000 to 10000 very small seeds. The panicle of each selected single plant is self fertilized by bagging and seeds from the self fertilized panicle are collected carefully by threshing the capsules. Well dried seeds are stored plantwise. A disease free and healthy raised bed nursery is prepared to grow single plant progenies adapting extra care and due precautions such as sterilizing the seed bed, maintaining seed bed soil by proper watering and standard nursery and seedling management practices. After 6 to 8 weeks, the seedlings are ready for transplanting. The seedlings are hardened before transplantation. The time of nursery raising is suitably adjusted so that in the proper season the seedling will be ready at the optimum time of planting. The planting time depends upon the regions where the tobacco is to be planted. Isolation distance of 10m is provided to the main transplanted field.

After establishing the transplanted crop with progeny wise plots, the field evaluation is made for the field standards like uniformity, true to type nature, morphological identity and genetic purity of the progenies. The progeny plot showing off types and deviant plants are uprooted. The progeny plots with deviant plants and off types found at the flowering and post flowering phases are uprooted along with the adjoining progeny rows.

In respect of resistant varieties, screening of progenies is done before transplanting for resistance against insect pests and diseases under artificial conditions and also before selection of plants for selfing. At flowering, the fully emerged panicle in each of the selected single plants is self fertilised by bagging. After seed maturity the entire panicle is collected and threshed separately. Careful steps are taken to avoid mechanical mixing or contamination of seeds. Seeds from true to type panicles are examined in the laboratory for uniformity and trueness. Another cycle of 1000 selected single plants from the previous cycle are raised in nurseries and field planted in 2 row progeny plots. Isolation distance of 10m is adapted. Progeny row evaluation procedures for trueness, uniformity, vigour, disease and pest free nature and conformity to the variety are followed upto seed separation stage. Seeds are packed from selected single plants. The seeds are then bulked to make the Nucleus seed and labelled. The remnant seeds are saved for developing nucleus seed in future cycle once in two years.

Breeder seed production: Nucleus seed is used for raising the nursery for breeder seed production. Good agronomic practices to raise a healthy nursery and transplanted crop with single seedling per hill are practiced (10m. isolation). Single plants in the breeder seed production plots are examined for the truness and genetic purity in the juvenile, vegetative, foliar, floral and matured capsule characteristics. Field evaluation at critical stages and harvesting and post harvest operations are followed as described in the nucleus seed production protocols. Off types and deviant plants not conforming to the descriptions of the variety are uprooted /discarded. Healthy looking true to type plants (1000 to 1200 plants) in the breeder seed crop are self fertilized and at maturity harvested, seed separated and dried with utmost care. Monitoring Team Visit is arranged at flowering/post flowering phase to asses for field standards and genetic purity.

Seeds of selected individual plants are separated with utmost care, cleaned, processed and bulked together to constitute the breeder seed.Samples are drawn from the breeder seed and tested for seed quality parameters. GOT is conducted with seed sample, and genetic purity is assessed and breeder seed tag is attached (Fig. 1).

Quality Assured Seedling Production from Breeder seed and Supply: The Breeder seed of the required tobacco variety is sown in the nurseries by Central Tobacco Research Institute (CTRI) at different locations in tobacco growing regions. The quality assured breeder seed seedlings are distributed to the identified farmers for production of foundation seed under the supervision and advice of the breeders (Deo Singh *et al.,* 2003).

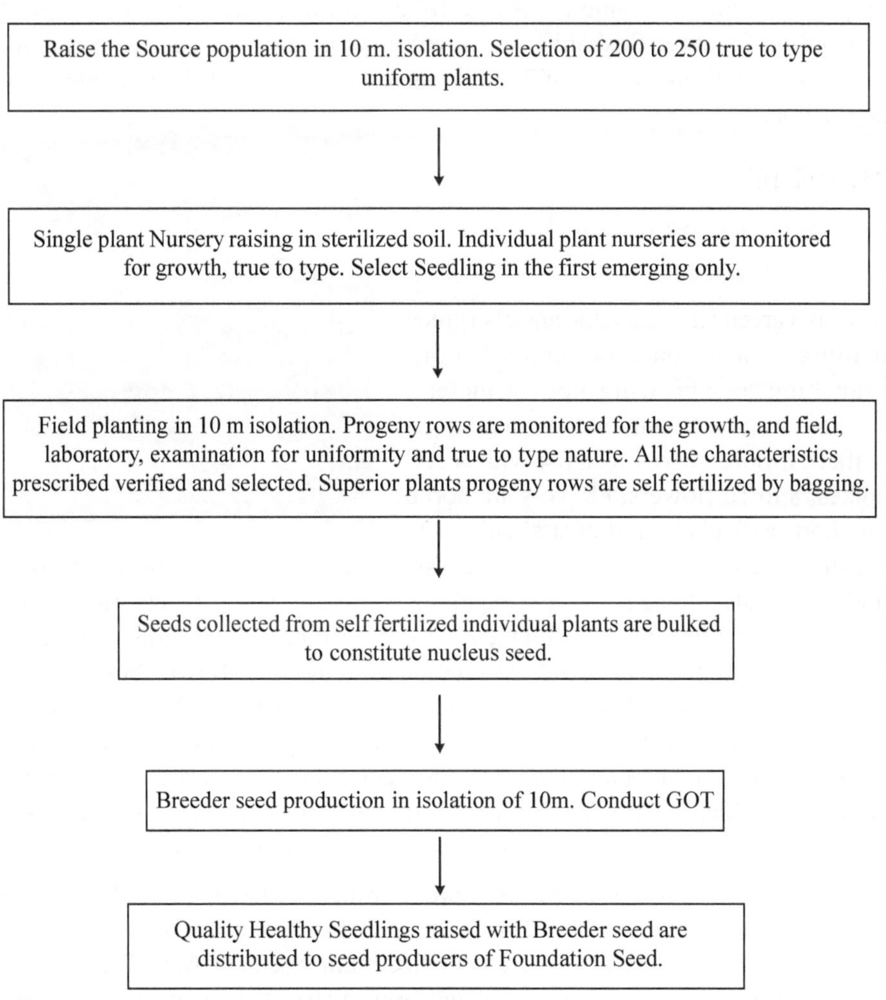

Fig. 1. Nucleus and Breeder seed production in Tobacco

Green Manure Crops

40. Daincha

(*Sesbania aculeata*) (Poirs.) (2n:12 and 24)

Family : Fabaceae

Daincha is a herbaceous green manure crop with soft stem and feathery like leaves and prominent racemose spike. It is tolerant to soil salinity and highly flooded situations. Daincha green manure crop improvement initiatives are seldom. The Haryana Agricultural University has developed and released an improved Daincha variety. Daincha is a self pollinated crop. Essentially the same procedure followed for nucleus and breeder seed production in other Sesbania sp are recommended in respect of *Sesbania aculeate*. Isolation distance of 20m. is suggested for nucleus and breeder seed production.

41. Sesbania

(*Sesbania speciosa* L.) (2n=12 and 24)

Family : Fabaceae

Sesbania green leaf vegetable and also green manure is a herbaceous annual, non-branching and growing upto 4 meters. Leaves prominent, pulvinate with 20 to 24 leaflets. Inflorescence racemose with lateral spikes, 8 to 10 flowers. Flowers are borne on short pedicels and almost sessile. The papilionaceous flowers show prominent standard petal and medium sized wings and keel petals. Stamens 10, in diadelphous condition. Stigmatic emergence and anther bursting are almost simultaneous events leading to self pollination. Insect visits are frequent and promotes pollination from the same flower. Sesbania is a self pollinated crop. Pods are flat and contain 10 to 14 seeds. The biomass production is almost high at 70 to 80 days after sowing. The root system produes nodules. Rhizobial inoculum mixed with seeds at the time of sowing promotes good and healthy root nodulation. The N content of the above ground biomass is 50 to 60 and the N content in the root biomass is 50 t0 60 kg N/ha.

Nucleus Seed Production: Seeds sourced from breeder stock seed are sown in isolation of 1600m. in 1000 sq.m. area adapting a spacing of 60 cm. x 60 cm. Before sowing, the seeds are treated with rhizobium inoculant. At pod developing stage, 1000 disease free, vigorous and uniform and true to type plants are selectd from the center of the field and omitting plants from border rows. At pod maturity, the selected plants are labelled and pods collected plantwise in separate packets. Simultaneously the selected plants are gently uprooted and judged for high root nodulation. Plants showing poor root nodulations are discarded along with the pods/seeds. In the laboratory, pods and seeds from the selected plants are examined for true to type and uniformity and the seeds showing deviations are discarded on plant basis.

In the next season, 400 progeny rows, each 4 m. long and in 2 rows are raised in 2 replications. The seeds of the reference variety are sown in 2 row progeny rows after every 20 progenies to facilitate comparison. Two meter wide border rows all around the plot are also raised. The plot is isolated by 20m. from other sesbania crops. Seeds were treated with rhizobium inoculants. Systematic observations are made on germination, initial vigour and healthiness, quality leaf or leafiness and disease free nature. Progeny rows with poor growth and vigour are uprooted. At different growth phases, leaf and high biomass development true to type nature and uniformity in the progeny rows are studied. Progeny rows showing deviant types of plants are rejected. At maturity, 20 uniform and true to type plants in each progeny row are selected and gently uprooted to assess the high degree of root nodulation ability. Plants are harvested progeny wise and leaf weight and biomass were estimated. Based on the performance, progeny rows showing performance above mean + 2SE are identified and pods and seeds are collected. In the laboratory, pods and seeds are examined and deviating types are discarded.

In the next cycle, progeny rows are raised with bulked seeds of selected plants within each progeny. Seeds are treated with rhizobium inoculants before sowing. Isolation distance of 20m. is adapted. Same procedure as outlined in the previous cycle was followed with emphasis on fast growth, high leaf biomass and high root nodulation. Deviating plants and off-types are uprooted. At maturity 1000 true to type plants are selected, gently uprooted and examined and plants showing high root nodulation alone are selected and pods collected. Laboratory examination of pods and seeds are made and only uniform and true to type seeds are finally selected. Part of the seeds from selected plants are pooled and the nucleus seeds is constituted. The remaining seeds of the selected plants are saved as such to initiate next cycle of nucleus and breeder seed production. Based on the requirement, NSI and NSII stages of seed production are organized.

Breeder Seed Production: Seeds from the nucleus seed stock are treated with root nodule rhizobium inoculation and sown in uniform spacing in an area of 2000 sq.m and provided with 20m. isolation distance. Border rows are raised around the breeder seed production plot. The plants are continuously examined for healthy, fast growth, uniformity and true to type nature, leafiness and biomass in each of the growth phase and deviating plants are uprooted. At flowering stage, Monitoring team visit is arranged to adjudge the genetic purity. At maturity stage, high root nodulation ability is verified and pods are harvested and seeds are seperated. Samples of seeds are drawn and seed testing is conducted for all quality parameters.

The quality assured seed is then packed and breeder seed label is tagged. Grow out test is conducted with samples drawn from the breeder seed and genetic purity is confirmed. The N content of root biomass and above ground biomass is estimated.

42. Stem Nodulating Thailand Agathi

(*Sesbania rostrata*)

Family : Fabaceae

It is an annual herb with some what ridged and angular stem showing characteristic stem nodules arranged in vertical rows. Leaflets are paired and pinnately arranged. Leaflets vary from 20 to 26 in number. Inflorescence is a loose racemose panicle, sometimes branched. Flowers are typical papilionaceous with corolla fused at base. Petals are 5 and free consisting of a large standard petal, yellow and pink or yellow coloured, two winged petals and two keel petals somewhat twisted around the bundle of diadelphously arranged ten stamens. Ovary superior, monolocular and 10 to 15 ovules present. It is a self pollinated species. Seeds are dark brown and cylindrical 4 to 5 mm. long. Anthesis occurs in early morning and pollen sheds on the stigmatic surface almost simultaneously with the flower opening and stigma exsertion. Efforts of genetic improvement of *S. rostrata* green manure are not well focussed so far. S.rostrata endowed with root and stem nodules is a native to Senegal and this species has been adopted as a green manure crop. In Tamil Nadu Agricultural University, breeding of improved stem and root nodulation line of S. rostrata has been successful with source seeds distributed from IRRI, Manila by the kind courtesy of Dr. M.S. Swaminathan, the then Director General of International Rice Research Institute, Manila. The improved selection has been released as Sesbania rostrata, CO 1 green manure in the year 1991 by the TNAU, Coimbatore.

Nucleus Seed Production: Seeds are sourced from the Breeder or the concerned institution. A wide spaced plant population with uniform spacing is raised in an area of 1000 to 1200 sq m. with isolation distance of 20 m. The seeds are inoculated with appropriate rhizobium culture before sowing. The crop at active growth phase around 20 days after sowing is sprayed with filtered water that is soaked with stems bits carrying stem nodules to facilitate the stems to be infected with the bacterium to develop stem nodules. As and when the crop establishes, weak unhealthy and disease affected plants are uprooted.

Around 1500 to 2000, fast growing true to type plants that show fast growth, high biomass, high expression of stem nodules and root nodules and uniform plants are tagged. At flowering stage, early and late flowering plants are removed to maintain uniformity and synchrony in flowering. At pod maturity, the tagged plants are gently uprooted and visually adjudged for high nodulation on stem as well as on roots. Finally 800 to 1000 such plants that fulfil all the criteria are selected and pods collected. In the laboratory, pods and seeds of individual plants are examined and true to type plants alone are selected and individual plant seeds are saved separately.

In the following season, around 500 plant progeny rows are raised with seeds saved from selected individual plants. Two row progeny row plots 4 m. long are designed and sown in 2 replications. For every 20 progeny rows, the seeds of the concerned variety are sown in two rows plots for comparison. The isolation distance provided is 20 m. The seeds are treated with rhizobium inoculant before sowing. Around 20 days after sowing; water suspension of filtered water that was soaked with stem pieces with stem nodules is sprayed onto the stem to facilitate infection for stem nodulation. Observations were made in the progeny row plots and weakly developing nonuniform and not true to type and diseases and pest affected progeny rows are uprooted. At flowering stage, early and late flowering plants and progeny rows are discarded. Healthy plants that are high in biomass and stem nodulation, true to type and uniform progeny rows are retained. At pod maturity uniform progeny rows are identified and labelled. The plants in the selected progeny rows are gently uprooted and examined for high leaf and biomass, high degree of root as well as stem nodulation. Matured pods are collected from 500 to 600 such plants that are showing high degree of root and stem nodulation and biomass. Pods collected from the selected plants are examined in the laboratory for pod and seed characteristics and single plant seeds are saved. Part of the seeds from each plant is pooled together to constitute Nucleus Seed and the rest of this seeds are utilised for continuing the nucleus seed cycle.

Breeder Seed Production: Seeds from nucleus seed stock are sown in well prepared field that is isolated by 20 m. from other *Sesbania rostrata* green manure crop. Border row of 3 meter wide are grown all around the field. Uniform spacing is provided and single plant per hill is retained. Inoculation steps for root and stem nodulation are fulfilled. Weak and disease and pest infested plants are removed. At different phases of crop development, fast and vigorous growth, high biomass, high expression of stem and root nodulation, uniformity in vegetative, flowering and other features and true to type expression and pod and seed set are critically looked into. Monitoring Team visit is arranged at midflowering stage to adjudge genetic purity. In the laboratory, pods and seeds of single plants are examined and plants showing poor pods setting are discarded. The seeds are bulked, dried and processed. Samples of the seeds drawn are tested in the seed testing laboratory for seed quality parameters. The quality assured seeds are then tagged with Breeder seed label. Grow out test in conducted as per the prescribed standards with seed samples drawn from Breeder seed to test the genetic purity. N content of root biomass and above ground biomass is estimated/ recorded.

43. Sunn hemp

Crotalaria juncea L. (2n=16)

Family : Fabaceae

Sunn hemp also known as Madra Hemp is a herbaceous plant growing upto 2 to 3 m. Flowers are bright yellow borne on terminal racemes. Branching is to a less extent at upper portions. Flowers are zygomorphic, consists of corolla, polypetalous petals namely standard petal, two wing petals and two keel petals, anthers 10 in number, dimorphic (globose and heart shaped) and gynoecium with hairy stigma.

Sunn hemp flowers show protandrous character. Breeding study confirmed the cross pollination behaviour of sunnhemp to be predominant (Nirmalaruban *et al.* 2020). Tap root is long and profusely nodulating. Plants are erect. Leaves are simple, elliptic, upto 4 to 12 cm. long. Vegetative growth is maximum during the long days of summer. Sunn hemp seeds can be collected 130 to 135 DAS. Sunnhemp yields 15 to 18 t/ha of green biomass and seed yield of 400 to 600 kg/ ha. Seeds are small and flattened. The sunnhemp is a multi purpose crop. The bast fibre quality is soft and lignified fibre used for strings, tissue paper, twines and like. As a fodder crop, it is highly relished by cattle. It is popularly known for green manure purpose for enriching the soil nutrient content. Sunnhemp as green manure crop adds 50 to 60 kg N/ha through biomass and in addition adds 50 to 60 kg of N/ha through biological N fixation by root nodules (Sarkar *et al.*, 2015). The natural cross pollination is very high in sunn hemp.

Nucleus Seed Production: Quality Seed of sunnhemp sourced from the Breeder/Institution are sown in a fertile field (1000 sq.m.) with isolation of 250m from other sunn hemp crops. The seeds are pre treated with rhizobium inoculants. Only one healthy vigorous plant per hill is retained at the age of 15 days. Continuous examination is made for true to type, uniformity and characteristics expression of the variety grown. Offtypes and deviating type of plants, weak plants, disease and pest affected plants are culled out. Very early and late flowering plants are discarded. At the time of pod maturity 1000 single plants are selected from the middle of the field. Pods from single plants are collected and saved plant wise. Seeds are separated and examined in the laboratory for trueness. Seeds of five hundred single plants are selected.

In the next favourable season 500 single plant progenies are raised in 2 row plots in a field isolation by 250 m. from other sunn hemp crop. Before sowing the seeds are treated with rhizobium inoculant. At 15 days after sowing, only one fast growing vigorous plant is retained per hill. Individual progeny plots are critically studied for good field establishment, fast growth and healthy stem and leaves. Off- types and deviating type of plants are discarded. Uniformity within the progeny row and between progeny rows is studied and less vigorous, less uniform, deviating type of progenies are discarded.

Nucleus and Breeder Seed Production in Sunn hemp

Raise source seed population in 1000 sq. m. in isolation of 250 m.

↓

Examine and select 1000 true to type single plants with high expression of biomass and root nodulation. Make laboratory studies on pods and seeds

↓

Select 500 single plant, self fertilise and save the seeds plantwise

↓

Raise 500 single row progenies and examine for true to type, stability, uniformity and biomass and pods. Also examine root nodulation by gently uplifting the plants at the time of harvest

↓

Select 200 progenies performing above mean value that are uniform true to type vigorous, high biomass and high nodulation. Seeds examined in the laboratory. Selected plants seed saved plants wise. Seeds are now pooled together to constitute Nucleus Seed. Breeder Seed Production (250m. isolation)

↓

Grow Out Test

Plant showing variation are discarded. Progenies are critically examined for true to type of stem, foliage, flower and pod setting at the corresponding growth phases and plants and progenies showing variation are discarded. At maturity, plants within each progeny are gently pulled up and examined for the nature and health of nodules and extent of nodulation. Plant showing poor nodulation are rejected. Performance of progeny rows for biomass and fast growth are assessed and pods collected from the selected progenies. Progenies that are equal to or higher than mean are identified for biomass. Seeds from superior plants in the identified progeny rows are saved progeny wise and bulked to constitute the nucleus seed. NSI and NSII can be organised based on the need.

Breeder Seed Production: In a well prepared field to fine tilth, rhizobium inoculated nucleus seeds of sunn hemp are sown adopting the recommended production practices of the crop. Isolation distance of 250 m. is maintained. Continuous monitoring is made. Weak, non branching plants are removed. So also deviating types of plant are culled out at pre and post flowering stages and plants surrounding the deviating type of plants. All the plants are examined for the characteristics of the sunn hemp variety. At mid maturity phase, visit of monitoring team is arranged. Matured plants are harvested after ascertaining the nodulating ability and their pods and seeds are verified for true to type and uniformity. Seeds are processed and seed quality is tested. Quality assured seeds are labelled with breeder seed label. N content of the biomass is estimated.

Forage Crops

44. Berseem

(*Trifolium alexandrinum* L.) (2n:16)

Family : Fabaceae

Berseem is also known as Egyptian clover and acknowledged as King of Fodder crop. It is a fast growing, N fixing, perennial, cross pollinated, drought tolerant, cut and feed rabi season fodder crop. Berseem is highly nutritious, highly digestable legume and does not cause bloating. It is erect growing upto 100 cm. tall, develop into multibranching crown at ground level; leaves are trifoliate with obovate leaflets that are 2 to 3 cm. broad and 4 to 5 cm. long. Flowers are 20 to 30 and are borne in compact racemes forming head like shape and are acropetally arranged. Flowers are sessile or small, pedicellate, pinkish to white coloured and papilionaceous in structure. Calyx is cup shaped and five petals arranged in 1+2+2 aestivation with respectively one keel, two free wings and two fused standards. Stamens are ten in diadelphous (9+1) manner. Stigma is globular. Berseem is a cross pollinated crop. (Malaviya *et al.* (2003).

Anthesis is normally in the early morning and the pollen grains are enclosed within the corolla as the flower opens at a later stage. The stamens are forced out of the keel under pressure by an insect tripping but return to original position when the pressure is released. Presence of bee coloy significantly increases the seed production. Bee colonies of *Apis meliferra* are maintained in artificially placed bee hives near the berseem fields.

Berseem can be harvested by first cut in 50 to 60 days time after sowing and then at every 30 to 40 days interval. Under irrigated cultivation, 5 to 6 cuttings are done. Seed rate of 25 kg. per ha. is recommended with a spacing of 50 cm. between rows and 5 to 10 cm. on the rows from seed to seed.

Nucleus Seed Production: Seeds of the variety are sourced from the breeder/ institution concerned responsible to maintain basic seed stock. The seeds are sown in 400 sq.m. area in isolation of 600m. Care is taken to sow in a fine tilth field not grown to berseem in the past 2 to 3 years. At the time of flowering, early and late flowering plants and off-types are uprooted. From the middle of the field, 800 true to type uniform plants are selected at maturity and after verifying true to type nature of plant type, the heads are harvested plant wise separately and saved. In the laboratory, individual heads and seed separated from them are verified for true to type and uniformity and the deviating types and non conforming type are discarded (Fig. 1).

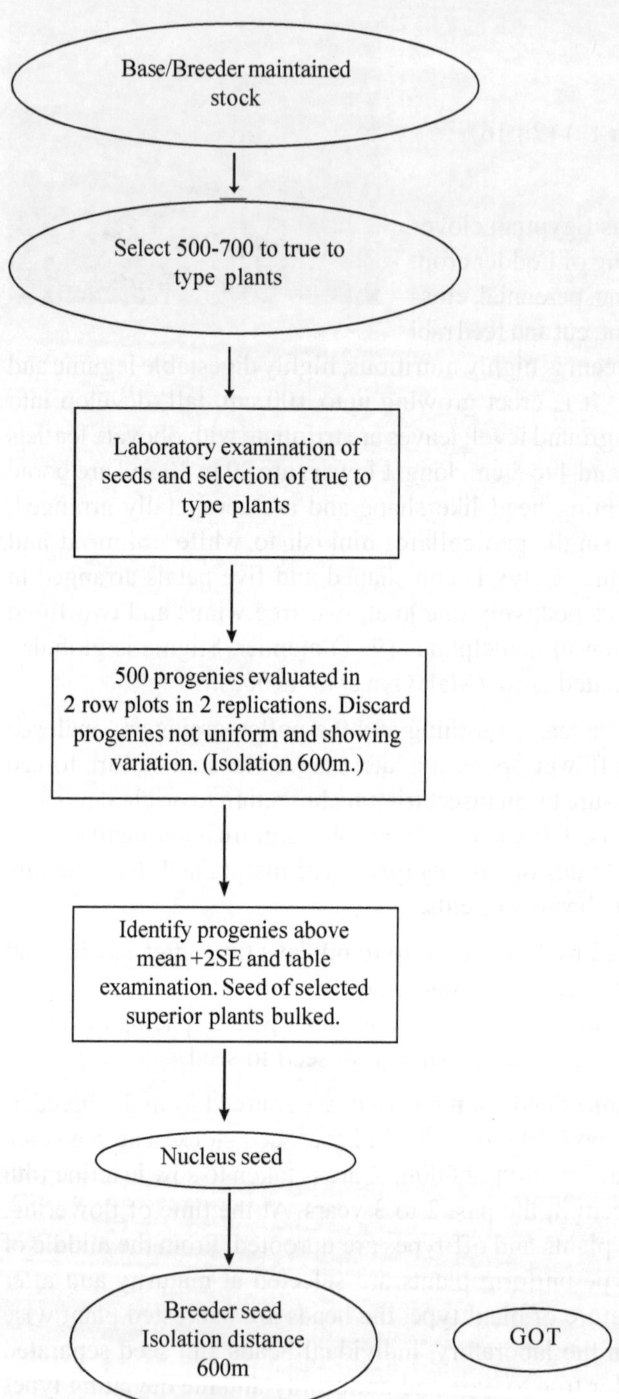

Fig. 1. Nucleus and Breeder Seed Production in Berseem

Seeds of 500 to 700 plants selected from the base population are sown in 2 row plots with check row once in 10 progeny rows in two replications. The row length is 4 m. In one block , 50 progenies are raised. The nucleus seed production plot is isolated by 600 m. from other berseem crop. The progeny rows are monitored for good germination and field establishment and fast growth and vigour. Poorly performing progenies are discarded. The observations are continued in different stages of crop development like juvenile, active growth stage till flowering and then upto pod and seed development and maturity phases. Individual plants are studied for stem, leaf, and flower characteristics in individual progeny rows. Progeny rows showing off-types and deviating plants not conforming to true to type nature and uniformity are rejected. At the time of flowering, very early and late flowering plants are discarded. Characteristics observed in individual plants are plant height, branching habit, leaf shape, size, colour, margin and hairiness, flowering, flower colour, head size and shape and maturity. Progeny rows with plants that are uniform and true to type for the prescribed characteristics alone are retained. Progeny rows with deviating plants, if any, are noticed, such progeny rows are discarded along with the adjoining progeny rows on either side. At multicut harvest, 2 to 3 cuts are done so as to ensure the ability of the progeny rows for regeneration and branching and to withstand multicuts. At maturity progeny rows showing performance higher than mean +2SE are identified /selected. In the selected progeny rows, 10 to 15 superior plants are harvested and labelled plant wise. In the laboratory, harvested single plants from the selected progeny rows are examined individually for plant type, pod and head shape and size and selected. Seeds are extracted plant wise and examined in the laboratory for seed shape, size, uniformity and trueness. Part of the seeds from each selected plant are bulked to constitute nucleus seed. The remainder of seeds are saved for future use.

Breeder Seed Production: Nucleus seeds are sown in rows in a well prepared field not grown to berseem crop in the past 2 to 3 years. The field is isolated by 600 m. from other berseem growing fields. Only single plant per hill is retained. Five border rows of berseem crop sown with the same check variety seeds are raised on all the sides. Individual berseem plants are monitored for good field establishment, initial vigour, growth, branching, leaf production and for true to type nature and uniformity. Off-types are uprooted. Deviating plants when noticed are uprooted along with surrounding plants to avoid vicinism and contamination. Very early and late flowering plants are discarded. Individual plants are monitored for true to type nature in stem, branching, foliage, head and flower characteristics and uniformity in head shape, size and maturity. At maturity, heads are collected on single plant basis and studied in the laboratory for head and seed characteristics. Visit of Monitoring Team is arranged at mid maturity stage. At maturity, single plants are harvested and examined in the laboratory for head and seed characteristics. The seeds of selected plants are pooled together and processed. Seed quality test is done and quality assured seeds are packeted and labelled with breeder seed label. Seed samples are drawn and grow out test is conducted to adjudge the genetic purity. Nutritional quality of berseem is estimated. The important diagnostic characteristics of berseem are given in table 1.

Table 1. Characteristics of berseem

Character	States	Stage of observation
Branching behavior	Secondary branching emerging from base/ emerging from upper nodes	Vegetative
Leaflets hairiness	Sparse/medium/dense	Vegetative/flowering
Stem thickness	Thick/medium thick	Vegetative/flowering
Leaf shape	Lanceolate/ovate/round	Vegetative/flowering
Leaflet apex	Tapering/round	Vegetative/flowering
Leaflet margin	Entire/serrate	Vegetative/flowering
Leaflet size	Bold/medium/narrow	Vegetative/flowering
Flower initiation	Early/medium/late/very late	Start of flowering
Inflorescence shape	Round/oval/elongated	Flowering
Flower colour	White/pink	Start of flowering
Maturity	Early/medium/late	Maturity
Seed size	Small/medium/bold	Maturity/after harvest

Ref: Malaviya *et al.* (2003); UPOV

45. Guinea Grass

(*Panicum maximum* Jacq.) (2n=18,36)

Family : Poaceae

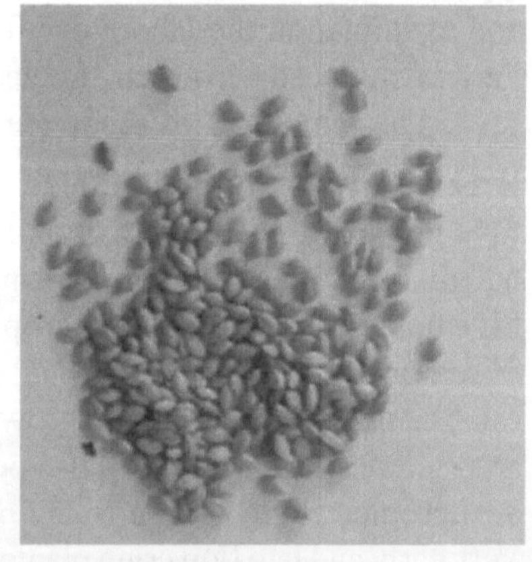

From slips or setts source: Stem setts method: Well established guinea grass multicut fodder production plot in clean field is selected and at least 500 healthy and disease free representative plants that are highly tilllering or ratooning plants are identified and tagged. From each such selected plants, slips with 3 to 4 buds are prepared and slips are pretreated with fungicide. Slips from selected plants are planted in multiplication plot to multiply the setts enough to plant them in individual progeny rows. Slips are planted in individual clonal progeny rows in two replications and good agronomical practices for multicut fodder grass production are adapted. Individual clonal progenies are assessed for uniformity, quick field

establishment, high tillering and leaf biomass. First cut is made after 40 to 45 days at the time of flowering. Subsequent cuts are made in 40 to 45 days interval. In the last cut, data on growth, height, size of leaves and stem and leaf softness are recorded. The progressive total of each cutting till the end of 7th and 8th cutting of the year is arrived at. Based on the progeny mean, the best performing clonal progenies above mean +2SE are identified and slips are collected and pooled to constitute nucleus seed. In the next step, the slips from nucleus stock are planted and breeder stock field is organized. Healthy setts/ Clonal progeny are multiplied and slips are prepared and distributed.

Nucleus and Breeder Seed (Seed to slip method): Well filled matured seeds collected from open pollinated panicles from healthy and high fodder biomass producing 1000 plants are sown in close spacing (10 cm. x 5 cm.) in a finely prepared raised bed nursery. Around 40 to 45 days after sowing, the seedlings are examined and vigorous, healthy, uniform, true to type fast growing seedlings are selected and then planted in secondary nursery with spacing of 45 cm. x 30 cm. and allowed to grow till 45th day after planting and first cuttings made. Vigorous, healthy fast growing, true to type and high tillering plants with high biomass are selected after 6 to 7 cuttings to verify the regeneration and stooling capacity. Slips are prepared from 500 selected plants and planted in the main field with 60 cm. x 45 cm. spacing.

Progeny rows are established and isolation distance of 20m provided. After the establishment, the first cut is made at the time of flowering around 70 to 75 days. Subsequent cuts are made in 40 to 45 days interval till the first year end. The data on plant growth, height, biomass of stem and leaf and the softness and nutritional qualities are studied. Based on the cumulative yield, best performing plant progeny rows are selected. Slips prepared from the plus performing plants are bulked together to constitute the nucleus stock.

Breeder Seed Stock: Slips from superior individual clonal progenies from the nucleus seed stock are field planted in rows with proper spacing and production practices followed for the multicut fodder crops. Isolation distance of 20m. is maintained. Characteristics like quick field establishment of slips, fast growth and development, leaf attitudes, non lodging, leaf and stem biomass and favorable leaf to stem ratio, suitability for 40 to 45 days cycle of fodder cutting, free from diseases and pests, stability in tillering and biomass production, nutritional content of fodder (green)are assessed across the multicut stages and recorded. Based on the performance considerations over multicut, superior clonal plants are selected. Slips propagules are bulked and distributed as breeder seed stock. The fodder quality and nutritive values of the breeder seed stock are assessed.

46. Lucerne

Medicago sativa L (2n=16, 32)

Family : Fabaceae

It is a premier forage crop and known as Queen of Forages. Lucerne is a highly nutritious crop with wide adaptability and fast growth and produces highly palatable fodder. It is grown as a multicut fodder crop; lucerne grows upto 65 to 100 cm. tall. Leaves are alternate borne on hollow stems. Leaflets are trifoliate and three and 10 to 45 mm. long and 3 to 10 mm. broad. Flowers are small and borne in axillary racemes. Flowers are papilionaceous, zygomorphic, with hairy tubular calyx. Petals length is upto 2 cm. long. Petals five comprising standard petal (8 to 12mm. long), two wing petals (5 to 8 mm. long) and two fused keel petals (5 to 8 mm. long). Stamens are 10 in (9+1) diadelphous manner. Pods are 2 to 5 in loose coils that are in anticlockwise spiral. Seeds are 2 to 6 per pod, cream to yellowish coloured. Lucerne is a cross pollinated crop adapted to entomophilious type of pollination. Tripping occurs when bees land on flowers. The flowers that are tripped set seeds.

Nucleus Seed Production: Base population of concerned variety of lucerne is raised with the seeds accessed from the breeder or from the respective Institution. Seeds are raised in 400 sq.m. area adapting a spacing of 45 cm. x 10 cm. The plot is isolated by 400m. from other lucerne crops. Fifteen to twenty days after sowing, seedlings are thinned to one seedling per hill. Individual plants are monitored for true to type and uniformity. Off-types are uprooted. Very early flowering and late flowering plants are discarded. Two to three cuts are made so as to verify the ability to withstand multicuts and regeneration of the plants. At maturity 1000 true to type uniform, healthy and vigorous branching representative plants are selected from the middle of the field and labelled. The selected plants are harvested as whole plant and examined in the laboratory for true to type of plant type, branching, stem, foliage, flower, and pod characteristics. Seeds from the selected plants are separated plantwise and examined for seed characteristics and 700 to 800 plants are selected and seeds saved single plantwise (Fig. 1).

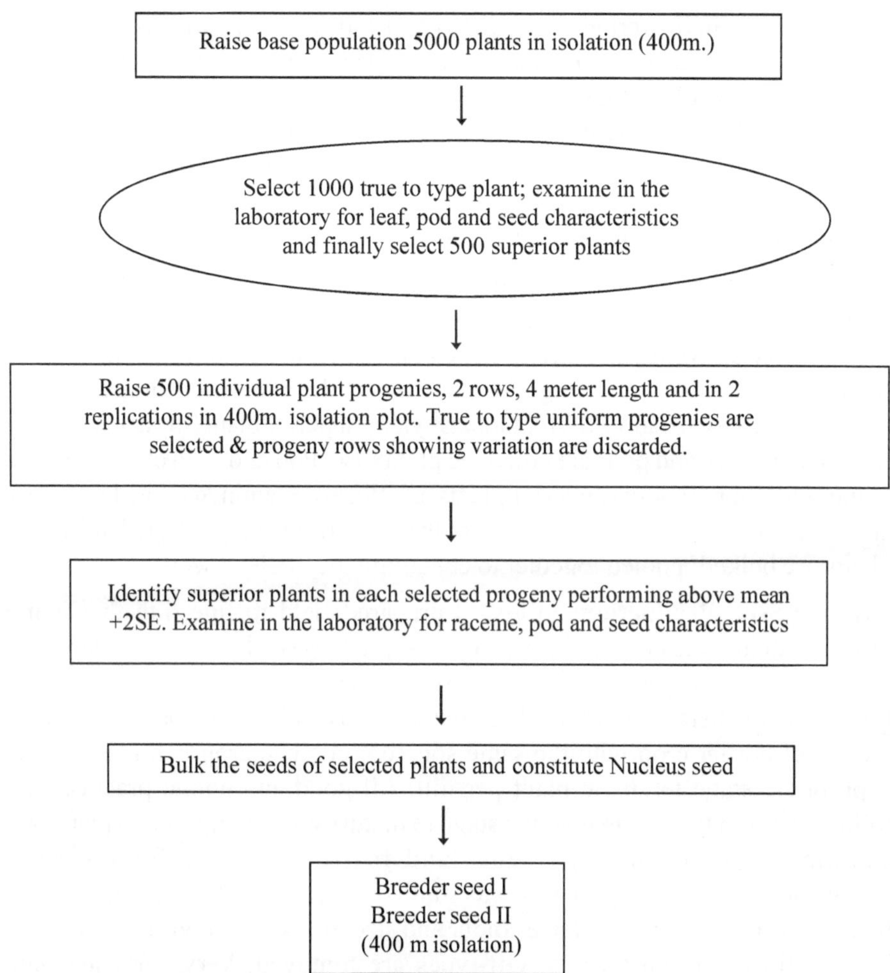

Fig. 1. Nucleus and Breeder Seed Production in Lucerne

In the next step, 1000 single plant progenies are raised in 2 row plots in 2 replications. Each row is 4 m. long. In each block, 50 progeny rows are sown in 2 rows plot with check seeds sown once in every 10 progeny plots. The field is grown with 5 to 10 border rows with the same variety seeds. Isolation distance of 400m. is provided. At juvenile stage, seedlings are thinned and only one plant per hill is retained.

Monitoring of individual progenies is initiated from juvenile stage onwards. Weak and poorly established progeny rows are rejected. Progenies showing vigorous and fast growth and uniformity with erect stem and healthy foliage are retained. At flowering, flowering uniformity within progeny rows and between rows are adjudged and very early and late flowering progenies and progenies showing non uniformity in flowering are uprooted. Off-types and deviating type of plants are rogued out. Progenies with non conforming and deviating types of plants are rejected as a whole progeny. If deviating plants are observed at flowering, progenies with such deviating type of plants are uprooted along with the progeny rows on either side to avoid vicinism and contamination of pollen. At maturity, one thousand progenies are studied for the pod characteristics and progenies conforming to the pod description are retained.

The first cut is made 90 DAS and later cuts are made at 40 to 40 DAS and observed for fast regeneration. Progeny rows showing high biomass yield performance above mean +2SE at the end of 3 cuts are identified and allowed for flowering and pod setting. At maturity of the identified progeny rows performing above mean +2 SE for biomass yield, superior plants in each identified rows are labelled and pods are collected plantwise. Pod and seed characteristics of the selected individual superior plants are closely examined in the laboratory and selected and saved plantwise. Part of the seeds from every selected individual plants are bulked/ pooled together to constitute the nucleus seed.

Breeder Seed Production: In a well prepared field to fine tilth that is not grown with lucerne crop in the past 3 years, nucleus seeds are sown adapting a spacing of 45 cm. between rows and 10 cm. between hills on the rows. The field isolation distance is 400m. The breeder seed production plot is grown with 5 rows of border crop with the same source seed. The crop is thinned at the appropriate stage to single plant per hill. All good agronomic practices are followed. From the young juvenile stage to maturity stage, individual plants are examined for uniformity, vigour and fast growth and for the field establishment.True to type nature was studied for the fast growth, vigour, stem, branching and plant type, foliage colour and size, flower and pod characteristics besides flowering uniformity. Off-types are removed. Very early and late flowering plants are rejected. Deviating / non conforming plants, if any, are up rooted, along with the adjoining plants. At mid maturity, visit of the Monitoring Team is organized. At maturity, individual plants are harvested with entire plants intact to facilitate examining of plants for plant type conformity. After selection of plants, pods are collected plantwise and verified for pod type. Seeds are separated and after conforming trueness, the seeds are pooled together to constitute the breeder seed. After drying and sampling, seeds are tested for seed qualities. Quality assured seeds are labelled with breeder seed tags. Samples drawn from the breeder seed are utilized for conducting GOT as prescribed to adjudge the genetic purity. Composite sample of the plant is analysed for nutritional quality and tested by animal feeding. The important diagnostic characteristics of lucerne are presented in Table 1.

Table 1. Characteristics in Lucerne

Character	State	Stage of observation
Plant growth	Dwarf/medium/tall	Vegetative
Growth habit	Erect/semi-erect/prostrate	Vegetative
Height after first cut	Dwarf/medium/tall	Vegetative
Stem girth	Thin/medium/thick	Vegetative
Stem pith	Hollow/semi-hollow/solid	Vegetative
Leaf shape	Lancoelate/ovate/obovate	Vegetative
Central leaflet length	Short/medium/long	Vegetative
Central leaflet width	Narrow/medium/broad	Vegetative
Days to flowering	Early/medium/;ate	Start of flowering
Flower colour	Purple/white/variegated	Start of flowering
Pod shape	Round/flat	Vegetative
Seed size	Small/medium/bold	Vegetative

Source: Yadarendra *et al.* (2003)

Sugar Crops

47. Sugarcane

(*Saccharum officianarum* L.) (2n=80)

Family : Poaceae

Sugarcane is tropical and subtropical in habit. It is bushy, profusely tillering with short to long internodes. Single buds on the nodes are alternately opposite. Nodes bear root primordial and intercalary meristematic tissue. Leaves are long and linear with hairy margin and leaf sheath often encircling the nodes covering the bud. Ligule is present, simple and ciliated. Inflorescence in sugarcane is a panicle, known as arrow. Panicle is very highly branched and branches end with paired spikelets that are subtended by callus hairs. In a pair of spikelets, one spikelet is sessile and the other is pedicelled. Spikelets have outer glumes 1 and glume 2 enclosing a sterile lemma.(L1). Palea (P1) is absent. L2 and P2 enclose a fertile flower which consists of three stamens, two lodicules, superior ovary with bifid style and feathery stigma. Flowering in sugarcane is location specific and influenced by environment. The list of the characteristics in sugarcane is given in Table 1. Production and maintenance of disease and pest free clonal seed crop of long duration is a challenging task. Sugarcane clonal seed production demands constant vigil and attention. Physical mixtures in sugarcane seed crop are difficult to search upon in the field conditions. Seed selection for seed production planting is the critical step for maintenance of genetic purity and disease and pest free planting material.

Clonal sett : Nucleus seed production: The first step in the nucleus seed production is to identify and select true to type single clumps in the basic seed source crop maintained by the concerned breeder in respect of newly released sugarcane variety and for the already released variety. The source clone seed material may be also from nucleus or breeder seed plot. The identified healthy single clumps (200 to 250 numbers) that are true to type and healthy are selected at tillering phase from the base population to maintain genetic purity, uniformity and conforming to the true to type of the characteristics of the clone variety in all aspects. Each clump is numbered serially and disease free clumps are harvested separately at proper maturity and harvest stage and utilized to raise Nucleus seed crop.

The nursery is raised following the procedure of Spaced Transplanting Method (STP). In STP method, healthy and uniform single sugarcane buds are selected from the base population and planted in a polybag nursery. The seedlings obtained after 6 to 7 weeks are verified for growth, vigour, free from diseases, uniformity and true to type nature and are transplanted in the well prepared, clean disease free field with a spacing of 90 cm. x 90 cm. from seedling to seedling for encouraging higher number of tillering per seedling. Isolation distance of 20m. is provided between the nucleus seed production plot and other sugarcane crops. The individual clumps are identified for true to type nature and uniformity and clumps are selected. The crop thus obtained from the STP is used as the base source of nucleus seed production.

The season of planting for nucleus seed crop is adjusted to the planting season for the breeder seed crop. Normally 8 to 10 month old crop is considered best for seed purpose.

The crop is isolated by 20m. from other sugarcane crops. The cane obtained from each clump at 8 to 10 months old crop are planted in paired rows (30 cm. between rows) and 120 cm. between paired rows. The seed rate for each paired row in the nucleus seed crop is four 3 budded setts and the row length is 3 m. and with 2 m. spacing between rows. The recommended crop management and plant protection measures for sugarcane crops are fully adapted. Control of important diseases are promptly and effectively done.Continuous field inspection and evaluation are carried out by the Breeder concerned.Paired rows showing off-types and disease and pest affected rows are removed. Each and every row is examined at every growth stage of the crop for disease infection and pest attack and for the presence of off-types. Special care is taken to identify virus infected rows and uprooting and destroying the paired rows showing virus infection. The nucleus seed crop is harvested at 8 to 10 months. The time of harvesting of nucleus seed crop is synchronized with the planting time of Breeder seed crop usually within 10 days time.

It is very important to closely examine the harvested nucleus seed canes and setts for the true to type and uniformity. Canes and setts showing the presence of infections for diseases and insect attack symptoms are detected and discarded.

There are as many as 5 important disease in sugarcane such as red rot, smut, wilt, grassy shoot and leaf scald and 8 insect pests (termite, top borer, internode borer, stalk borer, plassey borer, Gurdaspur borer, scale insects and mealy bug).

The nucleus seed canes that are tested for the disease free and insect attack free conditions are harvested. It is very important to closely examine the harvested nucleus seed cane/setts for the true to type and uniformity. After thorough examination, the nucleus seed canes that are true to type nature, uniformity and health are selected. The juice is analysed for brix value and other quality aspects in individual canes. If the seed canes need to be transported, the canes are cut into equal size setts, packed in gunny bags and insulated with paddy straw or wrapped with poly sheets to avoid bud damage (Fig. 1).

Breeder seed production: Healthy nucleus seed setts produced as already described are utilised for raising the Breeder Seed of newly released variety or/ and varieties already under cultivation. It is recommended to apply Moist Hot AIR Treatment (MHAT) to the cane setts used for planting Breeder Seed. MHAT treatment consists of moist hot air temperature at 54°C and 99% RH for 2.5 hours duration to control the majority of seed / sett borne diseases without affecting the germination. The spacing recommended is 90 cm. between rows and four budling per meter. The row length is 10 meters with 3 meters wide path between beds of 10 rows. All recommended sugarcane crop cultivation practices, crop cleaning steps and diseases, pest and weed management procedures are followed. Isolation distance of 20m. from other sugarcane crops is maintained. Roguing is carried out by the Breeder concerned at periodic intervals coinciding with 45 to 60 days and 120 to 130 days after planting and 15 to 30 days prior to harvest. During the roguing operations, disease affected plants, insect attacked plants, off types and deviant plants are uprooted. The physical purity, true to type nature, genetic purity and uniformity are thus ensured. The individual canes are studied for brix value and other juice qualities. Monitoring team visit is arranged at maturity stage.

The breeder seed cane is of utmost purity and collected from the true to type plant crop only with the said seed qualities with each node having only one sound bud. The number of nodes without sound bud shall not exceed 5% (by numbers) of the total number of buds per seed cane. Nodal roots must not be present. Ultimately, the breeder seed of sugarcane will have not less than 85% germination of buds, 98% physical purity and 100% genetic purity of the seeds.

Grow Out Test: Randomly, 50 true to type and healthy canes are selected and seed setts are prepared from individual canes and planted cane wise. Tillers of each cane are examined for purity and conformity of characters at different stages of development till cane maturity. The GOT plot is of one meter long, 50 rows, with 12 budlings per meter length in each row. Morphological characteristics, yield potential, sugar content and seed quality parameters are evaluated in addition to the brix value and other qualities of juice and test milling parameters.

Micropropagation: The micropropagation schematic details as recommended by the Department of Biotechnology, New Delhi for quality tissue culture plantlets to be followed by primary nursery, field planting and for GOT are given in Fig. 2.

Table 1. Descriptor of sugarcane

S. No.	Characters	States	Stage of observation
1.	Stool growth habit	Erect/semi-erect	Vegetative
2.	Number of millable canesper plant	Low/medium/high	Vegetative
3.	Adherence of leaf sheath	Self detraining/semi clasping/clasping	Vegetative
4.	Internode diameter	Thin/medium/thick	Vegetative
5.	Internode shape	Cylindrical/tumescent/ bobbin-shaped/conidial/ obconoidal/concave-convex	Vegetative
6.	Internode pithiness	Absent/present	Vegetative
7.	Internode color exposed to sun	Green white/green yellow /purple yellow/yellowgreen/ green/purple green/yellow purple/green purple/purple	Vegetative
8.	Internode appearance rind surface	Smooth/mottled/rough (corky)	Vegetative
9.	Node width of growthring (opposite bud)	Narrow/medium/broad	Vegetative
10.	Node shape of growth ring at bud	Horizontal/bent upward at bud side	Vegetative
11.	Node width of root bandopposite bud	Narrow/medium broad	Vegetative
12.	Node arrangement of root primordial	Staggered/regular	Vegetative
13.	Node shape of bud	Triangular pointed/oval/ obovate/pentagonal/rhomboid/ round/ovate/rect angular/beaked	Vegetative
14.	Node bud groove	Absent/shallow/deep	Vegetative
15.	Node pubescence on bud	Absent/basal/apical/lateral/entire	Vegetative
16.	Leaf sheath hairiness	Absent/slight/strong	Vegetative
17.	Leaf sheath shape of auricle	Incipient/deltoid/uniform/ calcariform/lanceolate/falcate	Vegetative
18.	Leaf sheath color of dewlap	Green white/green yellow/purple yellow/yellow green/green/ purple green/yellow purple/ greenpurple/purple	Vegetative
19.	Serration of leaf blade margin	Slight/ medium /strong	Vegetative

20.	Cane top color	Pale green/green/dark green/ purple green	Vegetative
21.	Cane height	Short/medium/long	Vegetative
22.	Cane harvest maturity (elaborate % in juice)	Early/late	Vegetative

Source: S.R. Misra and P.K Singh (2003); PPV & FRA (2001)

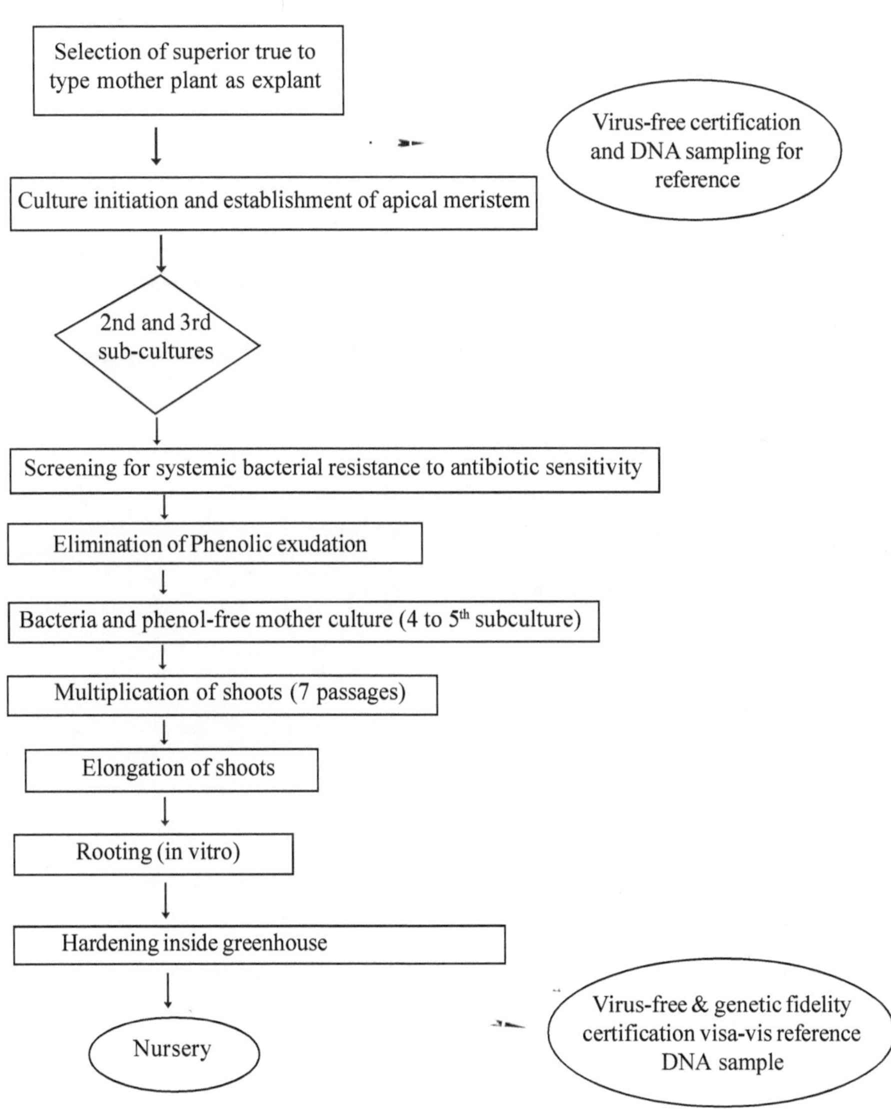

Fig. 2. Procedure and standard parameters for Sugarcane Tissue Culture
Source: Department of Biotechnology, New Del

13

Nucleus and Breeder Seed Production of Horticultural Crops

Vegetables

48. Agathi

(*Sesbania grandiflora* L.) (2n:24)

Family : Fabaceae

Agathi is a fast growing tree upto 10m. tall. Branching is drooping. The stem bark is deeply furrowing forming cork like plates. Leaves are compound, pinnae, 20 to 30 cm. long glabresence to pubescent. Leaflet oblong rounded 18 to 24 pairs of leaflets that are green to dark green mostly, bitter free. Inflorescence is a raceme 4 to 7 cm long and bearing 2 to 5 flowers that are white, red, yellowish and pink. Flowers are 5 to 10 cm long, 3cm wide, curved calyx bell shaped 2 cm long, 2 lobed, fine teeth like serration, fleshy. Petals 5 fleshy and large sized standard petal. Stamens 10 in 9+1 arrangement and pistil is narrow, slender styled. Pods are 20 to 60 cm. long, 6-9 mm. flat with 15 to 50 red brown seeds. Pods deliscent. Agathi is both self and insect pollinated. Agathi is known for food (leafy vegetables and flower vegetables), fodder and for wood. Agathi is planted as standard for vanilla, betelvine, pepper and as shade tree in young coconut plantations. Agathi is grown as wind break. Agathe is a medicinal and therapeutic use plant.

There are no improved varieties in Sesbania Agathi. It is a seed grown plant. Small plantations/ supports/border rows of Agathi plants are examined and at least 2000 plants are earmarked and seeds collected direct from open pollination. The seeds of 2000 plants are sown in isolation of 100m with a spacing of 90 cm × 90 cm. and plants are well looked after. Only one vigorous plant is retained in each hill. Individual plants are observed for vigour, stem girth, leaf and leaflet size leaf production, flower size, colour and each inflorescence on pod and seed set. At the maturity of crop, based on the examination, uniform looking, true to type for leaf and flower features, 500 plants are selected and seeds collected based on number of leaves, leaf biomass production, flower size and overall suitability for leafy vegetable and fodder. Laboratory examination of leaflets, flower, pod and seed is done and 200 selected true to type plants are self fertilised and seeds saved.

Nucleus Seed Production: In the following seasons, 200 plant progenies are raised, with 50 progenies in a block. Progeny rows are raised in single rows in isolation of 100m. Individual plants in the progeny rows are observed for uniformity with rows and between rows. The stem, leaf, flower, pod and seed traits are recorded progeny wise. Healthy and vigourous, true to type plants are selected in uniform looking progeny rows and 150 plants are self fertilised. At maturity, progeny rows are adjudged for their performance in terms of number of leaves and leaf biomass and flower yield. Progeny rows performing above mean +SE are identified and seeds from the self fertilised plants from the selected progeny rows are collected. The seeds of the self fertilised plants are bulked to constitute nucleus seed.

Breeder Seed Production: In the following season, breeder seed production is organised in a well prepared disease free field isolated by 100m from other sesbania crops. Early flowering and late flowering plants, off-types and plants showing deviations for any one of the characteristics are discarded. Visit of monitoring team is organised. Seeds harvested are tested in the seed testing laboratory. The quality assured seeds are labelled as breeder seed. Based on the need, NSI and NSII and BSP I and II can be organised. The market acceptability of leaf and flowers and their nutritive qualities are estimated.

49. Amaranth

(*Amaranthus* sp.) (2n:28)

Family : Amaranthaceae

Amaranth is a rich source of protein and nutrition. Amaranthus species are favored by human utilization as food grain, green leafy vegetable and forage also. Amaranth is herbaceous, erect stem, axillary and terminal inflorescences in cylindrical spikes. Amaranth is monoecious species. The sepals and petals are combined as tepal in a single floral whorl. Fruit is a capsule. Descriptor for vegetable and grain type are given in Table 1 and 2.

Table 1. Characteristics of Amaranth (Vegetable type)

1.	Cotyledon anthocyanin colour	Absent, present
2.	Seedlings-Cotyledon hypocotyl	Absent, present
3.	Young leaf length (cm.)	Short <10, medium 10-13, low <13
4.	Young leaf width (cm.)	Narrow <6, medium 6-8, broad >8
5.	Young leaf position of broad part	In the middle, lower base
6.	Young leaf prominence of veins	Weak, medium, strong
7.	Young leaf colour on upper side	Light green, green, medium green, dark green, purple, red
8.	Young leaf, distributions of colour on the upper side	Basal area coloured, central blotched colour, margin and vein coloured
9.	Colour of young leaf lower side	Green, red, purple
10.	Leaf margin	Entire, sinuate
11.	Leaf blade main colour	Light green, medium green, dark green, red
12.	Leaf blade- presence of blotches	Absent, present
13.	Leaf blade size of blotch	Small, medium, large
14.	Leaf blade colour of blotch	Green, silver, red, purple
15.	Leaf petiole anthocyanin colour	Absent, present
16.	Stem colour	Green, white, pink, red purple
17.	Plant height (cm)	Short <50, medium 50-80, full >80
18.	Stem shape cross section	Circular, undulated
19.	Plant- type of harvesting	Cutting, pulling
20.	Days to 50% flowering	Early <40, medium 40-50, late >50
21.	Colour of inflorescence	Green, purple, red, pink
22.	Plant growth habit	Determinate, indeterminate
23.	Seed colour	Brown, black
24.	Seed shape	Ellipsoid, discoid
25.	1000 seed weight (g.)	Low <0.6, medium 0.6-0.8, high >0.8

Source: PPV & FRA 2005

Table 2. Characteristics of GRAIN Amaranthus

1.	Seedling hypocotyle anthocyanin colouration	Absent, present
2.	Leaf blade length (cm.)	Small (<18), medium (18-22), high (>22)
3.	Leaf blade width (cm.)	Narrow (<10), medium (10-16), broader (>16)
4.	Leaf blade presence of blotch	Absent, present
5.	Leaf blade colour	Green, purple
6.	Petiole length (cm.)	Short (<14), medium (14-17), high (>17)
7.	Inflorescence colour	Light yellow, yellow, yellow green, orange, pink, pinkish, green, purple, red, reddish green, green.
8.	Days to 50% flowering	Early (<50), medium (70-80), late (>80)
9.	Inflorescence compactness	Lax, intermediate, loose

10.	Inflorescence length (cm.)	Short (<40), medium (40-70), high (>70)
11.	Inflorescence spininess	Absent, present
12.	Lateral spikelet length (cm.)	Short (<10), medium (10-15), high (>15)
13.	Inflorescence shape	Erect, semierect, drooping
14.	Plant height (cm.)	Short(<150), medium (150-200), tall (>200)
15.	Stem colour	Yellow green, pink, red
16.	Stem surface	Smooth, ridged
17.	Seed transparency	Translucent, opaque
18.	Seed colour	Cream, yellow, pink
19.	Seed shape	Elliptical discoid, oblate
20.	Seed weight (g/ml.) (g)	Low (<7), medium (7-8), high (>8)

(*Source:* PPV & FRA)

Nucleus seed production: Source seeds of Amaranthus variety are sown thinly in well prepared nursery with all precautions. On 21st day, well developed uniform and vigorous seedlings are selected from the nursery and transplanted at 20 cm. x 20 cm. spacing in a well prepared field. Ideal growing conditions like irrigation, weeding, fertilization and plant protection are provided. Isolation distance of 1600m. from other Amaranth crops is ensured. At vegetative phase, the plants are assessed for the true to type nature and uniformity in vigour, health and foliage characteristics. Off-types are removed. At flowering the panicle and floral characteristics are studied and plants deviating from the description are uprooted. At maturity, 1500 healthy true to type and uniform plants are labelled and seeds collected. In the next season well processed true to type seeds are sown in single plant nurseries with utmost care avoiding chance mixing of seeds. At proper stage of growth (21 days after sowing), the seedlings are transplanted with 25 cm. spacing between plants in 2 rows progeny plots with 25 plants in each row. The field is isolated by 1600m from other Amaranthus crops. The progenies are studied for the uniformity and conforming to the description in the vegetative stage. Progenies showing variation are uprooted as a whole. At preflowering stage, foliage in each progeny is studied for the shape, size, texture and softness. At flowering and seed maturity stage, the true to type and uniformity for panicle emergence and panicle characters are evaluated. Progenies showing deviating type of plants and features are discarded. Off-types and deviating plants are removed along with the progenies on either side to avoid pollen contamination. At harvest stage, superior five true to type plants in each progeny are selected and harvested. The seeds are examined in the laboratory for the uniformity and true to type.

In the next season, seeds of 500 such selected single plants are sown in separate single plant nurseries and seedlings selection made and planted in 2 rows progeny plots with isolation distance of 1600m. At the vegetative stage, stem and foliage characteristics are studied for the suitability for green leafy vegetable. Only vigorously growing uniform looking progenies suitable for green leaf market are retained. At the time of flowering, 250 superior plants selected in the uniform and true to type progenies for foliage and panicle characteristics and self fertilized by enclosing in large size bagging. Once in a day, gentle shaking of the plant is done. At maturity, self fertilized panicles are harvested and seeds extracted and after the laboratory examination seeds from each plant are saved separately.

In the third season, 250 self fertilised plant progenies are raised in two rows plots replicated twice by transplanting the seedlings in well prepared field that is isolated by 1600m. from other amaranthus crops. In each block, 50 progeny rows along with check are raised. All ideal growing conditions are provided. During the vegetative phase, the stem and foliar traits are assessed for suitability of fresh green soft leafy vegetable. Only highly uniform, fast growing and true to type progenies are retained and the rest of the progenies are cut and discarded. At flowering, 10 to 15 superior plants in each of the uniform progeny for stem, leaf and panicle characteristics are self fertilized. At maturity, self fertilized panicles are harvested and seeds separated and saved. After laboratory examination, part of the seeds from each self fertilized plants are bulked to constitute Nucleus seed.

Breeder seed production: Nursery is raised with the samples drawn from nucleus seed and well developed single seedling per hill is transplanted in the main field. Isolation of 1600 m. is provided from other Amaranth crops. All measures to maintain purity, true to type and uniformity as being followed in the nucleus seed production are followed. Only true to type , uniform , fast growing, healthy, disease free and vigorous plants are retained. In case of plants showing deviations, they are uprooted before flowering. Deviants identified at flowering are uprooted gently along with the surrounding plants to avoid vicinism. Visit of Monitoring Team to the field is arranged.Matured panicles are harvested and seeds extracted. The seeds are well dried, processed and tested for various seed quality parameters in the Seed Testing facility. The quality assured breeder seed is labelled with the Breeder seed label. Seed samples are drawn from the breeder seed lot and grow out test is conducted to adjudge the genetic purity (Fig. 1).

Amaranth grown for grain purpose is assessed for suitability and quality for cooking while nucleus and breeder seeds are developed. Nutritional quality of amaranth green leaves and grain are estimated

Source population from breeder source seed is raised with 1600 m isolation. The selected single plants are self fertilized. Select 100-150 true to type uniform vigorous, healthy single plants. Laboratory examination for leaf, panicle and seed features is done.

↓

Raise 100 single plant progenies, each progeny in 2 rows in raised beds (1600m isolation). Evaluate for field establishment, fast and vigorus growth and for leaf texture and tender stem and leaf yield at 25 to 30 days or at 40 days as the case may be. Select the lines and allow to grow to seed maturity. Select superior performing lines and identify 5 to 10 plants in each selected progenies. Collect the seeds and conduct laboratory examination in respect of grain types, Grain yield is evaluated and progenies identified.

↓

In the next season, raise progeny rows and evaluate the progenies for vigour, fast seedling growth and leaf production for green leaf vegetable .At maturity select superior plants 100 to 150 from the selected true to type uniform progeny rows. Constitute Nucleus seed by bulking seeds of selected plants.

↓

Breeder seed production (1600m isolation)

↓

GOT

Fig. 1. Nucleus and Breeder Seed Production in Amaranthus

Grain type

In respect of grain amaranths' the focused evaluation of single plants, self fertilized progenies and panicle seeds set, seed yield and seed quality's and nutritional aspect's is done. Isolation distance is 1600m.

Dual type

Evaluation for dual types (Vegetable greens and Grain purpose) is also done if specifically required.

50. Ashgourd

(*Benincasa hispida*) (Thunb.) Cogn.) (2n=24)

Family : Cucurbitaceae

A trailing or climbing herbaceous annual. Stem usually stout, thick, light green, less hairy and five angled. Leaves are large, alternate, inflorescence is cymose, axillary, flowers are monoecious, calyx 5, gamosepalous, petals 5 almost free, stamens 5, staminate flower is borne on long peduncle. Stamens single celled and in 2+2+1 arrangement. Pistillate flower with short thick peduncle. Ovary inferior, tricarpellary, style one, stigma 3, bilobed and curved. Fruit is a large pepo covered with ashy/waxy coating and flesh is white.

Nucleus seed production: From the breeder seed source or from maintenance seed stock, 300 to 350 plants are raised. The benincasa (ash gourd) plants are isolated by 1600m from other benincasa crops. Multistage evaluation is made on the stem and foliage traits. Plants that are not true to type are uprooted before flowering. Application of Ethrel 250ppm (2.5ml in 10 lit of water) four times commencing from 15 th day after sowing is made four times, at weekly intervals. At the time of flowering, each of the 100 selected superior plants is self fertilized. In the self fertilized plants, all other female flowers and developing fruits are removed to facilitate development of self fertilized fruit. Self fertilized fruits are examined in the laboratory and seeds are separated and labelled.

In the next generation, 50 to 60 single plant progenies are raised in isolation of 1600m. in two row plots in two replications. In each progeny, 20 to 25 plants are grown. The progeny rows raised in two replications are carefully examined for trueness expression and uniformity and stability of expression. Progenies showing deviations are uprooted. The progenies are well cared for in the plot that is isolated from other ash gourd fields by 1600 m. One entry from the original source of seed is also added for comparison from juvenile to pre flowering stages to flowering and fruiting phases. Individual plants from each progeny row over 2 replications are critically examined for stem, foliage, flower and fruit traits and progenies showing deviations for trait expression are uprooted. Growth, vigor, days to first female flowering, size of fruit , seed and yield data are recorded. All the progeny rows in the 2 replications are critically examined for true breeding and genetic uniformity and stability in traits expression. Based on the data, superior high yielding progenies recording mean + 2 SE or above are selected. The seeds form superior progenies are bulked together to constitute nucleus seeds. At the time of harvest, fruits are harvested from 10 superior plants in each selected progeny and seeds of single plants are separated and stored for future use as source seeds for developing nucleus to breeder seed chain.

Breeder seed production: Pregerminated nucleus seeds sprouting fast are sown in well prepared field not previously grown with the ash gourd crop. The field is isolated from other ash gourd crop by 1600 m. Only single healthy plant per hill is maintained. All recommended agronomic and horticultural practices are followed. Multistage examination of the individual plants are made continuously. Weak plants, off-types, diseases affected plants and deviating plants are uprooted .Two to three developing fruits per plant are retained from the beginning and the rest are removed at initial stage. At maturity, based on true to type nature for the overall plant features, vine and fruit characteristics, fruits are harvested after the visit of Monitoring Team. Laboratory examination of individual fruits for fruit shape, size, fruit wall appearance, flesh thickness, firmness, taste, keeping quality, seediness and market acceptability is made. Seeds are collected from selected fruits, dried, processed and tested for breeder seed quality assurance and Breeder seed label is tagged. GOT is arranged to test the genetic purity. The nutritional qualities are estimated in the marketable size fruits.

51. Bathua Keerai

Chenopodium album L. (2n=36, 54)

Family : Chenopodiaceae

Chakravarthi Keerai, Paruppu keerai, Lambs Quarters

Leaf Chenopodi

Paruppukeerai or Bathua is a fast growing dicot, upright in growth, herbaceous plant, annual herb. It is a minor leafy vegetable used like spinach. Bathua leaf extracts cure ulcer, antioxidant, antibacterial, purifies blood among many other medicinal properties Bathua possesses. India is one of the Bathua growing countries in large quantities commonly growing throughout year and consuming. It is commonly found growing with high intraspecific diversity and high degree of polymorphism. The characteristics are given in Table 1.

Table 1. Characteristics of *Chenopodium album*

1.	Plant -height (cm.)	80-100,150-180, 200>
2.	Plant-days to flowering	Early, medium, late
3.	Plant-days to maturity	Early, medium, late
4.	Stem-colour	Green, light purple, pink

5.	Stem-stripes	Absent, present
6.	Stem-shape	Angular, grooved, smooth
7.	Stem-intensity of pigmentation	Yellow, orange, pink, red
8.	Plant-Foliage gloucosity	Absent, present
9.	Plant- Foliage colour	Green, dark green, light purple, pink
10.	Plant-petiole base pigmented	Absent, present
11.	Leaf- angle at base	Acute, obtuse, wide
12.	Leaf-dentations	Absent, present
13.	Leaf-size	Small, medium, large
14.	Times of flowering	Early, medium, late
15.	Color of inflorescence	Yellow, orange, pink, red
16.	Panicle density	Sparse, medium, dense
17.	Panicle width	Narrow, medium, broad
18.	Seed color	Yellow, brown, red, black
19.	Seed (1000 no) wt.	1.4g

Nutrients content per 100g

1. Carbohydrate : 7.3
2. Fiber : 4.0
3. Fat : 0.8
4. Protein : 4.2
5. Calcium (mg) : 30.9
6. Iron : 1.2
7. Zinc : 0.4

An over view on the pharmacological effects and chemical constituents of chenopodium is given by Al snafi (2015). Sindhuja and Raja (2020) have described spacing, pinching and seed yield in chenopodium.

It grows upto 100 cm. in an upright manner and then becomes recumbent after flowering due to the weight of foliage and panicles unless support is provided. Stem is ridged and highly branched, The tap root system is strong. Leaves are opposite and leaves at basal part of stem is diamond shaped, toothed, 3 to 5 cm. long and 3 to 5 broad and leaves on the upper part of stem are entire and lanceolate, 1.5 to 2.0 cm long and 0.5 to 2 cm broad. Leaves are waxy coated. unwettable and mealy appearing with lower side of leaf white coated. Flowers are tiny small bisexual and female. Flowers are pentamerous, tepals 5 united at base to form cupshaped, anthers 5, placed opposite to tepal base, ovary superior. Self pollinalion is prevalent and upto 10% cross pollination occurs. Pollen is allergic and causes hey fever like.

Chakravarthi keerai Ooty1 was released in 2000 by the Tamil Nadu Agriculture University. It is a high yielding selection. Leaves are with pink tinge. The crop duration is 145 days and can be grown throughout year. The green yield is 28 t/ ha in high altitude, 17t/ha in the plains. It is grown with plant density of 100 to 150 plants per sq.m. It flowers in 50 to 60 DAS. Grain yield obtained is 0.2 to 0.6t per ha, Leaves and seeds are consumed. Pusa Bathua 1 with reddish green leaves and Pusa Green a late bolting variety are released in 2016. Kasthri Bathua 2 and Khasi. Bathua 4 are the high yielding varieties released and notified. They are leafy chenopods (Singh 2019).

Nursery is raised with 500 to 600g seeds in 2 to 3 mm depth. Healthy seedlings of 40-45 DAS are pulled out and planted in 60 cm. x 45 cm. spacing. Seed yield is 6 to 7 q/ha. Fresh and dried leaves are used. These varieties are multicut varieties. Fresh harvest of leaf at 40-45 DAS, subsequent 4 to 6 cuttings can be done at about 20 days interval. The leaves become unfit for consumption when the plants flower. In some Bathua varities, leaf cuttings are taken 45 days and 60 days after transplanting and then left for flowering and seed setting. It is found from studies reported from TNAU that sowing at a closer space (30 cm x 45 cm) and pinching on 30 days after transplanting results in a higher yield of seed (3680 kg/ha) (Sindhuja and Raja, 2020) with Ooty (C K) 1 variety which is rich in protein (2.2%), zinc (23 ppm), calcium (0.84%), mg (0.58%) and iron (474 ppm). Bathua can also be used as fodder. Bathua is cross pollinated by wind.

Nucleus and Breeder seed production: Seed of the Bathua/Paruppu keerai accessed from Breeders or institution source are sown in nursery. Seedling at 40 to 45 DAS are examined in the nursery. Vigorous growing true to type uniform size seedlings are selected and planted in a well prepared, weed free, contamination free field, not grown to bathua in the previous season are transplanted in 60 cm x 45 cm spacing. All good agricultural production practices are adopted. The plot is isolated by 1600 m from other bathua growing fields. From the plant establishment to seed setting stage individual plants are examined for fast grown, plant vigor, uniformity in growth and flowering and panicle size and shape. Off-types, deviating type of plants and disease affected plants are discarded as and when they appear at different growth phases of the crop. Deviating plants, if appear at flowering phase are uprooted gently without pollen dispersal. Plants around the discarded plants are uprooted to prevent pollen contamination. Early bolters are discarded to maintain flowering uniformity. True to type, uniform maturing plants conforming to the plant characteristics descriptor are now identified and 250 to 300 plants are tagged. They are self fertilized with butter paper bags and also secured to the stakes so that the selected plants do not lodge and recumbent. At seed maturity, panicles of individual plants are harvested with labeling, dried and examined in the laboratory for panicle and seed characteristics and seed setting.

In the following season, 100 plant progenies are raised by sowing the seeds in single plant nurseries. Selected seedlings of the individual plants are transplanted at 40 to 45 DAS in two row plots, with 45 cm spacing between the plants and 60 cm spacing between rows. Twenty five progenies each in two rows and one check (the same variety) in two rows are grown in one block. The field is isolated by 1600 m and is grown in field that is weed free contamination free and not grown to bathua in the past two to three seasons. The recommended bathua crop production practices are followed at different growth phases from transplanting. Observations are made on quick field establishments, fast growth, true to type for the stem, petiole, leaf, flowers and panicle and seed setting. In each progeny early bolters, off-types and deviating kinds of plants are uprooted from early growth phase onwards. One row in each progeny is allowed to grow up to flowering and seed setting. All the plants are staked. In the other row each progeny based on the crop duration, single or multicut harvest of the foliage are made commencing from 40 to 45 days after transplanting and then at about 20 to 25 days interval for 2 to 3 harvests of foliage. Once the plants flower, leaf quality is bitter and unsuitable for consumption. In the line in each progeny that was harvested for leaf, observations are made for the quick/fast regeneration of the plants with high leaf number and ideal stem development. After each harvest superior plants are identified in true to type uniform progeny and labelled (Fig. 1).

Source population grown in isolation of 1600m.
Select 300 true to type healthy vigorous plants and self fertilize.
Examine in the laboratory, select 100 plants

↓

Raise 100 progenies each progeny in 2 rows and one of the rows is for leaf harvest (single/multicut) and the other row for flowering and seed setting. Off-types and deviating plants are removed and true to type ascertained (Isolation distance 1600 m)

↓

Evaluate progenies for leaf yield and seed yield. Select progenies performing above mean +SE for leaf yield and seed yield. Seeds from Superior plants in selected progenies are now examined in the laboratory and bulked to constitute Nucleus seed

↓

Breeder seed production in isolation of 1600 m. Breeder seed tested for seed quality. Quality assured seeds labelled as a Breeder seed

↓

Nutritional quality of leaf and seeds estimated

↓

GOT

Fig. 1. Nucleus and breeder seed production in Bathua Paruppu Keerai

Based on the leaf harvest and seed setting extent in plants in row left without leaf harvest, progeny rows that are true to type and performing above mean +SE are selected. In each such selected progenies superior plants are identified and panicles are collected. Seeds of individual plants are examined in the laboratory and seeds of the selected plants now bulked to constitute nucleus seeds. The rest of the seeds are saved for use in need based of nucleus – breeder seed chain

In the subsequent favourable season for bathua growing, field that is free of weekly and contaminator and isolated by 1000m from bathua growing field is selected. Nucleus seeds are sown in raised bed nursery, 2 cm deep and well cared for. At 40 days after sowing growth healthy, vigorous true to type and uniform seedlings are selected and planted in 60 x 45 cm spacing. All recommended practices for raising a seed crop of bathua are adopted. From establishment to flowering, at each growth phases individual plants are examined for vigour, uniformity and true to type for characteristics and disease and pest free nature. All off types and deviating type of plants are uprooted. Deviating plants that are noticed at the time of flowering are discarded and also the plants surrounding the deviating type of plants to avoid contamination. Depending on the variety and duration, staking is provided to prevent recumbent attitude. At mid maturity, visit of Monitoring team is arranged. At maturity individual plants are harvested and panicles and seeds are examined in the laboratory for trueness and seeds from selected plants are bulked to constitute breeder seed. The breeder seed is tested in the seed testing laboratory for prescribed seed qualities. The quality assured seed is labelled with breeder seed label. Sample of the breeder seed is now raised in grow out test and the genetic purity is ascertained.

Samples of leaves collected at each leaf harvest and the seeds are tested for nutritional properties to adjudge the quality for consumption of leaves and grains.

Varieties released

1. Ooty 1 – Chakravarthi keerai (2000). Seed rate 50 kg/ha; leaves green with purple tinge, resistant to Cercospora, Colletotrichum and Macrophomina root rot. Harvest of leaf commences at 50days after planting and then once in 50 days leaf yield is 30 t/ha
2. Pusa bathua (2016) – leaf harvest 2 times at 45 and 60 days after planting. Duration – 150-160 days. Leaf yield – 35 t/ha
3. Khasi bathua 2
4. Khasi bathua 4
5. Pusa green

52. Beet Root

(*Beta vulgaris* ssp. *vulgaris*) (2n:18)

Family : Amaranthaceae

Beet Root is a biennial herb and to produce nucleus seed, two vegetative crop year cycle is needed. Beet root produces rosette of leaves in the first year and flowers and seeds in the second year. Panicle is a large sized spike. Flowers are small in size with calyx and without corolla. Flowers are bisexual or unisexual and occasionally dioecious also. Usually solitary flowers and sometimes two or more flowers grow together to form multiple flower structure known as utricle with multiple embryos. Calyx continues to grow after flowering and become corky and completely covers the seeds. True seeds are kidney shaped and brown (Kale and Masallkar 1993). Pollen is wind borne. Isolation distance is 1600 m for nucleus seed production. Beet root is cross compatible and cross pollinated crop. It is necessary that beetroot crop is also isolated from sugar beet, spinach and swiss chard. Beet root shows inbreeding depression upon selfing. The characteristics of beet root is given in Table 1.

Table 1. Characteristics of Beet Root

1.	Germity	Monogerm, multigerm
2.	Seedling-hypocotyl colour	Absent, present
3.	Leaf petiole- attitude	Erect, semi erect, horizontal
4.	Leaf blade attitude	Erect, semi erect, horizontal, pendulous
5.	Leaf length with petiole	Short, medium, long
6.	Leaf blade length	Short, medium, long
7.	Leaf blade width	Narrow, medium, broad
8.	Leaf blade shape	Narrow, elliptic, broad
9.	Leaf blade colour	Green, green red, red
10.	Leaf blade intensity of green colour	Light, medium, dark
11.	Leaf blade red colouration of vein	Absent, weak, medium strong
12.	Leaf blade lobing	Weak, medium strong
13.	Leaf blade blistering	Weak, medium strong
14.	Leaf petiole width	Narrow, medium, broad
15.	Root shape(L-S)	Narrow elliptic, medium elliptic, circular, obvate, narrow oblong
16.	Root length	Short, medium, long
17.	Root width	Narrow, medium, broad
18.	Root tip shape	Pointed, round, flat, depressed
19.	Root corkiness	Absent, weak, medium, strong
20.	Root external colour	White, yellow, reddish purple
21.	Root flesh colour	White, yellow, reddish purple
22.	Root-prominence of rings	Absent, weak, medium, strong, very strong
23.	Bolting tendency	Weak, medium, strong

Seed to seed and root to seed methods are followed for seed production. For nucleus and breeder seed production root to seed method alone is adopted. Seed is directly sown and after 15 days, seedlings are thinned to maintain one healthy seedling per hill. Good agronomic practices are followed to raise a healthy, disease and pest free beet root crop. During different stages of growth till matured root tuber formation, roguing is done. Critically stem and leaf characteristics are examined and off-types and the deviating plants are uprooted. Premature bolting plants and disease infected plants are removed. At this stage, matured roots are harvested/ lifted. Non conforming roots for size, shape, skin color, crown shape and surface corkiness and core colour are discarded. Uniform size and true to type roots alone are selected that conform to the variety description.

Root to seed: Before replanting the stecklings, verify and remove atypical roots, disease affected roots/plants. Root to seed is the method to produce basic seeds. The stecklings are prepared and planted in 60 cm x 30 cm spacing.

The schedule for seed production is seed sowing in July and good quality roots are produced when they take less time to grow. Root digging and examination for true to type and uniformity is done in November – December followed by steckling preparation by trimming the top without injuring the beet root plant crown, replanting early the selected stecklings in well prepared field at 60 cm x 45 cm spacing and harvesting the matured panicles in June – July. Isolation distance required is 1600 m.

Early bolting plants and off-types and deviating type of plants for leaf, panicle shape and floral characters are rogued out when the replanted stecklings are growing and bolting and at the seed maturity period. The panicles are cut at the base and harvested when the seeds at the basal portion of panicle is maturing as indicated by seed color turning to brown from green and ripe fruits are mealy and unripe fruits are milky. Nucleus seed is constituted by bulking of seeds from selected superior and uniform and true to type of plants.

In temperate condition growing, vernalisation is required for several weeks to initiate bolting. Stecklings prepared from one ha area is sufficient for planting 10 ha of seed production area.

Breeder Seed Production: Nucleus seeds are sown in well prepared field not previously grown to beetroot. The isolation distance followed in 1600m. Vigorous true to type single seedlings per hill are transplanted. Roguing is done in different stage of plant growth till the root reach edible maturity. True to type roots are collected and stecklings produced. Replanted stecklings produce spikes. In the selected true to type plants, selfing is done with muslin cloth bag. The seeds set are collected plant wise and progeny rows are grown in the following seasons at two stages. (i) seed to root and (ii) root to seed. At the second stage of root to seed, roots of selected true to type plants in each progeny are grouped together and raised to bolt and seed set. Progeny rows testing is carried out bestowing utmost care on true to type verification from seed to root and root to seed production stage. In the selected progeny rows, seeds from selected plants are pooled together to make the breeder seed.

53. Bitter Gourd

(*Momordica charantia* L.) (2n:22)

Family : Cucubitaceae

Plants are annual, viny, monoecious, staminate flowers small in size with slender and long thin pedicel. Pistillate flowers are solitary in nature, pedicel short, ovary oblong to long and green coloured. Petals yellow. Five calyx and 5 corolla. Stamens are three, two biloculed and one is uniloculed. Styles are short; stigma three, bilobed. Anthesis dehiscence occurs in the early morning. Bitter gourd is a cross pollinated crop. The characteristics of bitter gourd are given in Table 1.

Table 1. Characteristics of bitter gourd

1.	Cotyledon-green colour-intensity	Light, medium, dark
2.	Plant main vine length (m.)	Short (<2.0), medium (2 to 2.75), long (>2.75)
3.	Stem shape	Rounded, angular
4.	Stem-length of internodes of main stem (15to 20 th node) (cm.)	Short (<5), medium (5 to 8), long (>7)
5.	Stem- number of primary branches	Less (<10), medium (10-20), many (>20)
6.	Leaf blade length (cm.)	Short (<6), medium (6-9), long >9)
7.	Leaf blade width(cm.)	Narrow (<6), medium (96-10), broad (>10)
8.	Leaf blade margin	Entire, serrate, multifid
9.	Leaf blade shape	Obovate, cordate, oblong, reniform
10.	Leaf blade – number of lobes	Five, seven
11.	Leaf blade – depth of lobing	Shallow, medium, deep
12.	Petiole length (cm)	Short(<5), medium (5 to 8), long (>8)
13.	Flower colour	Light yellow, yellow, deep yellow
14.	Ovary length at anthesis (cm.)	Short (<1.5), medium (1.5 to 2.5), long (>2.5)
15.	Fruit length (cm.)	Very short (<5 cm), short (5 to 10), medium (10 to 15), long (15 to 20), extra long (>20)
16.	Fruit diameter (cm.)	Thin (<3), medium (to 4.5), thick (>4.5)
17.	Peduncle length (cm.)	Short (<5), medium (5 to 10), long (>10)
18.	Fruit- colour of skin	White, creamy white, light green, green,dark green, glossy green
19.	Fruit shape at peduncle end	Acute, obtuse, rounded flattened
20.	Fruit shape at apex at blossom end	Acute, obtuse, rounded flattened
21.	Fruit shape	Oblong, obovate, spindle shaped, club shaped triangular

22.	Fruit tubercles	Absent, few, medium, many
23.	Fruit ridge	Discontinuous, continuous
24.	Fruit bitterness	Mild, strong
25.	Fruit skin colour (ripe stage)	Yellow, orange, reddish, orange
26.	Seediness (number of seeds per fruit)	Very less (<10), less (10-20), medium (21 to 30), many (>30)
27.	Seed length (cm.)	Short (1.4), long (>1.4)
28.	Seed colour	Light brown, brown, dark brown, yellow, black
29.	Seed indentation of margin	Small, medium, large
30.	Seed surface	Smooth, rough

Source: PPV & FRA

Nucleus seed production: Select well filled and matured seeds, from the Breeder stock or maintenance breeding stock. Pregerminated and healthy well filled seeds are planted in pits of 30 cm x 30 cm x 30 cm size at a spacing of 2m. x 2m. in a plot isolated by 1600m. from other bittergourd crops. Only one fast growing healthy plant is retained per pit. Plants are allowed to trail on trellies at 2 meter height. Good vegetable crop growing practices including agronomic, fertiliser application and plant protection steps are followed. Spraying of Ethrel @ 250 ppm. for four times starting from 15 days after sowing is followed at weekly intervals. (2.5 ml. ethrel in 10 liters water). Regular and periodical observations are made to verify true to type nature and uniformity of the plants in respect of vine, foliar, days to first flowering, male and female flowers, fruit, fruit colour and fruit surface characteristics at respective stages of growth and development. All the deviating plants and off-types and disease affected plants are uprooted. At the time of flowering of the first female flowers, select 200 true to type, healthy, vigorous and disease free plants. One female flower and 2 to 3 male flowers on the same vine that will open next day are protected with paper bags. On the next day, aided pollination is made by collecting the anthers from the protected male flowers of the same plant that are dehiscing and gently rubbing on to the stigma of the female flower that is protected with paper bags. All female flowers and developing fruits in the vines are removed retaining the self pollinated flowers only. At ripening stage, self fertilized fruits are collected at maturity and just started ripening and stored for 2 to 3 days for full ripening. After laboratory examination, fruits and seeds are selected and seeds are separated on single plant basis and labeled (Fig. 1).

In the next generation, 100 self fertilised single plant progenies are grown in two replications with 20 plants in each progeny in the plot provided with 1600m isolation distance. Only single plant per pit is maintained. All the standard horticultural production practices and trelly support are followed. Ethrel spray (250 ppm.) is made 4 times commencing from 15 days after sowing. Individual plants in each progeny are observed in each growth/ development phase. The plants deviating from the prescribed vine, foliar and flower characters and fruit size and shape are uprooted. Progenies that are true to type, uniform, healthy and free from diseases alone are selected based on the vine, foliar, flower and fruit characteristics. Progenies showing performance above mean + 2 SE are selected. Within each selected progeny, superior plants are identified and self fertilised. At maturity, self fertilized fruits are collected progeny wise and seeds are separated, verified for conformity and stored.

In the following generation, the seeds from the superior 50 progenies are sown in isolated field with 1600 m. distance, in 2 row plots, 20 plants in each row. Monitoring the progenies for true to type breeding, genetic uniformity and stability is made at each crop stage until fruit maturity. The progenies showing deviating plants not in conformity are uprooted at preflowering and flowering commencement. Yield data are recorded. Fruits are collected from superior performing progenies. Seeds are separated from ripe fruits and examined for the size, shape and surface markings. Selected seeds from individual plants are bulked together to constitute the nucleus seed.

Breeder seed production: Pregerminated nucleus seeds are dibbled in pits of 30 cm x 30 cm x 30 cm size at a spacing of 2m x 2m in a well prepared field that is isolated from other bitter gourd growing fields by 1600 m. distance. Only fast growing healthy single seedling is retained in each pit.All good agronomical and horticultural production practices are adopted. Ethrel spray (250 ppm) is made 4 times commencing from 15 days after sowing. Individual plants are periodically examined and plants deviating from the prescribed vine, foliar and flower characters and fruit size and shape are uprooted. Early and late flowering plants are removed.

Plants that are healthy, true to type, uniform, and free from diseases alone are selected based on the vine, foliar, flower and fruit characteristics. Marketable fruits in each plant is examined for market acceptability and are finally retained. Monitoring Team visit is arranged to adjudge the field stand and genetic purity of the crop. Fruits are harvested at ripe maturity and seeds separated, dried and processed. Seeds are tested for seed qualities and quality assured seed is labelled with breeder seed label. Grow out test is conducted with seed samples drawn from the breeder seed and grow out test is conducted to check for the genetic purity. Nutrient contents of the fruits are estimated.

500 representative true to type uniform plants are grown adopting the gourd production system of trailing staking. From the middle of the field 300 representative true to type plants are tagged. The field is provided with 1600m. isolation distance At maturity 2 to 3 fruits that are uniform in size are harvested from each plant. Table and laboratory examination are conducted. 200 to 250 plants are selected and seeds obtained

↓

In the next season, progenies of selected single plants are raised at wide spacing. Off-types and deviating plants are removed. Uniform true to type 100 plants are identified at flowering phase and self fertilization of 2 to 4 female flowers is made in each plant and 2 to 3 developing self fertilized fruits are labelled. The fruits of single plants are allowed to ripe well and table and laboratory examination is made to verify the trueness of fruits and seeds of single plants and seeds are saved.

↓

Raise 100 single plant progeny plots each 2 rows with 10 m. long (10 plants per progeny) in two replications. 50 progenies in each block. Retain only one plant per hill.

↓

From seedling to active and fast vegetative growth, flower and fruiting stage evaluation is made for true to type, uniformity in flowering and other growth phase specific characteristics. At fruiting, suitability for fresh vegetable market stage is adjudged. Plant progenies showing variations (and disease infections) are discarded. Among the uniform progenies, superior plants from each progeny are selected and tagged based on prolific bearing nature and fruit size. At maturity fruits are collected from the tagged plants, allowed to complete ripening. Table and laboratory examinations are made for trueness. Seeds from 250 to 300 single plants are then packed and stored.

↓

Part of the seeds from each selected plants are then bulked to constitute Nucleus seed.

↓

Based on the demand NS II stage can be repeated (1600m. Isolation)

↓

Breeder seed Production BSP I and BSP II Storage (Isolation distance 1600m.)

↓

GOT

Fig. 1. Nucleus and Breeder Seed Production in Bitter Gourd, Ridge Gourd, Snake Gourd and Sponge Gourd. The protocol described is applicable to each of the four gourds. Isolation distance is 1600m.

54. Bottle Gourd

(*Lagenaria siceraria* sp.) (2n:22)

Family : Cucurbitaceae

It is an annual vine and monoecious in nature. Leaves are cordate-ovate to reniform shape. Flowers are solitary, white and open at night. Sepals five and petals five. Stamens are three, two of them are compound and one as single. Staminate flowers are borne on long pedicels. The pistillate flowers are single, supported on short pedicels. Ovary has 3 placenta and numerous ovules. Fruits are variable in size and shape. Flowers open in night. Selfing and self pollination may be done in the previous evening or next day morning. The crop duration is 135 to 140 days. The characteristics of bottle gourd are given in table1.

Nucleus seed production: Five hundred seeds are sourced from the maintenance breeding stock or from breeder stock. The source seeds are pregerminated and from them, vigorously germinating 300 seeds are selected and 2 to 3 seeds are planted in each pit of 30 cm. x 30 cm. x 30 cm. size at spacing of 2m. x 2m. or 2.5m. x 2m. Isolation of distance of 1600m. is provided. Only one plant is retained in a pit. All recommended production practices and ethrel application are followed for 4 times. Plants exhibiting genetic uniformity, true to type nature and stability alone are selected. Plants showing early and late to first flowering are discarded. At maturity, flowering 100 single plants are selected and 2 to 3 female flowers are self fertilised with male flowers of the same plant. Self fertilised fruits, one each from each plant, are harvested. The seeds are separated and selected based on laboratory examination and saved as single plant seeds.

Table 1. Characteristics of Bottle gourd

1.	Cotyledon length	Short, medium, long
2.	Plant length	Short, medium, long
3.	Leaf blade size	Small, medium, large
4.	Leaf blade green colour intensity	Light green, medium, dark green
5.	Leaf blade incision	Absent, slight, medium deep
6.	Male flower overlapping of petals	Free to slight overlapping, strong over lapping
7.	Male flower- corolla diameter	Small, medium, large
8.	Female flower- corolla diameter	Small, medium, large
9.	Female flower petals overlapping	Free, slight, strongly overlapping
10.	Fruit shape	Obovate, ovate, oblate round, elliptic, cylindrical, clovate
11.	Fruit length	Very short, short, medium, long, very long
12.	Fruit diameter	Very small, small medium, large, very large
13.	Fruit neck	Absent, short, medium long, very long
14.	Neck shape	Globose, fusiform cylindrical

15. Neck length	Very short, short, medium long, very long
16. Neck diameter	Small, medium, large
17. Fruit colour	Light green, medium green, dark green, very darkgreen
18. Fruit- no. of speckles	None, few, many
19. Fruit-size of speckles	Small, medium, large
20. Fruit-texture of skin	Smooth, slightly verrucose, medium verrucose, strongly verrucose, slightly corrugated, medium corrugated, strongly corrugated
21. Fruit pistil scar	Small, medium, large
22. Seed width	Narrow, medium, broad
23. Seed colour	Light brown, medium brown, dark brown, black

Source: UPOV. 2017(TP/313/1)

In the next generation, seeds of 50 single plants are sown in 2 rows plots with ten plants per row and replicated twice. The field is provided with isolation distance of 1600 m. Observations on individual plants are periodically made for the true to type, genetic uniformity and stability. Progenies showing deviations are uprooted before flowering. In each progeny, 3 to 5 plants are selected and in each selected plant three, flowers are self fertilized with pollen from male flowers borne on the same plant. Fruits are harvested at marketable stage and assessed for the various market and consumer qualities and fruit shape and other fruit charactertics.

Based on yield data, superior progeny rows performing above the overall mean value (mean + 2 SE) are selected. Matured fruits are collected from 50 to 60 single plants and the seeds from the well ripe fruit of superior progenies are processed and examined in the laboratory for trueness and selected and the seeds are pooled together to make the nucleus seed.

Breeder seed production: The nucleus seeds are pregerminated and from them, vigorously germinating 500 seeds are selected and 2 to 3 seeds are planted in each pit of 30 cm. x 30 cm. x 30 cm. size at spacing of 2m. x 2m. or 2.5m x 2m. All good agronomic and horticultural practices, fertilizer application and plant protection measures are properly done. At 12 to 15 DAS, only one healthy seedling is retained per pit. The breeder seed production plot is isolated by 1600 m. from other bottle gourds. From germination to maturity, the plants are examined continuously for uniformity and true to type nature conforming to the description in respect of vine, foliage, flower and fruit characteristics (Table 1). Non conforming plants are uprooted.

At the time of first female flower appearance, the female flower is self pollinated with pollen from the male flowers of the same plant. The stigma is profusely deposited with pollen to ensure good seed set. In each plant that is selected for early vigour, health and free from diseases, 2 to 3 flowers are self fertilized. Fruits are collected from true to type uniform individual plants at marketable stage and examined for fruit shape colour, flesh texture, non bitterness, slow fruit maturity, taste and flavor besides high female to male flower ratio. Visit of Monitoring team is arranged to judge the genetic purity. Self fertilized matured fruits are harvested from 100 to 150 selected single plants. Seeds are separated, examined in the laboratory and selected. Seeds from individual plants are pooled to constitute breeder seed. Samples of breeder seeds are tested for different seed qualities. The quality assured seed is labelled with breeder seed label. Grow out test is conducted with seeds sampled from breeder seed to adjudge the genetic purity. Nutritional quality is tested in the marketable stage fruits.

55. Brinjal

(*Solanum melongena* L.) (2n:24)

Family : Solanaceae

Annual, erect or semi spreading and cylindrical stem. Leaves are with oblong to obovate lamina. Inflorescence solitary or in clusters of 2 to 5 hermaphrodite flowers. Flower is complete with 5 calyx that are persistent, corolla 5 with prominent coloration, stamens five in number and free with apical pore dehiscence. Ovary hypogynous, bicarpellary and syncarpus. Fruit setting flowers are with long and medium style. Long styled flowers show 70 to 75 % fruit set while medium styled flowers show 12 to 55 % fruit set. Non fruit setting flowers are with short styles and smaller stigma and undeveloped papillae. Anthesis time is 6 to 8 am. in summer and get delayed upto 11 am. in cooler / winter period. Anther dehiscence and stigma receptivity are almost at the same time. Brinjal is a self pollinated crop but due to insect aided pollination, it is considered as often cross pollinated crop (Ram, 2005). The descriptors are given in Table 1.

Table 1. Characteristics of brinjal

1.	Seedling-anthocyanin of hypocotyl	Absent, present
2.	Seedling-intensity of anthocyanin colorationof hypocotyl	Weak, medium, strong
3.	Stem anthocyanin coloration	Absent, present
4.	Stem pubescence	Weak, medium, strong
5.	Leaf length (cm.)	Small (<10), medium (10-20), large (>20)
6.	Leaf width (cm.)	Small (<10), medium (10-20), large (>20)
7.	Leaf margin	Entire, dentate, sinuate
8.	Leaf blistering	Absent, present

9.	Leaf spininess	Absent, present
10.	Leaf blade colour	Green, purple
11.	Leaf blade colour intensity	Light, medium
12.	Leaf colour of vein	Green, purple
13.	Inflorescence number of flowers	1 to 3, >3
14.	Flower size	Small, medium, large
15.	Flower colour	Greenish white, light purple, purple, darkpurple
16.	Flowering time (days after sowing)	Early (<60), medium (60-80), late (>80)
17.	Fruit length (cm.)	Short (<10), medium (10-20), long (>20)
18.	Fruit diameter (cm.)	Small (<5), medium (5-10), large (>10)
19.	Fruit (general) shape	Globular, ovoid, obovate, pear shaped, clubshaped, ellipsoid, cylindrical
20.	Diameter of pistil scar	Small, medium, large
21.	Fruit – shape of apex	Indented, flat, round, pointed
22.	Fruit colour of skin at commercial harvesting	White, green, purple
23.	Intensity of green colour of fruit skin	Light, medium, dark
24.	Fruit stripes	Absent, present
25.	Fruit stripes-density	Sparse, medium, strong
26.	Fruit patches	Absent, present
27.	Fruit glossiness at maturity	Weak, medium, strong
28.	Fruit-size of-calyx	Small, medium, large
29.	Fruit-colour of-calyx	Green, purple
30.	Fruit spininess of calyx	Absent, weak, medium, strong
31.	Fruit colour of flesh	Whitish, greenish
32.	Fruit-length of peduncle (cm.)	Short (<1), medium (1 to 5), long (>5)
33.	Fruiting pattern	Solitary, cluster, mixed
34.	Plant growth habit	Erect, semi spreading, spreading, horizontal
35.	Plant height	Very short, short, medium, tall
36.	Plant spread	Narrow, medium, broad
37.	Fruit colour of skin	Yellow, orange, brown
38.	Fruit-time of ripeness (days after fruit set)	Early (<65), medium (65-75), late (>75)

(*Source:* PPV & FRA, 2009)

Nucleus Seed Production: Seeds are sourced from basic seed resource available with the breeder. Tray Nursery is raised with utmost care. The seedlings are fit for transplanting 30 to 35 days after sowing. One thousand vigorous disease free seedlings are selected from tray nursery and planted at a spacing of 90 cm x 60 cm. Isolation distance of 800 meters from other egg plants crop is adapted. Individual plants are examined from the seedling stage, active vegetative growth, at preflowering, flowering and fruit set stage to matured fruit stage. Individual plants that are uniform and conform to the true to type description alone (Table1) are retained and others showing deviations are uprooted. From the selected heathy plants, 100 plants are self fertilized and labelled. In the next generation, seedling from self fertilized seeds from 100 plants are raised as single plant progeny row. Verification and monitoring for the prescribed traits is done at every stage of crop growth. The progeny rows showing off-types and plants deviating from the descriptor and fruit qualities are removed. In the true to type uniform plants in each row, 5 to 6 plants are self fertilized. A total of 100 self fertilized plants are selected.

In the next season 100 single plant nurseries are raised. Healthy true to type seedlings from single plant nurseries are planted in 100 progeny rows plots, each row having 20 plants in 2 replications in randomized manner. Isolation distance of 400m. is provided. All the individual plants in each row/replication are critically examined for the vegetative, floral, fruit set and fruit characteristics. Twenty superior plants in each progeny are self fertilized. Yield data and traits expression are recorded. The fruits suitable for fresh market are harvested from each progeny and assessed for fresh fruit qualities, appearance, non bitterness, consumer appeal and shelf life and keeping quality. Final selection of progenies is made based on yield and fruit qualities including fleshy calyx and pedicel. Progeny rows performing above mean +2SE are selected. Part of the self fertilized seeds from each superior plant in the selected progeny rows are pooled to constitute the nucleus seed. The remnant seeds are saved for developing nucleus seed in the next cycle (Fig. 1).

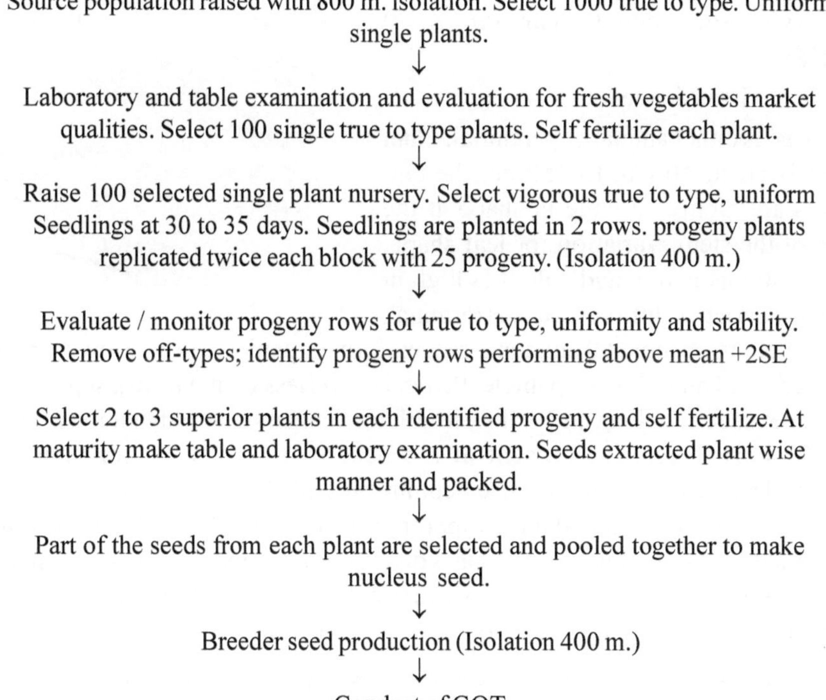

Fig. 1. Nucleus and Breeder seed production in Brinjal.

Breeder Seed Production: Seeds from the nucleus seed stock are sown and seedlings are transplanted in field with isolation of 400 m. distance from other egg plants. Continuous verification is made for observing for genetic uniformity, true to type description, health and vigour. Fresh vegetable stage fruit qualities are also assessed. Deviant plants and off-types are removed along with adjoining plants around. Visit of Monitoring Team is arranged to monitor the Breeder Seed Production plots and to assess the field and quality standards. Well matured fruits on the healthy plants that start showing yellow color are harvested and stored for ripening for seed extraction. Seeds are separated from well ripe fruit on plant basis. In the laboratory fruits and seeds of individual plants are examined to conform to true to type nature and then pooled together to constitute breeder seed. Breeder seed is tested for different seed qualities and the quality assured seed is labelled with the breeder seed label. Once in two years, the nucleus to breeder seed production cycle is followed. Seed samples drawn from breeder seed are evaluated in Grow Out Test to adjudge the genetic purity. Nutritional qualities are estimated in the marketable fruits.

56. Cabbage

(*Brassica oleracea* L. var. *capitata*) (2n:18)

Family : Brassicaceae

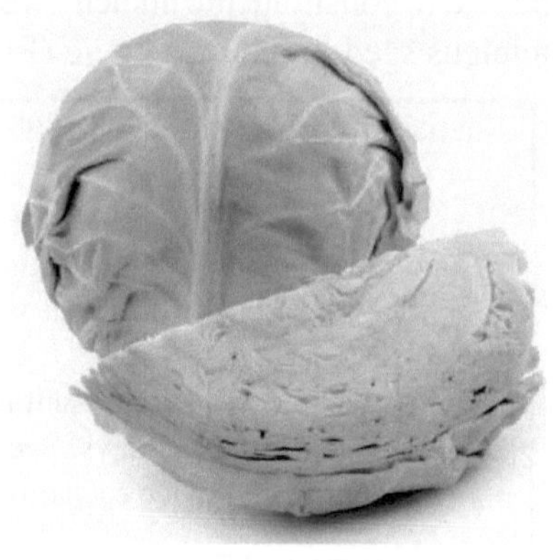

Cabbage is either annual or perennial. Plant habit is erect, 40 cm. to 150 cm. height; leaves are highly variable in shape at the base of the stem. Variation for leaf shape, size, glabrous nature and colour is high in cabbage varieties. Inflorescence is terminal, lax and elongate, upto 50 cm. long, racemeform or paniculiform. Flowers range from 25 to 40 numbers per panicle. Pedicel short (less than 0.8 cm), sepals four, erect; petals four oblong and yellow in colour. Stamens 6, tetradynamous, 4 long and 2 short. Nectaries present at the base of stamens. Ovary superior, bicarpellary, fruit is siliqua, false septum present. Flowers are protogynous. Cabbage is cross pollinated due to sporophytic incompatibility. Bees and other insects aid in cross pollination. Selfing is done by bud pollination. Cross pollination is accomplished by emasculation in one to two days prior to anthesis and pollinated with desired pollen. The descriptors are given in Table 1.

Seed Production: Seedlings are raised in a well prepared nursery with quality assured breeder seed or nucleus seed accessed from the Institution concerned. Seedlings are protected well with good plant protection care. Healthy and vigorous seedlings are transplanted in a well prepared field free of soil infection and not previously grown with cabbage or any Brassica crops. Isolation distance of 1600 meters is maintained. All prescribed good agronomic (GAP) practices for cabbage are adapted. Seed rate for cabbage is 600/g. per ha. Seed pretreatment or predrying the seeds at 40°C for 24 hours followed by air treatment at 70°C for 5 to 7 days is done to disinfect the cabbage seeds. Mid and late duration varieties are sown in June first fortnight and early maturing CVs are sown in July second or third week. Seedlings are transplanted in the second fortnight of August for early varieties and in the first week of August for medium and late duration varieties. Monitoring for true to type of the individual seedlings in the nursery and the individual transplanted plants in the mainfield, roguing of off-types and deviating plant types are carried out regularly. Roguing is continued in the pre head and head formation stages. Plants showing poor head size, loose heads and early and late head forming plants and poor health plants are removed. Plants are examined for uniformity, vigor, true to type and stability. The plants are allowed to overwinter *in situ* where they were originally transplanted. After vernalization and winter period exposure, any one of the following methods is adapted with heads of cabbage to fecilitate forming seed stalk and elongation:

1. Stump method
2. Stump with central core method
3. Head intact method

Seed stalk elongation commence from mid March when the temperature raises. Flowering and pod formation take place in April. Flowering flush in full and fruit formation are observed from mid of April to mid of May. Pods start ripening in mid of June. In the second week of July, pods are harvested in two lots from true to type plants only that are left after through rouging. Off-type removal was initiated from flower opening in first week of April onwards.

In the first season of cabbage growing, the heads are produced and in the following season, seed production is done. The seed crop is continuously monitored in the nursery stage and then in the seedlings planted field in situ and in transplanted field during autumn for nucleus and breeder seed production. In situ crop is allowed to over winter and produce seeds in the original position where they are first planted in the seedling stage. In the transplanted method, the selected plants are carefully uprooted and after removing whorls of leaves, the plants are reset immediately in a well prepared new field.

1. Stump method: When the curd produced in the first season is fully matured, the curd/ head is cut just below the curd by a sharp knife keeping the stem with the outer whorls of leaves in tact. The beheaded portion is termed as

stump. The stumps of the selected true to type plants are left *in situ* in their original positions or replanted in the following season. After over wintering and dormancy breaking, the buds sprouts from the axils of all leaves and leaf scars on the stem. The flower stalks that develop subsequently from the axillary shoots become decumbent and require heavy staking. In this method, the crop matures 10 to 15 days earlier than in the head in tact method.

2. Stump with central core in tact method: In this method, the crop is allowed to mature fully. Off-types and plants not conforming to true to type are rogued out. Central core of the head is now made by carefully chopping the head on the sides only with perpendicular cuts without damaging the central core. The true to type plants with central core in tact are allowed to over winter. After over wintering, the core will start bursting and now two cuts are given to facilitate bursting in the core. Care is taken not to injure the central growing point in the core. If no such cuts are made, the heads burst irregularly and sometimes the growing tip is broken. In this method, the shoots arising from the central core are not decumbent and heavy staking is not required. Seed yield realized in this method is high.

3. Heads in tact method: In this method, the cabbage heads are allowed to fully mature. Off-types and deviating types of plants are removed. Keeping the heads in tact, a cross cut is given in each head to facilitate the emergence of flower stalks. In this method, heavy staking is not required. Seed yield is slightly lower than in the other two methods. In all the three methods, the isolation distance adapted is 1600 m. The seeds produced in the true to type representative individual plants are collected as single plant seed. Before saving the seeds, laboratory examination of seeds is carried out. In situ method or seed to seed method is adapted for foundation and certified seeds production.

Nucleus Seed Production: Single plant seeds nursery is raised in well prepared seed beds, raised by 5 to 8 cm. height. Individual seedlings in each progeny nursery are examined at 2 weeks age, for true to type and off-types are removed. Progenies showing less uniformity and weak health are discarded. Selected seedlings in each of the progeny nursery beds alone are transplanted in progeny rows in a well prepared field free of soil infection and isolated by 1600m. from other Brassica crops. Individual plants in each progeny row are critically examined for true to type and off-types are discarded. Good agronomic production practices are adapted. In each progeny, early and late heading plants are discarded to maintain uniformity in heading. Plants showing loose and mini sized heads are rejected. At head maturity, in each true to type progeny, superior plants alone are retained and allowed for vernalisation and over wintering. Seed production is done following the head intact method in the autumn. The head of the individual true to type plants is kept intact and only a cross cut is given to facilitate the emergence of the flower stalk. The stalk and the flowers and the pods set are critically examined in the individual plants in each progeny row for trueness. Pods are harvested at maturity in one or two times. The pods and seeds are examined in the laboratory and progeny wise selected and bulked to constitute nucleus seed. After drying to 8% moisture, the seeds are tested for seed quality assurance and labelled as nucleus seed.

Breeder Seed Production: Healthy seedlings are produced and planted in Breeder Seed Production field provided with isolation of 1600m. The field is not grown with cabbage crop in the past 2 to 3 years. Good agronomic production practices and maintaining genetic purity protocols are adapted. Head in tact method is followed for seed production. Seeds are harvested from the selected true to type plants. Seeds are tested for quality and Breeder Seed labeling is done. Nutritional qualities are estimated.

Table 1. Descriptor of cabbage

1.	Plant height		Very short, medium, tall or very tall
2.	Plant	Maximum diameter	Small, medium or large
		Length of outer stem	Short, medium or long
		Attitude of outer leaves	Erect, semi erect or horizontal
3.	Outer leaf		Small, medium or large
		Shape of blade	Broad elliptic, broad ovate, circular, transverse broad elliptic or broad ovate
		Profile of upper side of blade	Concave, plane or convex
		Bilstering	Absent or very weak, weak, medium strong or very strong
		Size of blisters	Small, medium or large
		Colour (with wax)	Yellow-green, green, grey-green, blue-greenor violet
		Intensity of colour	Light, medium or dark
		Green flush (red cabbage only)	Absent or present
		Waxiness	Absent or very weak, weak, medium strong or very strong
		Undulation of margin	Absent or very weak, weak, medium strong or very strong present
		Incisions of margin	Absent or present
		Reflexion of margin	Absent or present
4.	Head	Shape of longitudinal section	Transverse narrow elliptic, transverse elliptic, circular, broad elliptic, broad obovate, broad ovate or angular ovate

		Shape of base in longitudinal section	Raised, level or arched
		Length	Short, medium or long
		Diameter	Small, medium or large
		Position of maximum diameter	Towards top, at middle or towards the base
		Cover	Uncovered, partially covered or covered
		Blistering of cover leaf (savoycabbage only)	Absent or very weak, weak, medium, strongor very strong
		Colour of cover leaf	Yellow-green, green, grey-green, blue-greenor violet
		Intensity of colour of cover leaf	Light, medium or dark
		Anthocyanin coloration of cover leaf (white & savoy cabbages only)	Absent or very weak, weak, medium, strongor very strong
		Internal colour	Whitish, yellowish, greenish or violet
		Intensity of internal colour (redcabbage only)	Light, medium or dark
		Density	Very loose, loose, medium, dense or verydense
		Internal structure	Fine, medium or coarse
		Length of interior stem (in relationto length of head)	Short, medium or long
5.	Time of harvest maturity (in relation to length of head)		Short, medium or long
6.	Time of bursting of head after maturity		Early, medium or late
7.	Resistance to race 1 of Fusarium oxysporum		
8.	Method of seed production		Open-pollinated or hybrid
9.	Axillary Heading		This is undesirable and is assessed by splitting the head vertically through the core.
10.	Storability		Longer storability is desitable/ It is positively correlated with dry matter contents and late maturity. Dry matter could be assessed by sampling about 200g portion of head excluding core tissue and measuring the fresh and dryweights of the sample.

11.	Head Compactness	By applying pressure on head by thumbs, compactness of the head can be judged to some extent. Following 4 methods are more useful (Swarup, 1991) i) Examination of the position of the uppermost wrapper leaf indicates compactness. If it covers two third or more area of head, the head is considered to be compact. ii) Measuring the length of core inside the head by cutting a longitudinal section also indicates compactness. A compact head has relatively small core. iii) Compactness of the head can also be judged by the following formula given by Pearson. $Z = \frac{C \times 100}{W^3}$ Z an index of compactness C net weight of the head W average of the lateral and polar diameters of the head. A higher value of Z indicates more compact head. Net weight of head (without stalk and non-wrapper leaves) also gives a fair idea of compactness. More the weight, higher is the compactness
12.	Frame of the Plant	This is the maximum spread of plant at a maturity. Usually smaller frames are preferred.

Note: Description based on DUS guidelines (1992)

57. Carrot

(*Daucus carota* L. subsp. *sativus*) (2n=18)

Family : Apiaceae

There are two distinct types of carrots, namely, Tropical or Asiatic and Temperate or European. The temperate type forms roots both under tropical and temperate conditions but set seed only under temperate conditions as they need low temperature of 5 to 8°C for 40 to 60 days before putting forth the seed stalks. The Asiatic types are high yielding and produce seeds under tropical conditions. Carrot is an annual herb for root production and a biennial for flowering and seed production. Stem is branched, erect arising from a thick fleshy tap root. Stem grows upto 100 cm. while the tap root is from 10 to 30 cm. long. Root consists of outer peridermous layer and inner core. The root shape ranges from short conical to long conical and the root flesh color may be white, yellow, orange, red, purple or dark purple. The inflorescence is a terminal compound umbel. Umbels are of primary, secondary, tertiary and quarternary order. Primary umbels show around 1000 flowers per umbel. Seed setting is 100% in primary umbel and 90 to 100% in secondary and 82% in tertiary umbel (Hort.vt.edu). The umbels are centripetal. Flowers are epigynous, each flower consisting of 5 sepals, 5 petals, 5 stamens and two carpels. Flowers are protogynous and cross pollinated. Carrot exhibits high degree of inbreeding depression upon selfing. The characteristics of carrot are given in table1.

Table 1. Characteristics of Carrot

1.	Foliage width of crown	Narrow, medium, broad
2.	Leaf attitude	Erect, semi erect, prostrate
3.	Leaf length	Short, medium, long
4.	Leaf division	Fine, medium, coarse
5.	Leaf-intensity of green colour	Light, medium, dark
6.	Root length	Very short, short, medium, long, very long
7.	Root width	Narrow, medium, broad
8.	Root shape (LS)	Circular, obovate, medium narrow, narrow oblong
9.	Root shape of shoulder	Flat, flat round, conical
10.	Root tip	Blunt, slightly pointed, pointed
11.	Root colour	White, yellow, orange, purple, red
12.	Root-intensity of exterior colour	Light, red, dark
13.	Root anthocyanin colour	Absent, present
14.	Root diameter	Very small, small , medium, large
15.	Root colour of core	White, yellow, orange, pinkish, red, purple
16.	Root-intensity of root colour	Light, medium, dark
17.	Root-tendency for bolting	Weak, medium, strong

Seed production in Himalaya region: Seed to seed: Seed sowing is done in September to October. The plants are verified for true to type in stem, leaf, flower and seed characteristics and off types and deviating types are removed. Root to seed: Matured roots are lifted and examined for trueness to shape, size, color, root surface smoothness, absence of root forking and pithiness and deviating in any of the root characteristics. Off-types and deviating types are discarded. Early blooming or late bolting plant roots are rejected. The matured roots are lifted and one third of the top and one fourth of the root are cut and uniform sized stecklings are prepared and replanted in 60 cm x 30 cm spacing in well prepared field. Isolation distance of 1600 m. is provided for nucleus seed production plots. Good agronomical practices are adapted. Roguing is done and only true to type plants are retained. At full bloom stage, late and early bolting plants and atypical plants are uprooted. Plants showing uniformity in flowering, umbel shape and seed maturity and free from pests and diseases are retained. The ideal time for harvesting is when the secondary umbels are fully ripe and tertiary umbel begins to turn brown. Harvest is made by 2 to 3 pickings from selected superior true to type healthy plants. After drying, heads are threshed and seeds separated. Seeds are saved individual plant wise. Seed moisture content should be reduced to 8%.

In the following season, seeds are sown progeny wise in progeny rows in an isolated plot (1600 m.) to produce seed to root. Progenies that are uniform, vigorous and true to type in respect of stem and leaf characteristics are identified. Progeny row showing deviating plants and off types and early and late bolting are discarded. At the tuber maturity stage, tubers are lifted progeny row wise and examined for true to type and uniformity. Tubers showing variation for size, shape, skin color and core size are rejected along with the other plants/tubers in the same progeny. In continuation of the tubers/root selection, stecklings are prepared and planted in progeny rows. True to type nature of individual plants in each progeny is examined and progenies showing off-types and deviant plants are discarded. Early blooming plants are discarded. At maturity, umbels are collected from selected plants within each identified progeny rows. Seeds are extracted single plantwise and stored separately. Part of the seeds from selected single plants are bulked to constitute Nucleus seed.

Breeder Seed Production: Seed to Root and from Root to Seed production stages are followed on single plant progeny basis with utmost care to maintain trueness and genetic purity of the breeder seed. Isolation distance of 1600 m. is adhered to. Off-types and deviating plants, atypical plants are culled out in different phases of plant growth and development in seed to root and root to (stecklings) seed production respectively in each progeny row. Umbels are harvested 2 times from selected superior plants in the identified progeny rows. Umbels are dried and seed separated and part of seeds are bulked to form Breeder Seed. Seed quality assurance test is conducted. Grow out test is organized to adjudge the genetic purity. Nutritional qualities of carrot are estimated.

58. Cauliflower

(*Brassica oleracea* L. var *botrytis* L.) (2n:18)

Family : Brassicaceae

Flower: Sepals 4, petals 4, standing opposite to each other in a square cross. Stamens 6 (2 short + 4 long) in tetradynamous arrangement. Ovary superior, carpels two, ovary with false septum and in two rows of campylotropous ovules. Petals are in bright yellow colour, 15 to 25 mm. long and 10mm. wide. The flowers become fully expanded during the following morning and anthers open a few hours later. There exist slight protogynous nature. Nectaries are situated between the base of short stamens and the ovary. Nectaries are also present outside the base of the long stamens (Ram, 2005). Flowers are borne on long recemes on the stem and branches. The inflorescence may attain 1 to 2 meters length. Fruits are siliqua, 10 cm long and 4 to 5 mm wide. Seed number varies from 10 to 30 per siliqua. When the siliqua is ripe, pod dehiscence takes place through two valves breaking away from below upwards leaving the seeds attached to the placenta. The descriptors are given in Table 1.

Self incompatibility (SI) is present and the population exists in a highly hetrozygous manner. Bud pollination in SI varieties can be done successfully, 2 to 4 days before flower opening. The self compatible varieties can be self fertilized within nylon cages or in isolated planting if the SI is weak.

Difference between Indian and European Cauliflower (Gopalakrishnan 2007)

S.No.	Indian types	European types
1.	Tolerant to heat	Not tolerant to heat
2.	Curd formation at about 20°C	Curd formation at 5 to 20°C
3.	Annual	Biennial
4.	Yellow curds with less flavour	Snow white curds with or no flavour
5.	Early duration/maturity	Late maturity
6.	Short juvenile phase	Long juvenile phase
7.	No need of vernalisation but needs cold treatment for 6 weeks at 10-13°C	Need vernalisation at 7°C for 8 to 10 weeks.

During vegetative phase, temperature should be a few degrees higher than required for curd initiation and development. Cauliflower transform to curding at temperature ranging from 5 to 20 to 30°C depending on varieties. Nearly four weeks of vegetative growth is sufficient to curd development stage in snow ball types. Typical Indian cauliflower produces curd above 20°C and curding upto even 35°C.

Cauliflower being a biennial, the formation of curd/head is completed in first phase when the reproduction phase is complete and in the second phase only flowering is induced after getting necessary stimuli of 4°C to 10°C for about 60 days to break dormancy. Scooping is done (removal of the central portion of the curd) to facilitate uniform emergence of flowering shoot. In case of very compact curd producing varieties, longitudinal half cutting proved easier for bolting and early flowering besides high yield of seeds. Harvesting of seed is done 2 to 3 times.

Depending upon the varieties, plants may transform to curding at temperature 5°C to 28-30°C. All cole crops require vernalisation. But Cauliflower is an exception. In Indian or early CV, temperature range of 20 to 25°C is needed for transition to the reproduction stage while late or snow ball type needs 10°C to 16°C.

Seed Production: In cauliflower seed to seed production in situ is followed in early types. Very late type of cauliflower withstand transplanting shock. Seeds are produced by bagging of individual plants with muslin cloth bag followed by hand pollination among the selected plants. For transplanting method, plants are selected with good size curds and uprooted with ball of earth. The uprooted plants are verified for trueness and planted in a solid block for seed production.

In cauliflower, different maturity groups such as early, mid early and late varieties (Ram Singh and Sharma, 2001) are available. The sowing time of different maturity group is given below:

Maturity group	Sowing time	Transplanting time	Temperature requirement for curd initiation and development
1. Early maturity: Sept.	Mid May	July beginning	20° to 27°C
2. Early maturity : Oct.	May end mid June	Mid July	20 to 25°C
3. Mid early maturity: Nov. /Dec.	August	Sept. beginning	16 to 20°C
4. Mid late maturity: Dec.	Aug. end	Sept. end	12 to16°C
5. Late maturity	Sept. end, Oct. mid	Oct. end to Nov.	10 to16°C

Nucleus Seed Production: Seeds of breeder /nucleus seed source origin are sown in a well prepared fine tilth soil in raised beds. All precautions are taken to protect the young nursery plants. Each individual seedling is examined for vigour, fast growth / disease free nature and true to type expression. Deviant types and off-types are continuously monitored and discarded. Five to six week old heathy seedlings are pulled from the nursery and transplanted with recommended spacing in a well prepared disease free field that is isolated by 1600 m from other Brassicas. Prehardened seedlings are considered to withstand transplanting and in subsequent good growth and development. All good agronomical practices are adapted. Regular observations on true to type nature are made to cull out deviating type of plants, off types and disease and pest affected plants. Growth regulators are sprayed (NAA 10 ppm, GA_3 at rate of 80 mg/litre for 2 applications of GA_3 spray). Early and late bolting plants are discarded by uprooting.

At curd development completion stage or just prior to it, compact curds are selected from superior and true to type plants and central portion is removed by scooping without damaging the apical portion. The cut portion (wound) is applied or swapped with fungicides. Portions not flowered yet are removed. Disbudding is completed. Staking is provided for each plant. High quality healthy seed buds develop in the flowering season. When the fresh flower stalk develops, early and late bolters are removed. Seeds/pods are collected from selected true to type plants that are uniform. Seeds/pods collected from individual superior plants are examined in the laboratory for trueness. In the following year, single plants nurseries are raised in well prepared nursery beds free of soil borne infection. During the seedling development, critical observation is made on fast growth, vigour, true to type nature of seedling. Off-type and deviating type of plants are removed. Single plant nurseries that are poor in vigor and health are rejected. At 6 to 8 week age of seedling, the seedling are transplanted in progeny rows in a well prepared field free from soil borne infection. At every 10 rows, seedling of the source seeds are planted for comparision. The field should be isolated by 3000 m. from other Brassica fields. From the field establishment to bolting, in each of the progeny rows, care is taken to examine for the uniformity, stability and true to type nature of individual seedling. Uniformity within each progeny and between progeny are assessed.

Early and late bolting plants and disease /pest affected plants are discarded. At curd formation stage, progeny rows showing trueness and good performance are identified and superior plants within each such identified lines are tagged. The tagged plants alone are retained and other plants are cut and removed. Care is taken to maintain 400 to 500 selected plants so as to avoid inbreeding and to promote cross pollination among them. The selected cauliflower plants are labelled. When the curd firmness is at appropriate stage , suitable for scooping, removal of portion of curd is done. After scooping, the cut surface is protected to prevent infection. As and when individual plants show flower stalk and flower formation, critical care is taken to verify true to type nature for flower and developed pods and to discard the deviating types and off-types. Matured pods are collected from individual plants and examined in the laboratory for pod and seed characteristics and then selected. Seeds from selected plants are stored/ saved plantwise. Seeds from each selected plants are bulked to constitute nucleus seed.

Breeds Seed Production : Seedling of nucleus seed origin are produced under ideal conditions and after discarding off-types, the well developed seedlings are planted in a well prepared field free of soil infection, and not grown to cauliflower in the past 2 to 3 years. The seed production field is isolated by 1600 m. from other Brassica fields. Individual plants are verified for conformity to the descriptor, uniformity and true to type nature. Early and late curding plants are discarded. The plants with true to type stem and leaf characteristics and size and firmly developing flower heads are selected and scooped to remove the central portion the curd. The shape of inflorescence, the colour and size and shape are compared and verified to match with trueness. At flowering and pod setting stage again trueness of individual plants are verified and selected. Seeds/pods of individual plants are examined in the laboratory and selected and stored. The seeds are then labelled to constitute breeder seed. Breeder seed quality is tested in the laboratory and seed quality assurance in made. Breeder seed is labelled. Grow out test is conducted to adjudge the genetic purity.

The nutritional qualities are estimated.

Table 1. Descriptor of Cauliflower

1.	Type of location and specific season forgrowing the market crop	
2.	Seedling anthocyanin	Absent or present
3.	Plant height, at time of harvest for fresh market	
4.	Outer stem : length (upto insertion of first leaf)	Short, medium or long
5.	Leaf : attitude	Erect, semi-erect or horizontal
	Length	Very short, short, medium, long or very long
	Width	Very narrow, narrow, medium, broad or very broad
	Shape	Narrow elliptic, elliptic or broad
	Lobing	Absent or present
	Colour (with wax if present)	Green, grey-green or blue-green
	Intensity of colour	Light, medium or dark
	Cross-section of midrib	Strongly flattened, weakly flattened or rounded
	Torsion of tip	Absent or very weak. weak, medium, strong orvery strong
	Distribution of blisters	At tip only or on whole leaf
	Crimping near main vein	Absent or very weak, weak, medium, strong orvery strong
	Undulation of margin	Absent or very weak. medium, strong or very strong

6.	Curd : Covering by inner leaves	Not covered, partly covered or covered
	Height	Short, medium or tall
	Diameter	Small, medium or large
	Shape in longitudinal section	Circular, broad transverse elliptic, transverse elliptic, narrow transverse elliptic or triangular
	Doming (excluding cultivars with triangular curds)	Weak, medium or strong
	Colour	Whitish, yellow, orange or greenish
	Knobbling	Very fine, fine, medium, coarse or very coarse
	Texture	Fine, medium or coarse
	Anthocyanin coloration after harvest maturity	Absent or present
7.	Flower :Colour	White or yellow
	Earliness in specific flowering season (50% at harvest maturity)	Very early, early, medium or late

Note: Description is based on George (1999).

59. Chilli Hot Pepper

(*Capsicum annuum* L.) (2n: 24,36)

Family : Solanaceae

Annual, shrub with a tendency to be perennial. Plants are branching. Foliage light green to dark green. Flowers usually single per leaf axil and are pentamerous. Calyx 5, corolla 5 stamens 5, alternate with petals. Fruit is a bicarpellary berry which is pendant to radiating to erect fruits. Fruits mature in about 45 days after pollination. Calyx is persistent, pedicel short to medium straight or curved. Fruit colour varies from light red to dark red. Pericarp fruit wall is thin to thick. Cross pollination occurs in chilly from 7 to 36% depending on the variety, season and insect prevalence (Ram, 2005). The descriptors are given in Table 1.

Table 1. Characteristics of Chilli (Hot pepper), Bell pepper (Sweet) and Paprika (*Capsicum annuum* L.)

1.	Seedling hypocotyl anthocyanin	Absent, present
2.	Plant habit	Spreading, semi spreading, upright
3.	Main stem length (cm.)	Short upto 5 cm, medium 5-10, long >10
4.	Anthocyanin colouration of nodes	Absent, present
5.	Intensity of node coloration	Weak, medium, strong
6.	Plant height (cm.) hot pepper and paprika	Short (<40), medium (40-80), tall (<80)
7.	Plant height capsicum (cm.)	Short (<40), medium (40-80), tall (<80)
8.	Plant spread- (hot pepper and paprika (cm.)	Narrow (50), medium (50-100), broad (>100)

9.	Plant spread Bell pepper (cm.)	Narrow (50), medium (50-100), broad (>100)
10.	Stem pubescence (hairiness) intensity	Sparse, medium, strong
11.	Stem shape	Round, angled, flat
12.	Leaf length of blade (cm.)	Narrow <3, medium 3-5, broad (>5)
13.	Leaf colour	Green, purple, green with purple, tinge, purple with green tinge
14.	Leaf shape	Lanceolate, ovate, broadly, elliptic
15.	Leaf margin undulations	Weak, medium, strong
16.	Leaf pubescence	Absent, present
17.	Leaf pubescence intensity	Sparse, medium, heavy
18.	Flower petal colour	White, yellow green purple, others
19.	Flower anther colour	Yellow, pale blue, purple, others
20.	Days to 50% flowering from date of sowing	Early (<70), medium 70-90, late 90>
21.	Flower/fruit orientation	Drooping, semi drooping, erect
22.	Fruit bearing habit (number of fruits per node)	Solitary, 2 to 3, >3 (Cluster)
23.	Fruit colour unripe	White, green cream purple
24.	Fruit length (cm.) hot pepper and paprika	Very short (<2), short (2-5), medium (5-10), long (10-15), very long (>15)
25.	Fruit length (cm.) bell pepper	Short <3 medium (3-6), long (>6)
26.	Fruit diameter (cm.) hot pepper and paprika	Narrow (<0.8), medium (0.8-1.5) and broad (>1.5)
27.	Fruit diameter (cm.) bell pepper	Narrow (3<), medium (3-5), and broad (>5)
28.	Fruit shape (longtitudinal) section	Oblate, circular cordate, square, rectangular trapezoid moderate triangular narrow triangular, horn shaped.
29.	Fruit curvature intensity	Low, medium, high
30.	Fruit neck at basal end	Absent, present
31.	Fruit (cs) at placental level	Round, slightly corrugated, strongly corrugated
32.	Fruit sinuation of pericarp	Weak, moderate, strong
33.	Fruit surface texture	Smooth, slightly rought,rough
34.	Fruit colour (ripe shape)	Yellow, orange, red, brown other
35.	Fruit colour transition	One stage, two stages, > two stages
36.	Fruit glossiness	Weak, medium, strong
37.	Fruit shape at base	Acute, round, sunken
38.	Fruit shape at apex	Acute, blunt, depressed and acute
39.	Fruit number of locules (Bell pepper)	2, 2 to 3, 3, 3-4, >4
40.	Fruit pericarp thickness (mm)	Very thin (<1), thin (1-2), medium (2-3), thick (3-4), very thick (>4)
41.	Fruit stalk –length (cm)	Very short (1.5), short (1.6-2.5), medium (2.6-3.5), long (3.6-4.5), very long (>4.5)
42.	Fruit pedicel attachment	Weak, strong
43.	Fruit days to ripening	Early <100), medium (100-120), late >120
44.	Seed 1000 seed weight (g)	Low <4, medium (4-6), high (>6)
45.	Seed recovery %	Low <30, medium (30-50), high >50
46.	Seed colour	Light yellow, yellow, orange yellow others

(*Source:* PPV & FRA)

Nucleus Seed Production: Seeds are sourced from the breeder stock seed or from the maintenance breeding stock. Nursery is well prepared and maintained free from fungal and insect infestation and protected against damping off disease. Vigorous and healthy 40 days old, seedlings with good leaf development are selected and roots pretreated with fungicide. Seedlings are planted singly in 60 cm. x 30 cm. spacing in well prepared field isolated by 800 m. from other chilli crops. Around 2000 seedlings are transplanted and all ideal growing environment is provided and the crop is fertilized well. Need based plant protection is done against aphids, thrips, mites, nematodes, fruit borer, leaf spot diseases, powdery mildew, fruit rot and chilli mosaic disease. NAA growth regulator at, 10 ppm. on 60th and 90th day after planting and triacontanal at 1.25 ppm at 20, 40, 60 and 80 days after planting are applied.

The individual chilli plants are assessed for vigour and development, plant type, canopy shape and size, stem, branches, petiole and leaf features such as leaf shape, size, laminar surface and thickness, free from diseases and pest infestation. Plants showing deviations in any one of the characteristics are uprooted. Days to flowering is recorded and very early flowering plants and late flowering plants are removed. From flowering to fruit set, the individual plants are monitored and deviant types for flower colour, orientation, pedicel shape and behaviour, changes in shape, size and length of fruit characteristics and poor or low fruit bearing are studied critically and plants showing deviations and nonconformity with the variety description are removed along with the plants around. The plants finally retained are assessed for the health, vigour, high fruit bearing, uniformity and true to type and 100 superior true to type plants with high yield are tagged and self fertilised by enclosing in net cages. Matured selfed pods are collected. Laboratory examination of pods and seeds, flavor, pungeny, skin colour, skin surface, extend of seediness, seed size and shape are studied. Finally 50 superior plants are selected and seeds saved plantwise.

In the next season, 50 single plant nurseries are raised with all due precautions to avoid mixing of seeds. At 40 days after sowing, 50 well developed two to type seedlings from each progeny nursery are selected and planted in the main field in 60 cm x 30 cm spacing in 2 rows progeny plots. Single plant is maintained per hill. The progeny rows were looked after with ideal growing conditions for maximising the plant expression. The plot is isolated by 800 m from other chilli crops and well protected from diseases and pests. Periodical observations are made critically on every plant in every progeny row. Before the flowering phase commences, progeny rows showing off-types and deviant plants from the varietal description are uprooted as a whole. Among the progeny rows retained, critical assessment is made for very high uniformity for habit and growth rate, canopy size and shape, stem, branch and petiole features and all aspects of leaf characteristics and for fresh fruit (green chilly) qualities. Plants that flower very early and late are also discarded. At active flowering and at early fruit formation, 60 superior plants across very uniform true to type progeny rows are tagged and the fruits already set and flowers opened that day are all removed to ensure removal of already fertilized pollinated flowers. Such selected plants are then enclosed in Net Cages of required size to enclose the selected individual plants to facilitate self pollination only. Once in a day, the plants inside the cages are gently disturbed. After 45 to 50 days, the ripe pods from each self fertilized plants are collected and stored for complete ripening. The pods are then dried. Laboratory examination is made on the developing, unripe green pod and ripe pods and later on all aspects of dry pods and seeds as described earlier. Self fertilized seeds from each of the plants that fulfill trueness to all the prescribed characteristics are forwarded to the next season.

In the third year, 50 progeny seeds are raised in single plant nurseries and on 40th day, well developed healthy and vigorous seedlings are selected and planted in 2 rows progeny plots in two replications with 800 m. isolation distance from other chilli fields.Single seedling is maintained per hill. All other observations and selecting uniform superior progeny rows are carried out as described earlier. In each progeny, one to three selected plants are individually self fertilized by enclosing each selected plant in a separate Net Cage adopting all the precautions and steps as followed in the previous generation. Periodical harvesting for green pods and ripe pods is made and yield recorded. Progenies performing above mean ± 2SE are selected. The seeds from ripe pods of the self fertilized plants from the identified progenies are then composited to constitute the nucleus seed. Little remnant seeds from each of the superior self fertilized plants are saved for the next cycle of nucleus seed development.

Breeder Seed Production: The nucleus seed is raised in well drained raised nursery. Forty days old vigorous healthy and true to type seedlings are transplanted in well prepared main field that is isolated by 800 m. from other chilli crops. Frequent monitoring and roguing are done at vegetative, flowering and pod development and ripening stages. True to type, uniformity and stability of the individual plants are ensured. Very early and late flowering plants are discarded to maintain flowering uniformity. Monitoring Team visit is organized to adjudge the genetic purity of Breeder Seed crop.Green and ripe pods and seeds of individual plants are examined in the laboratory for uniformity besides qualities like flavor and pungency, pigmentation and colour and market traits and acceptability. The seeds from the selected plants are pooled together to constitute breeder seed. Samples drawn from the breeder seed lot are then tested for the seed qualities. Quality assured seeds are labelled with Breeder Seed Tag. Seed samples from the breeder seed are grown in grow out test as per the standard procedure to test the genetic purity. Nutritional qualities of green fruits and dry fruits are estimated for pungency and other qualities (Fig. 1).

Source seeds raised with 800m isolation, Select 1000 true to type plants. Table and laboratory examination and adjudging fresh vegetable market /dry pod qualities.

↓

Identify 100 uniform true to type plants and raise in wide space. Each selected plant is covered by nylon net cage, after removing already opened flowers and developing fruits, collect fruits and seeds and examine plant wise.

↓

Evaluate progeny rows for uniformity true to type and stability. Reject progenies showing variations/deviations, Identify progenies performing above mean +2SE and select superior 2 to 3 plants in each such progeny. After laboratory examination seeds are selected plant wise. Part of the seeds from each plant is bulked together to constitute nucleus seed.

↓

Nucleus Seed (Isolation distance 800m.)

↓

Breeder seed (Isolation distance 800m.)

↓

GOT

Fig. 1. Nucleus and Breeder seed production in Chilli / Hot Pepper.

Bell Pepper/ Sweet Pepper/ Capsicum (*Capsicum annuum*) and Paprika (*Capsicum annuum* var. *longum*): In Bell Pepper and Paprika the method of nucleus and breeder seed production is similar to that in Chilli Hot Pepper. Characteristics are shown in Table 2.

Table 2. Descriptor of chilli – Bell Pepper/ Sweet Pepper

1.	Seedling	Anthocyanin colouration of the hypocotyls : absent or present
2.	Plant	Shortened internode in upper part : absent or present
3.	Leaf	Length of blade : short, medium or long
		Width : narrow, medium or broad
4.	Flower	Attitude of peduncle : erect or non-erect
5.	Fruit	Colour before maturity : greenish-white, yellowish, green or purple
		Length : very short, shprt, medium, long or very long
		Predominant shape of longitudinal section : flattened round, heart- shaped, square, rectangular, trapezoid, triangular, narrow triangular or horn shaped

		Colour at maturity : yellow, orange, red or brown Stalk cavity : absent or present Predominant number of locules : only two, two and three, Thickness of flesh : Thin, medium or thick Capsaicin in placenta : absent or present
6.	Resistance to Tobacco-virus, specific pathotypes	Absent or present
7.	Resistance to potato virus Y (PVY), specified pathotypes	Absent or present
8.	Resistance to *Phytophthora capsici*	Absent or present
9.	Roguing	9.1 Before flowering 9.2 Early flowering and first fruit 9.3 First fruit at market maturity
10.	Seed yield and 1000 seed weight	5-5.0 g. seed / kg. of fruit 1000 seed weight 5g. Seed yield 100-200 kg/ha.

Note: The description is based on UPOV and George (1999)

60. Chow Chow – Chayote

(*Sechium edule* (Jacq) SW.) (2n:28)

Family : Cucurbitaceae

Chow chow or chayote is the only member in Cucurbitaceae that is viviparous. The vivipary condition does not occur in any wild relatives. It is the domestication that has suppressed the dormancy mechanisms. Chow chow plants are vineyard climbers. Leaves triangular 5 to 8 lobed. Flowers are small, white or yellow and five lobed. Stamens 3. Connate filaments at base. Chow chow is having pistillate and staminate flowers. It is pollinated by insects. Because of insect pollination, it is difficult to maintain genetic integrity. In chayote, seeds mature in 25 to 30 days after pollination. Seed embryo has the ability to sprout from the seeds while still the fruit is attached. The characteristics of chow chow are given in Table 1.

Table 1. Characteristics of Chow Chow (Chayote)

1.	Stem colour at young stage	White, yellow, light green, green, dark green
2.	Pubescence of internode	Absent, few, medium, many
3.	Stem-pubescence of nodes	Absent, few, medium, many
4.	Stem colour-mature stage	Dark green with brown stripes, green, lightgreen, yellow with brown stripes

5.	Tendril length	Short, medium, long
6.	Tendril branches	2, 3, 5 or more
7.	Tendril colour	Light green, green, dark green
8.	Leaf blade size	Very small, small, medium, large
9.	leaf blade shape	Angular, cordiform palmately lobed, tripartite, deltoid, sectate
10.	Leaf blade colour	White, light green, dark green
11.	Leaf blade- colour variation	White, light green, dark green
12.	Leaf blade- no. of nucrone	Few, medium, many
13.	Petiole length	Short, medium, long, very long
14.	Petiole colour	White, light green, green, dark green, very dark black
15.	Female inflorescence : no. of flower per node	One, two, three or more
16.	Female flower colour	White, green, greenish yellow
17.	Female flower-calyx colour	Light green, green, dark green
18.	Male flower colour	White, green, green yellow
19.	Male inflorescence- no. of flowers	Few, medium, many
20.	Peduncle length	Short, medium, long
21.	Fruit size	Very short, short, medium, long, very long
22.	Fruit length	Very short, short, medium, long, very long
23.	Fruit shape	Conical, pyriform, sphere oblong, cylindrical, obovoid ellipsoid, broadly ellipsoid
24.	Fruit shape in cross section	Flattened, oval, round
25.	Fruit profile at base	Depressed, flat, raised, very raised
26.	Fruit profile at apical	Depressed, flat, raised
27.	Fruit size of apical fissure at apical part	Absent, very small, small, medium, large
28.	Fruit groove	Absent, present
29.	Fruit colour of skin	White, yellow green, light green, green, darkgreen
30.	Fruit skin smooth texture	Absent, present
31.	Fruit-main colour of flesh	White, cream, light green, green, dark green
32.	Fruit-flesh thickness	Thin, medium, thick
33.	Fruit fibrous flesh	Absent, very few, medium, many
34.	Fruit spines	Absent, present
35.	Fruit spines, if present	Very few, few, medium, many
36.	Fruit flavour	Neutral, sweet
37.	Seed size	Very small, small, medium, large
38.	Seed length	Short, medium, long
39.	Seed diameter	Short, medium, long
40.	Seed shape	Conical, pyriform, epheroid, oblong, cylindrical, obovoid, broadly obovoid, ellipsoid, bradly, ellipsoid
41.	Seed adorn	Absent, present
42.	Seed colour	Whitish, cream
43.	Seed flavour	Neutral, sweet, bitter
44.	Root flavour	Neutral, sweet, bitter

Nucleus seed production: From the source plot grown in isolation of 1600m distance from other chow chow fields, healthy and virus free highly productive plants are selected. Depending upon the population size of the garden, 150 to 200 plants that are uniform free from mosaic virus infestation are selected. At least 10 to 15 well matured fruits that conform to the specified variety are selected from the selected individual vines. Fruits from individual plants are maintained as an unit or progeny and allowed to sprout. Well matured and sprouted fruits are also collected on individual plant basis. Fruits showing deviations in appearance, shape, size and colour are rejected.

In the following season, each plant progeny is established maintaining the identity. Isolation distance provided is 1600m. All standard good agronomical and horticultural practices are followerd. Two meters raised trailing structure ,manuring and plant protection measures are adapted. Control of aphids, fruitfly, scale insects and mealy bugs is properly taken up. Prompt control measures are initiated to control mosaic virus disease. Observations are recorded on progenies for the vine, foliage, flower and fruit traits conforming to true to type nature in the first year at the time of active flower and fruit formation stage. The deviating plants whenever occur in a progeny, the entire progeny plants are uprooted. Soon after all the fruits set, matured fruits on the vines are discarded to ensure that the fruits are not set owing to the cross pollination from the deviants type of plants. The selected chayote plants are well protected from cross pollination from other plants/gardens of chow chow by maintaining isolation of 1600m distance. In the end of the first year of the selection of the true to type plants, 20 to 25 matured fruits from each plant progeny are collected and stored for viviparous germination.

In the next generation, in a well prepared germinated matured fruits are planted in progeny rows. At least 80 to 100 progenies are maintained in plots that are provided with isolation of 1600m. In this generation of progeny rows, continuous monitoring is done to verify the true to type nature. Progenies having all the plants conforming to the true to type and genetic uniformity for foliar and fruit traits are retained. All other progenies are uprooted as a whole. At this point of time, the immature and mature fruits borne on all the plants and already fertilized fruits are removed ensuring that there is no fruits set owing to cross pollination from deviating type of plant. At full maturity fruits harvested from true to type individual lines are saved for next generation growing. The fruits thus saved from identified true to type progenywise lines constitute the Nucleus seed.

Breeder seed production: In the third generation as in the previous generation, at least 80 progenies (10 to 15 plants per progeny) are grown in 1600m isolation plot in rows maintaining the progeny identity. At the time of flowering and fruit development and maturation, true to type nature for foliage, flower and fruit traits are assessed. Only such of those progenies that are true to type and genetically uniform nature and stable and disease free are retained and all other progenies showing deviating type or variation are discarded. The foliage, flowers and fruits in the plants of the retained progenies/families are again examined and verified for genetic uniformity and true breeding nature. At this stage of breeder seed crop stand, visit of Monitoring Team is arranged to assess the field purity and genetic purity of the crop.

The selected chow chow/chayote plants constitute the breeder seed source population. This is achieved at the end of the first year of the third generation. Matured fruits constitute the Breeder Seed. The quality assured seeds or fruits are labelled with breeder seed tags. Grow out test is conducted with samples from breeder seed lot. Fruit nutritional quality aspects and fresh vegetable market acceptance are estimated.

61. Coleus/Chinese Potato

Chinese potato (*Coleus parviflorus* Benth) Syn. *Solenostemon rotundifolius*/ (2n:56,64) *Plectranthus rotundifolius* (Poir.) Spreng.

Family : Lamiaceae

Annual herb, succulent stem, flowers arranged in racemose cymes. Calyx two lipped and corolla uniformly greenish brown. The flower is tubular and 2-lipped, measuring 7 to 12 mm. long, light to dark violet, velvety and gland-dotted. The tube is strongly curved. The upper lip is very short and with 4-lobed while the lower lip is boat-shaped. There are 4 stamens that are curved within the lower lip. Anthers are up to 2.5 mm. long, shortly united at the base and envelope the style. The style is bifid and slightly exceeding the stamens. Flowering: September-October.

Flowers are completely sterile. Hybridization is not possible. Coleus bears heteromorphous tubers. The crop matures in 5 months. Vine cuttings are used as propagules. Tubers are first planted in flat beds and in 7 to 10 days tubers putforth sprouts and in 2 weeks time the herbaceous sprouts grow to 20 cm. Sprouts are planted in the main field with a spacing of 60 cm. x 30 cm. (55000 plants/ha). In 5 months time, tubers develop and mature and 15 to 20 tons of tubers per ha. can be harvested in 120 days. Coleus can be propagated by single node cutting (TNAU – Agric Tech Portal).

In coleus there are flowering and non flowering types. Tubers of coleus are used as vegetables. The small tubers are aromatic in smell and round to oval shaped. Roots are of medicinal value and economic importance. Leaves are also used for medicinal purpose.

Harvested tubers are stored under room temperature for two months. Sprouted tubers are then planted in the nursery beds. Herbaceous sprouts cuttings (10 cm length) from the tubers are collected from nursery beds and planted in the main field at a spacing of 30 cm × 60 cm. Yield obtained is from 15 to 20 t/ha in 120 days. With maturity of tubers, leaves start falling from stems indicating crop maturity.

Table 1. Characteristics of Coleus- Chinese Potato

1.	Stem succulence	Less, medium, more
2.	Stem attitude	Decumbent, ascending
3.	Stem pubescence	Sparse, medium, dense
4.	Leaf thickness	Thick and juicy, less thick and less juicy
5.	Leaf arrangement	Opposite, subopposite
6.	Leaf stalk (cm)	Short (<1.5), long (1.5 to 3.0)
7.	Leaf blade shape	Ovate, oblong, near round
8.	Leaf blade size	2 to 6 cm × 1.5 to 4 cm
9.	Leaf shape at base	Wedge, straight
10.	Leaf margin	Low dentate, highly dentate
11.	Inflorescence spike length (cm)	Short (<5), medium (5-10), long (10-15)
12.	Number of flowers per whorl	4 to 6 numbers
13.	Flower stalk length	1 to 2 cm
14.	Flower tube length	7 to 12 mm, 2 lipped dark velvetty and gland dotted
15.	Stamens attitude	Curved, less curved
16.	Stamens size	2.5 mm long, united at base and envelope the style

Nucles seed production: Uniform true to type individual single plants grown in source nursery are examined and 3000 tubers are selected from single plants. The tubers are stored for 2 months in room temperate and planted in raised beds. Sproutlings / sproutings from individual tubers that are from single individual plants are planted in tuber progeny rows in the main field. The field is isolated by 40 m. from other coleus crops. At harvest, all the tubers originating from the same progeny are examined separately as tuber progenies for trueness, uniformity and vigour. In case deviations are noticed in any of the plants in the plant progeny for plant stature, foliage, flower and tuber traits, the entire single plant progenies (all the tuber progenies arising from the tubers of the orginal individual plants) are discarded. Tubers that are selected conforming to the trueness of the variety are thus maintained separately plant progenywise. In the next cycle, developing sprouts from individual plant progeny tubers or the sprouts or the herbacous cuttings are planted in the main field in plant progeny plots, in two rows plot .The plant progeny rows are periodically observed for true to type nature, genetic uniformity and free from diseases and pests. Based on productivity, progenies are selected and then harvested and tubers are examined in the laboratory and final selection of tubers and progenies are made. Healthy and uniform sized tubers are pooled to constitute nucleus seed (Fig. 1).

Breeder Seed production: Single node cuttings from selected individual plants, true to type and superior in performance in the nucleus seed source are utilized to develop single plant progenies in every generation till the third generation of clonal crop seed is reached. The essential steps enumerated in the previous paragraphs in respect of tuber progenies will be meticulously followed here also. The breeder seed crop is grown in the field that is not grown to coleus crop in the past 2 to 3 years. This field is isolated by 40m. In the third generation, tubers from the selected plants that are in conformity are pooled together to constitute Breeder Seed tubers. Visit of Monitoring Team is arranged to assess the breeder crop. GOT is conducted to adjudge the genetic purity. Nutritive content is estimated in the tubers.

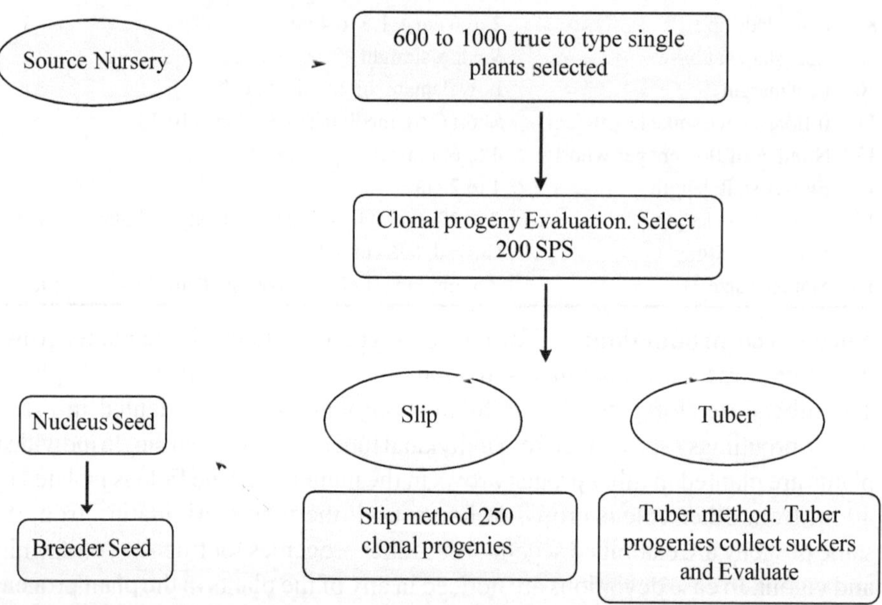

Fig. 1. Nucleus and Breeder Seed Production in Coleus

62. Coriander

(*Coriandrum sativum* L.) (2n:22)

Family : Apiaceae

An annual herb. Leaves are alternate. Stem erect and hollow and sympodial branching. Each branch terminates in an inflorescence. Inflorescence is a compound umbel. In every umbel, the peripheral umbellate and peripheral flowers are first to flower. Flowers are hermaphrodite. Calyx five and petals five. Inner flowers are stamiante. Hermaphrodite flowers are protandrous. Ovary inferior, Stigma is receptive for pollination for maximum of 5 days. Pollination behavior is often cross pollinated and facultative cross pollination. Inbreeding depression is not observed upon self fertilization. The characteristics of coriander is given in Table 1.

Table 1. Characteristics of Coriander

1.	Number of leaves	Low (1-3), medium (4-6), high >6
2.	Length of the longest leaf (cm)	Short (<6), medium (6-10), long (>10)
3.	Habit of basal leaf	Flat, raised, very erect
4.	Leaf lusture of lowest basal leaf	Non shiny, shiny
5.	Leaf margin of longest basal leaf	Deeply serrated, serrated
6.	Leaf-colour of longest basal leaf	Green, dark green
7.	Stem colour pigmentation	Absent, present
8.	Nodal pigmentation	Absent, present
9.	Involucres	Absent, present
10.	Growth habit	Erect, semierect, spreading
11.	Primary branches number	Less<3, medium (4-6), more(>6)
12.	Angle of primary branch	Narrow (35°C), medium (36-40°), wide (>40°C)
13.	Number of secondary branches	Less (20), medium (21-30), more (>30)
14.	Angle of secondary branches	Narrow (<35°C), wide (>35°C)
15.	Plant height upto tip (cm.)	Short (<30), medium (31-40), tall (>40)
16.	Umbellate per umbel	Low (<4),medium (4-5), high (>5)
17.	Seeds per umbellate	Average (<5), good (>5)
18.	1000 seed weight (g.) (7 to 8% moisture)	Low (<15), medium (15-20), high (>20)
19.	Seed colour	Light, brown
20.	Seed shape	Oblong, oval, slightly round

Nucleus Seed Production: Breeder stock seed, maintenance breeding stock or the AVT stage are the sources for seeds to develop nucleus seed. Coriander source population is raised by sowing split seeds in 20 cm x 20 cm spacing in an area of 400 sq.m Isolation distance of 400 m is followed. All good agronomic practices and plant protection care are adopted. Critical and frequent observations are made on individual plants for growth, vigour, health and disease free nature. At the stage of 25-30 days after sowing, stem and leaf characteristics are studied. At flowering and umbel/seed development stage, seed setting, uniformity and true to type nature are studied. Based on observations, 1000 superior true to type uniform plants are selected, harvested and seeds extracted. In the next season, 1000 single plant progeny rows are raised adopting the recommended crop production practices and plant protection in respect of aphid, powdery mildew and grain mould. Isolation distance of 400 m is ensured. Observations are continued from germination to emergence to leaf development, flowering and seed set. Progeny rows having off-types and deviant type of plants are uprooted. Progeny rows showing uniformity and true to type nature conforming to the description are retained. During active flowering, 1000 superior plants are selected and all the peripheral umbellate flowers in each plant are self fertilized. Self fertilized seeds are collected on plant basis. In the following generation, split seeds are sown in 1000 progeny rows. Progenies are examined for genetic uniformity, true to type breeding and stability. The progenies are evaluated for leaf characteristics, aroma and flavour at 30 to 35 days growth. At maturity, seeds are separated following the harvest practices and examined in the laboratory for shape, size, colour , uniformity in development and aroma. Seeds from the selected self fertilised plants are then bulked to constitute nucleus seed. If sufficient quantity of nucleus seed is not available, Nucleus Seed stage II can be initiated by one more season growing. Evaluation and selection of superior plants can be continued and sufficient quantity of nucleus seeds obtained.

Breeder Seed Production: Isolation distance recommended is 400 m. Breeder seed production plot is raised in clean field with sowing of spit seeds and following all standard and ideal agronomic and seed production practices. The crop is well protected from pests and diseases. Field evaluation of the breeder seed crop is made at every stage of plant growth and development including leafy stage and grain maturity stage. The off-types and deviant plants are removed as and when noticed. When deviant plants are located at flowering, they are uprooted along with the surrounding plants to avoid vicinism.Early and late flowering plants are discarded. Monitoring Team visit is arranged at mid flowering / maturity stage. The crop is harvested at maturity and seeds are separated, well dried and processed. Seed quality is tested in the Seed Testing Laboratory. Breeder seed label is tagged to the Breeder seed packets. Seed samples are drawn from the breeder seed lot and tested in Grow Out Test to adjudge the genetic purity. Fresh leaf and seed qualities are adjudged in laboratory tests.

63. Cucumber

(*Cucumis sativus* L.) (2n=14)

Family: Cucurbitaceae

Annual, trailing or climbing monoecious and day neutral plant. The vine is angular, leaves mostly triangular, indistinctly 3 lobed, staminate flowers occur with slender and short pedicels. The pistillate flower is solitary borne on short pedicels. Calyx and corolla are five lobed. Staminate flowers have three stamens with 2 of them having two locules each and the third one uniloculed. Filaments are free but united by their anthers. Pistillate flowers are epigynous. Carpels are mostly three and ovary is correspondingly three loculed. Pistillate flowers usually exhibit five stigmas. In the monoecious type cucumber, the main stem is characteristically showing 3 phases: producing staminate flowers only in the first phase followed by the second phase of irregularly alternating female and male flowers and finally a phase of production of only pistillate flowers. Sex expression is generally influenced by the environment. Under long day and high light intensity, staminate flowers predominate whereas under short day and low light intensity pistillate flowers production predominates (Ram, 2005). The characteristics of cucumber are given in Table 1.

Table 1. Descriptor of cucumber

1.	Plant	Growth type : determinate or indeterminate Vigour : weak, medium or strong
		Total length of first 15 internodes : short, medium or long
		Length of internodes of side shoots : short, medium or long
2.	Leaf	Size of blade : small, medium or large Intensity of green color : light, medium or dark
		Blistering : absent or very weak, weak, medium, strong or very strong
		Length of terminal lobe : short, medium or long
		Width of terminal lobe : narrow, medium or broad
		Ratio length/width of terminal lobe : less than 1, equal to 1 or more than 1
3.	Plant	Sex expression : male and female flowers approximately equally present, mainly female flowers, or almost exclusively female
		Number of female flowers per node: one to three, or more than three
4.	Young fruit	Type of vestiline : hairs only, prickles only or hairs and prickles Density of vestiline : sparse, medium or dense
		Color of vestiline : white or black
		Size of warts : absent or very small, small, medium, large or very large
5.	Parthenocarpy	Absent or present
6.	Fruit	Length : very short, short, medium , long or very long Diameter : small, medium or large
		Ratio length/diameter : small, medium or large

		Core diameter in relation to diameter of fruit : small, medium , or large
		Predominant shape of stem end at market stage : necked, acute or obtuse
		Length of neck : short, medium or long
		Shape of calyx end at market stage : acute or obtuse
		Ground color of skin at market stage : white, yellow or green Intensity of ground color of skin : light, medium or dark
		Ribs : absent or present
		Prominence of ribs : weak, medium or strong
		Coloration of ribs compared to ground color : lighter, equal or darker
		Vestiline : absent or very sparse, sparse, medium dense or very dense
		Warts : absent or present
		Stripes (ribs excluded) : absent or present Length of stripes : short, medium or long Mottling : absent or present
		Predominant type of mottling : small and round, or large and irregular
		Intensity of mottling : weak, medium or strong
		Length of peduncle : short, medium or long
		Thickness of peduncle : thin, medium or thick
		Ground color of skin at physiological ripening : white, yellow, green, orange, or brown
7	Time of development of female flowers (80% of plants with at least one female flower):	Medium or late
8	Cotyledon bitterness	Absent or present
9	Fruit : bitterness	Absent or present
10	Resistance to	scab : absent or present powdery mildew : absent or present downy mildew : absent or present curcumins mosaic virus : absent or present

Note: Descriptions are based on UPOV (1993) as described by George (1999).
(*Source:* Ram 2005)

Nucleus Seed Production: Well prepared field previously not grown with cucurbitaceous crops in the past 2 to 3 years are selected to avoid the carry over of foliar and root diseases in cucumber. Isolation distance of 1000 m. is maintained. Seeds sourced from the Breeder stock seed or from the maintenance breeding stock are sown in rows drawn 1.5 m. apart. On the row, pregerminated seeds are sown in pits at one meter spacing. Before sowing, the seeds are treated with fungicide. The seedlings in each pit are thinned, leaving only one healthy and disease free seedling. All horticultural practices, fertilisation and plant protection measures are duly followed. During the crop growth upto flowering, the vine and foliar characteristics are examined and plants deviating from true to type are uprooted. Very early flowering and late flowering plants are uprooted to maintain flowering uniformity. At early flowering and active flowering and fruit development, flower characters, ovary size, shape and colour are noted and plants deviating from trueness are uprooted. At flowering initiation / commencement, 500 to 600 uniform single plants are selected based on the vigour and trueness in trait expression. One to two female flowers in each plant are self pollinated with pollen from the flowers of the same plant. Because of slender and delicate nature, female and male flowers are twist tied with cotton threads on the previous day to anthesis to protect from cross pollination. On the next day, the female flowers are opened up and pollen from male flowers are pollinated on to the stigma by dusting the pollen of the dehiscing anthers. At maturity, matured fruit characters are collected and examined in the laboratory. Fresh market stage fruits from the corresponding single plants are also examined for fruit colour, taste and texture, uniformly cylindrical shape, colour, non bitterness, absence of separation or hollowness in the fruits and other market acceptable features. Based on the final evaluation of plant traits and fruit qualities, 100 single self fertilized plants are selected and matured and ripe fruits are harvested. Seed separation is made separately for each plant, examined for the trueness, selected and stored plantwise.

In the next season, the plant progenies, 100 in number, are raised in rows with 1m. x 1m. spacing and 30 plants per progeny in two replications. The plant progenies are raised in the field that is isolated by 1000m from other cucumber crops. Periodical observations are made on individual plants in each progeny row at early vegetative, pre-flowering and flowering and marketable fruits and at fruit maturity stage as described in the previous paragraph. Progeny rows with plants deviating from true to type are uprooted. Progeny rows performing above mean +2SE are identified for yield performance. Superior plants in the selected progenies are labelled. Part of the seeds from each selected plant are bulked together to constitute nucleus seed and remainder of the seeds from each selected plant are saved for the next nucleus and breeder seed cycle.

Breeder Seed Production: In the next generation, seeds from 100 selected single plants are sown in 2 row plots at a spacing of 1.0 m x 1.0 m. In each row, 10 plants are raised. Isolation distance of 1000 m is maintained. Observations are made at each growth stage on all the plants and progenies showing off-types and deviations are uprooted. Very early and late flowering plants are discarded. Yield data and other quality traits are recorded in each of the rows. Fresh market stage fruit and its quality are assessed in the laboratory. Field visit of Monitoring Team is arranged Fruits from high performing superior plants are selected and harvested. The seeds are separated, dried and processed. Seed quality is tested in the loabotatory and quality assured seeds are labelled with breeder seed label. Sample seeds are drawn and grow out test is conducted as per the guidelines to declare the seed quality. Nutritural quality of the market grade fruits is evaluated.

Cucumber Cultivar Types:

1. Field cucumbers, with prominent black or white spines.
2. The greenhouse or forcing type (often referred to as 'English cucumber'). They have spineless fruit which can be produced parthenocarpically; there are also short-fruited cultivars in this group.
3. The Sikkim cultivars (originating from India) with reddish – orange fruits.
4. The small-fruited cultivars frequently used as 'gherkins' for pickling.
5. The 'apple' fruited cultivars which have ovoid-to spherical-shaped fruit, from diverse areas of the world including the USA and the Far East.

64. Curry Leaf (Sweet Neem)

(*Murraya koenigii* L.) Sprengel) (2n=18)

Family : Rutaceae

Curry leaf is widely distributed in India. It is a popular leaf spice. Curry leaves are slightly pungent, bitter and feebly acidic in taste. The leaf flavor is retained in dried leaves as well. The leaves are used in Ayurvedic and Unani prescriptions. The volatile oil extracted from the leaves and flowers is used in industrial applications. In homesteads, usually one or a few curry leaf trees are maintained for easy access to fresh curry leaves for daily culinary use. In commercial field plantations, shoots that can grow upto 10 m. are cut back and well trained into compact bushes of nearly 1.0 to 1.5 m. tall with 15 to 20 well developed stem branches with full of current season branchlets or coppices bearing fresh leaves which are harvested with the terminal twigs with as may as 15 to 20 marketable fresh leaves in each twig. These terminal twigs are then loosely bundled or packed for transit to fresh vegetable markets. Quality of curry leaf is narrated by Jagadeesh *et al.* (2017) and genetic variability by Shoba *et al.* (2020).

The leaves are exstipulate, bipinnately compound. 30 cm. long with 20 to 24 leaflets. Main stem is greenish to dark brownish. Inflorescence is terminal cyme with 60 to 90 flowers. Flowers are bisexual, white, funnel shaped, sweet scented, pedicellate, regular and pentamerous. Sepals 5 with lobes persistent, petals 5, free polypetalous, stamens 10 in circles of 5 each, 5 long (6 mm) and 5 short (4 mm); gynoecium 5 to 6 mm long with sticky and bright stigma. Style is short. Ovary is superior. Fruits round to oblong, 1 to 6 cm long and 1.2 cm thick weighing around 800 mg. Polyembryonic fruits are produced upto 14 to 16% of fruits which are usually 2 seeded. Flowering time is February to May.

Distinct varieties based on leaf colour and size have been identified. A perceptible variation is evident for leaflet shape, size, thickness, glanded nature and flavor. Seedling progenies are variable and true to type plants are not realized from them (Joseph and Peter, 1985).

Root suckers are produced from the matured roots of adult trees and form propagules developing into true to type plants. Only a few root suckers per plant are produced and are quite inadequate for raising large number of plants. Stem cuttings dipped in 5000 ppm of IBA respond well for rooting. This method helps in acceleration of efficient and rapid cloning method of true to type plants. Field survival is very high with the rooted cuttings. Stem segments from selected mature curry leaf plants are the ideal explants that respond more to rooting and upto 85% field survival.

Nucleus Seed Production: A critical survey of individual bushes is made in curry leaf plantations production gardens maintained by bush management system. Bushes that are healthy, vigorous and suitable for multicut, showing high leaf productivity and fresh leaf qualities are selected. Stem cuttings with 3 to 4 buds are prepared from each selected bush. Individual clonal progenies rows are raised with the IBA treated rooted stem cuttings by planting in progeny rows. When grown upto 2 meters, the adult clonal plants in each progeny are assessed for health, fresh and dry leaf qualities and foliage productivity. Selected elite clonal progeny rows that are true to type, uniform and stable are utilized as base clonal stocks for multiplication by stem cuttings to develop nucleus source plantations to ensure continuous supply of quality planting material. The selected plants in the nucleus source nursery are trained by pruning regulation to bush shape for maintenance of nucleus source nursery.

Breeder Seed Production: Stem cuttings prepared from the Nucleus Stock Plants/bushes are rooted in polybags and planted in the Breeder seed plot that is free from contamination and other nutrient deficiencies. After planting, with 5m. isolation the young plants are looked after well and trained properly to maintain bushy nature.True to type of bushes and good quality leaf production and quick rejuvnevation and leaf production stability are assessed. Monitoring Team visit is arranged. Stem cuttings are selected from the identified bushes/ stock plants are distributed as breeder stock for raising high quality curry leaf plantations for leaf production and commercial utility. The qualities of curry leaf are analysed.

65. Drumstick

Moringa oleifera Lam (2n:28)

Family : Moringaceae

Moringa is perennial to annual in duration. It is a multipurpose nutritious and mutiutility tree. The leaves are pinnate and leaflets are numerous and arranged into pinnae and pinnarulae. Flowers are zygomorphic. Gullet type flowers. Flowers are fragrant, white to creamy white, pedicelled borne on panicles that are large and spreading. Flowers are with 5 sepals and 5 petals and stamens 5, yellowish; ovary superior, tricarpellary, locular with biseriate pendulous ovules on three parietal placentation. Pods pendulous, triangular. 30 to 120 cm. long, elongated and uniloculed. Dry pods dehiscent or longtitudinal splitting. Seeds are dark brown and 3 winged. Anthesis commences and completes between 4.30 am and 6.30 am. Anther of longest stamen dehisces first followed by anthers of descending length of filament. Stigma becomes receptive a day before flower opening and continues to be receptive on the day of flower opening and pollination occurs through honey bees. The bisexual, oblique, stalked, axillary and heteromorphic flowers are highly cross- pollinated due to heteromorphism. Flowers are frequented by the insect pollinators. Out crossing is estimated to be upto 76 % and selfing 24 %. Presence of selfing and early sexual maturity (6 months to 1 year) offers scope to develop inbreds. There is significant biparental inbreeding (Ritland, 1990). Inbreeding depression is reported. Hence the orchard planting is organised to minimize the distance between related individuals to maximize cross fertilization among unrelated clones. There exisits not much genetic diversity. Fruits split along each angle to expose the rows of rounded blackish oily seeds, each with 3 papery wings. The characteristics of moringa are given in Table 1.

Table 1. Characteristics of moringa

1.	Plant growth habit	Upright, spreading, drooping
2.	Seedling-hypocotyl colour	White, green, pink
3.	Basal shape of the first leaf blade	Acute, obtuse, rounded
4.	Apex shape of first leaf blade	Acute, obtuse
5.	Size of leaf in primary branches	Small in length and width, small in length but wider, large length but narrow
6.	Number of primary branches	Few, medium, many
7.	Colour of leaf	Green, dark green
8.	Colour of petiole	Light green, medium green
9.	Petiole length	Short, medium, long

10.	Days for flower initiation	Early, medium, late
11.	Inflorescence length	Short, medium, long
12.	Petiole anthocyanin colouration	Absent, present
13.	Number of branches in inflorescence	Low, medium, high
14.	Flower bud colour	White, pink, dark pink
15.	Flower size	Small, medium, big
16.	Pod length (cm)	Short (<40), medium (40 to 60), long (>60)
17.	Pod girth (cm)	Thin, medium, big
18.	Pod skin colour	Pink, greenish, pink
19.	Pod-texture of surface	Smooth, slightly, rough
20.	Pod shape	Straight, curved
21.	Pods per cluster	One, many
22.	Seed size	Small, medium, large
23.	Seed colour	Light, medium, dark
24.	Pod maturity	Early, medium, late
25.	100 seed weight (g)	Low, medium, high

(*Source:* PVP & FRA 2015)

The variability of characters in Moringa perennial type is explained by Balaguru *et al.* (2020). The presence of selfing as well as early sexual maturity (6 months to 1 year) might provide an alternative breeding programme for this species as compared with traditional directional selection. If inbreeding depression is weak, then it should be possible to use breeding schemes involving inbred lines and hybridisation (El-Kassabya *et al.*, 2007). In designing seed orchards, randomisation and minimum distance between related individuals will need to be worked out to maximise cross-fertilisation among unrelated clones and to minimise selfing or mating among related ramets.

Moringa types are broadly perennial and annual types in addition to many land races and ecotypes that exists. Many ecotypes exist in perennial and annual types. Quality and productive individual trees are also commonly grown in home gardens. The range and variation for flowering, fruits bearing, fruit length, thickness, tasty flesh and seed texture and softness and taste of leaf and fruits upon cooking is more or less similar in perennial and annual types.

Perennial Moringa Types: A new clonal source nursery plot is raised with 400 to 500 clonal cuttings sampled across productive elite and superior perennial Moringa trees bearing quality fruits that are in second or third year of fruit production. Isolation distance of 20 m. is maintained. The source clonal garden is well looked after with good horticultural and sound plant protection protocols. Individual trees are monitored for their vigour and growth, regular productivity of fruits, leaf and fruit quality traits over the past two or three years life span. Poor, weak, unthrifty trees are removed. At the 2nd or 3rd year cycle, 200 elite true to type, uniform and genetically stable and productive trees are selected and limb cuttings are prepared from individual trees and planted in the clonal seed orchard that is isolated by 20 m from other Moringa plantations. The orchard is well maintained with pruning/cutting back of individual trees to 1 to 1.5 m. height to encourage branching. Individual trees are verified and selected for uniformity and trueness for stem, branches, leaf colour and leaflet size and shape and panicle, flower, fruit, fruit setting and fruit quality characteristics. This orchard will be the nucleus source plantation for stem cuttings.

Breeder source orchard. Isolation of 20m is maintained. Stem cuttings are prepared for raising the next generation orchards for commercial fruit production. To speed up the fast multiplication, innovative methods of preparing stem cuttings with 3 to 5 buds and raising them in mist character and greenhouse could be done for mass multiplication. Standards for the cuttings may be evolved to assure quality rooted cuttings production, that will have a good field establishment.

Annual Moringa types: Annual types are raised from seeds sourced from elite plantation sources. Well developed seeds are sown in polybags. Only vigorous growing seedlings are retained and 35 to 40 days old seedlings are transplanted in well prepared pits. Plants commence flowering and fruiting after 6 months.

A few well developed and well maintained annual Moringa varieties plantations are available in Tamil Nadu, India. PKM 1 (Periyakulam 1) and and PKM 2 (Periyakulam 2) are bred varieties evolved and released by the Tamil Nadu Agricutlural University. PKM 1 is an inbred moringa variety developed by continuous selfing and selection for 6 generations. PKM 1 bears 70 cm long fruits that are fleshy and tasty. A very popular and seed propagated Moringa. PKM 2 (Periyakulam 2) moringa is developed by hybridization and selection and bear 125 cm.long fleshy fruits. PKM 1 and PKM 2 are very popular and stable seed propagated annual moringa varieties, flowering in 110 to 120 days and yielding harvestable immature edible and marketable fruits in another 60 days and matured fruits with well developed seeds after another 60 to 70 days. The trees are cut back to 1.0 to 1.5 m. at the end of first season and allowed to ratoon. Ratooning can be practised for 3 year period. After 3 year cycle, fresh planting need to be done. On an average, annual types bear 250 to 300 pods per tree per year; Pod length varies from 25 to 30 cm. to 100 to 120 cm. in annual types. Each fruit is having 10 to 15 seeds.

Experiments have shown that natural cross pollination (xenogamy) from different plants gave 100% fruit set, 81% seed set and 9% fecundity while self pollination (geitonogamy) resulted in 62% fruit set, 64% seed set and 6% fecundity. (Jyothi *et al.,* 1990). Natural pollination exhibited 11 to 15% fruit set only. Phenology and floral biology studies have been reported by Kanthasamy (2006). Out crossing rate in moringa using AFLP was estimated (Mulvi *et al.,* 2004).

Nucleus Seed Production

Nucleus seed production in annual seed propagated Moringa: Seeds are sourced from highly productive annual Moringa plantations. Nursery is raised and healthy vigorous seedlings are selected and planted in 2 m x 2 m spacing in a field plot with isolation of 1600m.. Single plant per hill is maintained. Individual plants are examined and variant types for pigmentation in stem, petiole and foliage are uprooted. Only uniform flowering plants at first flowering 200 to 250 in number, are retained. Plants showing vigour and uniformity in leaf, flower and fruit traits at the fruiting phase are earmarked and at the time of harvest, highly productive superior and elite plants are selected and retained. Open pollinated seeds collected from the selected individual superior plants are raised in separate single plant nurseries and vigorous and healthy seedlings are selected and planted as progeny rows with at least 20 to 25 plants per progeny. Monitoring of the progenies is done from first flowering to fruiting. Progeny rows that are uniform true type alone are retained. Progeny rows showing deviating type of plants are uprooted along with adjoining rows. At maturity, superior individual trees are earmarked after testing for leaf and fruit qualities. The seeds from selected plants are forwarded as separate progenies in the next generation. One hundred progeny rows are raised. High density planting could be adapted. Progeny rows are selected in this generation for uniformity, vigour, true to type, leaf quality, nature of fruits and stability in fruit bearing. Early blooming and late blooming plants and progeny rows showing deviant plants along with the adjoining rows of plant are uprooted. Highly productive fruit bearing uniform true to type plants alone are retained. Seeds collected from matured fruits are extracted, graded and tested for fruit and seed qualities. The quality assured pooled seed from the selected plants are constituted as Nucleus seed and labelled.

Breeder Seed: Annual Moringa: Quality seeds from Nucleus seed are sown in poly bag nurseries. Vigorous seedlings alone are selected and planted at single seedlings per hill in an isolated plot by 1600m. from other moringa plants. Systematic examination in made on each and every established plants for vigour and stem, leaves, flowering, percent pod setting, pod fleshiness and market acceptable leaf and pod qualities. Plants, that are weak, less productive, bearing bitter leaves and poor pod bearing and deviating plant types are removed. The fresh green leaf market acceptability and non bitterness of leaves and high palatability of immatured marketable stage drumstick pods are critically appraised in each and every tree in the breeder stock. Matured pods are collected and seeds extracted. After laboratory examination of pods and seeds, the seeds are bulked to make Breeder Seed. Seed quality parameters are tested for quality assurance. Breeder seed labeling is done.

Nutritional qualities of moringa leaf and fruit pods are estimated. Moringa dried leaf powder is becoming popular health powder.

66. Elephant Yam

Amorphophallus campanulatus. Roxb. (20:28)

Family : Araceae

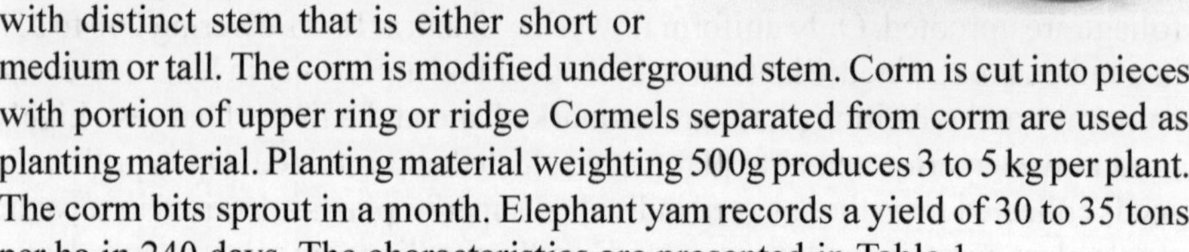

Elephant yam is an erect herbaceous plant with distinct stem that is either short or medium or tall. The corm is modified underground stem. Corm is cut into pieces with portion of upper ring or ridge Cormels separated from corm are used as planting material. Planting material weighting 500g produces 3 to 5 kg per plant. The corm bits sprout in a month. Elephant yam records a yield of 30 to 35 tons per ha in 240 days. The characteristics are presented in Table 1.

Elephant yam is recommended to be planted in pits with mounds resulting in better yield. In 7 to 8 months after planting, the crop matures indicated by yellowing and drooping of leaves.

Table 1. Characteristics of elephant yam

1. Plant height (cm.)	Short (<50), medium (50-100), tall (>100)
2. Plant type	Upright, semi upright, horizontal
3. Number of leaflet per branch	Low (10-15), medium (15-20), high (>20)
4. Leaflet shape	Round, intermediate, long
5. leaflet size	Small, intermediate, large
6. Leaflet colour	Yellow, light yellow, light green, dark green
7. Leaf waxiness	Low, medium,
8. Petiole colour	Light green, light pink, pink
9. Size of petiole patches	Small, intermediate, large
10. Distribution of patches	Absent, spotted, contiguous
11. Petiole colour	Dark green with purple patches, green with purple patches
12. Rachis pattern	V shaped, y shaped
13. Corm shape	Round, round elliptical, oval
14. Number of corm wrinkles	Few (1-2), intermediate (2-5), many (>5)
15. Corm epidermal colour	Light brown, brown, dark brown
16. Corm flesh colour	Pinkish yellow, yellow, deep yellow, saffron, orange
17. Skin thickness	Very thin, thin, thick
18. Number of cormlets	Absent, few, more
19. Cormlet shape	Globular, club shaped
20. Bract colour of main bud	Light purple, purple, pink, purple pink
21. Leaf emergence days	Within 30, more than 30,
22. Maturity days	Early (150-160), medium (150-180), late (180)
23. Corm weight (g.)	Low (510-700), medium (700-1000), high (>1000)
24. Softness	Very soft, soft, medium, hard, very hard
25. Taste	Sweet, tasteless
26. Mealiness	Highly coarse, medium coarse, low coarse, very low course
27. Acidity	Very low, low, no acidity

67. Greater Yam (Peruvalli)

Dioscorea alata L. (2n:40) esculentum (Siruvalli)

Family : Dioscoreaceae

Dioscoreacea is most widely cultivated yam in India. It is dioecious and a herbaceous climber. Land races, mutant, hybrids are prevalent. Leaves are large, ovate or cordate in shape. Tubers vary in size and shape, usually 1 to 3 per plant. Tuber flesh is white, cream, yellow or purple. Many cultivars of dioscorea produce aerial tubers or bulbils in the leaf axils. It is vegetatively propagated using seed or whole yam. Recommended size of setts is from 200 to 250 g in weight. Spacing adopted is 90 cm × 90 cm. As soon as yam vines emerge, staking is done on poles or live trees. Tubers mature and vine started drying in 8 to 10 months after planting. The yield potential is 20 to 25 t/ha in 240 days. The characteristics are given in Table 1.

Table 1. Characteristics of Dioscorea Yam

1.	Young stem colour	Green, greenish yellow, purple reddish purple
2.	Young fully opened leaf colour	Yellow, dark green, purple
3.	Colour of wing	Greenish yellow, green pigmented
4.	Petiole colour	Greenish yellow, green pigmented
5.	Petiole length (cm)	Short (<5), medium (5-10), long (>10)
6.	Mature leaf colour	Pale green, dark green
7.	Leaf shape	Cordate narrow, cordate broad, saggitate narrow, saggitate broad
8.	Leaf lobe	Non overlapping, overlapping
9.	Flowering	Absent, present
10.	Sex	Male or Female
11.	Aerial tuber	Absent, present
12.	Aerial tuber colour of cortex	Yellow, purple
13.	Aerial tuber cortex colour	Yellow, purple
14.	Aerial tuber flesh colour	Green, yellow, light purple
15.	Tuber shape	Linear, oval, digitate, cylindrical, irregular
16.	Tuber cortex colour	Green, yellow, light yellow, purple, dark purple
17.	Tuber flesh colour	White, yellowish, yellow orange, light purple, purple violet, mixed colour
18.	Hairs or roots on tubers	Sparse, dense
19.	Tuber – cross section appearance	Amorphous, granular
20.	Tuber – oxidation browning	Absent, present

68. Ivy Gourd

(*Coccinea grandis* L.) Voigt.
(*C. indica* L.)

Family : Cucurbitaeae

IVY gourd is also known as little gourd is a dioecious plant. Matured and immatured fruits are used for vegetable purposes. Ivy gourd is distributed widely in tropics. Ivy gourd has long tuberous roots. Fruits are striped, ovoid or elliptic 4 to 6 cm long and about 2.5 cm. diameter. This vegetable crop is propagated by vegetative cuttings of 20 to 25 cm. long. stem cuttings from old stems that are thick and better in rooting. Ivy gourd is a dioecous plant. Flowers are dull white with 5 calyx and 5 petals. In male flowers, anthers are three (Table 1).

Table 1. Characteristics of IVY gourd *Coccinea grandis* (L) Voigt

Dioecious, Perennial	
Stem	20 to 30 m. long
Stem	Ribbed longtitudinally
Leaf	Simple, alternate, trifoliate 1 to 5 cm.
Leaf shape	Ovate to orbicular 3 to 12cm× 3 to 12cm. size
Leaf lobing	Shallow, 3 to 5, entire margin
Flowers	Axillary unisexual, 5 merous, tubular
Sepals	5, linear 6mm. long
Petals	5, campanulate with lobes
Male flowers	Solitary or paired in short raceme, pedicelled
Stamens	Three, united at base
Female flowers	Solitary, pedicel 2 to 5 cm.
Ovary	Inferiors, cylindrical, 1.5 cm long
Stigma	3 lobed, style 3mm. long
Fruit	Ellipsoidal, 3 to 7 cm × 1 to 3.5 cm, size green with white stripes,many seeded at maturity
Planting ratio	10 female to one male plant

Nucleus and Breeder Seed: Planting material: Stem cuttings, 20 to 25 cm. long from healthy highly vigorous and productive female plants are prepared. One each cutting is planted in a well prepared pit and 10% of the pits are planted with healthy vigorous stem cuttings from male vines. Plants are trained and grown on a bower system suitable for the growth of vine. Plants are supported with stakes to enable them to reach the bowers and spread. Periodical pruning, training and fertilization are done to encourage profuse flowering and good fruit production. Ivy gourds produce fruits throughout the year.

Based on the vigour and growth rate, and true to type foliage cover, health, flowering and fruiting, individual plants are identified 50 to 100 individual female plants are selected based on the productivity, fruit (mature/immature) size and quality. Stem cuttings of 25 cm. length are prepared from each vine that are 2 years old and known to yield high and planted as single plant clonal progeny with 10 plants in each progeny. One to two budded cuttings are planted on raised beds with utmost care and sprouted cuttings are transplanted after 20 to 25 days and 150 to 200 plants are planted in a block isolated with 10m wide path. True to type plants alone are retained. Off-types and deviating plant are discarded. At maturity, fruits are harvested for 5 times and based on the mean performance in fruit, 100 plants are finally selected for fruit yield and quality and disease free nature of the vines. Stem cuttings are prepared from the selected female vines and planted in a vine stock nucleus block and maintained with utmost care. In another block, cuttings from 25 well developed and performing male type vines are prepared and raised separated by 10m. wide path to maintain isolation of the male stock vines.

Breeder Seed linc stock is raised by planting the selected stem cuttings ofNucleus Stock Plant. Ten percent of the plants are male parents. Isolation distance of 10m is provided. Individual plants are verified for trueness and conformity to the prescribed characteristics. Deviating plants are uprooted. The plot which is isolated by 10m. is well cared for with proper pruning, training and fertilizer schedules. Monitoring team visit is arranged at the 1 st or second year of peak fruiting. The fresh market acceptability and nutritional qualities are assessed. Cuttings from the plus performing female vines of Breeder stock are supplied to the growers and nurseries along with the cuttings of the male parent.

69. Knol Kbol (Kohlrabi)

(*Brassica oleracea* var. *gongylodes* L.) (2n:18)

Family : Brassicaceae

Knol khol is a biennial crop tend to remain vegetative in the first year and flower and set seeds in the second year. Knol khol plant is characteristic in exhibiting knob which is swollen hypocotyl and stem immediately above. Commonly it is light green knobs and internally having creamy yellowish color flesh. Planting season need to be mild winter (August to September). The crop attains height upto 25 to 30 cm. Seeds take 7 to 12 days for sprouting. Thirty days old seedlings are transplanted in row (in 30 cm x 25 cm), 10 plants per sq. meter. Fleshy knob develops above ground. The crop is harvested when the fleshy knob attains 5 to 7 cm. dia. Early knol khol varieties attain harvestable stage around 40 to 45 days while that of medium maturing and late maturing varieties are harvested respectively in 60 to 65 days and 70 to 100 days after transplanting. Knol khol can also be propogated by leaf cutting or tip cutting. The descriptor is given in Table 1.

Table 1. Descriptor of Kohilrabi (Knolkhol)

1.	Seedling Colour of cotyledons	Yellow-green, light green, dark green, blue green or violet-green
2.	Foliage : Formation of crown leaves Petiole crossing of one of two leaves	Weak, medium or strong Absent or present
3.	Leaf : Attitude	Erect, semi-erect or horizontal
4.	Leaf blade : Size	Small, medium or large
	Shape	Narrow ovate, ovate or broad ovate
	Colour	Yellow-green, grey-green or blue-green
	Intensity of the colour	Light, medium or dark
	Shape of the tip	Narrowly acute, acute, obtuse, rounded or broadly rounded
	Lobing (division to or almost to midribs)	Absent or very weak, weak, medium, strong or very strong
	Crenation of margin	Absent or very weak, weak, medium, strong or very strong
	Dentation of margin	Absent or very weak, weak, medium, strong or very strong
	Sinuation of margin	Absent or very weak, weak, medium, strong or very strong
	Cuping of upper side	Concave, plane or convex
	Blistering	Absent or very weak, medium, strong or very strong
	Colour of veins	White-green, yellow-green, pale violet or dark violet
5.	Petiole Length	Short, medium or long
	Thickness	Thin, medium or thick
	Colour	White-green, yellow-green, pale violet or dark violet
6.	Kohlrabi Size	Small, medium or large
	Shape (in longitudinal section)	Transverse narrow elliptic, transverse elliptic, circular or broad elliptic
	Shape of apex	Concave, plane or convex
	Colour of skin	White-green, yellow-green, pale violet or dark violet
7.	Start of harvest maturity (when 50% of kohlrabi reach 7 cm. diameter)	Very early, early, medium, late or very late
	Start of lignification	Early, medium or late

Note:

1. UPOV description refers to marketable part of the plant as kohlrabis while seeds men refer it as bulb.
2. Description based on the guidelines for DUS tests produced by UPOV (1980) (George, 1999).

Knol khol inflorescence is a terminal raceme. Flowers are hermaphrodite and with short pedicels. Sepals 4, petals 4, anthers are 6, didynamous, 2 short and 4 long filamented anthers. Ovary superior, tricapellary, stigma protogynous, being receptive 5 days before and 4 days after the anthesis. Pod is a siliqua. Sporophytic self incompatibility is inactive. Bud pollination is successful for self fertilization (Gopalakrishnan, 2007).

Nucleus Seed Production: In situ method of seed production: Seeds are sown on ridges. Seed rate is 1.5 kg/ha. Isolation is 1600m from other cruciferae crops, Brassicas, is a must. However, in the in situ method critical evaluation of true to type nature is not thorough in respect of knol khol knob.

In the transplantation method of seed production, the crop is grown with isolation distance of 1600m. The crop is lifted when the knob is well developed. The knobs are lifted and individually examined in the laboratory for free from disease infection and for true to type for colour, size and shape of the knob, the leaf and leaf base attached to the knob and the leaf features. After table and laboratory selection, selected knobs are transplanted/ replanted in 60 cm x 45 cm spacing. Shoots develop and flower and when the pods turn to yellow brownish, harvest is made. After 7 to 10 days plants are threshed and seeds are collected. In the following winter season, seeds of, individual true to type plants are sown in progeny rows. Seedlings are transplanted at 3 to 4 weeks age in the individual progeny rows. At the knob maturity, plants are lifted. Individual plants in each progeny row are verified for true to type nature in respect of foliage, leaf base attachment to the knob, knob colour, size, shape and free from deviations. Off-types for knob or leaf characters are discarded. In the next step, the knobs in the same progeny are selected after table examination and planted in progeny rows in a well prepared field isolated by 1600m from other Brassica crops. The developing shoots, leaves and the raceme and developing fruits are very closely examined for true to type nature. Atypical and deviating plants along with the knobs are discarded. Early bolting and very late bolting plants are removed. At maturity, the matured siliqua and seeds of each plant are examined progeny wise in the laboratory. Uniform and superior performing progeny rows are identified and seeds of superior plants in each identified progeny are now tagged. Seeds are collected from the tagged plants, dried to 7% to 8% moisture and examined in the laboratory for size, shape and colour. Seeds of selected plants are saved plantwise. Part of the seeds from each plant are bulked to constitute Nucleus seed.

Breeder Seed Production: Seeds of nucleus seeds are sown and transplanted in a well prepared clean field that was not grown to any Brassica crop in the immediate past 2 years. The field is isolated by 1600m from other Brassica crops. Seedlings are examined for trueness and uniformity and transplanted to raise knol khol crop knobs. Individual plants are examined for true to type. Off-types and atypical plants are discarded. In the next step, matured true to type knobs are lifted and transplanted in another well prepared clean field with isolation of 1600m. from other Brassicas. The developing shoots, racemes and siliqua are monitored at their respective development crop stage and verified for trueness of characteristics. Off-types and deviating plants are rejected. At maturity, single plants are harvested, seeds separated and dried to 7 to 8% moisture. Seeds of single plants are examined for true to type nature and seeds from all selected plants are bulked to constitute Breeder seed. The breeder seed is tested for seed quality of breeder seed standard. The quality assured seeds are labelled with Breeder Seed label. Grow out test is conducted. The nutritional content of the knolkhol vegetable is estimated.

70. Lettuce

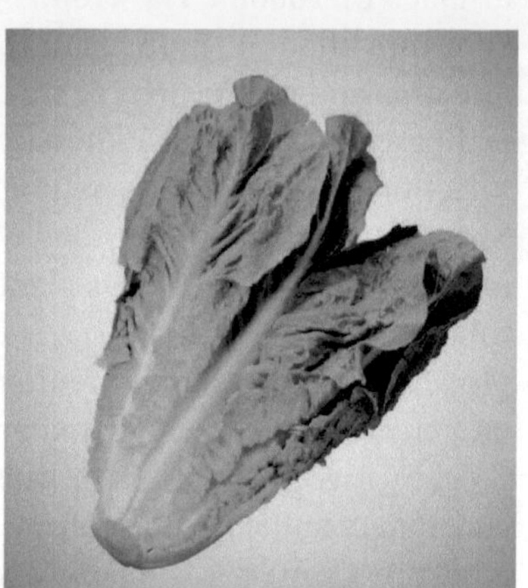

(*Lactuca sativa* (L.)) (2n:18)

Family : Asteraceae

Lettuce is consumed throughout the world as salad. In US, lettuce is the second largest fresh vegetable crop next to the potato. China is the largest producer and exporter of lettuce. Lettuce leaves are crisp to succulent and rich source of vitamins K and A and moderate source of iron and folate. Lettuce is a temperate annual or biennial crop. The characteristics are given in Table 1.

Table 1. Characteristics of lettuce

S.No.	Characteristics
1.	Plant diameter : low, medium, high
2.	Plant : degree of overlapping of upper part of leaves : absent, present, wide
3.	Plant : number of leaves : few, medium, many
4.	Leaf attitude : erect, semierect, horizontal
5.	Leaf anthocyanin colouration: absent, present
5.1	Leaf : size of anthocyanin colouration
5.2	Leaf : area covered by anthocyanin
5.3	Leaf colour : green, yellowish green, greyish green
5.4	Leaf : intensity of green colour
6.	Leaf : glossiness of upper surface
7.	Leaf : thickness
8.	Leaf : blistering : absent, present
9.	Leaf : blistering size
10.	Leaf : undulation of margin
11.	Leaf : type of incisions on the margin

12.	Leaf : density of incisions on the margin
13.	Leaf : venation
14.	For stem type vars: stem length : short, medium, long
15.	For stem type vars: stem width : narrow, medium, broad
16.	For stem type vars: stem colour : whitish, light green, green, medium green, greenish purple, purple, red
17.	Time of beginning of bolting : very early, early, medium, late, very late
18.	Fasciation of bolting stem : Absent, very weak, weak, medium, strong, very strong
19.	Colour of seed: white/ yellow / brown/ black
20.	Resistant to mosaic virus : absent, present

(**Ref:** UPOV. TG/13/11. Lettuce 2017-04-05)

Lettuce plants are herbs, spreading to 15 to 25 cm. Colour of the leaf is commonly green and also varies to red, yellow and golden colours. Root system is typically tap root habit. Duration of lettuce crop ranges from 65 to 135 days. Lettuce leaves become bitter on bolting and not preferred by the consumers. Flower stalks grow upto 100 cm. bearing yellowish flowers consisting of ray florets and tubular florets in each flower. Multiple heads are borne on the same inflorescence. Flowers exhibit five modified calyx into pappus hairs and 5 petals shaped into ligules. Anthers are five, syngenesious forming tubular shape enclosing style and stigma. The ovary contains only one ovule. Primarily self pollination takes place. Fruits have 5 to 7 ribs on each side with 2 rows of minute pappus hairs on the upper tip.

Lettuce was cultivated in Egypt for production of seed oil. Lettuce is also used as tobacco for cigarettes.

Breeding improvement in lettuce is focused on developing late bolting or delayed bolting, large succulent leaves, large heads with better texture leaves and taste, lower latex content, different leaf shapes and colours, high leaf biomass and large seeds. Being a low caloried nutrient source, it is a popular diet/ salad crop. Also cooked and eaten. Leaves are harvested before bolting. Lettuce is sown in Sept/Oct.

Seeds are sown in nursery and 4 to 6 weeks old seedlings with 2 to 3 leaves are transplanted in 30 to 45 cm. spacing between rows and 20 to 25 cm. on the rows. If lettuce seeds are directly sown in the field, thinning at 4 to 6 weeks is done. In higher altitude, lettuce in sown in March - June. Seed rate varies from 2 to 3 kg/ha. The first harvest of leaves commence in 45 to 60 days after sowing. Head type of lettuce yield 10 to 11 tons/ha while leaf types yield more.

Lettuce types are: 1. Leaf, 2. Head and 3. Romaine (cos) are most common. In leaf types, leaves are loose, cut and bunched. In Romaine types, leaves are long, upright and used in salads and sandwiches. Butter head and leaf types are known as cabbage lettuce. Iceberg types are grown in U.S., Butter head (round lettuce) in U.K. and summer crisp head in France. Summer/crisp types are with well flavoured large leaves and bolt resistant. Stem type lettuce is the seed stalk type grown in China and used in cooking. Oilseed type produces few leaves, quick bolting and produce seeds, 50% higher seed yield than other lettuce types. Besides, baby types of lettuce are also known. Lettuce seed ripes in 11 to 14 days after flowering. Lettuce seed is an achene type, 4 to 5 cm. long, beaked, 1000 seed weight ranges from 0.6 to 1.0 g. Lettuce is dry seeded vegetable crop. Lettuce seed is susceptible for thermodormancy. Isolation of 400m is recommended to avoid cross pollination in lettuce crop.

Nucleus Seed Production : Stock seeds of selected lettuce type/ variety that are genetically refined and maintained by the breeder are sourced. The stock seeds are raised with wide spacing in isolation of 400 m. in a healthy clean fertile field plot. Individual plants are monitored for quick germination, field establishment, initial and sustained seedling vigour, leaf production, leaf texture and crispness, taste, head/leaf type, plant stature, free from diseases and pests and resistance to premature bolting. Leaves are harvested 3 to 4 times and tested for suitability of fresh leaf vegetable market. True to type nature, uniformity and stable expression of characteristics are finally verified. Off-types and deviant type of plants are rejected and 250 to 300 true to type single plants are tagged. Seeds of the single plants are separated and examined in the laboratory for uniformity and trueness and finally seeds of 200 single plants are saved. In the next favourable season, in a clean disease infection free, fertile field, seeds of single plants are raised in 2 row progeny rows, each row 4 m. long and in 2 replications. Isolation distance of 400 m is provided. The progenies are closely examined for quick germination, vigour of seedlings and fast growth in critical phases of growth.

Individual plants in each row in each progeny are studied for trueness and uniformity and leaf traits expression. Off-types and deviating plants for growth habit, and leaf texture and premature bolting are discarded. Detailed examination and verifications for the characteristics of stem, leaf and flower stalk are carried out. Leaves in one of the rows in each progeny in each replication is harvested periodically and yield recorded besides the assessment of leaf characteristics, texture, succulence; taste and suitability as salad. At maturity, leaf yield data and leaf quality features are compiled and the progeny rows performing above mean +SE are identified. Superior plants in each identified progenies are tagged and seeds collected plantwise, dried and examined in the laboratory for trueness and uniformity in seed development. Part of the seeds from each selected plant is then bulked to constitute nucleus seed.

Breeder Seed Production : In a well prepared, clean and disease infection free field not grown to lettuce in the previous 2 to 3 seasons, nucleus seeds of the concerned variety of lettuce are sown in somewhat wide spacing. Only one plant per hill is retained. All recommended good agronomic practices are followed. Critical examination of the crop for germination, seedling vigour, quick leaf emergence, healthy succulent and tasty leaves and other fresh vegetable features was made. Off-type and deviant types are removed/ uprooted at prebolting stage. Isolation distance of 400m is provided.

Premature bolting plants are discarded. If deviating plants are noticed at post flowering, such plants are uprooted along with the plants around. Leaves from each plants are collected and tested for suitability of salad qualities. Plants that do not confirm to the ideal salad leaf quality are uprooted. This procedure is continuously followed. Composite samples of lettuce leaves are analysed for nutritional and other properties. At maturity phase, visit of Monitoring Team is arranged to verify field stand and crop uniformity. At maturity, individual plants are harvested, seeds separated and saved plantwise and examined in the laboratory. Deviating types of seeds are discarded plantwise, seeds are then pooled, processed and breeder seed constituted. Seeds are tested for seed quality assurance and breeder seed is labelled. Samples drawn from the breeder seed are adjudged by Grow Out Test for genetic purity. Nutritive quality of lettuce leaves are estimated (Fig. 1).

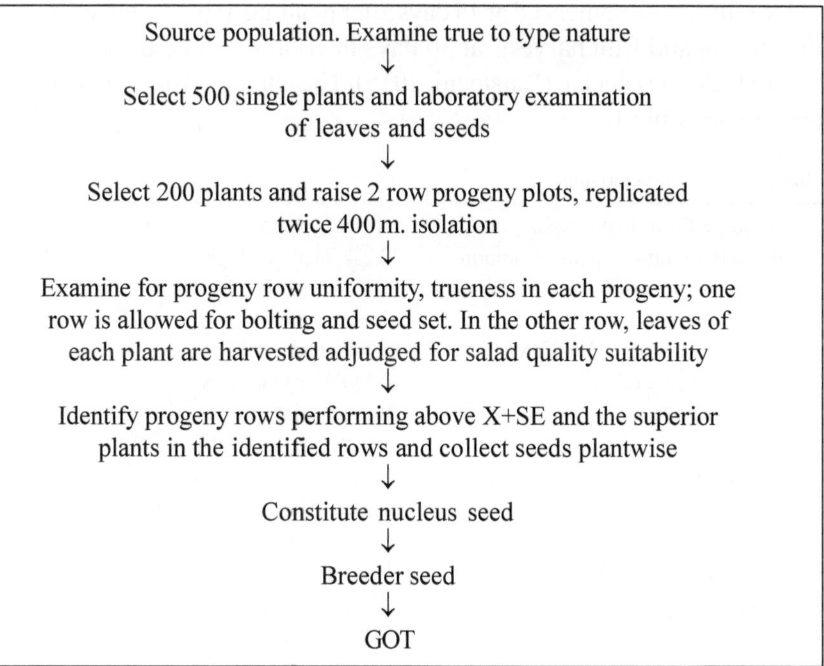

Fig. 1. Lettuce Nucleus and Breeder Seed Production

71. Manathakkali, Makoi, Black Night Shade

Solanum nigrum L. (2n=24)

Family : Solanaceae

Solanum nigrum is distributed widely and also occurs in semi wild form and higly polymorphic. It is a medicinal plant being used as a rejuvenant, tonic and in a range of remedial alternant.

Plants are mostly erect, herbaceous nature. Stem angular or cylindrical. Leaves are alternate spreading to tendency to be erect and green colour. Leaves are borne in petioles 5 to 7 cm long. Leaf vein is thick and prominent. Leaf margin straight. Flowers are in clusters of mostly 3 to 5 or more in number and in axillary cymes. Flowers are symmetrical, Calyx 5, petals 5 and stamens 5. Ovary superior. Fruit is a fleshy berry. Seed is flattened. Manathakkaali is a minor vegetable. Both leaf and fruits are edible. It is a highly self pollinated plant and autogamous.Seeds are sown in nurseries and 30 to 35 days old seedlings are transplanted. Leaves along with the tender stem are harvested at 3 to 4 times. First leaf harvest commenced at 45 days after planting and is followed by second, third, fourth and fifth harvests at 30 days interval.Fresh herbage yield will be from 15 to 20 tons per ha. (Rajamani, 2015). The characteristics of manathakkali are given in Table 1.

Table 1. Plant Characteristics

1.	Anthocyanin in hypocotyl/seedlings	Absent, present
2.	Intensity of anthocyanin coloration in hypocotyl /seedlings	Weak, medium, high
3.	Stem anthocyanin coloration	Absent, present
4.	Stem anthocyanin colour intensity	Weak, medium, high
5.	Stem at cross section	Angular, round, wavy
6.	Stem pubescence	Present, absent
7.	Leaf length	Small, medium, high
8.	Leaf width	Small, medium, broad
9.	Leaf shape	Lanceolate, ovate, obovate
10.	Leaf margin	Straight, dentate, wavy
11.	Leaf anthocyanin - intensity	Weak, medium, high
12.	Leaf pubescence	Absent, present
13.	Leaf upper surface	Smooth, blistered
14.	Leaf lower surface	Smooth, blistered
15.	Leaf blade colour	Light greens, green, dark green
16.	Leaf pigmentation	Absent, present

17.	Stem node junction pigmentation	Absent, present
18.	Stem node pigmentation	Light, medium, strong
19.	Leaf colour of vein	Absent, present
20.	Inflorescense –no.of flowers	1-3, more than 3
21.	Flower colour	White, light green
22.	Flower pedicel length	Short, medium, long
23.	Flower shape / diameter (mm.) at base	
24.	Fruit shape	Globose, oval , round
25.	Fruit girth mm.	
26.	Fruit length mm.	
27.	Fruit skin colour	Light purple, purple, blackish
28.	Fruit skin colour intensity	Dark, medium dark
29.	Fruit glossiness at maturity	Low, medium, high
30.	Fruit – patches on skin	Absent, present
31.	Fruit – calyx size	
32.	Fruit flesh colour	
33.	Seed colour	
34.	Seed shape (l x b)	
35.	Fruit bearing	Cluster of 2 to 3, more than three
36.	Plant growth habit	Erect, non-lodging, semi erect, spreading bushy
37.	Plant leaf habit	Erect, spreading, flat
38	Plant branching	3-4 and more
39	Plant height (cm)	60 to70, 70 to 90, 90>
40	Plant- days to flowering (50%)	45 to 60
41	Plant- days to maturity	(seed to seed) 160 to 180
	Plant Nutraceutical / nutritive composition	
	Ascorbic acid 21.60 mg / g	
	Fiber content 0.81%	
	Total soluble protein 1.4 mg / g	
	Total alkaloid content 0.38%	
	Solasodine content 0.05%	
	Iron content 6.10 mg / 100 g	

Source : Rajamani (2020)

Raise single plant progenies in 2 row plots with isolation distance of 400m. Remove early and late flowering plants. Progenies rows showing deviating type of plants are uprooted. One row in each progeny rows is left growing as such. In the another row systematic leaf harvesting is made at appropriate stages. Individual plants are harvested for leaves with stem at first, second, third,fourth and fifth harvests are made at 30 days interval commencing 45 days after planting. The fast regeneration ability and leaf development stability are critically assessed on individual plants. Progenies showing poor regeneration are rejected. In another corresponding row that is not harvested for leaf, uniform maturing, true to type progeny rows for good vegetative growth, floral and fruit characteristics are selected. In the selected progenies, superior individual plants are identified,labelled and fruits from them collected plant wise.

Seeds from the selected plants are examined in the laboratory and bulked to constitute Nucleus seed.

Breeder seed production: Raise breeder seed crop by sowing nucleus seed in the field not grown to the crop in the past 2 to 3 seasons. Isolation of 400m is provided. The field should be weed free and contamination free. Border rows 2 to 3 are raised around the breeder seed plot with the seeds of the same variety. Single plant per hill is maintained. At different growth phases-seedlings, juvenile, vegetative, flowering and fruit setting, plants are verified for true to type nature, uniformity in day to flowering. Early and late flowering plants uprooted. At appropriate stage, individual plants are harvested for leaves with stem. First, second, third, fourth and fifth harvests are made at 30 days interval commencing 45 days after planting. The fast regeneration ability and leaf development stability after each harvest are critically assessed on individual plants. Plants showing poor or low production of fresh shoots and leaves, die back and poor health are uprooted. Fruits and seeds from selected individual plants are examined in the laboratory for trueness. Seeds from individual plants are now bulked to constitute Breeder seed. Visit of Monitoring team is arranged. The seed qualities are assessed in the seed testing laboratory. Breeder seed labelling is done. Meanwhile, the nutritional qualities are estimated. Breeder seed is grown in Grown Out Test (GOT) to assess the genetic purity (Fig. 1).

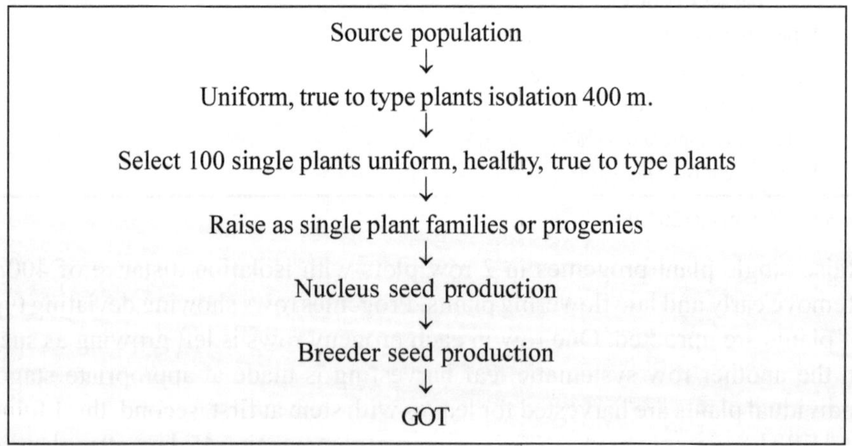

Fig. 1. Nucleus and Breeder seed production

72. Minor Tuber Vegetables

Nucleus and Breeder Seed Production in Taro (Colacasia), Dioscorea Elephant Yam and Coleus (Chinese potato)

Seed tuber multiplication: Survey the productive cropped area to identify and select healthy, vigorous, disease free, representative true to type plants nearing maturity and earmark 1500 plants. The plant habit, stem, stem pubescence, leaf and petiole colour, size and the petiole pubescence are examined in the individual plants. At the harvest time, the selected individual plants are gently lifted with tubers intact and washed to understand the root system, tuber size, colour, number, shape and free from contamination. The tubers are then separated and studied plantwise for tuber shape, size, tuber surface colour and texture, tuber flesh colour, texture and flesh development, free from pithiness and other defects. Based on the observations, 800 to 1000 plants with tubers in tact are selected that are true to type, uniform and for root tuber traits and plantwise labelled and from them 300 to 500 plants are saved.

In the following crop, the tubers or corm are cut into pieces suitable for the source of respective tuber crop. Cormlets are also selected and included as seed. The seed material comprising tuber pieces or cormbits, or cormlets are planted in ridges that are 90 cm. apart from each other on the rows, 30 cm or 60 cm space on the rows is provided as per the crop. All recommended production practices for the crop are followed and a productive crop is raised. Only one healthy plant per hill is retained 25 to 30 days after planting. Earthing up is done after 30 to 40 days. Crop specific agro or horticultural technology is followed. In respect of dioscorea, support is provided to each plant/vine 45 days after planting and the climbing stem vine is fixed./guided on the support. Continuous monitoring is made to verify for uniform, true to type nature, stem, leaf and petiole characteristics of size, thickness, colour, pigmentation and pubescence. Healthiness and fast growth are also recorded. Based on the observations the individual clonal progenies are established.

Progenies showing offtypes and deviating plants are rejected. At maturity, the selected individual plants in each clonal progeny are gently lifted up with tubers in tact. The entire root system true to type with tubers are adjudged for trueness, free from diseases and contamination. The shape, size, colour and texture of skin of the tubers, tuber flesh colour, texture, flavour and uniformity in flesh development, free from sponginess are evaluated and individual plants are selected in each progeny and saved. At the end of the second crop, selected plants now represent individual progenies for further evaluation for trueness, characteristics and uniformity suitable for nucleus seed development in each of the four tuber crops.

Nucleus and Breeder Seed Production: The tuber crop is raised with the seed material free of contamination in a clean field following the recommended production practices specified for each of the tuber crops. In each crop 100 progeny rows are raised. Isolation distance of 5m is provided.

Nucleus Seed Production: True to type and uniformity for the plant type, leaf, petiole and stem characteristics are examined in the progeny rows. Progenies showing deviating plants are discarded. At maturity, the plants are gently uplifted, and tubers washed free of soil. Plants (tubers) showing disease infection and contamination are rejected. The tuber characteristics such as size, shape, colour of skin, flesh colour, texture, fragrance, free from mealiness and sponginess are studied plant wise and plants showing deviations and malforned tubers are discarded. After completion of the harvest and tuber examination and collection plantwise, the performance of each progeny is studied with reference to tuber yield and other tubers traits. The progenies performing above mean +SE are identified and the tubers from the superior plants within the identified progenies are collected and tuber samples are pooled together to constitute the nucleus seed. The remnant seed tubers are stored for further use in nucleus-breeder seed cycle in the next season.

Breeder Seed Production: In a well prepared, disease free, fine tilth soil, the rhizome bits or cormels of the required size for seed purpose are planted adopting the space prescribed for the crop. In case of dioscorea, support system is arranged for stem to climb. Earthing up is done properly. All recommended production practices are adapted. Continuous monitoring is made to verify the uniformity and trueness of all the plants for each of the stem and leaf characteristics. At the time of maturity, monitoring team visit is arranged. The individual plants are gently lifted up with tubers in tact. The tubers are washed, examined for uniformity and free from contamination and disease infections denoting a clean crop. The tubers are studied in the laboratory for the prescribed tuber characteristics such as tuber shape, size, tuber surface, colour and texture, flesh colour, flest texture, crispness and flavour and selected. The selected tubers from individual plants are bulked to constitute Breeder seed. Quality of breeder seed is assessed. Breeder seed labelling is done. The tuber are examined for nutritional quality and pharmaceutical quality.

The description and production practices of individual crops. Coleus, Taro, Dioscorea and Elephant yam are presented in pages 330, 344, 345 and 400 respectively.

73. Okra Bhendi

(*Abelmoschus esculentus* L.) (2n:72 to 133)

Family : Malvaceae

Okra is an annual vegetable crop grown commercially in large area in tropical countries. Stem is green with or without reddish pigmentation. Leaves are alternate. 5 to 7 lobed with serrate margins. Flowers are solitary and axillary. Epicalyx is upto 10 number, petals 5, yellow coloured with prominent crimson petal spot at base. Stamens are many in the staminal column united at the base of the petals. Stigma 5 to 9 lobed. Fruit is a dehiscent capsule. Okra is primarily a self pollinated crop but due to insect pollination it is treated as often cross pollinated crop (Thamburaj and Singh, 2001). The descriptors are given in Table 1.

Table 1. Characteristics in okra

1.	Stem colour	Green, red
2.	Leaf blade-depth of lobing	Shallow, medium, deep
3.	Stem-number of nodes at first flowering	Few (<5), medium (5-8), many (>8)
4.	Flowering time (50% of plants with at least oneopen flower) (Days)	Early (<35), medium (35-45), late (>45)
5.	Leaf blade length	Small, medium, long
6.	Leaf blade width	Small, medium, large
7.	Leaf blade serration of the margin	Weak, medium, strong
8.	Leaf blade- colour between veins	Green, red
9.	Vein colour	Light green, purple
10.	Petiole length	Short, medium, long
11.	Flower petal colour	Cream, yellow, purple
12.	Flower petal base colour (purple)	Inside only, both sides
13.	Flower length (cm)	Small (<3), medium (3 to 5), large (>5)
14.	Fruit colour	Light green, green, light red, red, purple
15.	Fruit length at marketable stage (4 days after anthesis)	Small, medium, long
16.	Fruit diameter (cm) at mid length	Small (<1.5), medium (1.0 to 1.5), large (>1.5)
17.	Fruit surface between ridges	Concave, flat, convex
18.	Fruit pubescence	Absent, weak, medium, strong
19.	Fruit-constriction at base	Absent, weak, strong
20.	Fruit shape of apex	Narrow acute, acute, blunt
21.	Fruit number of locules	<6, 6 to 7, >8
22.	Plant-number of branches	Few (<2), medium (2 to 4), many (>4)
23.	Stem diameter (at 10 cm above ground level) (cm)	Small (<1), medium (1.0 to 1.5), large (>1.5)

24. Plant height (cm)	Short (<90), medium (90 to 120), tall (>120)
25. Fruit length (cm) at physiological maturity	Short (<10), medium (10 to 15), long (>15)
26. Fruit diameter (cm) at mid length	Small (<1.5), medium (1.5 to 2.5), large (>2.5)
27. Seed colour	Green, brown
28. Seed hairiness	Absent, present

Source: PPV & FRA (2009)

Nucleus Seed Production: Seeds from the source population or from Breeder stock are sown in an isolated plot with 600 m isolation. The crop is raised with good agronomic and production practices in the field not grown previously with okra crop. Healthy and vigorous plants conforming to the description of the variety characteristics are labelled. Individual plants are critically examined at different stages of growth juvenile to pre flowering, flowering till fruit maturity. Three hundred healthy single plants that are true to type in pod and seed features are selected and seeds collected from such plants are grown in progeny rows in the next generation. The field is isolated from other okra crops by 600m. The progeny rows are critically examined. Single plants are selected in progeny rows that show uniformity and true to type nature and self fertilised. The progenies that show deviant plants and off-types are uprooted as a whole progeny row. In the second season, the same steps are followed and 200 single plant selections are made, self fertilized and seeds carried forward. In the third season, the self fertilized seeds are sown in progeny rows plot in 2 replications. Superior plants are monitored from juvenile, pre-flowering, flowering and fruiting stage and as many as 200 plants are selected and self fertilized. The marketable tender fruits of the selected plants are evaluated in the laboratory for various fruit traits, tenderness, slenderness, taste, flavour and consumer appeal. Based on the field performance, progenies performing above mean +2SE are selected.The final selection of single plants is made. Based on the laboratory examination, seeds from 100 to 150 selected plants are bulked to constitute the Nucleus seed, Plants that are free from nematode, powdery mildow, white fly and YMV alone are selected (Fig. 1).

Breeder Seed Production: Nucleus seed is sown in well prepared clean field not grown with okra in the previous 2 to 3 years isolation provided is 600m. Only single plant is maintained per hill. Individual plants are monitored from seedling emergence to maturity for vigour, free from virus disease, pests and diseases like powdery, yellow mosaic virus and enasion virus. Plants in conformity with the variety description for stem and leaf pigmentation, leaf size, lobing and margin, petiole length, petal color, petal spot and pollen color, tender and mature fruit shape, color, ridges and tenderness and market quality of tender fruits are retained. Off-types and plants showing deviations are uprooted along with the plants surrounding them. Matured fruits are harvested in first 3 pickings only. Monitoring Team visit is arranged before or after the first picking of fruit. Seeds are extracted, dried and processed. Seed testing for seed quality aspects is performed. Quality assured Breeder seed is labelled. GOT is conducted with seeds samples drawn from the Breeder seed to ascertain genetic purity. Vegetable and nutritional qualities are estimated in marketable stage fruits.

Raise source population, Provide 800 m. isolation,
Evaluation and selection of 2000 uniform true to type plants.

↓

Laboratory and table examination for market stage fruits,
extent of fruit set, fruit colour and texture and slow maturity.

↓

Select 200 true to type plants and self fertilize. Raise two rows Progeny plots replicated twice (600m. isolation). Only single vigorous seedling is retained. Monitor progeny rows for uniformity and stability and true to type nature. Discard progeny rows showing variation Evaluate and identify uniform progeny rows performing above mean +2SE, Select 3 to 5 plants in each of the identified progeny.

↓

Collect fruits and seeds plant wise. After laboratory examination, selected seeds are packed plant wise. Part of the seeds from each plants are bulked to make nucleus seed.

↓

Breeder seed and nutritional quality of fruits assessment.
(Isolation distance 600m)

↓

GOT

Fig. 1. Nucleus and Breeder seed production in Okra – Bhendi

74. Onion

(*Allium cepa* L.) (2n=16)

i) Onion. *Allium cepa* var. *cepa* and

Small onion or Multiplier onion : *Allium cepa* var. *aggregatum* (2n=16)

Family : Alliaceae

Bulbs of *Allium cepa* are a popular vegetable everywhere. It can be used raw, sliced for seasoning salads, boiled with other vegetables or fried with other vegetables. They are an essential ingredient in sauces. It is consumed in small amounts for their pungency and is considered as a condiment. Onion is a biennial glabrous herb, usually grown as an annual from seed or bulbs. Grows upto 100 cm tall. Stem is very short formed at the base of the plant in the form of a disk, with adventitious roots at base. Bulbs are formed by the thickening of leaf-bases at short distance above the true stem, solitary or in clusters, depressed globose to ovoid or oblate, upto 20 cm in diameter, variously coloured. Leaves 3 to 8, distichously alternate, glaucous, with tubular sheath; hollow, upto 50 cm long, acute at apex. Inflorescence a spherical umbel upto 8 cm in diameter on a long, erect and hollow scape upto 100 cm. long, usually inflated below the middle; Umbel initially is surrounded by a membranous spathe splitting into 2 to 4 papery bracts. Flowers bisexual, stellate; pedicel slender, upto 4 cm long. Each flower comprises six calyx, six petals, 6 stamens and 3 carpels united and each carpel has 2 ovules. Flowers are protandrous and anthers shed pollen for 3 to 15 days prior to the time when style elongates to full length. Stigma becomes receptive 3 to 4 days after pollen shedding and this protandry nature favouring cross pollination. The characteristics of onion are given in Table 1.

Table 1. Characteristics in Onion

Common for onion varieties

1.	Number of leaves per plant	Few (<10), medium (10-15), more (>15)
2.	Foliage/ pseudo stem to leaf tip (cm)	Short (<30), medium (30-45), long (>45)
3.	Bulb time of maturity (days)	Early (<140), medium (140-160), late (>160)
4.	Bulb height (cm)	Short (<3), medium (3-5), tall (>5)
5.	Bulb diameter (cm) - Multiplier onion -	Small (<4.5), medium (4.5-6.0), large (>6)
6.	Total number of leaves per hill	Few(<10), medium (10-15), high (>15)
7.	Total number of shoots per plant	Few(<10), medium (10-15), high (>15)
8.	Foliage length (Pseudo stem to leaf tip) (cm)	Short(<30), medium (30-45), long (>45)
9.	Bulb time of maturity (days)	Early(<65), medium (65-80), late (>80)
10.	Bulb- compound bulb polar diameter (cm)	Small(<3), medium (3-4), large (>4)
11.	Bulb- compound bulb equatorial diameter (cm)	Small(<3.5), medium (3.5-5.0), large (>5.0)
12.	Bulb- compound bulb polar diameter (cm)	Small(<1.5), medium (1.5-2.0), large (>2.0)
13.	Bulblet equatorial diameter (cm)	Small(<2), medium (2.0-2.5), large (>2.5)
14.	Number of bulblets per bulb- for all varieties	Few(<6), medium (6-8), high (>8)
15.	Foliage attitude	Erect, semi erect
16.	Leaf diameter (cm)	Small (1.0), medium (1.0 to 1.5), large (>1.5)
17.	Foliage waxiness	Absent, present

18.	Foliage intensity of green colour	Light, medium, dark green
19.	Foliage- crinkling	Absent, medium, strong
20.	Pseudo stem length (upto last green leaf) (cm)	Small (<7), medium (>10), long (>10)
21.	Pseudo stem dia (cm)	Small (<0.5), medium (1.5 to 2.0), large (>2.0)
22.	Bulb thickness of neck (dia) cm	Thin (<0.5), medium (0.5- 1.0), thick (>1.5)
23.	Bulb L.S. Shape	Elliptic, ovate, globose flat globose, flat
24.	Bulb – colour of dry skin	White, yellow, pink, light medium red, dark red,brown, purple
25.	Bulb adherence of skin after harvest	Weak, medium, strong
26.	Bulb thickness of ring (av) (mm)	Thick (<3), medium (3.0-3.5), thick (>3.5)
27.	Bulb – Firmness of flesh (lbf)	Weak (<70), medium (70-100), strong (>100)
28.	Bulb – colour of epidermis	White, yellow, reddish purple
29.	Bulb position of root disc	Inserted, at surface, flat, exerted
30.	Bulb – predominant number of axis	Single., multiple
31.	Bulb – cross section	Asymmetrical, symmetrical
32.	Bulb – degree of splitting	Absent, medium , high
33.	Bulb – T.S.S. %	Low (<10), medium (10-15), high (>15)
34.	Male sterility to be tested under microscope	Absent, present

(*Source:* PPV & FRA)

Fruit is a globular capsule 4 to 6 mm. in diameter, splitting loculicidally and upto 6-seeded. Seeds are around 6 mm × 4 mm black. Flowers 50 to 600 in number per umbel. Secondary umbels will also occur by branching from primary umbel. In the large onion, varieties and hybrids have been developed and utilised in commercial production. CMS based male sterile line (A), maintainer line (B) and corresponding R line (pollen fertility restorer line) are the component parent lines in the hybrid seed production system. It is necessary to develop and maintain high quality assured parent lines in Breeder Seed Production Onion is highly cross pollinated and an isolation distance of 1600 m. between onion fields is maintained.

Nucleus Seed Production: In onion in the source maintenance breeding plot raised with genetically uniform seeds of the concerned variety, 2500 to 3000 true to type, disease free, healthy and vigorous plants are selected at the time of flowering. At harvest, the bulb is lifted up along with the panicle stalk. The mother bulbs are examined for colour, size, shape, uniformly developed single bulbs, panicle stalk and true to type nature and disease free conditions. Based on the matured bulb examination, 1500 to 2000 mother bulbs are finally selected for seed production and stored with ventilation. In the next season, bulbs weighing

60 to 70 g. are prepared for sowing by removing top one fourth of the bulb and treating the cut ends with fungicide for 10 minutes. The treated bulbs are sown in isolation plot (1600m) adapting wide spacing of 50 cm x 40 cm. The plants are examined for vigour and fast growth, healthiness and true to type and uniformity. Early bolters are removed. The umbels developed from uniform plants are self fertilized at the time of flowering in all the plants. The self fertilised seeds are saved on single plant basis. This process completes one cycle of seed to bulb and bulb to seeds (Fig. 1).

Source population : Access well developed uniform medium sized bulbs that are disease free. Select 3000 to 3500 true to type healthy bulbs and make table and laboratory examination. Reject malformed and split bulbs.

↓

Select 500 true to type healthy uniform and representative bulbs. Remove top ¼ th of the bulb and apply fungicide. Space plant the bulbs in 50 cm x 20 cm providing 1600 m Isolation. Monitor the growth of germinating bubs for healthiness and true to type nature till flowering. Self fertilize the selected 200 plants. Raise 200 progeny rows in 2 row plots.

↓

Monitor the growth, vigour, flowering and seed maturity. Plants showing variation and deviation in character expression are discarded. Harvest the seeds plant wise from the superior plants in each selected progeny. After laboratory examination part of the seeds are bulked to constitute Nucleus Seed to be utilized in seed to bulb and bulb to seed operation cycle. Based on the requirement. NS II Stage can be repeated.

↓

Breeder seed stage I and II are organized on need based manner (1600m Isolation is provided)

↓

GOT is conducted

Fig. 1. Nucleus and Breeder Seed production in Large (Common) Onion

In the next season, seeds of single plants are raised in well prepared nursery as single plant progenies. At 35 to 40 days, the seedlings are monitored for vigour, health and expressing uniformity and true to type. Progenies showing poor or weak expression , deviations in uniformity and disease infections in the seedlings are rejected. Healthy seedlings are transplanted in the progeny rows in the main field with 20 cm x 20 cm spacing. Progenies showing deviations, non uniformity and not true to type and poor growth and development and vigour are rejected well before the bulb harvest stage , usually at 110 to 120 days after transplanting. Well matured bulbs, from each plant progeny are examined in the laboratory for uniformity, true to type nature, colour, size, shape and non split bulbs and disease free naure. Bulbs of the progenies that are not true to type , showing split bulbs and non conforming to the descriptor are discarded. Healthy bulbs are stored with ventilation.

In the next cycle of mother bulb to seeds production, the well developed bulbs, that are selected and carried forward from the previous cycle are prepared for the progeny wise bulb planting to produce seeds. Field isolation of 1600m is maintained. In every selected bulb, top one fourth portion is sliced and removed. Bulbs are then pretreated with fungicide for 10 minutes before field planting with spacing of 50 cm x 20 cm in progeny rows. Examining for vigour and true to type nature is continuously made and roguing done to discard deviant progenies and off-types. At the time of flowering, progenies are allowed to intermate. At the time of seed harvest, only healthy and vigorous true to type and productive plants as evident from bulb development in all its aspects are selected. Seeds are collected from single plants separately in each progeny. Seeds collected from single plant progenies are pooled in part or in full to constitute nucleus seed.

Breeder Seed Production: Seedlings raised from the nucleus seed and nursery examination was made to discard less vigorous and deviant types seedlings. Well developed true to type seedlings alone are transplanted. Isolation distance of 1600m is maintained. Field observations on true to type, uniformity and stability and removal of bolters and disease free nature are made till the harvest of bulbs (around 110 to 120 days after transplanting). Thorough removal of plants showing split bulbs and deviations for true to type characteristics and poor in vigour and disease and pest infested plants is done. The harvested mother bulbs are verified for uniformity in colour, shape, size, development and nonsplit bulbs. The matured bulbs after harvest are stored. In the next season, well matured bulbs are trimmed in the top one fourth and treated with fungicide and sown in1600m isolated plot. All the bulbs from selected plants are planted and seed production is taken up.Careful monitoring is done for verifying true to type,uniformity and pest and disease free conditions of the individual plants. Off-types, deviant plants, very early and late flowering plants are uprooted. Monitoring Team visit is arranged. Seeds collected from the selected plants are bulked together and breeder seed is constituted. The breeder seed samples are tested for seed qualities and the quality assured seeds are labelled with Breeder seed labels. GOT is completed with seed samples drawn from the quality assured breeder seeds. At this stage, seeds from 1000 to 1500 single plants are collected and maintained as single plant selection for further nucleus and breeder seed production cycle. Bulb characteristics and seed quality are assessed in every stage of production in bulb to seed and seed to bulb cycles. Nutrients composition is analysed/ evaluated.

Small or Aggregatum or Multiplier Onion: Aggregatum onion group is characterized by production of smaller, several to many (upto 5 or more) bulbs form in an aggregated cluster originating from single mother bulb. In the aggregatum onion grown commonly in Tamil Nadu state there will be 4 to 5 well developed bulbs per clusters. In multiplier onion production, mother bulb to bulb crop is commonly practiced and bulb seed material is utilized as planting material. Isolation distance 5m. is prescribed. In multiplier or aggregatum onion production the steps are 1. Seed to bulb and 2. Bulb to bulb as the same in large onion. In the recent year, the quality assured true seed production cromprising the cycles of seed to bulb and bulb to seed are being followed by the TNAU, Seed production Agencies and the individual seed producers. Seedling raised aggregatum onion in on the increase in TN state. Nucleus and Breeder Seed production is shown in Fig. 2.

Seed production of parents in Cytoplasmic Genic Male Sterility (GMS) syatem for hybrids in onion: Hybrid seed production in onion based on CGMS system requires three parent lines (i) male sterile (female fertile) line (A), (ii) fertility maintainer pollen fertile line (male parent) (B) and (iii) fertility restoring pollen fertile (male parent) (R) line. In this system, seed to seed method or bulb to seed method could be followed. The first step is to make A x B paired crosses to maintain and multiply the male sterile parent stock line (A)that will be the female parent of the hybrid. The next step to cross the A line with the male fertility restorer line (R) that will be a very good general and specific combiner resulting in highly heterotic A x R hybrid. (genotypes of A=S msms, B=N msms, R=N MsMs)

Nucleus Seed Production of Parents Lines (A, B and R): Stock production of seeds of A, B and R lines require utmost care and are produced in netlon caged condition.

Bulb to Seed Method: In raised beds of 6 m x 2 m size, selected quality assured bulbs of A and B lines that are true to type are planted in alternate rows with 60 cm x 20 cm spacing maintaining single plant per hill and 1600m isolation. In A and B rows, single plants developing from the bulbs are critically examined for true type nature upto the commencement of flower initiation. The 6 m x 2 m size bed being grown with A and B lines in 1 to 1 row ratio is covered with a nylon net cagé well before flowering commences and the four corners were secured to strong poles that are fixed in the four corners of the plot. Care is taken not to allow the umbels of A or B line to touch the walls of the net cage so as to avoid pollen contamination. The cagé is provided with a nylon flexible door. The door is opened and bee hive is placed and honey bees are released.

Plants in B line are gently shaken so as to disburse the pollen to A line umbels.Uniformity and synchronization in flowering in A and B lines are ensured by timely and stringent removal of early and late flowering plants, off-types and deviant plants. At maturity, matured umbels of the B line plants are harvested 3 to 4 days earlier and dried and seeds separated. The matured umbels of A lines are collected 4 to 5 days later and seeds separated. The dried seeds are processed, germination tested and stored as the Basic stock seeds of A line and B line. Similar procedure is followed to produce R line seeds in netlon caged condition in isolated plot (1600m) to maintain genetic purity (Fig. 3).

Bulb to bulb

Disease free source bulbs are accessed. 2000 bulb clusters of uniform size with uniform bulblets, that are true to type are identified. The bulb clusters, each from a single plant, are table and laboratory examined and 500 clusters are finally selected.

↓

Bulb to Seed

Raise 500 single plant by planting only one well developed bulb in each selected plant Isolation of distance of 1600 m adapted. Monitor the plant establishment, growth, true to type nature, vigour, free from disease. Self fertilize the plants and at maturity self fertilized umbels and seeds are examined in the laboratory for trueness

↓

Seed to Bulb

Single plant wise nurseries are raised by sowing self fertilized seeds in clean raised seed nursery. Each nursery is critically examined for uniformity, health, true to type and vigour of seedings. Nursery that are not uniform and non conformity are rejected, Vigorous healthy seedlings at 25 days are removed. 300 single plant progeny rows are transplanted. The progeny rows are examined for establishment, vigour, healthiness, trueness and uniformity till the bulb formation and bulb maturity. Bulbs are harvested progeny wise and table and lab examination made for trueness and selection. Bulbs selected from each progeny are then bulked to constitute Nucleus seed. Nucleus seed stage II can be formed based on the raising clean healthy bulb to the bulb crop.

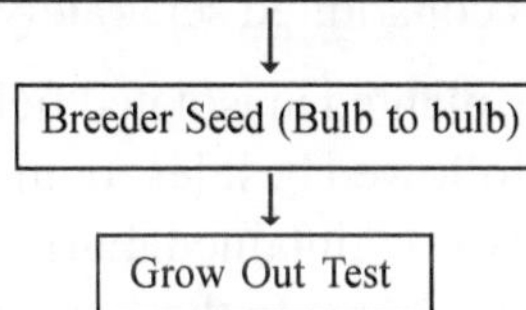

Fig. 2. Nucleus and Breeder Seed Production in Aggregatum Onion. (small onion) Bulb to bulb **Note:** I. Seed to seed cycle: II. Bulb to seed cycle can be adapted in case of small aggregatium onion TNAU Onion CO.5 and CO.6. Seed forming other genotypes of aggregatum onion also responds by flowering to adapt bulb to seed and seed to seed cycle.

Seed to Bulb Production: Seedlings are raised in separate raised nurseries for A,B and R lines with at least one meter space between the different nurseries. Nurseries are well protected by proper spraying of agrochemicals against pests and diseases. Well developed 30 days old healthy and vigorous true to type seedlings are transplanted in well prepared field. Single seedling is planted per hill. Seedlings of A, B and R are planted in separate three blocks each spaced at

2 m. apart which is sufficient isolation for bulb production in onion. Individual plants in each block of A, B and R are continuously examined for trueness to type and vigour and free from diseases. Deviant plants and disease infected plants are uprooted. At bulb maturity, when the tips of shoots commence drooping, plants are lifted and bulbs are separated. In the laboratory, individual bulbs are examined for size, shape, colour, non split nature and free from disease and pest infection. Bulbs of A and B lines are critically compared for gross similarities by virtue of their isogenic nature and, if any, deviant types are found, such bulbs are discarded. Bulbs of A, B and R lines are thus produced from seed to bulb and the bulbs are separately stored in ambient temperature and around 12% moisture.

Bulb to Seed Production: In A, B and R lines, uniform size bulbs are planted with recommended spacing and an isolation distance of 1600m is maintained from other onion crops. Bulbs are vernalised in cold temperature before planting or stored in ambient temperature and planted in cold season. Tropical adapted onion lines do not need vernalisation. Bulbs of A line and bulbs of B line are planted in alternate rows. Continuous examination is made to study the true to type nature and free from diseases and pests. At the flower initation onwards, plants are observed for synchronisation in A and B lines. Very early and late flowering plants are also uprooted and uniformity in flowering is maintained in A and B lines, Pollen shedders in the A line and pollen sterile plants in B line, if any, are uprooted. Overall synchrony for flowering, flower opening and seed maturity is thus maintained. At maturity, the matured umbels in A line (AxB crosses) are harvested first. Matured umbels of B line are harvested 2 to 3 days later. Seeds separated from individual umbels of A and B are examined in the laboratory for trueness and saved plantwise. Due care is taken to avoid mixing the seeds of A and B lines. Nucleus seed is constituted separately for A line and B line.

In respect of R line, bulbs are planted in isolated fields 1600 m apart. All steps in bulb to seed production are followed besides roguing the plants that are not true to type. Flowering uniformity is maintained. Early and late flowering plants are uprooted. It is advisable to note days to the flowering in R line in comparison with A line to decide maintaining the time staggering gap to plant A and R lines in hybrid seed production plots. At maturity of R line, matured umbels are cut with 15 cm. flower stalk, thinly spread, dried and seed separated. Laboratory examination of seeds of individual plants are made for trueness. Nucleus seed is constituted by pooling together of seeds from selected individual plants in the R line.

Breeder Seed Production in A and B lines: Bulb to seed method : Uniform size true to type bulbs in A, B and C lines are planted with recommended spacing and an isolation distance of 1600 m. is maintained from other onion crops. Tropical adapted onion lines do not need vernalisation. Bulbs of A line and bulbs of B line are planted in alternate rows. Continuous examination is made to study the true to type nature and free from diseases and pests.Off-types, deviating plant types and early and late flowering plants are uprooted to maintain synchrony in flowering and true to type nature. Bee hives are placed to promote cross pollination and seed set. At flowering, uniformity and synchrony between A ans B line is ensured by removal of early and late flowering plants in A and B line. Pollen shedders in the A line and pollen sterile plants in B line, if any, are uprooted. Monitoring Team visit is arranged at mid flowering to completion of flowering. Fully matured umbels of the B line plants are harvested 3 to 4 days earlier and dried and seeds separated. The matured umbels of A lines are collected 4 to 5 days later and seeds separated. The dried seeds are processed, germination tested. The seed quality is tested in seeds of A line and B line and the quality assured seeds are labeled with Breeder seed label. Seed samples drawn from the breeder seed lot are tested in GOT.

Breeder Seed Production in R line: The procedure to produce seeds in R line is similar to the steps followed in onion variety for seed to bulb and the bulb to seed Isolation distance of 1600m. is followed.

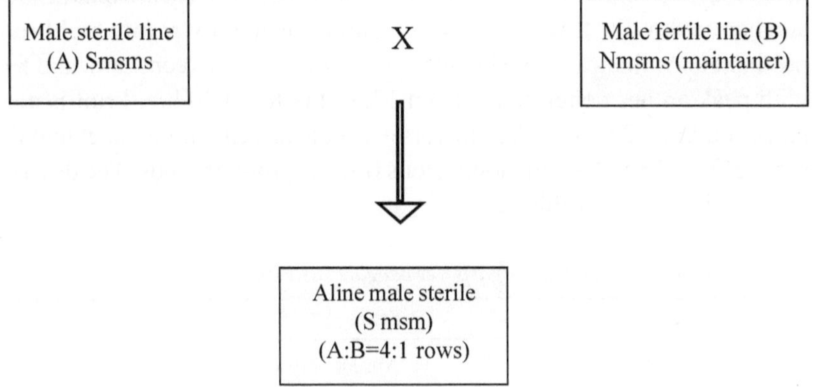

Fig. 3. Parent Seed Production using CGMS line

75. Palak (Indian Spinach)

(*Spinacea oleracea* var. *bengalensis*)
(2n:18)

Family : Amaranthaceae

Palak: Spinach Beet

Beta vulgaris L. var *bengalensis* Roxb.
(2n:18)

Family : Chenopodiaceae

1. **Palak Spinach Beet:** Palak also known as spinach beet is a herbaceous annual for leaf production and biennial for seed production. Palak is a nutritions leafy vegetable crop and a perishable vegetable. It is also suitable for hydroponic culture system production. Stem is short, small and thick. Leaves are turgid and are arranged in rosette fashion, Leaf petiole is longer. Leaf margin entire. Plants are with either reddish midrib and leaf veins or greenish midrib and leaf veins. Stem elongates till 75 DAS and produces a raceme inflorescence. Individual flowers are cyme type. Flowers are small, sessile and hermaphrodite flowers. Calyx is symmetrical. Flowers are highly cross pollinated. Anthers are five in number. Anthesis is between 7 am. and 5 pm. with peak anther deheiscence from 12.30 hrs to 14.30 hrs. Fruit is a seed ball containing 2 to 3 seeds. Flowers are male or vegetative male, female or monoecious. Females and monoecious flowers produce seeds. The describtor of palak is given in Table 1.

Table 1. Descriptor of palak (*Beta vulgaris* var *bengalensis* Roxb)

S.No.	Characters	State
1.	Seedling anthocyanin color	Absent, present
2.	Seedling anthocyanin intensity	Weak, medium, strong
3.	Leaf length (cm.)	Short (<20), medium (20-30), long (<30)
4.	Leaf attitude	Erect, semi erect, prostrate
5.	Leaf blade length (cm.)	Short (<10), medium (10-15), long (>15)
6.	Leaf blade width (cm.)	Narrow (8<), medium (8-9), long (>9)
7.	Leaf blade-green color intensity	Light, medium, dark
8.	Leaf blade glossiness	Dull, medium, bright
9.	Leaf blade blistering	Weak, medium, strong
10.	Leaf blade lobing	Very weak, weak, medium, strong
11.	Leaf blade anthocyanin	Absent, present
12.	Leaf blade shape	Triangular, medium, ovate, broad ovate, medium elliptic, broad, broad elliptic

13.	Leaf blade apex shape	Acute, obtuse, rounded
14.	Petiole color	White, green, pink, purple
15.	Petiole length (cm.)	Short (<10), medium (10-12), long (>12)
16.	Petiole width (cm.)	Narrow (<0.6), medium (65-70), late (>70)
17.	Petiole attitude	Absent, present
18.	Time of bolting (in days)	Early (<65), medium (65-70), late (>70)
19.	Seed spines of harvested seeds	Absent, present
20.	Resistance to cercospora species 1 to 9 scale	Very low, low, intermediate, high, very high

(*Source:* PPV & FRA)

High extent of variation exists in succulent green leaves that are crispy. Savoy types are for fresh market. Virginia savoy produces prickly seeds. Early smooth variety produces smooth seeds.

Palak is an annual plant (rarely biennial), which grows up to 30 cm tall. Palak spinach may survive over winter in temperate regions. The leaves are alternate, simple, ovate to triangular, and very variable in size from about 2 to 30 cm. long and 1 to 15 cm broad, with larger leaves at the base of the plant and small leaves higher on the flowering stem. The flowers are inconspicuous, yellow-green, 3 to 4 mm in diameter, maturing into a small, hard, dry, lumpy fruit cluster, 5 to 10 mm size containing several seeds. Isolation distrance of 1600 m is to be maintained from other spinach varieties, beet root, sugar beet, Swish chard. Seed yield is 600m to 800 kg/ha (Gopala Krishnan, 2007).

Seed Production: Seeds are soaked overnight in water before sowing to improve the germination. Seeds are sown with spacing of 60 cm x 25 cm. Seven to ten days after sowing seedlings are thinned and single seedling per hill is maintained. Harvest of leaves is done after 30 to 35 days after sowing and then the plants are allowed to grow and flower. Seed stalks are produced after 75 to 80 days and seeds harvested in 180 to 185 days after sowing.

Desirable traits are late bolting, fast regeneration and growth, Leaf harvest in 30 to 35 days, turgid succulent crisp leaves, freshness and attractive appearance.

2. **Indian Spinach or Vilayatti Palak** ***Spinacaea oleraceae*** **(2n=12)** Leaves are with entire margin and flowers are hermaphrodite.

   There are four sex forms in Spinach.

Extreme males:

- Produces only staminate flowers with minimum foliage.
- Small with very little vegetative development and tend to bolt quickly.
- Flowers early and die soon after flowering.
- Eliminate the extreme males from commercial strains by selection.

Vegetative males

- Vegetative males produce only staminate flowers, more foliage and flower later than extreme males.

Female plants

- Female plants produce only pistillate flowers and have well developed foliage with very late tendency to flower.
- Gibberellic acid plays an important role in sex expression of spinach.
- Female and vegetative male plants are preferred because they are larger, slower bolting and higher yielding.
- Pollination is mostly by wind.
- Fruit is usually an achene. The fertilized ovary form a one seeded fruit called utricle. Parthenocarpic fruits also occur in spinach.

  Vegetative males and females: Are slower to flower and produce considerably more foliage, making them the preferred plants type for commercial cultivation.

Monoecious plant

- Monoeciousplants produce staminate and pistillate flowers, well developed foliage and are slow to flower.

There are two types of seeds

1. Smooth round seeded types (considered as summer type)
2. Irregularly prickly shape (considered as winter type)

S.No.	Spinach Beet, Palak	Spinach
1.	Botanical name: *Beta vulgaris* var. *bengalensis* Hort	Botanical name: *Spinacea oleracea* L.
2.	Chromosome number 2n=18	Chromosome number 2n=12
3.	Leaves with margins	Leaves with lobed leaf margin
4.	Produce hermaphrodite flowers	Produce staminate, pistillate and hermaphrodite flowers
5.	Tolerates high temperature and grow well in hot weather	Purely a cool season crop and cannot tolerate high temperature. In warm weather and long days it quickly tend to flower
6.	Common name Palak, Saag, Indian Spinach, Indian Palak, Desi Palak	

Palak

Origin: Indo-Chinese region, especially Bengal region of Indian subcontinent.

Botany: Spinach be*et al.*so known as Palak or Palong and is closely related to beet root and Swisschard. Botanically it is known as *Beta vulgaris* var. *bengalensis* (2n=18) and belongs to the family Chenopodiaceae. Plant of spinach is a herbaceous annual for the edible leaf production while it is a biennial for seed production. Leaves are long with entire margin and long petioles. Flowers are small, sessile, bisexual, perfect and are borne in the axil of leaf in a group of 2 to 3, with 5 stamens. Flowers produce abundant, small and light pollen grains which are carried by wind. Hence the crop is highly cross-pollinated (a wind pollinated crop). Thinning and roguing is to be done for several times. Thinning will start as the leaves of seedlings start touching each other and continue till the plant to plant distances stands at about 15 to 20 cm. Before bolting, thinning and roguing for off-types will include plants in terms of colour of plants, shape of leaves, and stoutness in establishment. Early and late bolters should be removed as soon as observed. The isolation distance of 1600m. from other palak is maintained.

Land should be kept clean through weeding and mulching. Plants of *Chenopodium album* must not be allowed to grow in and around the field. Irrigation is to be provided as and when required.

Harvesting, Threshing, Drying etc: Spinach-beet seeds mature in sequence even in the same plant. First opened/formed inflorescence seeds mature first. Seeds after maturation may get discoloured/damaged due to foggy weather/rain. As such sequential harvesting should be done. When the seed stalks and seeds turn yellow and dry, harvesting is to be done by cutting the branch. About 2/3 harvests will be necessary before leaving the crop for seed production. Harvested branches should not be kept heaped which may cause damage to seed due to fermentation. Harvested branches to be dried in the sun over drying floor/canvas. Threshing is done by hand beating with a stick after being properly dried. Seeds are cleaned by winnowing and dried to 9% seed moisture content for temporary open storage. Seed Yield: Seed yield is about 600 kg/ha.

76. Pointed Gourd (Parwal)

(*Trichosanthes dioica* Roxb.) (2n:22)

Family : Cucurbitaeae

A dioecious climber, annual with perennial rootstock. Not much systematic research work is being done to develop improved varieties (Thamburaj and Narendra Singh, 2001). It is vegetatively propagated through vine cuttings and root suckers (2000 to 2500 cuttings per ha). In seed grown progenies, the seedlings segregate in 1:1 ratio for male and female sex due to its dioecious nature. Seed propagation is not done to avoid problem in selecting female plants at early stage of crop. However, in vegetable parwal production gardens, the male plants are maintained in 10% of the population to facilitate pollination of the female plants. While planting the vegetative cuttings for every ten female vine cuttings, one male vine cutting is planted. Parwal plantings reach harvest in 80 to 90 days after planting and immature and tender fruits are harvested. The yield increases to two fold in the second year.

One hundred plus performing female plants (above 2 to 3 years) that are in regular fruiting are identified and from each such plant 10 cuttings are prepared. True to type for stem, leaf, flower and are verified before selecting the female vines. In every 10 th row of female vines a single row male plants are planted. The support system and other crop production practices are followed. Uniform flowering (first flowering) plants alone are retained. Observation on individual plants are made for vine, leaf, flowers (male and female), developing, marketable stage fruits and matured fruits. Deviating type of plants are discarded. In each progeny selected plants are tagged. Harvest data for 2 to 3 harvests are compiled and plant progenies performing above mean +SE are selected. Superior performing female plants are selected and fruits are harvested and assessed for market qualities. Seeds are extracted from selected single plants and examined in the laboratory for trueness.

Each selected female plant is then cloned by standard size cutting. Clonal progeny rows are raised and individual clonal progenies are verified for trueness, uniformity and productivity and stability in female flower production and fruit set. The progeny rows performing better over mean performance of female rows are identified. Superior plants are selected. Three budded stem cuttings from each selected plants are then bulked to constitute Nucleus clone stock. Likewise cloning is made with the male plants that were raised along with the female progeny rows. Isolation distance followed is 10m.

Breeder Stock: Healthy well prepared 3 budded setts are prepared and planted in raised ridges adopting standard crop production practices for pointed gourd.

Plants are selected for vigour of the budlings and good field establishment. Selected plants are planted in the clean field. Male plants are planted in 10% proportion to the total plants. Regular examination is made from field establishment to trailing on supports, plant vigour, leaf development and healthiness, days to first female flower, profuseness of female flowers and fruit set. Isoaltion distance followed is 10m. Trueness for all the prescribed characteristics are noted. Deviating plants are removed. Yield data are recorded. The visit of Monitoring Team is arranged. Stem cuttings from selected plants are pooled together to constitute breeder seed clonal stock for female and male lines. The market acceptability for the fruits and nutritional qualities are analyzed and recorded. The list of characteristics of pointed gourd is presented in Table 1.

Table 1. Characteristics of pointed gourd *Trichosanthes dioica*

1.	Stem shape	Round, angular
2.	Leaf shape	Auriculate, cordate
3.	Leaf margin	Entire, undulated
4.	Stem number of secondary branches upto 20th node	Few <5, medium 5 to 9, many >9
5.	Node number at which first female flower appears on the main stem (indicate earliness)	Earlines <12 th node, medium maturity 12-15th node, late maturity >15 th node
6.	Stem intensity of pubescence	Sparse, dense
7.	Leaf blade length (cm)	Small >7, medium 7-10, long >10
8.	Blade width (cm)	Narrow <6, medium 6-9, broad >9
9.	Leaf lobe	Present, absent
10.	Leaf blade depth of lobing	Shallow, medium, deep
11.	Petiole length (cm)	Short <2, medium 2-4, long >4
12.	Flower sex type	Dioecious, gynomonoecious
13.	Fruit	Peduncle attachment- soft, hard
14.	Fruit shape	Club shaped, cylindrical, oval, spindle, elongate spindle, ovate, spheroid,taperning
15.	Fruit skin primary colour	Light green, green, dark green
16.	Fruit surface colour pattern	Uniform mottled striped
17.	Fruit glossiness	Non glossy, glossy
18.	Fruit length (cm)	Small (<5), medium (5-10), long (>10)
19.	Fruit diameter at widest portion (cm)	Small <4, large >4
20.	Fruit size (l/b ratio)	Small <1.5, medium 1.5-2.5, large >2.5
21.	Fruit shape at apex blossom end	Depressed, flattened, rounded, pointed
22.	Plant vine length (m)	Short <3, medium 3-5, high >5
23.	Number of seeds per fruit	Lessthan <5, less 5 to 10, medium 10-15, many >15

(*Source:* PPV & FRA, 2020)

77. Potato

(*Solanum tuberosum* L.) (2n:48)

Family : Solanaceae

Potato is a self pollinated crop. It is heterozygous and vegetatively propagated by tubers. Stem is angular. Inflorescence is a cyme and many flowered. Calyx 5, campanulate; corolla 5, gamopetalous. Stamens are five with pores at the tips. Stem tuber is short small or enlarged; it is a shortened apical portion of the stolon. Fruit is a berry, two celled and many seeded. Usually small tubers at multisprouting stage is preferred as seed material to produce strong stems. To get maximum yield, 30 stems per sq.m. are maintained. Yield: 15 to 20 tons of tubers / ha. in 120 days. The characteristics of potato are given in table 1.

Potato seed production is presented by Chakrabarthi (2018)

Table 1. Characteristics in Potato

1.	Light sprout- predominant colour	White green, pink, red, purple, black
2.	Light- sprout-shape	Spherical, conical, cylindrical
3.	Light- sprout anthocyanin intensity	Light, medium, dark
4.	Light sprout, pubescence	Absent, weak, strong
5.	Light sprout, length of apical portion (cm)	Small (<2), medium (2-4), long (>4)
6.	Plant foliage structure	Compact, semi compact, open
7.	Stem solidarity	Solid, hollow
8.	Stem cross section	Round, angular
9.	Plant height main stem (cm)	Hills; Small <70 Medium 70-90 Tall >90 Plains; <5050-70>70
10.	Stem predominant colour	Green, red brown, purple, dark purple
11.	Plant – wing	Poorly developed, highly developed
12.	Plant – wing type	Straight, wavy
13.	Leaf appearance	Open, intermediate, close
14.	Leaf anthocyanin colouration	Absent, present
15.	Leaf anthocyanin- midrib colouration	Absent, present
16.	Leaf length (cm)	Small (<16), medium (16-20) long (>20)
17.	Leaf width (cm)	Narrow (<11), medium (11-15), broad >15
18.	Lateral shape	Narrow, lanceolate, ovate, ovate lanceolate
19.	Leaflet glossiness upper side	Weak, medium, strong
20.	Flower- anthocyanin of bud	Absent, present
21.	Flower anthocyanin colouration pedicel	Absent, present
22.	Flower pedicel articulation position	Below middle, at the middle, above middle portion

23.	Flower- corolla colour	White, red, violet, blue
24.	Flower corolla size (dia.) (cm)	Small (<3), medium (3-4), large (>4)
25.	Inflorescence size (number of flowers)	Small (<10), medium (10-20), large (>20)
26.	Flower anther colour	Greenish yellow, yellow, orange
27.	Flower anther cone	Normal, irregular
28.	Flower pedicel type	Normal, irregular
29.	Flower style length in comparison to stamen column	Shorts, equal, lorge
30.	Flower stigma shape	Unilobed, bilobed, trilobed
31.	Plant maturity days	Hills Plain Early <100 medium 100-20 l Late >120
32.	Tuber predominant skin colour	Whitish, cream, yellow, orange, brown, pink, purple red, reddish purple, dark purple, black
33.	Tuber secondary skin colour	Absent, white cream, yellow pink, red, purple, black
34.	Tuber skin type	Smooth, rough
35.	Tuber shape	Flattened, round, ovoid, oblong, pear shaped long oblong, reniform, irregular
36.	Tuber depth of eye	Protruding, shallow, medium deep, deep
37.	Tuber predominant colour of flesh	White, cream, yellow, reddish, reddish purple, dark purple
38.	Tuber secondary colour of flesh	Absent, white, cream, yellow, reddish, purple, purple dark purple
39.	Tuber- distribution of secondary colour of flesh	Outer cortex, inner cortex, outer medulla, innermedulla, vascular, ring, mottled

(*Source:* PPV & FRA, 2006)

Nucleus seed production: Quality and clean clonal seed production that is true to type, genetically uniform, stable and free from diseases and pests is done in the nucleus seed production plot once in 3 years. The nucleus seed production steps are:

1. Select 2000 disease and insect infestation free, healthy, vigorous highly productive true to type potato plants conforming to the plant and tuber description from the source seed stock or maintenance breeding seed stock.
2. Store the heathy and clean potato tubers from true to type, genetically uniform, stable individual plants for 10 to 15 days. At the end of storage, grade the tubers as small (25 g), medium (25 to 30g) and large (25 to 50g).
3. After grading, the tubers are dipped in 3% boric acid solution for 30 minutes. In the plains, tubers are stored at 2 to 4°C till sowing.
4. Seed rate adopted is 3.0 to 3.5 t./ha. At multisprouting stage, the tubers are sown in clean fertile field. Single plant progeny plots are raised adapting spacing of 45 cm x 40 cm between tubers to get 30 well developed main stems per square meter. Isolation distance of 20m is provided.

5. Continuous observations are made during the crop period and the clonal plant progenies showing deviant plants and off-types are uprooted as a whole progeny. Those plant progenies that are true to type conforming to the varietal description alone are selected.
6. Harvest of tubers is made after discarding plants infested with nematodes, fungi, bacterial and viral diseases. Tubers are saved individual plantwise within each selected clonal progeny. Tubers are stored, graded, treated with fungicide and insecticide and sown as individual plant wise as described in steps 2 to 5.
7. In the second clonal generation, single plant clonal progenies are sown in a clean field maintaining 0.5 meter spacing between the progenies for easy identification, monitoring and avoid mixing during harvesting. During the crop development and growth till harvest all the procedures and observations are completed as in step 5. The tubers from each selected clonal progeny are then stored in separate bags, pretreated with 3% boric acid and stored for sowing the next generation crop.
8. In the third clonal generation, tubers weighing 40 to 50 g are sown in 45 cm x 45 cm spacing on the rows. Between clonal progenies, 0.5m space is provided to avoid mixing of tubers and to monitor the progenies easily . Each progeny is sown in two rows plot and replicated twice. Single plant progenies with plants showing deviations in varietal characteristics are uprooted. Plants showing virus infection, lack of vigour, pathogen and insect infestations, nematode affected plants and those showing micronutrient deficiency and physiological disorders are rogued out. Single plant progenies that are true to type, vigourous, stable, genetically uniform and free from diseases and pests alone are retained and evaluated for productivity. Progeny rows showing performance higher than mean are identified and tubers are harvested from them.
9. The harvested tubers are examined in the laboratory, as it was done in previous clonal generations for tuber size,shape and other tuber characteristics and healthiness. Progenies showing deviations in any of the characteristics for tubers are discarded as a whole. The selected tubers from superior plants from the progenies at the end of third clonal generation are bulked to constitute the nucleus seed in the potato variety.

Breeder Seed Production: Uniform sized true to type heathy potato tubers free from virus infection from nucleus seed stage are sown in well prepared clean fields that is not grown to potato in the previous two years. The field is also free from diseases and insect inoculums. Spacing of 45 cm x 40 cm is adopted. Single plant per hill is maintained. The field is isolated from other potato crops by 20 m. distance. Good agronomic and horticultural practices are adopted to raise clean, biotic and abiotic stress free healthy crop. Thirty well developed stems per sq.m. are maintained. Critical examination and monitoring are done on individual plants in different crop development stages from the crop emergence to tuber harvest for true to type, uniformity, health, vigour, virus and diseases infection free nature. Off- types and deviating plants are uprooted and discarded. Visit of Monitoring Team to adjudge the Potato crop breeder seed plot is arranged. At harvest, single plants are lifted with the stem and tubers intact and are examined in the laboratory for trueness and uniformity. Plants showing deviations for tuber traits are discarded. Sound tubers are separated from the selected individual plants and bulked to make the Breeder seed of the concerned Potato variety. Nutritional qualities of tuber are tested in the samples from breeder seed. Quality assured seeds are labelled with the Breeder seed label. Grow out test is conducted with tuber samples from breeder seed to ascertain the genetic purity. Based on the requirement for breeder seed potato, second stage of breeder seed crop can be raised.

Tuber indexing Detection and elimination of virus is regularly followed in each clonal generation in the nucleus and breeder seed production cycle. Slow multiplication and high field exposure increased the viral infection.

Seed Plot Technique envisages growing of the seed crop in nil or very low aphid prevalence conditions. Regular application of granular insecticide at planting and earthing up and removal of disease and pest affected plants and spraying of systemic insecticides are recommended in the seed tuber production plots.

Aeroponic system: Mini/Microtuber production: This is described in ICAR. News letter April to June 2010 (Vol. 16). This system facilitates 3 to 4 times multiplication of minitubers. Aeroponics is the process of growing plants in an air or mist environment

In vitro **micropropagation**. In this method, buds from seed tubers are grown in tissue culture medium under sterile conditions. The potato seed tubers, true to type are pretested for virus and diseases free nature. The buds are collected from indexed seed tubers free of virus infection. The axillary buds are separated with nodes and cultured to produce disease free tubers. *In vitro* micropropagated plants produce minitubers and microtubers. They are in turn grown in greenhouse for production of normal tubers. Microtubers are also obtained from meristem culture. *In vitro* plants and plants developed from minitubers and microtubers do not differ significantly for yield and other traits. Procedure for micropropagation in potato is outlined by the Department of Biotechnology (Fig. 1).

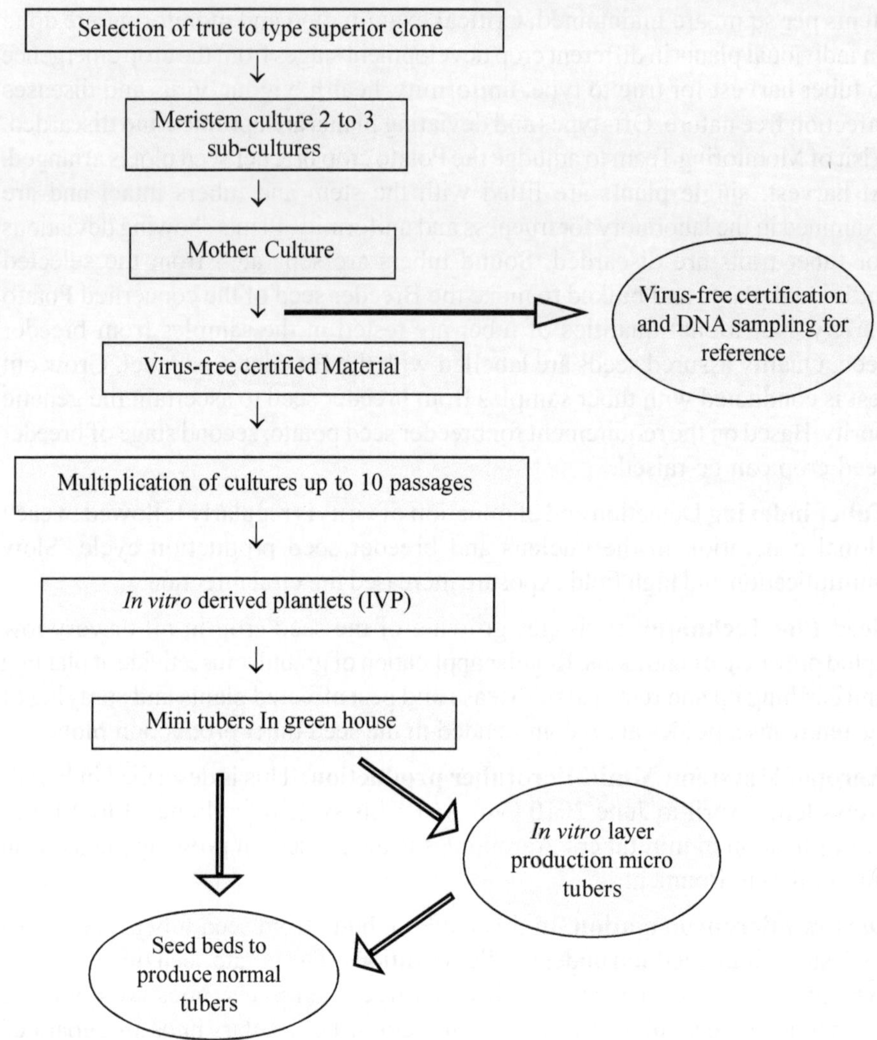

Fig. 1. Procedure and standard parameters for Micro Propagation in Potato
Source: Department of Biotechnology, New Delhi.

True Potato Seed (TPS): TPS is used to grow commercial potato crop by two methods; (i) transplanting of seedlings and (ii) use of seedling tubers. In the former method, potato seedlings are raised in nursery beds and transplanted in the field at 4 to 5 leaf stage. (Sarjeet Singh *et al.,* 2000). Potato crop raised from seedlings is harvested at maturity. This method requires about 125g. TPS and 75 sq.m. nursery bed area for growing seedlings sufficient for transplanting in one-hectare area.

In the second method, small size tubers are produced by allowing the seedlings to mature in nursery beds. Seedlings are dehaulmed (cutting of plants at ground level) at maturity and harvested in another 10 to15 days when the tuber skin becomes firm and well developed. These seedling tubers are stored and used as planting material for growing the commercial crop in the following crop season. About 50 to 60 g. TPS and 250-300 sq.m. nursery bed area are required for producing seedling tubers sufficient for planting one hectare. The disadvantages with TPS technology are that the seeds are not genetically homozygous and exhibit high heterogeneity; the crop is late in maturity as compared to the crop grown from seed tubers; and the technology is labour intensive. Quality clean seed system in potato consists of:

1. Tubers indexing – virus detection
2. Virus elimination
3. Clonal multiplication as clonal progenies in successive generations
4. Tissue culture based system of quality seed tubers production. *In vitro* plants and microtubers are planted in insect proof greenhouses for minituber production. Hardening is not needed for minitubers and microtubers. Increase in productivity of 20% is reported.
5. Minitubers are small tubers harvested from the plantlets. Virus enters only through old cells and in shoot tips or meristem virus get precluded.
6. Microtubers are miniature tubers developed *in vitro* under tuber inducing conditions. They are very small (average weight 100 to 150 mg) and convenient for handling, storage and long distance transportation. The general microtuber production procedure consists of mass multiplication of *in vitro* plants in liquid propagation medium in flasks/magenta boxes.
7. Stem segments (each with 3 to 4 nodes) obtained from 5 to 6 *in vitro* plants are cut and inoculated in each container containing 20 ml. liquid propagation medium and the containers are incubated stationary under normal potato propagation conditions (22 to 25°C temperature and 50 to 60 μE/m /s light intensity for 16 h).
8. Minituber production from *in vitro* plants Minituber is an intermediate stage of potato seed production between laboratory micropropagation and field multiplication. The simplest method of producing minitubers from *in vitro* plants particularly suited to North Indian plains has been described by Naik and Karihaloo (2003) and Sharma *et al.* (2013). In this method, 15 to 20 day old cultures are kept in glasshouse or polyhouse for 8 to 10 days for hardening without removing culture tube plugs or lids of magenta boxes.

78. Pumpkin and Squash

(*Cucurbita* sp.) (2n:40)

Family : Cucurbitaceae

a) *Cucurbita pepo* L. (Squash)

b) *Cucurlbita moschata* Poin. (Pumpkin)

Pumpkin is cooked edible fruit and pepo is eaten raw. Herbaceous, annual trailing plants, stem angular and tendrils branched.

Plants are monoecious; flowers are large, showy, yellowish coloured and unisexual. Staminate flowers are with long pedicel and pistillate flowers having short and stout pedicels. Calyx and corolla, five each, campanulate. Anthers 3 united partly. Stigma 3 lobed; Ovary inferior tricarpellary. Placentation is parietal. Style short thick with 3 to 5 stigma. Highly cross pollinated species. Inbreeding depression is not reported in this crop. The descriptor of pumpkin are given in Table 1.

Table 1. Descriptor of pumpkin

1.	Plant	Shape of cotyledon : elliptic, broad elliptic, or obovate Growth habit : bush, semi – trailing or trailing
2.	Branching	Absent or present
	Degree of branching	Weak, medium or strong
	Bush cultivars only	Attitude of petiole (excluding lower external leaves) : erect, semi-erect or prostrate
3.	Stem	
	Green color	Light alone, dark alone or light and dark
	Tendrils	Absent or rudimentary, or well developed
4.	Leaf blade	
	Size	Small, medium or large
	Incisions	Absent or very weak, weak, medium, strong or very strong
	Green color of upper surface	Light, medium or dark
5.	Marbling	Absent or present Area covered by marbling : very small, small, medium, large, very large
6.	Petiole	
	Green color	Light, medium or dark
	Length	Short, medium or long
7.	Thickness (at base) of petiole	Thin, medium or thick
	Shape of cross-section	Round or angular
	Number of prickles	Few, medium or many
8.	Female flower	
	Ring at inner side of corolla	Absent or present
	Color of ring at inner side of corolla	Green, yellow or green and yellow
	Intensity of ring color inner side of corolla	Weak, medium , strong

	Character	Description
	Length of sepal	Short, medium or long
	Color of pistil (before opening)	Yellow or orange
9.	Male flower	Expression of colored ring at inner side of corolla : absent or very weak, weak, medium, strong or very strong
	Color of ring at inner side of corolla base	Green, yellow or green and yellow
	Length of pedicel	Short, medium or long
	Diameter of pedicel	Small, medium or large
	Color of pedicel	Light green, medium green or dark green
	Grooving of pedicel	Weak, medium or strong
	Hairiness of pedicel	Weak, medium or strong
	Length of sepal	Short, medium or long
10.	Young fruit	
	Length	Short, medium or long
	Maximum diameter	Small, medium or large
	Ratio of length/maximum diameter	Small, medium or large
	Neck	Absent or present
	Curving of neck (on normally set fruits)	Absent or present
	Number of colors	One, two or three
	Major color	Whitish, yellow, orange, green, whitish and yellow, whitish and green or green and yellow
	Intensity of major color	Light, medium or dark
	Shape of cross-section	Round, ribbed, angular or lobed
	Warts	Absent or present
	Glossiness	Weak, medium or strong
	Color of ribs compared to color of rest of fruit	Same or different
	Color of linear bands on ribs	Whitish, yellow, orange or green
	Mottling	Absent or present
11.	Type of mottling	Only diffuse ; diffuse and in linear bands ; or diffuse, in patches and in linear bands
	Color of mottling of skin	Whitish, yellow, orange or green
	Conspicuousness of mottling of skin	Weak, medium or strong
	Size of flower scar	Small, medium or large
	Protrusion of flower scar	Absent or present
12.	Length of peduncle	Short, medium or long
13.	Main color of peduncle	Only green, only yellow or green and yellow
	Intensity of main color of peduncle	Light, medium or dark
	Mottling of peduncle	Absent or present
	Conspicuousness of mottling of peduncle	Weak, medium or strong
14.	Fruit	
15.	General shape	Scallop-shaped, transverse elliptical, globular, top-shaped, ovoid, elliptical, cylindrical, club-shaped, pear-shaped or crook-neck

	Fully developed in size, major color	Whitish, cream, yellow, orange or green
16.	Ripe fruit	
	Intensity of major color	Light, medium or dark
17	Fruit fully developed in size	Whitish, yellow orange or green
18	Ripe fruit	
	Color of stripes on ribs	Whitish, yellow orange or green
19.	Fruit	
	Maximum length	Very short, short, medium, long or very long
	Ratio length/maximum diameter	Very low, low, medium, high or very high
20.	Seed	
	Size	Small, medium or large
	Shape	Narrow elliptic, elliptic or broad elliptic
	Color	Whitish or yellowish

Note: Descriptions are based on UPOV (1988) which considers vegetable marrow and squash together and as described by George (1999), PPV & FRA (2001)

Nucleus seed production: Seeds secured from the breeder source are sown in the field (isolation of 1600m) that are not grown in the pervious two seasons with pumpkin or squash crop. Plants are raised with pregerminated seeds in pits of 30 cm x 30 cm x 30 cm with 2 m x 2 m spacing. Only one well developed healthy plant is retained in each pit. Ethrel 250 ppm (2.5 ml per/10 liters of water) is sprayed four times at weekly interval from 15 days after sowing. Plant protection measures are taken up against downy mildew, powdery mildew and fruit fly. Observations are made on individual plants for true to type and uniformity for the stem, foliage, tendrils and floral characteristics. Deviant types are discarded. Individual vines are tagged for the days to first flowering and very early or late flowering vines are uprooted to maintain uniformity in flowering. When the female flowering phase commences, individual vines are examined and all the female flowers already opened are pinched. From the next day onwards, female flowers and male flowers about to open the following day are covered with selfing paper covers in the previous day evening. In the next day morning, each female flower is dusted with pollen from 2 to 3 male flowers from the same plant by gently rubbing in rotating manner so as to cover all the sides of stigma of the pistllate flower. The self pollination on to the stigma is done before 9 am. and after the self pollination, the female flowers are again bagged and labelled. At that location, coloured sticks may be pegged down for easy locating the self fertilized pumpkin fruits. Two to three fruits are thus self pollinated in each plant and at maturity, self pollinated fruits from each plant are harvested and stored separately to raise single plant progeny.

In the next season, at least 100 single plant progenies are raised in two replications and with 20 plants in each progeny. All the agronomic and horticultural production practices as described earlier are followed. Isolation distance of 1600m. is practiced. The individual plants in the progeny row plots are examined and the progeny row plots showing any off types or deviation types are uprooted. From among the true to type, uniform and stable progeny row plants, 100 healthy vigorous and uniform plants are identified. At least one female flower in each such selected 100 plants is self pollinated in the following morning and the self pollinated female flowers are protected by butter paper covers. At maturity, the self pollinated fruits are collected and stored for 30 to 35 days. At the end of storage, the fruits are examined for size, shape, ridges on the fruit surface, flesh colour and thickness, consistency ,flavour of flesh taste and extent of seediness. The yield of fruits in each of the progeny in each replication is compiled and progenies showing performance above mean +2SE are identified. At least 50 to 60 fruits are selected and seeds extracted and dried .Seeds from individual plants are examined for trueness and uniformity and saved plant wise. Part of the seeds from each plant sample is drawn and pooled to constitute nucleus seed. The rest of the seeds are stored plant wise for future nucleus seed stock development.

Breeder seed production: Pregerminated nucleus seeds are planted in pits of 30 cm x 30 cm x 30 cm size in 2m x 2m spacing. Isolation distance of 1600m is maintained. Good agronomical and horticultural practices are adhered to. Ethrel application of 250 ppm is made four times at weekly interval commencing from 15 to 20 days after sowing. Roguing and removal of off-types and deviant plants are regularly done and only true to type plants for vines, foliage, floral and fruit characteristics are retained. Very early flowering and late flowering plants are discarded. Monitoring Team visit is organized at mid maturity stage. At maturity, fruits are harvested in each plant and stored for 30 days. At the end of storage, fruits are examined for trueness and uniformity. Selected fruits are cut open and assessed for fruit thickness, flesh colour, texture and other market acceptable traits. Seeds are then extracted from each fruit and dried separately. Laboratory examination is made for seed features. After final examination, the seeds are bulked together, dried and samples drawn for seed quality tests. Samples seed of the quality assured breeder seeds are then grown in GOT to adjudge the genetic purity. The seed lots are labelled with Breeder Seed label. The fruit quality and nutrients contents are estimated (Fig. 1).

Raised in wide spacing by planting pre graminated 500 seeds. Select uniform true to type plants. Population monitored for healthiness, trueness and uniformity. The fruits are retained in the field till maturity. Harvest the fruits and seeds and examine for trueness and uniformity. Select 100 representative lines (plants). Self fertilize 1 to 3 flowers in each plants and collect one fruit in each selected plant and conduct table and laboratory examinations of fruits and seed characteristics. In the next season grow 100 single plants and examine for trueness and self fertilize 1 to 2 flowers in each plant.

↓

100 Plant progenies are raised in 2 row progeny plots. Monitoring and evaluation of single plants in each progeny row for true to type and uniformity for the characteristics specific for each crop. Off-types, unhealthy and less vigorous and disease affected plants are discarded. Fruits in progeny rows are adjudged for vegetable market features. At maturity based on evaluation for productivity, progeny rows performing above mean +SD are identified and 2 to 3 plants in each identified progeny are harvested.

↓

Table and laboratory examination of fruits and seeds and keeping quality of fruits is assessed. Seeds are extracted from fruits on progeny plant wise.

↓

Seeds of superior plants in the identified rows are separated and stored plant wise. Portion of seeds from each plant are bulked to constitute Nucleus seed.

↓

Breeder seed production

↓

GOT

Fig. 1. Nucleus and Breeder Seed production in Pumpkin and Squashes, Bottle gourd & Ash gourd
Note: Isolation distance of 1600m. is maintained in all the generations

79. Radish

(*Raphanus sativus* L.) (2n:18)

Family : Brassicaceae

Radish is a biennial for seed production and requires 4 to 8°C chilling for 40 to 60 days to initiate flowering. The flowers are borne on terminal inflorescences. Flowers are small usually white coloured. Sepals are four and erect and petals are four and coloured. Ovary superior. Fruits is a capsule. Radish flowers are protogynous and is cross pollinated by virtue of sporophytic system of incompatibility. It shows inbreeding depression upon self fertilization. It is insect pollinated mainly by bees. Stigma receptivity is maintained upto 4 days after anthesis. Selfing is performed by bud pollination and buds are pollinated 2 days prior to opening by their own pollen freshly obtained from already bagged flowers of the same plant. The edible portion of radish develops from the primary root and hypocotyl. Isolation distance required is 1600m for nucleus and breeder seed production.

In respect of nucleus and breeder seed production, root to seed method is followed so as to ensure trueness is tested, verified and established in respect of underground roots and aboveground vegetative parts, flowers and seeds. Seeds sourced from breeder institution, or nucleus seed and breeder seed are sown in October in plains and September in locations in hilly regions. Systematic roguing is done 2 to 4 times covering all individual plants in respect of stem, leaf, root, inflorescence, flower, fruit characteristics at edible maturity of roots.

Matured roots are carefully lifted and examined and selected for the shape, size, colour, absence of forking and pithiness, root surface smoothness, slow bolting and high degree of uniformity and true to type. Underdeveloped, atypical root, roots showing sponginess and pithiness, early bolting and disease infected roots are discarded. Nearly 1000 roots are selected and examined on the table and in the laboratory. Stecklings are prepared from the selected matured typical true to type roots by cutting/removing 1/3 of the shoots top and 1/4th of the root. The stecklings are planted at 60 cm x 30 cm spacing in a well prepared field with fine tilth. Isolation distance of 1600 m is ensured. Good agronomic practices are adapted to raise a disease and pest free crop. In each progeny, pods are harvested, dried and seeds separated and saved plantwise with pedigree details. Seeds of all the plants are now examined in the laboratory for trueness and selected for conformity. Part of the seeds from each plant are bulked together to constitute the nucleus seed.

Breeder Seed Production: Nucleus seeds are sown in a finely prepared field isolated by 1600 m in the proper season. Single plant is maintained in each hill. The radish crop is grown till the roots attain edible root maturity. During this field growing, due examination is critically done in individual plants to verify for trueness and conformity in all characteristics. Early bolting as well as late bolting plants and off types are discarded. At this stage of seed to root production, the roots are lifted from the soil and critically examined for the shape, size, uniformity, skin color and texture, absence of pithiness and deformities in root. Non conforming roots are rejected. The selected roots are examined for absence of root infection and other diseases. Selected roots are examined on the table and stecklings are prepared in the prescribed manner.

Stecklings are planted in the following season in the well prepared field isolated by 1600 m. The field should be free from disease infection and not grown with radish crop in the previous 2 seasons. Stecklings are well cared for and put forth new shoots and reach flowering stage. During this period, continuous and critical monitoring is done to remove off-types, deviating plants, early bolters, disease and pest affected plants. Trueness of the retained plants is established. At flowering and after flowering, if deviating plants are noticed, such plants and plants around them are gently removed to avoid accidental cross pollination. At pod maturity, again the crop is verified for conformity of pods. Whole plant is cut and removed and in the laboratory examined for pod and seed characters. Selected plants are threshed, seeds extracted, cleaned and dried to 8% moisture level.

Samples of seeds are tested for breeder seed quality assurance. Seeds are labelled with breeder seed tag. GOT is conducted. Nutritional qualities of radish are estimated.

80. Ridge Gourd

(*Luffa acutangula* L. Roxb) (2n:26)

Sponge Gourd

(*Luffa cylindrica Roem*) (2n:26) (Syn. *aegyptica*)

Family : Cucurbitaceae

Ribbed gourd: Plants are monoecious, annual. Stem vines angular, leaves 5 to 7 lobed. Tendrils branched. Flowers yellow and showy, five petals, staminate flowers in racemes, pistillate flowers solitary with short to long peduncle. Pistil has 3 placenta. Stigma three and bilobed. Flowers start anthesis in the evening and remain open in the night and are ready for pollination in the early morning. Fruit is oblong or cylindrical. Tip of the fruit is club shaped or acute tip.

Sponge gourd: Vigorous and strong vine with angular stem. Leaves orbicular and with pointed apex, 3 lobed, fruits are nearly cylindrical, 30 to 60 cm. long, mostly straight, with light furrows or stripes but not ribbed. Seed black, flat, smooth without margin, 10 to 15 mm long. Flowers of sponge gourd open in early morning hours and selfing can be done throughout the day.

Nucleus seed production: Source 500 well matured seeds from Breeder source stock or from maintenance breeding stock. Seeds are pregerminated and fast germinating seeds are selected and planted in pits with 2 m spacing between plants on the rows (2m x 2m). Single plant per hill is maintained. Isolation distance of 1600m. is provided. Plants are well cared with good vegetable crop production practices and plant protection measures. Ethrel spray (250 ppm) is applied four times commencing from 15 days after sowing at weekly interval. Periodical observations are made on the vigour of vines, fast growth rate, foliage, flower and fruit characters (Tables 1 and 2) besides tolerance to powdery and downy mildew diseases affecting ridge gourd and sponge gourd. On the basis of observations, 100 vigorous and healthy, true to type and uniform plants are selected and 2 to 3 female flowers in each selected plant are self pollinated by profusely dusting pollen from the male flowers borne on the same vine. The female and male flowers are already protected by selfing bags. Fruit characters such as shape, size, flesh thickness, length and breadth of the fruit, number and prominence of ridge, nature of surface between ridges, colour of the fruit skin, slow maturity of seed and fruits and free from bitter taste and seed colour, shape and size are recorded.

Table 1. Characteristics of ridge gourd

1.	Cotyledon- intensity of green colour	Light, medium, dense
2.	Stem shape	Rounded, angular
3.	Stem pubescence	Absent, present
4.	Leaf margin	Entire, serrated
5.	Leaf shape	Ovate, orbicular, reniform
6.	Leaf length	Small, medium, large
7.	Leaf width	Small, medium, large
8.	Leaf pubescence	Absent, present
9.	Leaf pubescence nature	Soft, hard
10.	Leaf blade, no. of lobes	Three, five
11.	Stem internode length	Short, medium, long
12.	Stem –number of primary branches	Less (<10), medium (10-15), more (>15)
13.	Petiole length	Short, medium, long
14.	Flower colour	White, light yellow, yellow
15.	Ovary length	Short, medium, long
16.	Peduncle length	Short, medium, long
17.	Fruit length (cm.)	Short (<20), medium (20-30), long (>30)
18.	Fruit girth (cm.)	Small (<10), medium (10-14), large (>14)
19.	Fruit shape	Oblong, elongate, elliptical
20.	Fruit skin colour	Light green, green, dark green
21.	Fruit shape at blossom end	Round, pointed
22.	Fruit shape at base	Round, pointed
23.	Fruit ridge (rib) shape	Pointed, superficial, medium, deep
24.	Fruit skin luster	Matt, glossy
25.	Fruit skin softness of marketable fruits	Mild, soft, hard
26.	Flesh texture	Smooth, spongy, fibrous, gelatinous
27.	Flesh colour	White, cream

(*Source:* PPV & FRA)

Table 2. Characteristics of sponge gourd

1.	Cotyledon-intensity of green colour	Light, medium, dark
2.	Stem shape	Rounded, angular
3.	Stem pubescence	Absent, present
4.	Leaf margin	Entire, serrate
5.	Leaf shape	Ovate, orbicular, reniform
6.	Leaf length (cm)	Small (<10), medium (10.1 to 15.0), long (>15)
7.	Leaf width (cm)	Small (<15), medium (15.1 to 20), large (>20)
8.	Leaf pubescence	Absent, present
9.	Leaf blade number of lobes	3, 5, >5 lobes
10.	Stem length of internode (cm)	Short (<10), medium (10.1 to 15), long (>15)
11.	Stem number of primary branches	Less(<3), medium (5), more (>7)
12.	Petiole length (cm)	Short (<0), medium (8.1 to 10) long(>10)
13.	Flower colour	White, light yellow, yellow
14.	Ovary length at anthesis (cm)	Short (<4), medium (4.1 to 6), long (>6)
15.	Peduncle length (cm)	Short (<8), medium (8.1 to 10), long (>10)
16.	Fruit length (cm)	Short (<20), medium (20.1 to 30), long (>30)
17.	Fruit girth (cm)	Short (<10), medium (10.1 to 14.0), large (>14)
18.	Fruit shape	Oblong, elongate, elliptical
19.	Fruit skin colour	Light green, green, dark green
20.	Fruit shape at blossom end	Rounded, pointed
21.	Fruit shape at base	Rounded, pointed
22.	Fruit ridge (rib) shape	Superficial, medium, deep
23.	Continuity of ridge	Continuous, discontinuous
24.	Fruit skin luster	Matt, glossy
25.	Skin softness of marketable fruits	Mild soft, soft, hard
26.	Flesh texture	Smooth, soft/ spongy, fibrous gelatinous
27.	Flesh colour	White, creamy
28.	Time of marketable maturity of fruits (days)	Early (<60), medium (60.1-70), late (>70)
29.	Plant growth habit (m)	Short viny (3.5), medium viny (3.5-5.5), longviny (>5.5)
30.	Seediness (number of seeds / per fruit)	Low (<100), medium (101 to 200), high (>200)
31.	Seed length (cm)	Small (<1), medium (1.1 to 1.2), large (>1.2)
32.	Seed colour of testa	Black, grey
33.	Seed luster	Matt, glossy
34.	100 seed weight (g)	Low (<10), medium (10.1-14.0), high (>14)

Source: PPV & FRA (2014)

In the next generation, 100 selected progeny rows are raised in 2 replications. Isolation of 1600m. is provided. In each replication, each progeny is grown in 2 row plots, 20 plants on a row. All ideal crop production and plant protection practices are followed. Ethrel @ 250 ppm. is applied as spray for four times at weekly interval commencing from 15 DAS. The progeny rows are monitored for true to type breeding, genetic uniformity and stability of the vine, foliage, flower and fruit characteristics. Off-types and deviant type plants and disease infected plants are uprooted as and when they occur. Plants around them are uprooted. Plants showing poor vigour and low fruit set are culled out. Based on the yield and fruit size and quality traits, 20 to 25 superior true to type progenies showing performance above mean + 2SE are selected. Fruits are collected from individual plants selected in each identified progeny. Fruits harvested at marketable stage are evaluated in the laboratory for non bitterness, uniformity, thick flesh, taste, cylindrical shape and slow maturity. Progenies are selected after evaluating the fruit traits and seed features. Seeds are separated from fruits of individual plants and bulked for utilizing as nucleus seed. This cycle is repeated once in 2 to 3 years.

Breerder seed production: Seeds from nucleus seed source are pregerminated and fast germinating seeds are sown in pits of 30 cm x 30 cm x 30 cm size at 2.0m x 2.0m spacing. After seeds have geminated, only one vigorous seedling is retained in each pit. Isolation distance of 1600m. is provided. All recommended agronomic and horticultural production practices are applied to the crop. Trailing support system at 2m. height is provided. Ethrel 250 ppm. is administered four times commencing from 15 DAS at weekly intervals. Individual plants are continuously monitored for trueness, uniformity in plant expression and in days to first flowering, vine, flower and fruit characters besides number of female flowers production, slow fruit maturity and high fruit set in each plant. Very early and late flowering plants are removed. Off-types and deviant plants and disease infected plants are uprooted as and when they occur and plants around them are also uprooted. Plants showing poor vigour and low fruit set are culled out. At mid maturity, visit of Monitoring Team is organized to adjudge field stand and genetic purity. At maturity, well developed 3 to 4 fruits are harvested from each vine and allowed to ripen. After laboratory examination, true to type fruits alone are selected. Seeds from selected individual fruits are separated, dried and after verifying for the conformity to the description, seeds are selected, graded and pooled together to make breeder seed lot. Samples of seeds is labelled with the Breeder seed label. Grow out test is conducted with breeder seed sample to test the genetic purity. Nutritional quality of the fresh vegetable marketable fruits are estimated.

81. Snakegourd

(*Trichosanthes cucumerina* L.) (2n:22)

Family : Cucurbitaceae

Plants are monoecious, vines are annual carrying branched tendrils. Leaves glabrous, 5 to 7 lobed. Flowers are yellow coloured. Staminate flowers are in racemes, female flowers are solitary. Anthers are three in number. Pistil is with three placenta and many seeds. Highly cross pollinated crop.

Nucleus seed production: Five hundred well developed seeds are sourced from the Breeder seed resource or maintenance breeding stock. Pregerminated 300 vigorously sprouting seeds are selected and planted in pits in rows with 2 m spacing between plants and 2 m between rows. Only one plant is retained in each pit 15 days after sowing. Plants are trained on trelley support at 2 m in height. The plot is isolated by 1600m from other snakegourd crops. Ethrel @ 250 ppm spray is applied four times at weekly interval commencing from periodical observations at different stages of crop growth and development are made on vine, foliage, days to first flower, flower colour and size and fruit length and thickness, fruit colour and other fruit characteristics. Uniformity in flowering is maintained by removing very early and late flowering plants. Plants not conforming to the description are uprooted as and when noticed. At flowering stage, 100 single plants that are vigorous, healthy and true to type are identified, selected and self pollinated. The female flowers and 2 to 3 male flowers that will open next day are selected in the same plant and covered by paper bags. The next day, the male flowers is the same vine coverd by papper bags are removed and the dehiscing anthers are gently rubbed to profusely release the pollen on to the stigma on the female flower. At maturity of the crop, all the single plants are assessed for the vigour, disease free nature and for productivity. From among them, 50 superior single plants that are self fertilized are selected and fruits harvested. Laboratory examination is made for fruit and seed characteristics.

In the next generation, 50 plant progenies are raised in 2 row plots, each row with 10 plants and in 2 replications with a spacing of 2.0 m x 1.0 m. Single plant in each pit is maintained. The plot is grown with isolation of 1600 m from the same crop. All horticultural practices of crop production and plant protection measures are followed. Ethrel @ 250 ppm. spraying was done four times at weekly interval commencing from 15 days after germination. Observation is made on different characteristics at periodical intervals at different stages of crop growth upto fruiting. True to type breeding nature, genetic uniformity and stability of the progenies are recorded. Progeny rows showing deviations are uprooted. True breeding progeny rows alone are considered. Yield data and productivity are recorded. Progeny rows having values above mean + 2SE are identified. Superior performing 20 to 30 plants in the identified progenies are selected and fruits are harvested and seeds separated and examined in the laboratory. The seeds from the superior plants are pooled together to constitute the nucleus seed. Once in 2 to 3 years, the nucleus seed development cycle need to be repeated for initiating fresh nucleus to breeder seed chain.

Breeder seed production: Pregerminated seeds from nucleus seed are sown in pits of 30 cm x 30 cm x 30 cm size at 2m x 2m spacing. At 15 to 20 days after sowing, only one vigorous growing seedling is maintained in each pit. All good agronomic and horticultural production practices and training on staking support at 2m. in height are followed. The breeder seed production field is isolated by 1600m. from other fields with snakegourd. Ethrel 250 ppm. spray is done four times commencting from 15 DAS at 7 to 10 days interval. Examination of individual plants is made at seedling, juvenile, preflowering. flowering and fruit setting stages for expression of true to type nature, uniformity and stability of the characteristics of stem , foliage, tendril, inflorescence, flower, fruits and fruit ripening. Very early and late flowering plants are removed to maintain flowering uniformity. Weak slow growing and diseases affected plants, off-types and plants showing deviating characteristics are uprooted. At mid maturity, Monitoring Team visit is arranged to visti the breeder seed production plot to adjudge field stand and genetic purity. From the selected superior plants, 3 to 4 fruits showing ripening are harvested from each plant and allowed to ripe fully. Seeds are separated, dried and pooled to constitute breeder seed. Samples of breeder seed are tested for seed quality parameters and the quality assured breeder seed is labelled. Seed samples drawn from breeder seed is tested in grow out test to adjudge genetic purity. The market stage fruits are tested for the nutrient contents and other qualities.

Table 1. Characteristics of snake gourd *Trichosanthes cucumerina.* L

1.	Stem shape	Round, angular
2.	Leaf shape	Auriculate, cordate
3.	Leaf margin	Entire, undulated
4.	Stem: number of secondary branches upto 20thnode	Few (<5), medium (5 to 9), many (>9)
5.	Node number of which first female flower appear on the main stem (indicate earliness)	Earliness <12 th node, medium maturity12-15th node, late maturity >15 th node
6.	Stem: intensity of pubescence	Sparse, dense
7.	Leaf blade length (cm.)	Small (>9), medium (10-12), long (>15)
8.	Blade width (cm.)	Narrow (<6), medium (6-9), broad (>9)

9. Leaf lobe	Present, absent
10. Leaf blade depth of lobing	Shallow, medium, deep
11. Petiole length (cm)	Short (<2), medium (2-4), long (>4)
12. Flower sex type	Dioecious, gynomonoecious
13. Fruit	Peduncle attachment- soft, hard
14. Fruit shape	Cylindrical, spindle, elongate spindle, taperning
15. Fruit skin primary colour	Light green, green, dark green
16. Fruit surface colour pattern	Uniform mottled striped
17. Fruit glossiness	Non glossy, glossy
18. Fruit length (cm)	Small (<25), medium (40-50), long (>50)
19. Fruit diameter at widest portion (cm)	Small (<4), large (>4)
20. Fruit size (l/b ratio)	Small (<1.5), medium (1.5-2.5), large (>2.5)
21. Fruit shape at apex blossom end	Depressed, rounded, pointed
22. Plant vine length (m)	Short (<3), medium (3-5), high (>5)
23. Number of seeds per fruit	Less than <10, less 10 to 15, medium (15-20), many (>20)

82. Sweet Potato

(*Ipomoea batatas* (L.) Lam) (2n:90)

Family : Convolvulaceae

Sweet potato is self compatible but many cross compatible cultivars are available. Herbs or shrubs, often twining, sometimes prostrate, Leaves petiolate, entire, lobed, or divided. Inflorescences mostly axillary, cymose, one to many flowered, rarely paniculate; bracts various. Flowers borne on solitary or in cymose inflorescence that are vertically raising from leaf axils. Flowers small to large. Sepals 5 persistent, equal to unequal,. Corolla variously colored, rarely yellow, funnel form, campanulate 5-lobed to entire, Stamens 5 included or exserted; filaments filiform, usually unequal, anthers ovate or linear, longitudinally dehiscent, not twisted; pollen globular, pantoporate, finely spiny. Ovary 2 to 4-loculed, 4 or 6-ovuled. Style one, filiform, included or exserted; stigma capitate or 2- or 3-globulose. Capsule globose or ovoid, 4- or 6-valved. Seeds 4 to 6 or fewer, glabrous or pubescent. Flowers open in the early morning hours (4am. to 5am.) and close and wilt before noon. Insects aid in cross pollination. Fruit is a dehiscent capsule. Tubers are close to the base of the main stem or at the nodes of the trailing stems. Sweet potato is usually propagated by terminal vine cuttings. (80000 per ha). Each cutting is 10 to 15 cm. long with 2 to 3 nodes. The cuttings are collected from matured vines aged 3 months and above. The characteristics are given in Table 1.

Table 1. Characteristics of sweet potato

1.	Plant growth habit	Semi erect, spreading
2.	Vine internode length (cm)	Short (<5), medium (5-10), long (>10)
3.	Vine Pigmentation	Yellow green, green with purple spot, mostly purple
4.	Vine tip pubescence	Sparse, moderate, dense
5.	Mature leaf shape	Round, reniform, cordate, triangular, hastate, lobed, almost divided
6.	Leaf lobe type	No lateral lobes, very slightly teeth, moderate teeth, deep teeth
7.	Shape of central leaf lobe	Triangular, semicircle, elliptic, lanceolate, oblanceolate, linear
8.	Mature leaf size (cm)	Small (<8), medium (8-15), large (>15)
9.	Abaxial leaf vein pigmentation	Absent, medium, strong
10.	Immature leaf colour	Green, green with purple, slightly purple, mostly purple
11.	Mature leaf colour	Yellow green, green, green with purple, purple
12.	Petiole length (cm)	Very short (<10), short (10-20), medium (21-30), long (>31)
13.	Petiole colour	Green, green with purple, purple
14.	Presence of flower	Absent, present
15.	Flower colour	White, pale purple, purple
16.	Flower length (cm)	Short (<3), medium (3-5), long (>5)
17.	Shape of flower limb	Semi stellate, pentagonal, round
18.	Colour of stigma	White, purple
19.	Stigma exsertion	Embedded, same height as anther, exserted
20.	Length of peduncle (cm)	Short (<8), medium (8-15), long (>15)
21.	Tuber shape	Rounded ovate, elliptic, obovate, oblong, irregular
22.	Tuber cortex thickness (mm)	Thin (<1.5), medium (1.5-3.0), thick (>3)
23.	Tuber skin colour	Cream, Yellow, light pink, purple, dark purple.
24.	Tuber flesh colour	White, cream, yellow, orange, dark orange, strongly pigmented (purple, violet)
25.	Tuber latex	Absent, present

(*Source:* PPV & FRA)

Nucleus seed production: Source 4000 terminal cuttings from the matured veins of the variety concerned from the maintenance breeding block. Each terminal cutting is from a single individual plant. Plant the cuttings in a well prepared field previously not grown with sweet potato crop in the past 2 years. This field is isolated by 20 m. from other sweet potato crops. Alternatively source 4000 tubers of 120 to 150 g. size and plant them in 60 cm x 20 cm spacing in a well prepared primary nursery (100 sq.m.). Tubers are pretreated with fungicide and insecticide. After 45 days, vines emerging from the tubers can be cut at a length of 20 to 30 cm and with 3 to 5 nodes for further multiplication in the secondary nursery for tubers production. The tubers used for planting in the primary and secondary nursery are from single individual plants and represent single plant progenies. At the time of harvest, the single plants raised by cuttings or the single plants raised from tubers are critically examined for the sweet potato varietal characteristics, true to type nature, uniformity and stability of traits expression. All the deviant plants are uprooted and discarded. After the harvest, the tuber characteristics were examined in the laboratory and selection made for true to type.

In the next season, the selected individual tubers along with the vines are retained and carried forward for second year cycle. The field is not grown to sweet potato in the past two years and is isolated by 20m from other sweet potato crops. In this second year generation, either tuber progeny or terminal vine cuttings progeny is chosen and single plant progenies are raised in progeny plots. Periodical observations are made for the uniformity, true to type nature and vigour. All deviant types of plants along with the rest of the plants in the single plant progenies are uprooted. At the time of crop maturity, selected plants within the uniform looking single plant progenies are harvested. After harvest, laboratory examination is made for tuber features for final selection.

In the following season, the selected plants are planted with vine cuttings in two rows plots as plant progenies replicated twice. The field is not grown to sweet potato in the past two years and is isolated by 40m from other sweet potato crops. Systematic and periodical examination of all the plants in each progeny are made for true to type, uniformity and stability. Deviant plants are removed along with the rest of them in that plant progeny. The retained plant progenies in the field are harvested and evaluated for tuber yield. The progenies showing performance above mean + 2SE are identified and their tubers are examined in the laboratory for the size, shape, colour, uniformity flesh characters and other prescribed features and selected. Nutritional and starch contents are estimated in the samples. After satisfying true to type nature and uniformity, the tubers and corresponding vines of superior plants are bulked to constitute clonal nucleus seed stock.

Breeder Seed Production: Single tubers and the corresponding vines are selected from nucleus clonal stock and verified for conformity and vines are utilized for planting in the clonal breeder seed production. All good agronomical practices are adapted to raise healthy and pests and diseases free sweet potato crop. The field is not grown to sweet potato in the past two years and is isolated by 20m from other sweet potato crops. Individual plants are verified for conformity for all the traits of vine and tubers and deviating plants are uprooted along with the tubers. The vines are lifted and turned on 50th and 75th day after planting but before earthing up to prevent root formation at nodes and to form the originally formed roots larger. Ethrel at 250 ppm is sprayed five times at 15 days interval commencing 15 days after planting. Monitoring Team visit is arranged before harvest. Pests and disease affected vines and tubers are removed and destroyed. After harvest, the tubers are verified for conformity to trueness, uniformity and stability. Biochemical and nutritive contents are analysed in the samples and starch content is also assessed. After laboratory examination is done, the selected vines are bulked and so also the tubers and used as the breeder seed source stock. Tubers are examined for nutritive qualities.

Aeroponics technique is also adopted for sweet potato. Foliar cuttings from the foliage of selected mature plants and shoots from the storage roots are used. Vines are dipped in monochrotophos to ensure pest free planting material.

In china 30% increased yield is obtained from disease and pest free quality clones. In UC, Davis, clean seed production and maintenance of disease free seed stocks are accomplished by heat treating and sterilizing the disease free shoot tips and using them in micropropagation. *In vitro* shoots rooted in sand beds are transplanted in main field to establish genetic purity, true to type, stability and disease and pest free stock that are to be released for seed purpose.

The following protocols are suggested in Horticultural Technology (Dec, 2010)

1. Virus testing
2. G1 Seeds grown after year 1 (Virus testing)
3. G1, G2, G3,
4. VT (virus tested) seeds

83. Tapioca (Cassava)

(*Manihot esculenta* Grantz) (2n:36)

Family : Euphorbiaceae

Tapioca is a perennial shrub with tuberous roots. Stem glabrous and variously branched. Leaves large, palmate, 5 to 9 deep lobed. Leaf life duration is longer. The main stem or branches terminate into an inflorescence. Tapioca is a monoecious plant. The female flowers are located on the lower part of the inflorescence and are fewer in number than male flowers which are numerous on the upper part of the inflorescence. On the same inflorescence, the female flowers open 1 to 2 to weeks before the male flowers (protogyny). Cross pollination is common. In female flowers, sepals are 5 and united and petals absent. In male flowers, ten stamens are with free filaments arranged in 2 whorls of five stamens in each whorl. In female flowers, ovary is mounted on glandular disc. Locules three, each locule with single ovule. In each inflorescence, female flowers open first and male flowers open 8 to 10 days later. Seed mature 3 to 5 months after pollination. Swollen tap root is the economic edible part in cassava. The characteristics of cassava are given in Table 1.

Table 1. Characteristics of Cassava

1.	Pubescence on apical leaves	Glabrous, medium, high
2.	Colour of first fully expanded leaf	Light green, dark green, purple green, purple
3.	Predominant shape of central leaf lobe	Linear, lanceolate elliptic, ovoid, lanceolate elliptic
4.	Petiole colour	Yellow, green, reddish cream, reddish green,greenish red, red, purple
5.	Mature leaf colour (dorsal)	Light green, dark green, purple green
6.	Predominant number of leaf lobes	Three, five, seven, nine
7.	Length of central lobe (cm)	Short (<10), medium (10-15), high (>15)
8.	Width of central lobes (cm)	Narrow (<2), medium (2 to 3), broad (>3)
9.	Leaf lobe margin	Smooth, winding
10.	Petiole length (cm)	Short (<25), medium (25-35), long (>25)
11.	Leaf vein colour	Green, partially reddish, red
12.	Petiole orientation	Included upward, horizontal, inclined downward, irregular
13.	Young stem colour at top 20 cm	Light green, dark green, greenish red, greenish purple
14.	Prominence of foliage scar	Semi prominent, prominent, highly prominent
15.	Colour of mature stem	Silver grey, orange, light brown, dark brown, golden red
16.	Stem distance between leaf scars (cm)	Short (<3), medium (3-6), long (>6)
17.	Growth habit of stem	Zig zag, straight
18.	Plant type	Erect, semi spreading, spreading
19.	Plant height (cm)	Short (<100), medium (100-200), tall (>200)
20.	Plant branching habit	Single, dichotomous, tri-dichotomous, tetrachotomous
21.	Plant canopy	Compact, open
22.	Fruiting	Absent, present
23.	Crop maturity (months)	Early (<6), medium (6-8), late (>8)
24.	Tuber peducde	Absent, present
25.	Tuber shape	Conical, conical cylindrical
26.	Tuber peelability	Easy, medium, difficult
27.	Tuber skin colour	Cream, light brown, red brown, dark brown
28.	Tuber rind colour	Cream, cream with pink, patches, yellow, pink
29.	Tuber flesh colour	Purple, white, cream, light yellow, dark yellow
30.	Mealiness of cooked tuber	Fairly mealy, moderate mealy, very mealy

(*Source:* PPV & FRA)

Propagation is by vegetative setts taken from middle portion of the mature stem. Setts 25 to 30 cm long and 2 to 3 cm thick are field planted for getting better field establishment, better yields and proper tuber size. Virus (SMD) elimination can be achieved by micropropagation. The *in vitro* plants are tested for virus free nature, harvested and planted in greenhouse before field planting. Usually slow multiplication rate and virus disease infection are the limiting factors in tapioca. Clonal multiplication after virus elimination or virus free breeder stock for at least 4 to 5 successive generations will be very effective in Basic or breeder seed clones establishment in tapioca. Single budded/single node cutting can be established in polybags, pretreated with fungicide/insecticide and then transplanted in locations free from the virus infections. Control of whitefly vectors and adapting IPM measures will be helpful in preventing the spread of virus infection.

Nucleus Seed Production: Individual plants that are healthy are selected from disease free source plants or from the disease free maintenance plot and 2000 to 2500 setts are prepared by selecting single sett per plant. Each sett is pretreated with fungicide and planted in 90 cm x 90 cm spacing in plots isolated 20 m away from other tapioca growing fields. Periodical observations are made to verify the true to type and genetic purity of individual plants. Off-types and deviant type of plants are uprooted The main stem alone is retained and branches are trimmed at maturity and tubers are harvested and examined for productivity and the true to type description nature. Rigorous selection is made for the typical individual plants. As many setts as possible are prepared from the middle portion of the main stem of each selected plant after discarding the basal and top portions of the stem.

In the next generation, setts from each selected plants are planted as progenies in progeny plots isolated by 20m. distance from other cassava crops. Continuous observations are made for trueness and uniformity within each plant progeny and between progenies. Deviant type plants noticed are uprooted along with the rest of the plants in the progeny showing deviant type plants. Selected progeny rows that are true to type for foliage and stem characteristics are harvested with the tuberous roots in tact. The tuberous edible roots of individual plants are examined for conformity, trueness and for flesh colour and starch content. Based on the characteristics of foliage and tubers, the plants are selected and from each such progeny setts are prepared.

In the following generation, setts from each of the selected progenies are planted in two rows progeny plots for evaluation of genetic purity, true to type nature and stability. Progeny rows showing deviations for the prescribed traits are discarded as a whole. Progeny rows are identified and selected based on the productivity above mean. Setts are prepared from the selected plants in each progeny.

Tuber characteristics are compared. Starch content and other tuber quality aspects were finally evaluated. Progenies that are fulfilling all the standard description and yielding higher than mean are finally selected. Stem cuttings from all such plus plants are pooled together and the clonal seeds of nucleus seed stock are thus constituted.

Breeder Seed Production: Setts prepared from the clonal progenies in nucleus seed stage stock are planted with 90 cm x 90 cm spacing in well prepared field that is isolated by 20 m from other fields being grown with tapioca crop. High degree of field and crop sanitation are maintained to raise a clean crop. Whitefly vector control and IPM practices are regularly done. Individual plants are verified and selected for free from virus infection and conformity of traits of main stem and leaf and uniform maturity and free from pests and diseases. After the harvest with the tubers in tact, the tubers are examined for trueness and for starch content. Cyanide content (CG) content and physical deterioration in post harvest life are estimated and plants showing desirable expressions are finally identified. Monitoring Team visit is organized. The setts prepared from the main stems of the selected plants are bulked to constitute breeder seed stock. Based on the requirement, another 1 to 2 stages of multiplication of breeder seed can be followed. It is reported that around 25 to 30% increased yield is obtained by this process. GOT is conducted to verify /adjudge the genetic purity:

All cassava organs, except seeds, contain cyanogenic glucoside (CG). Cultivars with <100 mg/kg fresh weight are called 'sweet' while cultivars with 100 to 500 mg/kg are 'bitter' Juice extraction, heating, fermentation, drying or a combination of these processing treatments aid in reducing the HCN concentration to safe levels. Cassava roots have the shortest postharvest life of any of the major root crops (Ghosh *et al.*, 1988). Roots are highly perishable and usually become inedible within 24 to 72 h after harvest due to a rapid physiological deterioration process, in which synthesis of simple phenolic compounds that polymerize occurs, forming blue, brown and black pigments (condensed tannins). It is suggested that polyphenolic compounds in the roots oxidize to form coloured pigments that are deposited in the vascular bundles (Ghosh *et al.*, 1988).

84. Taro (Sembua) (Aravi)

Colacasia esculenta L. (2n:24)

Var. *esculenta*

Var. *antiquorum*

Var. *stoloniferum*

Family : Araceae

An upright shrub, leaves are erect or spreading, green to dark green, borne on sturdy grooved petiole. Suckering is moderate to high. Small suckers are removed from the mother plant. The crop is raised by planting 25-30 g size cormels in Feb- March and June-July in spacing of 45 cm in furrows. Earthing up and desuckering are done allowing 3 suckers per plant. Before harvesting all suckers are wrapped up around the base and earthed up to arrest sprouting.

The plant matures and harvesting is done in 6 to 8 months after planting by uplifting the plants. Cormels are separated from mother cormels. The yield obtained is 8 to 10 t/ha in 180 days. The characteristics of taro are presented in Table 1.

Table 1. Characteristics of Colocasia

1.	Plant height	Dwarf, medium, tall
2.	Growth habit	Dwarf, medium, tall
3.	Plant type	Erect, intermediate, spreading
4.	Postion of leaf	Cup shaped, erect, apex downwards
5.	Leaf number	None, few, many
6.	Leaf length	Small, medium, large
7.	Leaf colour	Light green, green , dark green
8.	Leaf vein colour	White, orange, green, pink, red, brown
9.	Leaf vein pattern	V type, Y type
10.	Leaf blade colour	White, yellow, green, dark green, pink, red, brown
11.	Leaf blade margin	Sinuate, undulate
12.	Leaf tip shape	Pointed, slightly pointed, round
13.	Petiole length (cm)	Small (<16), medium (16-30), long (>30)
14.	Leaf sheath length (cm)	Low (<10), medium (10-20), long (>20)
15.	Number of stolons	Absent, few (1-5), intermediate (6-10), many (>10)
16.	Number of suckers	Absent, low, medium, high
17.	Corm shape	Conical, round, cylindrical, ellipsoid, dumbled, elongated,flat, spindle shaped
18.	Corm length (cm)	Short (<8), medium (9-12), high (>12)
19.	Corm skin surface	Smooth, fibrous, scaly
20.	Corm skin colour	Light brown, brown, dark brown
21.	Corm branching	Clustered, dispersed, branched, unbranched
22.	Number of secondary corms	Few, intermediate, many
23.	Sprouting from side corm	Absent, intermediate, rare
24.	Corm weight (g)	Low (<100), medium (100-250), high (>250)
25.	Corm flesh colour	White, yellow, cream, orange, pink, pinkish red, purple, other
26.	Number of cormels	Low (5-10), medium (10-15), high (>15)
27.	Corm storability	Low, intermediate, high
28.	Edibility of petiole, leaves cooked corms and cormels	All are edible
29.	Taste	Sweet, light, salty, tasteless
30.	Softness	Very soft, soft, medium hard, hard, very hard

85. Tomato

(*Lycopersicon esculentum Miller*) (2n=24)

Family : Solanaccae

Tomato is a highly self pollinated crop. It is a herbaceous annual with determinate to indeterminate growth habit. Leaves alternate variously shallow lobed. Glands sparsely distributed. Stem roundish to angular. The inflorescence is compound type with 4 to 8 bisexual flowers. Sepals 5, petals 5 and anthers 5 forming cone surrounding the stigma leading to self pollination when anthers dehisce on the days of anthesis. The descriptors are given in Table 1.

Table 1. Characteristics of Tomato

1.	Seedling- Anthocyanincolouration of seedling	Absent, present
2.	Leaf: intensity of green colour	Light, medium, dark
3.	Plant growth type	Determinate, indeterminate
4.	Stem- Pubescence	Absent, present
5.	Stem- anthocyanin colouration of upper portion	Absent, weak, medium, strong
6.	Stem length- length of internode between first and fourth inflorescence (in cm) for indeterminate varieties	Short (<25), medium (25-40), long (>40)
7.	Stem: length of inter node between first and fourth inflorescence (in cm) for determinate vars	Short (<20), medium (20-30), long (>30)
8.	Leaf length (cm)	Short (25<), medium (25-30), long (>30)
9.	Leaflet- length (cm)	Short (<5), medium (5-10), long (>10)
10.	Leaf width (cm)	Short (<15), medium (15-20), long (>20)
11.	Leaflet width (cm)	Narrow (<4), medium (4 to 6), broad (>6)
12.	Leaflet – Serration	Absent, less, highly serrated
13.	Leaflet Structure	Open, intermediate closed
14.	Leaf – Attitude of petioles of leaflets	Semi erect, horizontal semi drooping
15.	Leaf attitude in relation to mainstem	Semierect, horizontal drooping
16.	Inflorescence type (second and third truss)	Uniparous, intermediate, multiparous
17.	Plant – number of inflorescenceson mainstem	Few (<4), medium (4 to 8), many (>8)
18.	Flower fasciation (first flower of inflorescence)	Absent, present
19.	Flower- pubescence of style	Absent, present
20.	Flower colour	Yellow, orange
21.	Flower-anther colour	Green, yellow
22.	Flower-stigma nature	Non exserted, exserted
23.	Flower stigma	Unilobe, bilobe, multilobe
24.	Flower calyx size (cm)	Short (<1), medium (1 to 1.5), long (>1.5)
25.	Peduncle-abscission layer	Absent (jointless), present (Jointed)
26.	Time of flowering (days)	Early (<65), medium (65 to 80), late (>80)
27.	Fruit-intensity of green colour (before maturity)	Light, medium, dark
28.	Fruit-green shoulder (before maturity)	Absent, present

29.	Fruit size (g)	Very small (<100), small (100-200), medium (201-700), large (701-1000), very large (>1000)
30.	Fruit length (cm)	Very short (3.0), small (3 to 5), medium (5 to 7), large (7-9),Very large (>9)
31.	Fruit width (cm)	Very short (<3), small (3 to 5), medium (5 to 7), large (7-9),Very large (>9)
32.	Fruit shape (L.S)	Flattened, slightly flat, circular, rectangular, cylindrical, heartshape, obovoid, ovoid, pear shaped
33.	Fruit ribbing at peduncle end	Absent, weak, medium, strong
34.	Fruit cross section	Not round, round
35.	Fruit depression at peduncle end	Absent, shallow, medium deep
36.	Fruit – size of scar around peduncle	Small, medium, large
37.	Fruit size of blossom scar	Small, medium, large
38.	Fruit shape at blossom end	Intented, intented to flat, flat, flat to pointed, pointed
39.	Fruit size of core in cross section	Small, medium, large
40.	Fruit-pericarp thickness (cm)	Thin (<0.3), medium (0.3-0.6), thick (>0.6)
41.	Fruit- number of locules	2, 3 to 4, >4
42.	Fruit colour at maturity	Yellow, orange, pink, red
43.	Fruit colour of flesh	Yellow, orange, pink, red
44.	Fruit firmness (kg/cm^2)	Soft (<3), medium (3-6), firm (>6)
45.	Time of maturity (from seed sowing) (days)	Early (110), medium(110 to 130), late (>130)
46.	Fruit total soluble solids (Brix)	Low (<3), medium (3 to 4), high (4.1-5.0), very high (>5)

(*Source:* PPV & FRA (2009)

Nucleus Seed Production: Seeds from breeder stock source population are sown in seed trays and 25 days old healthy seedlings are planted in well prepared field that is isolated by 50 m. from other tomato fields. Five hundred single plants are transplanted. All good agronomical, horticultural and plant protection practices are adopted. Individual plants are critically examined for true to type nature and uniformity with other plants. The individual plants are examined from seedlings to fruit maturity stage for plant type, foliage, inflorescence, flower, pedicel, fruit set and uniform maturity besides fruit characteristics such as shape, size, number of fruit locules, firmness, fruit wall thickness and fleshiness. Hundred true to type plants are selected and self fertilized. In the next season, seeds of single plants are sown in single plant nurseries or seedling trays and 50 plants are transplanted in each progeny row. The plot is isolated by 400m. distance from other tomato crops. Single plant is maintained per hill. Monitoring of the progeny rows is done for true breeding, true to type and uniformity from juvenile stage to pre- flowering, flowering and fruit development stages. Progeny rows showing plants with deviations are uprooted as a whole progeny. In the true breeding progenies, 100 superior plants are selected and self fertilized Fig. 1.

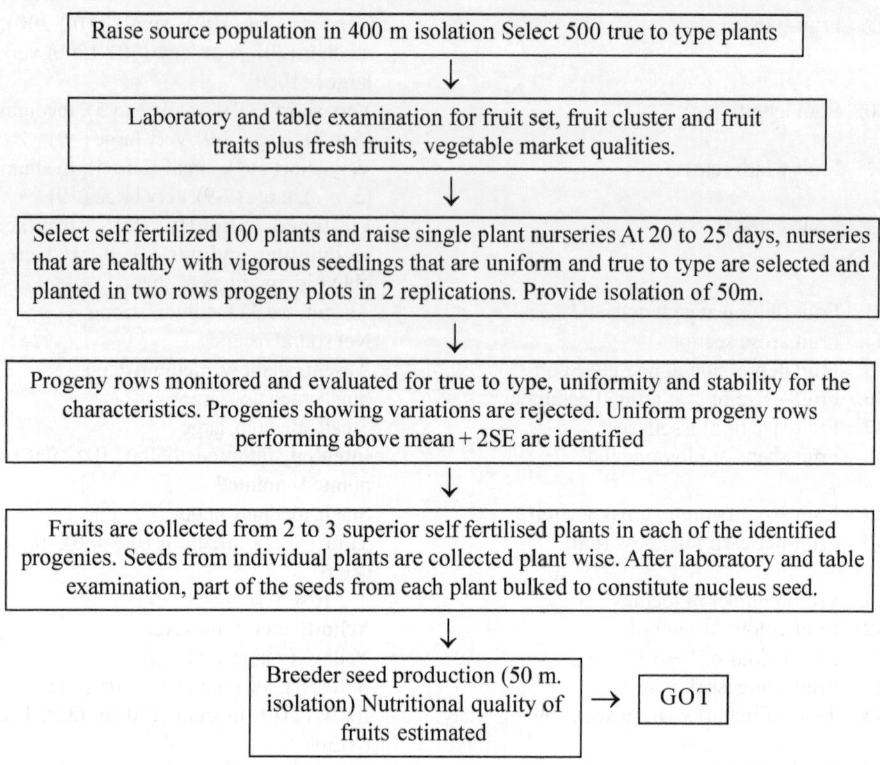

Fig. 1. Nucleus and Breeder Seed Production in Tomato

Seeds from 100 selected self fertilized plants are raised in small plot nursery and 25 days old healthy seedlings in each progeny are transplanted in two rows plots with 2 replications and 25 plants in each replication. Isolation distance of 50m. is maintained from other tomato crops. The progeny rows are examined critically and progenies showing deviations are rejected. In each selected progeny, 10 plants are selected and self fertilized. Yield and related data like number of flower clusters, fruits per cluster, locules per fruit, fruit weight, seediness, taste, flavor ,juiciness, shelf life and other quality aspects such as fleshy or juicy fruits are studied with marketable fresh fruits collected from each progeny row. Based on yield and marketable fruit quality traits, progenies performing above mean +2SE are selected. Seeds from superior plants in the identified progeny rows are pooled together to constitute the nucleus seed of the given tomato variety or inbred parent.

Breeder Seed Production: Breeder seed crop is raised from self fertilized nucleus seeds of the selected plants in nucleus seed plot. Nursery is raised and plants are transplanted in rows and examined for the true breeding, genetically true to type and uniform nature in respect of stem, foliar, flower and fruit related characteristics. Isolation distance of 50m from other tomato crops is ensured. Selected plants are self fertilized. Off-types and deviant type of plants and very early and late flowering plants are removed. Monitoring Team visit is arranged. Seeds from each plant are bulked. Seed samples are tested for seed quality. Quality assured seed is labelled with the Breeder Seed label. Each time, self fertilized seeds from true to type plants should be produced and used as breeder seed as per the requirement. Grow out test is conducted with seed samples drawn from the breeder seed. Nutritional qualities of marketable fruits are assessed/estimated.

86. Turnip

(*Brassica campestris* var. *rapa* L.) (2n:20)

Family : Brassicaceae

Turnip is a cross pollinated crop due to sporophytic system of incompatibility. Usually crosses are made by enclosing male and female flowers heads in nylon net cages after removing opened flowers, if any. Turnip seed production requires two seasons of growing namely seed to root and root to seed. In Turnip, two types, biennial temperate and annual Asiatic races exist. They are two distinct races.

Turnip is a long day plant. Turnip inflorescence is a raceme. Flowers have short pedicel. Flowers are small, pale yellow colour. Sepals 5, petals 5 and persistent, stamens 5, ovary superior, 3 celled, large lobed disc at bottom and stigma is capitate. Fruit is a siliqua. Seeds are blackish or reddish brown. Flowers are protogynous and varying degree of cross pollination takes place. Insects aid in pollen transfer. Anthesis usually takes place in the forenoon from 7.30 am to 11.30 am and then anther dehiscence continues. In turnip, stigma become receptive 2 to 3 days after anthesis. The characteristics of turnip are given in Table 1.

Table 1. Characteristics of Turnip

1.	Leaf attitude	Erect, semi erect, horizontal
2.	Leaf-reflexing of top	Absent, weak, medium, strong, very strong
3.	Leaf green colour	Very light, medium, dark green
4.	Leaf type	Lobed, entire
5.	Leaf number of lobes	Few, medium, many
6.	Leaf margin undulation	Absent, weak, medium, strong, very strong
7.	Leaf length	Short, medium, long
8.	Leaf width	Narrow, medium, broad
9.	Leaf hairiness on upper side	Absent, weak, medium, strong
10.	Leaf anthocyanin colouration	Absent, weak, medium, strong

11.	Root thick cork layer around skin	Absent, present
12.	Root skin colouration	White, green, yellow, orange bronze, red purple, scarlet
13.	Root skin colour-below ground	White, yellow, purple, red
14.	Root flesh colour	White, yellow
15.	Root anthocyanin colour	Absent, present
16.	Root shape (LS)	Narrow elliptic, circular, obovate, square, oblong obtriangular
17.	Root length	Very short, short, medium, long, very long
18.	Root diameter	Small, medium, large
19.	Root position of widest part	Above middle, at middle, below middle
20.	Root curvature of main axis	Absent, present
21.	Root-time of harvesting maturity	Early, medium, late

(*Source:* UPOV)

Seed Production: Isolation required is 1600m for nucleus and breeder seed production.

Seed to seed method: In this method the crop is allowed to over winter in the field and produce seed in the following spring in their original position. For nucleus and breeder seed production, root to seed method is followed. This method consists of seed to root or steckling production and followed by root to seed production. Seeds sourced from nucleus / breeder / institution origin are sown in the last week of August or first week of September. The field is prepared to fine tilth and all good agronomic practices are followed in raising the seed crop in both the stages. After 70 to 80 days of sowing, the roots are well developed and lifted up without damage. During the seed germination to root harvest, care is taken to verify individual plants for true to type nature and plants showing off-type and precocious bolting are rogued out.

The uprooted Turnip roots are critically examined for size, shape, colour, skin surface, uniform filling and free from pithiness, shape of root tail and only true to type roots are selected. Off-types, a typical and pithier roots are discarded. The top of the selected roots are clipped to 2/3 and the root tails are primed. The stecklings thus prepared are replanted or reset in the well prepared field in 60cm x 30cm spacing. Stecklings are planted in December second half. Stecklings putforth vegetative shoots and then bolt. Every individual plant is verified for trueness to stem, leaf and panicle, flower and seed characteristics. Plants showing deviations are removed. Off-types are culled. The deviating plants, not conforming to true to type that are noticed at the time of flowering are discarded and also the plants surrounding the deviant plants are removed carefully to avoid accidental cross pollination. Early bolting and late bolting plants, plant bearing deviations in flower and pollen colors are also culled out. The disease affected and insect affected plants are discarded. The crop is harvested by cutting the flower stalk at the base when the seed pods show reddish brown color. Seeds are collected from 1000 to 1500 superior individual plants and saved plantwise to raise nucleus seed.

Seed to Root Stage: In the following season, individual plant progeny rows are raised by sowing the seeds. Around 20 days, plants are thinned to retain only one plant per hill. The individual plants in progeny rows are critically examined upto root edible maturity stage for true to type nature. Deviating/ atypical plants are discarded. Off-types are removed. At root maturity, the roots are lifted and carefully examined for trueness in respect of root characteristics in each progeny row. Progeny rows showing deviations are rejected.

Root to seed stage: Roots from superior plants in the identified uniform progeny rows are selected after examination of conformity to true to type nature. Stecklings of roots saved from individual progeny rows are prepared and planted in progeny rows. The shoots and flowers / spikes developed from the stecklings are examined on individual plant basis in each uniform progeny. Deviating plants are discarded. Nonconforming plants for flower and seed characters are rejected. Seeds are collected from the superior plants in each progeny and stored plant wise. Part of the seeds from each plant are bulked together to constitute Nucleus seed.

Breeder Seed Production: Nucleus seed is sown in well prepared field that is isolated by 1600m. from other Turnip growing fields. One plant per hill is retained. Individual nature of plants are monitored for true to type till root maturity. The matured roots are lifted and roots are examined for the characters conforming to the prescribed description. Stecklings are prepared from the selected roots and planted in the following season to produce seeds. All steps are taken to verify the true to type nature of vegetative, floral features and spike features. Roguing of off-types and deviating plants are carried out at different growth phase of the crop. Early bolting and late bolting plants are culled out to get flowering and seed maturity uniformity. Seeds are collected, processed and tested for seed quality assurance. Breeder seed labelling is done. Nutritional qualities of Turnip are estimated.

Fruits

87. Acid Lime, Sweet Orange and Mandarin Orange

Acid lime : (*Citrus aurantifolia* Swingle) (2n:18)

Sweet orange : *Citrus sinensis* (L.) Osbeck

Mandarin orange : *Citrus reticulata* Blanco

Family : Rutaceae

Trees upright or erect, semispreading or spreading. Shoots spiny, leaves alternate, elliptic to ovate; petiole winged or not. Flowers axillary, single or in clusters of 2 to 6 in corymbs. Flowers are mostly sauce shaped. Calyx five, pointed and petals 5, oblong, white. Stamens, many, upto 20 to 25. Anthers yellow coloured. Fruit is a hesperidium modified berry, globate to subglobate to oval or oblate shape. Outer rind of the fruit is differently coloured. Mesocarp is spongy and white. Endocarp (pulp) yellow to varying colour, enclosing 10 to 14 wedge shaped compartments that could be separated to individual segments, each with 2 to 4 seeds that are irregular in shape. Sometimes oranges are seedless too. Citrus is propagated by seeds / seedlings. More often, in many varieties budding onto appropriate root stock is practised. Budded trees are earlier than seedlings planted trees. Budded trees are more uniform in maturity. Plants could be grown in close planting with appropriate pruning and training.

Citrus species is self compatible and self pollinated. Occasionally, in some of the oranges, it is cross pollinated by bees. In addition to zygotic embryony, nucellar embryony is common leading to the production of maternal clonal embryos. Nucellar clonal seeds are used for propagation and multiplication. Nucellar embryos are polyembryonic or multiple viable embryos within each seed. Polyembryony is conducive to very high germination upto 100%.

The characteristics of Sweet orange, Mandarin orange and Acid lime are presented in tables 1, 2 and 3 respectively.

In acid lime (*C. aurantifolia*) nucellar seedlings are utilized for planting and plants commence bearing from third year after planting. In sweet orange (*C. sinenies*) budded plants onto Rangpur lime root stock and in Mandarin orange (*C. reticulata*) budded plants are utilized for planting. Bearing commences from 5th year after planting in sweet orange and in mandarin orange from 3 to 5 years after planting the budded plants and 5 to 7 years after planting the seedlings.

Table 1. Characteristics of sweet orange (*Citrus sinensis*) (Sathgudi, Mosambi)

1.	Tree - growth habit	Erect, semi spreading, spreading
2.	Leaf length (mm.)	Short (<70), medium (70-90), long (>90)
3.	Leaf width (mm.)	Narrow (<40), medium (40-50), broad (>50)
4.	Fruit weight (g.)	Light (150<), medium (150-200), high (>200)
5.	Fruit dia (mm.)	Small (<60), medium (60-70), large (>70)
6.	Fruit length (mm.)	Short (<60), medium (60-70), long (>70)
7.	Shape at fruit base	Convex, truncate, concave
8.	Shape of fruit at apex	Rounded, truncate
9.	Fruit rind colour	Greenish yellow, yellow, dark yellow, orange, dark orange
10.	Number of segments per fruit	<11, >11
11.	Fruit rind thickness (mm.)	Thin (<4), moderate thick (4-5), thick (>5)
12.	Fruit juiciness (%)	Low (<35), medium (35-45), high (>45)
13.	TSS (Brix)	Low (<10), medium (10-12), high (>12)
14.	Titrable acidity (% citric acid)	Low (<0.5), medium (0.5-0.8), high (>0.8)
15.	Number of seeds per fruit	<5, 5-10, >10

(*Source:* PPV&FRA) **Note:** In sweet orange twenty budded plants on region specific root stock plants – Rangpur at - 5th year after planting are observed in two independent identical seasons.

Table 2. Characteristics of mandarin orange (*Citrus reticulata*)

1.	Tree-growth habit	Erect, semi erect, spreading
2.	Length of leaf (mm)	Short (<70), medium (70-80), long (>80)
3.	Leaf width (mm)	Narrow (<30), medium (30-40), broad (>40)
4.	Fruit weight (g)	Light (<110), medium (110-140), heavy (>140)
5.	Fruit dia (mm)	Small(<60), medium (60 – 70), large (>70)
6.	Fruit length (mm)	Short (<55), medium (55 – 65), long (>65)
7.	Shape of fruit at base	Truncate, concave
8.	Shape at fruit apex	Truncate, depressed
9.	Fruit rind colour	Greenish yellow, light orange, orange, dark orange
10.	Fruit rind peelability	Easy, moderate
11.	Fruit rind thickness (mm)	Thin (<2), moderate (2-3), thick (>3)
12.	Fruit juiciness (%)	Low <40, medium (40-45), high (>45)
13.	TSS, (Brix)	Low (<8), medium (8-11), high >11
14.	Titrable acidity (% citric acid)	(<0.5), medium (0.5-0.7) high (>0.7)
15.	Number of seeds per fruit	<5, >5
16.	Boldness of seed in g (weight of 20 seeds)	<1.10; >1.10

(*Source:* PPV & FRA) **Note:** Twenty observations are made in mandarin orange in two independent identical seasons in 5th year after planting. Budded plants on region specific root stock are considered.

Table 3. Characteristics in acid lime (*Citrus aurantifolia*)

1.	Tree-growth habit	:	Erect, semi spreading, spreading
2.	Spine density on adult tree (number of spines on one year old spring shoot of 30 cm length).	:	Low (<10), medium (10-15), high (>15)
3.	Spine length (mm)	:	<5, 5-15, >15
4.	Leaf lamina length (mm)	:	Short (<60), medium (60-70), long (>70)
5.	Leaf lamina width (mm)	:	Narrow (<35), medium (35-40), broad (>40)
6.	Petiole wing	:	Present, absent
7.	Fruit weight (g)	:	Light (40), heavy (>40)
8.	Fruit diameter (mm)	:	Small (<40), medium (40-45), larger (>45)
9.	Fruit length (mm)	:	Short (<40), medium (41-45), long (>45)
10.	Albedo colour	:	Greenish, white, yellow
11.	Fruit axis	:	Solid, hollow
12.	No. of segments per fruit	:	8-10, >10
13.	Fruit rind thickness (mm)	:	Thin (<2), (thick >2)
14.	Fruit juiciness (%)	:	Low (<40), medium (41-50), high (>50)
15.	Total soluble solids (TSS)	:	Brix low (<5), medium (5-6), high (>6)
16.	Titrable acidity (% citric acid)	:	Low (<5), medium (5-6), high (>6)
17.	Seediness-number of seeds per fruit	:	<4, 4-10 , >10

In acids lime, twenty nucellar plants (six months old) are planted. Observations are made in two independent ideal fruiting seasons at 3 to 5 years after planting.
(*Source:* PPV & FRA)

Nucleus Stock Production: In each of the species of Citrus, namely, *Citrus aurantifolia*, *C. sinensis* and *C. reticulata*, in a well established garden, superior performing plants that are true to type, uniform and highly productive are selected and labelled. They are examined for free from diseases and pests. Ten to twenty superior elite plants are selected and labelled and cloned by budded plants on to root stocks following species specific method. At appropriate age of the cloned plants, they are planted adopting crop specific standard good fruit production practices. Minimum of 20 clonal progenies are planted in 2 row progeny plot at rate of 10 clonal plants per row. All around the nucleus seed production plot, two to three border rows are planted with the same source of planting material . The nucleus seed production plot is isolated by 5m. from fields grown with the same species.

Observations are made on individual plants for the prescribed vegetative and floral / fruit characteristics (refer tables 1 to 3) to verify for true to type nature, uniformity and disease and pest free nature. The flowering uniformity and fruit set and fruit production are recorded over 2 to 3 seasons from the commencement of first flowering after planting. Based on the productivity and trueness to fruit and juice characteristics, true to type clonal progenies are identified and selected (at 5 years after planting). The identified plants are tagged and retained as nucleus stock. The unselected plants are uprooted.

Breeder Stock Production: Budwoods or nucellar seedlings as the case may be are selected or produced from selected true to type healthy uniform clonal plants from the nucleus stock plot. They are verified for free from diseases and pests and two to three plants cloned from each of the selected plant from the nucleus plot are planted in Breeder Seed Production plot adopting good horticultural production practices. Two to three border rows are raised with the same source plant material as that of nucleus stock. Isolation distance of 5m. is maintained. True to type conformity with the prescribed characteristics are verified periodically. At flowering and fruiting the trueness is verified. Based on the productivity and the fruit quality over 2 to 3 normal seasons, deviant plants / off-types, if any are culled out. Individual selected plants are labelled and tree record is maintained on single tree basis. High degree of phytosanitary standards are ensured. Single plants in breeder seed plot are verified for virus free status. Budwoods from the superior trees are labelled and distributed as breeder stock for orchards raising foundation stock in mother block. Visit of monitoring Team is arranged at the stage of budwood collection in the Breeder stock plants.

88. Amla (Aonla)

(*Phyllanthus emblica* L.) (2n:98,104)

Family : Euphorbiaceae

Amla, Aonla or Indian gooseberry (synon. *Emblica officinalis* Geartn.) is an indigenous fruit to the Indian subcontinent. AMLA or Aonla tree is small to medium sized deciduous tree, usually grows up to 7.5 m; trunk often crooked and gnarled, up to 35 cm. in diameter. The bark is thin, smooth, grey, peeling in patches, with numerous knobs. Leaves alternate, distichously and densely crowded along lateral twigs, simple and entire, glabrous, sessile; stipules triangular; blade narrowly oblong, 5 to 25 mm long , rounded and more or less oblique at base, acute or obtuse and mucronate at apex. Flowers are unisexual, fascicled in axils of leaves or fallen leaves. The male flowers are numerous at base of young twigs and the female flowers are solitary and higher up the twig; male flowers pedicellate, with 6 pale green perianth lobes 1.5 to 2.5 mm long and 3 entirely connate stamens; female flowers sessile, with 6 somewhat larger perianth lobes, a cup like disk, and a superior 3 celled ovary crowned by 3 styles, connate for more than half of their length and deeply bifid at apex. Fruit a depressed globose drupe up to 4 cm in diameter, pale green changing to yellow when mature and with 3 slightly dehiscent compartments, each usually containing 2 seeds. Seeds trigonous, 4 to 5 mm × 2 to 3 mm. The characteristics are given in Table 1.

Table 1. Characteristics of AMLA

1. Vigour and fast growth
2. Tree size, shape and habit such as erect, semi erect spreading
3. Nature of stem and girth
4. Leaf and leaflets size, shape and number
5. First flowering period and duration (month of flowering)
6. Fruit maturity period and months: early. medium, late
7. Fruit setting extent percentage: heavy, prolific regular
8. Fruit shape: Round, oval, oblong, flattened oval, oval round, flattened round, triangular
9. Fruit size: large, medium, small, 40 g. and less. 40 to 50 g., more than 50g.
10. Fruit colour: greenish, yellow,
11. Fruit shining, smooth skin
12. Fruit segments: 6 or 8 or 6 to 8
13. Fruit fiber content: fibrous, less fibrous, fleshy, pulpy, pickling, juice, other products making
14. Fruit usefulness/suitability for medicinal purposes
15. Fruit free from necrosis
16. Polleniser from other varieties needed or not
17. Consumer appeal for fresh fruit market, storability
18. Seed size and shape
19. Suitability for high density planting (HDP)

Source: SHODHGANGA and FRUITPEDIA

Amla is raised through seeds. However ,there is prolonged juvenility and wide variability in the seedlings. It need to be propagated through budding which is an efficient and an ideal method of true to type propagation in Amla. A high percentage of rooting (84%) has been reported from semi hard wood cuttings collected from the middle portions of vigorous shoots of young trees and planted in beds. The tree coppices well and pollards moderately well. In its natural distribution area *Phyllanthus emblica* is a light demanding species. It is photosensitive, producing flowers only at a day length between 12 and 13.5 hours. It occurs from almost sea level up to 1500 m altitude. The tree is slightly tolerant of alkaline soils and rather drought resistant, but some cultivars are sensitive to drought and frost. It is fire resistant and is one of the first trees to recover after a fire.

Amla is a deciduous tree and the emergence of new shoots starts in the beginning of April. Leaf fall is in November-December, flowering in March-May & fruiting in November-February. The fruits are globose, fleshy, spherical, light greenish yellow, quite smooth and hard on appearance, with 6 vertical stripes or furrows.

Scion selection: The scion should be selected from the healthy representative true to type superior mother plants that are prolific, high yielding and known for regular bearing. and show low incidence of pests and diseases. Patch and modified ring budding techniques are used for commercial propagation. Budded plants start bearing from year 4 and attain full bearing in 8 to 10 years, giving a production of 80 to 100 kg fruits/ tree.

Root stock selection: Six months to one year-old seedlings obtained from amla trees are being used as rootstock. Mature amla fruits are collected during January-February and their seeds are extracted after drying. The seeds are graded by soaking in water. Floating seeds are discarded and sunken seeds are dried and kept for sowing. For better germination, collected seeds are treated with 200 to 500 ppm. gibberelic acid. Treated seeds are sown on raised beds during March-April. They can also be sown in poly bags for better handling. Seeds sown during March/April germinate within 2 weeks and seedlings of about pencil thickness are obtained after 90 to 100 days and these seedlings can be used for budding. One year old seedling is highly suitable for budding (TNAU).

A budded/grafted aonla tree starts bearing third year/ fourth year onwards after planting, whereas a seedlings tree may take 6 to 8 years. Vegetatively propagated plants attain full bearing within 10 to 12 years and may continue to bear for 60 to 75 years of age under well managed conditions. An aonla tree may bear 300 kg. per tree, giving 15 to 20 tons/ha. High yielding amla varieties are suited for closer planting with a spacing of 3 m. by 3 m. By adopting this spacing and following hexagonal or triangular method of planting as many as 1200 plants can be accommodated in a hectare.

Scion Collection: Scion is collected from individual healthy trees of 10 years and above age that are prolific and regular in bearing. The scion shoots are serially numbered and stored hygienically and cured for a week. Root stock seedlings that are one year old, healthy and being grown in poly bags are prepared for grafting and patch budding is performed. Budded plants from each selected tree origin are planted in a well prepared field. As many as 25 to 30 budded clones, with 25 trees per clones are thus planted adopting the recommended horticultural production protocols. Periodical training and pruning are adopted. Budded plants are planted in 3m. x 3m. spacing. All around the field, suitable pollinator clones of Aonla are planted in border rows. Continuous care and management practices are followed to raise a good clonal orchard that is well protected from pests and diseases. Each progeny that consists of 25 plants are regularly monitored for different characteristics in the juvenile, pre flowering, flowering and fruit development stages. Early and late flowering progenies are discarded. The productivity of individual trees and the fruit characteristics in each clone are recorded in permanent records. Characteristics of Amla to be recorded are shown in **Table 1**.

Nucleus Stock: Individual plants in each of the clones are regularly verified for the expression of characteristics. The productivity and the fruit traits are studied each year plant wise, within each clone and based on the progressive performance and period for reaching economic productivity, the clones performing above mean are identified. Systematic efforts are bestowed to rejuvenate the selected individual uniform plants from true to type clones that are in complete conformity to the characteristics to facilitate collection of highly healthy scion shoots for collection of buds. The selected elite plants are maintained with ideal attention and management and training and pruning regulations. The scion bud woods constitute nucleus source capital. Isolation distance 10m is provided

Breeder Stock: Meanwhile a dedicated breeder stock clonal bank with the buds sourced from the selected nucleus superior plus trees of the clonal orchard (grafted on to the root stock) is established as mother Block and true to type plants maintained in a highly healthy and hygienic conditions with suitable pruning and training regulations in disease and pest free conditions. The breeder stock is planted in 3m to 3m spacing. Border rows of pollinator trees are planted. Isolation distance 10m. is provided. Regular and prolific bearing individual trees are retained. Pruning is done periodically. This breeder plant service could be supported by modern protected nursery and propagation facilities and freeing from pests and diseases and adopting improved phytosanitary measures. A dedicated root stock source nursery will be also developed and maintained to supply compatible root stock specific for the scion source. Visit of Monitoring Team is arranged for visit to the Breeder Stock Plants in the Mother Block.

89. Ber (Indian Jujube)

(*Zizyphus mauritiana* Lam.) (2n:48)

Family : Rhamnaceae

Tree: Spreading habit, growing upto 10m tall. Branches vine like and thorny. Branchlets zigzag and thornless or thorns with short sharp straight or hooked spines. Leaves dark green, upper side green and lower side densely felted. Leaves are alternate, ovate, oblong. Leaf apex rounded. Leaf margin finely toothed. Foliage evergreen or deciduous depending on the climate. Flowers 3 to 5 or more, borne in cyme inflorescence in leaf axils, yellow in colour, 2 to 3 cm long, tiny flowers, bisexual, pentamerous, 5 petals, Calyx is persistant, ovary 2 loculed, pyrene. Fruit is a juicy drupe. Fruit contains a rough central stone that consists of two elliptic brown seeds, 5 to 6 mm. long. Fruit colour, shape, juiciness, keeping quality and fruiting season varies according to the agroclimatic zones. Flowers are protandrous. Anthesis occurs in the morning and in some varieties, late in the day. Pollen is thick, heavy and not airborne. Ber is predominantly insect pollinated. Cross incompatibility is prevalent and varieties are to be matched for good fruit set. Some cultivars set fruit parthenocarpically and fruit set is good. Fruit skin is smooth to glossy and light. Fruit flesh is white and crisp. Fruiting season is from October to November and from March to April depending upon the region. Viability of the seed is for 2 to 2½ years.

Budding: Ring budding or shield budding is done to propagate. In budded plants, the buds sprout after 20 to 22 days. Budded plants will be ready after 2 months (National Horticultural Board). Spacing of 5 m x 5 m is advocated. Training and pruning need to be given. In the Ber crop, in the 2nd and 3rd year, fruit matures in 150 to 170 days from flowering. Preharvest spray of 750 ppm. of Ethephon induces early maturity. Ber fruit matures in February to April in Tamil Nadu, December to January in Gujarat and January to March in Rajasthan. Usually bearing commences under rainfed condition from second year. Prime bearing occurs in 10 to 20 years (Chadha 2001).

Root stock: Well matured seeds of ber from selected ber trees grown in the Root Stock orchard that are healthy, vigorous, resistant to diseases and pests and highly tolerant to moisture stress are scarified or pretreated with water and raised in polybag nursery. Watering and proper care and management are bestowed. Only healthy vigorous true to type single plant is retained in each polybag. The raising of root stock plant is planned and programmed to coincide with very ideal and proper season for grafting with selected buds.

Scion stock: Scion bud woods of true to type plus trees are identified from disease free well developed pencil thickness stem of past season growth. Bud woods with well developed axillary buds are selected and properly cured. Care is bestowed to collect bud woods from previously cloned plus trees from scion stock resource exhibiting, true to type and uniform characteristics, high and regular bearing with high degree of repeatability for flowering and fruiting with quality ber fruits.

Nucleus stock production: Sixty to ninety days old true to type root stock seedlings in polybags are systematically prepared for ring or shield budding with buds from scion donor trees and kept in partial shade and well protected after budding operation. After 60 to 65 days, grafts showing good sprouting of buds are selected and closely examined for the characteristics of bud, pigmentation and young leaf traits of donor tree. Grafts, 20 in number made with donor buds from each of 20 to 25 selected true to type mother trees in ber scion orchard are planted in 2 row clonal progeny plots with 10 grafts in each row. The clonal orchard is isolated by 5 meters from other ber trees to prevent mixing up of fruit products and scion shoots when harvested. It is necessary to plant 2 rows of pollinator variety all around to ensure proper pollination and fruit set in the clonal progenies. Good agronomic and horticultural production protocols, pruning and training regulations and frequent attention to health care are ensured. Periodical observation on true to type and uniformity in expression of the prescribed characteristics are noted and progeny rows showing deviating plants are uprooted. Clonal progenies are verified for the foliar, stem, branch, thorny nature, type of inflorescence, flower buds, opened flowers shape and size and for the extent of fruit set and fruit maturity within the well defined seasons specific for the region concerned where the clonal progenies are evaluated. The fruits set on the individual plants in the clonal progeny are verified for characteristics like fruit size, shape, skin glossiness and ripe fruit colour at harvest. In the laboratory, fruits from each plant in each clonal progeny are verified for skin colour, fruit shape, flesh firmness, colour, consistency, taste and seed traits. The clonal progeny evaluation is continued upto fourth year or for at least 2 to 3 good bearing seasons. The individual superior plants in each uniform true to type clonal progenies are identified based on the overall fruit productivity consistent with quality parameters. The rest of the clonal plants that are not selected are uprooted. The selected true to type, uniform and superior performing clonal plants are tagged and properly trained, regulated and maintained by horticultural protocols in Nucleus stock plot. Fruit quality parameters are adjudged.

Breeder stock production: A well prepared polybag nursery is raised and sown with pretreated seeds of selected vigorous individual ber trees identified for raising seedlings for root stock purpose. Sixty days old seedlings that are vigorous and healthy are selected one each in a polybag. Healthy scion stems with well developed buds free from diseases and pests are separated from true to type identified superior plus trees and tagged in the nucleus seed clonal progeny plot. Healthy and quality grafts are raised by ring or shield budding on sixty days old root stock seedlings. Well established grafts that are healthy and vigorous with sprouted buds are selected and critically examined for bud sprouts, fresh young leaf and stem characteristics in conformity with the scion donor tree. After verifying for the trueness, the clonal plants are planted in well prepared pits adapting good agronomic and horticultural practices. Isolation of 5m is provided. Periodical verification is made for conformity to the characteristics and deviating clonal plants, if any, are uprooted in the Breeder Stock Plot. Pollinator variety is also planted in two rows around the breeder stock plot as per the need. Periodical pruning and training regulations are practised. At the time of flowering and at fruit maturity period, due attention is made to observe for trueness and conformity to the prescribed characteristics. Matured fruits / ripe fruits are harvested tree wise and yield data recorded besides assessing the fruit qualities. Laboratory testing of fruit quality is completed on single tree basis. Superior plus trees at the end of fifth year are identified and tagged for bud wood collection of Breeder seed status. It is suggested to maintain performance record and duly filled descriptor of the characteristics for each tree in the breeder stock. Visit of Monitoring Team is organised when the trees are in full bearing.

Table 1. Descriptor of indian jujube –ber (*Zizyphus mauritiana* Lamk)

S.No.	Characters	State
1.	Growth habit	Erect, semierect, spreading
2.	Shoot surface	Smooth, thorncutose
3.	Leaf apex	Acute, obtuse
4.	Leaf base	Acute, cordate, round, oblique
5.	Leaf shape	Ovate, oval, cordate, obovate
6.	Leaf curving	Absent, present
7.	Leaf pubescence in lower surface	Smooth, sparse tomentose, dense tomentose
8.	Leaf blade length (cm)	Short (<7), medium (7-9), long (>9)
9.	Leaf blade width (cm)	Narrow (<6), medium (6-7), broad (>7)
10.	Thorniness of branch	Less, medium, high
11.	Thorn shape	All curved, alternate curved
12.	Thorns persistence	Absent, present
13.	Immature fruit anthocyanin	Absent, present
14.	Fruit maturity group	Early, medium, late
15.	Fruit bearing habit – bunching	Absent, present
16.	Fruit cracking	Absent, present
17.	Fruit shape at apex	Flat, rounded, pointed
18.	Mature fruit shape	Oblong, oval, ovate, oblate, round, falcate
19.	Mature fruit length (mm)	Short (<20), medium (20-30), long (>30)
20.	Mature fruit width (mm)	Narrow (<15), medium (15-25), broad (>25)
21.	Mature fruit weight (g)	Very low (<5), low (5-10), medium (11-20), high (21-30), very high (>30)
22.	Mature fruit color	Yellow, greenish yellow, chocolate brown
23.	Mature fruit surface	Plain, ridged, wart
24.	Pulp texture	Soft, medium, hard
25.	Pulp cavity	Absent, present
26.	Pulp content at stylar end	Absent, present
27.	Pulp cavity at stem end	Absent, present
28.	Stone shape at apex	Oblate, acute
29.	Stone shape	Oblong, oval, spindle, club, falcate
30.	Stone length (mm)	Short (<15), medium (15-25), long (>25)
31.	Stone width (mm)	Narrow (<7), medium (7-9), broad (>9)
32.	Stone weight (g)	Very low (<1), low (1-2), medium (2.1.-3), high (>3.1)
33.	Pulp to stone ratio	Very low (<5), low (5-10), medium (11-15), high (16-20), very high (>20)
34.	Pulp TSS (Brix)	Low (<10), medium (10-20), high (>20)
35.	Pulp acidity (%)	Less (<0.3), medium (0.3-0.5), high (>0.5)
36.	Ascorbic acid content of pulp (mg/100 FW)	Very less (<10), less (50-100), medium (101-150), high (150-200), very high (>200)

(*Source:* PPV & FRA)

90. Custard Apple, Sugar Apple, Sithaphal

(*Annona squamosa* L.) (2n:14,16)

Family : Annonaceae

Custard apple is grown in more than five states in India. It is a semiarid minor fruit crop capable of tolerating drought and heat. High atmospheric humidity during fruit set is ideal for this crop. There are few superior germplasm identified for fruit production. The extent of genetic variation is considerable for fruit shape, size and skin colour in addition to number of aerioles and number of seeds per fruit. Fruit colour is green or yellow. Green colour skin is common. Plant habit and foliage attributes help to recognise the different germplasm.

Leaves are alternate, ovate to obovate, thin and darkgreen. Young leaves pubescent, smelling peculiar strong odour on crushing. Flower solitary or seldom 1 to 3 in number borne on stem of current season's growth. Sepals three, short and deciduous. Petals are 6 in biseriate, outer 3 being lanceolate and inner 3 petals alternate with outer ones. Stamens numerous attached to the cushion around gynoecium. Pistils are numerous and dark found above the level of stamens. Stigma sessile. Stamens and pistils form a cone shaped structure. Fruit is syncarp due to fusion of ovaries. Exterior polygonal tubercles are corresponding to the carpels forming fruit. Fruit flesh is soft, white and juicy. Seeds dark brown or black, elliptical, shiny and are embedded in the edible pulp inside the fruit. Custard apple is a cross pollinated crop. Flowers are protogynous, dichogamy, stigma becoming receptive a day before the anther dehiscence. Insects aid in cross pollination. (Kundan *et al.,* 2012; Jenkins *et al.,* 2015).

Custard apple varieties need to be vegetatively propogated to maintain the identity, trueness, genetic virtues, yielding ability, uniformity and stability in qualitative and quantitative traits. Cleft or veneer grafting is mostly adapted. Root stock seedlings are raised by sowing the water pretreated seeds of custard apple in raised beds. In three weeks time, germination is completed. When seedlings grow to 12 to 15 cm. they are transplanted in suitable size polybag nurseries and maintained with good moisture level till they grow to 30 to 35 cm. height in 18 months. Well developed healthy buds from the selected scion shoots from the high performing identified mother donor trees are grafted onto the custard apple seedlings root stock of their own species that will ensure high graft compatibility. Budded plants ensure true to type nature, uniformity and precocious bearing.

Grafted plants are planted in pits of 60 cm x 60 cm x 60 cm size in 4 m x 4 m spacing that would provide close planting of custard apple plants to promote better fruit set by improving pollination which is a problem in custard apple. Isolation of 5 m distance is provided. Good agricultural and horticultural production practices are applied. Training and pruning are done to promote single stem and efficient compact canopy to get better yield over a longer period. Yield commences 4 to 5 years after planting and continues till 15 years. Fruits are harvested in 4 to 5 pickings in a year.

Neglected plants and severely water and nutrient stressed plants set in dormancy in advance and fruits turn brown and remain as stone fruits. Good care, irrigation and fertilizer application help the plants to flower and fruit before the plant enters into dormancy and without stone fruit formation. Attraction of pollinators by using lures can increase fruit set. Artificial pollination leads to improve the fruit set upto 85% or more in Arka Sahan hybrid custard of apple of IIHR (Sreenivasamurthy, 2012). Without hand pollination, the fruit set is 1 to 2% only. Hand pollination is performed from 6.30 am to 9.30 am with pollen collected from custard apple variety concerned. Commercial lures are also used (Jenkins *et al.,* 2015).

Nucleus Stock Production: Superior true to type mother plants of custard apple developed from budded plants authenticated by the breeder or the institution are identified and tagged. Good agronomic and horticultural practices are bestowed in preflowering to fruit set stage prior to dormancy of shoots. In the appropriate stage of current shoot growth, well developed buds are isolated from individual plants and grafted onto root stock seedlings of the same genotype. From each mother plant, 15 to 20 budded clonal plants are developed. In the favourable season, 15 to 20 clonal progenies from each selected superior true to type mother plant are raised in two row progeny plots. Once in 10 progeny rows, 2 rows of the original variety, each row with 10 plants, are planted for comparison and for aiding pollination. Isolation distance of 5m. is ensured. Critical observations are made at periodic intervals covering different phases of plant establishment, canopy development following training and pruning and faster shoot and foliage development. Deviating plants for stem and leaf characteristics are culled. Plants are monitored for uniformity in dormancy setting in and recovery to active growth after dormancy. Plants showing early flowering and late flowering are uprooted. At the age of first flowering, true to type and conformity of the flowers to the descriptors on flower, fruits and pulp are verified and deviating plants are removed. At active flowering, steps are taken to attract pollinators with appropriate lures and also by hand pollination. Individual trees are then examined for number of flowers produced, fruit set, fruit colour, shape, uniformity and extent of fruit set under natural and aided pollination. Fruits collected at ripening are studied for the number of aerioles, seed number, seed traits, flesh colour, juiciness and time for ripening of fruits. Data on various aspects collected over 4th to 5th years are again supplemented with one more year of comparable fruiting season. On the basis of individual trees that conform to the trueness and performance over 2 years, superior individual trees of custard apple are identified and labelled. These identified elite plus trees constitute the nucleus stock.

Breeder Seed Production: A well established superior breeder seed garden is organized by planting grafted/budded clones derived from budding of the superior plus plants from nucleus clone plot onto the compatible root stock. The field is isolated by 5m. All ideal planting, pruning, training, fertilizer inputs, canopy management practices are done. Continuous monitoring for trueness and conformity of stem, branches, foliage, flower, fruit setting and fruit trait characteristics upto two years of fruiting is taken up. Matured fruits after harvest are studied in the laboratory for conformity. Dormancy period setting and following period of short growth are observed. True to type selected plants alone are retained and breeder seed garden is established and maintained for bud sourcing. Border planting with the same genotype is raised around Breeder clones.

91. Fig

Ficus carica L. (2n=26)

Family : Moraceae

Deciduous, small sized tree or shrub, leaves stipulate, petiolate; leaf blade ovate, palmately lobed, leaf margin irregularly dented. Fig is gynodioecious, from a functional point of view. It is dioecious with two tree morphs. Pollination is performed by wasp. Flowers are lining inside the synconium, a kind of inflorescence. Flowers are tiny and unisexual. Plants are pistillate or staminate, trees are propagated by cuttings of 2 years old dormant matured wood, the cuttings are 20 cm to 30 cm long and are planted in nursery for one month period and replanted in poly bag containers and allowed for 6 to 9 months. The rooted well established seedling are planted in 4m × 4m spacing. Adopting HDP with 2.5 m × 2.5 m spacing is also advocated. Fruiting on the current year growth will commence in less than a year from planting. Two times fruit bearing occurs in south India ie in July – September and Feb- March. Commercial scale harvests can be had from 3rd year onwards. The characteristics of Fig are presented in Table 1.

Table 1. Characteristics of Fig. *Ficus carica* L.

1.	Plant growth habit	Upright, semi spreading, spreading
2.	Plant - secondary weeping shoots	Absent, present
3.	Plant - no. of basal suckers	Few, medium, many
4.	Plant - density of branching	Sparse, medium, dense
5.	Plant - bark tubers	Absent, present
6.	One year old shoot colour	Orange, brown, greyish brown
7.	Terminal bud size	Small, medium, large
8.	Terminal bud colour	Yellow green, orange, brown, greyish brown
9.	Shoot- no.of leaves	Few, medium, many
10.	Leaf- predominant type	Entire, 3 lobed, 5 lobed
11.	Entire leaf shape	Cordate, triangular, lanceolate, oblong
12.	Leaf shape of central lobe	Triangular, narrow, broad, spatulate, linear
13.	Leaf length	Short, medium, long
14.	Fruits- number of fruits first crop	Few, medium, many
15.	Fruits- number of fruits in the main crop	Few, medium, many
16.	Fruit shape	Spherical, cucurbit form, ovoidal, pyriform, urceolate
17.	Fruit size	Small, medium, large
18.	Fruit width	Very narrow, narrow, medium, broad, very broad
19.	Fruit neck fruit length	Absent, very short, short, medium, long
20.	Fruit stalk length	Short, medium, long
21.	Fruit ostiole size	Small, medium, large
22.	Fruit outer colour of skin	Yellow, red purple, purple
23.	Fruit ribbing	Absent, weak, medium, strong
24.	Fruit –pulp colour	Yellowish, brown yellow, pink, purple orange red, red, light brown, dark brown
25.	Fruit internal cavity	Absent, very small, small, medium, large
26.	Fruit- no. of achenes	Few, medium, many
27.	Fruit- size of achene	Small, medium, large
28.	Fruit juiciness	Low, medium, high
29.	Fruit – time of maturity	Early, medium, late, very late

(*Source:* UPOV – Draft, 2009)

Survey the actively and regularly bearing Fig trees aged 4 to 6 years and 25 regular and early bearing trees are identified. At the time of fruiting, fruit production and fruit qualities are studied. Selected plants high yielding are cloned by hard wood cuttings to raise at least minimum of 15 clonal progenies from each selected tree. The clonal progenies are raised under ideal condtions of growing. Isolation of 5m. is provided. In each progeny 10 plants are selected and raised in 2.5m × 2.5m spacing. The progenies are evaluated in the pre flowering stage for true to type nature, health, vigor and fast growth. True to type healthy plants are retained and fruiting and fruit yield and qualities of fruits are studied/ monitored for at least 2 to 3 fruit bearing periods/ seasons commencing two year after planting. Based on the performance, progenies performing above mean + Sd are selected. Superior performing trees in the selected progenies are identified, cloned and the nucleus stock is constituted. Selected plants are now cloned and planted in the Breeder Stock/ Mother seed garden. The breeder seed garden is maintained with recommended training/ pruning and other horticultural practices. Fruit qualities are assessed in fresh and dried fruits and the garden is well protected from biotic and abiotic stresses. At appropriated stage of plant development and fruiting the clone/planting material is supplied to the nurseries and growers.

92. Grapevine

(*Vitis vinifera* L.) (2n:38)

Family : Vitaceae

In grapes, rooted cuttings are commonly used for propagation. In view of the importance of this vine crop in fresh fruit market and in food industry, quality of the grape berry is very important in the source variety. Based on the past performance for growth, productivity, fruit yield and fruit quality traits as many as 500 vines that are well maintained, disease free and withstanding pruning and training regulations are earmarked in the well maintained variety source plantation. Among the vines, 100 vines of good performance and fruit and juice quality are selected and 50 to 60 rooted cuttings are prepared from each selected vine.

Nucleus Stock: Rooted cuttings which are clones are planted with a spacing of 4 m x 3 m at the rate of 20 rooted cuttings per selected vine. All recommended grapevine production practices are followed. Periodical observations are made among individual clones of each entry and the clonal progeny showing variants arising will be discarded. In the individual clones responding to training and pruning and flowering and possessing high fruit set and very good berry quality traits are continuously studied and evaluated. After 3 crop seasons performances, yield data, quality and other horticultural performances are recorded in individual vines. Best and superior performing 50 clonal selections are identified tagged and from each selection, 20 to 25 or more number of cuttings are selected, pooled and maintained as nucleus stock. Isolation of 5m distance is followed. Genetic uniformity, trueness, stability and distinctness are verified in all the selected clones. This will form the authentic and productive nucleus stock source nursery and maintained in very healthy state and diseases, pests and nematode free condition.

Breeder Stock: Rooted cuttings are prepared from the individual tagged vines in the nucleus stock and are planted in the breeder source nursery. Isolation distance of 5m. is adapted. The vines from rooted cuttings are verified for the response to training and the foliage, flower blossom and berry trait qualities, seasonal performances. The performance of the individual plants in the breeder stock vine yard are recorded and verified. The deviating type of vines are discarded. Only healthy, productive true to type vines are retained as breeder stock and stem cuttings from the breeder stock nursery are supplied as breeder seed.

Meriestems or shoot tips culture in *in vitro* is practised in University of California, Davis, USA. Breeder stock plants are developed from meristem or shoot tip cultured for 6 months followed by 4 weeks in rooting medium and roots are trimmed. The *in vitro* plants are transferred to the closed containers. After a month, the plants are transferred to green house. Virus free test is conducted by ELISA or PCR method. After testing for virus free nature, the herbaceous shoots or cuttings are used for mass propagation in mist chambers.

Green grafting or *in vitro* grafting is adapted for rapid multiplication. The quality control of the propagules is to be maintained. Indian Institute of Horticulture Research (IIHR), Bangaloru has formulated combined *in vitro* and *ex vitro* techniques for providing cleaned planting material in grapevines.

Root stock: Root stocks provide genetic and physiological strength to the grafts (scion – clone) to confront the soil and water problems besides resisting abiotic (drought) and biotic stresses and to overcome physiological disorders and to provide compatibility for fruitfullness of the variety /clone. Root stocks should be identified based on its compatibility with the scions and sustaining and maintaining the vegetative growth ,flowering and fruiting abilities of the scion and the quality of the marketable produce. It is of paramount importance to select and to maintain the distinctness, genetic uniformity and stability of the rootstocks while multiplying. In grapes, as many as 11 distinct root stocks have been listed by Nookaraju (2011). Most of them are interspecific hybrids or originating from such hybrids.

Table 1. Descriptor of Grapevine

1.	Time of bud bursting days after proving (in days)	Early, very early, medium late, very late
2.	Young shoot form	Closed, half open, fully open
3.	Young leaf colour (upper side)	Green, green with bronze, yellow with brown spots, copper yellow, copper reddish
4.	Time of full bloom number of days after fruit pruning	Very early (<25), early (25-30), medium (31-36), late (37-42), very late (42>)

5.	Average number of inflorescence per shoot	1, 1-2, 2-3, 3 or more
6.	Shoot growth habit	Erect, semi erect, horizontal, semi erect, horizontal, semi drooping, drooping
7.	Number of lobes in matured leaf	1, 3, 5, 7, 9
8.	Mature leaf anthoyanin colour of main vein on lower side of leaf	Absent, present
9.	Time of venaison after days after pruning	Early (<70), medium, (70-90), late (>90)
10.	Physiological maturity of berry days after fruit pruning	Early (<110), medium (110-130), late (131-140)
11.	Bunch weight (g) with peduncle in wine grapes	Small (<150), medium (150-200), large (>250)
12.	Bunch length of table grape (mm)	Short (<120), intermediate (120-200), long (>200)
13.	Bunch length of wine grape	Short (90), intermediate (150), long (>150)
14.	Bunch berry density compactness of table grape	Loose, medium, compass
15.	Bunch-peduncle length (mm)	Short (<50), medium (51-70), long (>70)
16.	Bunch shape	Globular, cylindrical, winged cylindrical, polywinged, double laterals
17.	Berry diameter (mm)	Small (<14), medium (14-18), large (>18)
18.	Berry shape	Oblate, globose, round, short elliptical, long elliptical, cylindrical, ovate obovate, finger shape
19.	Berry skin colour after removal of bloom	Green, yellow, rose, red, purple, blue, black, others
20.	Berry skin thickness	Thin, medium, thick
21.	Berry mesocarp anthocyanin	Absent, present
22.	Berry firmness of mesocarp	Soft, firm
23.	Berry attachment with pedicel	Loose, firm
24.	Berry – formation of seeds	Seedless, medium maturity, well developed
25.	Sugar content of must (%)	Low (<16), medium (16-20), high (>20)
26.	Total acid content of must (g/l) tartaric acid	Very low (<3), low (3-6), medium (6-9), high (9-12), very high (>12)
27.	Wood shoot cross section	Circular, elliptic, oblate
28.	Colour of wood shoot	Yellow, brownish, red violet, grey

(*Source:* PPV & FRA)

Schematic diagram for foundation stock in grapes (UC. Davis)

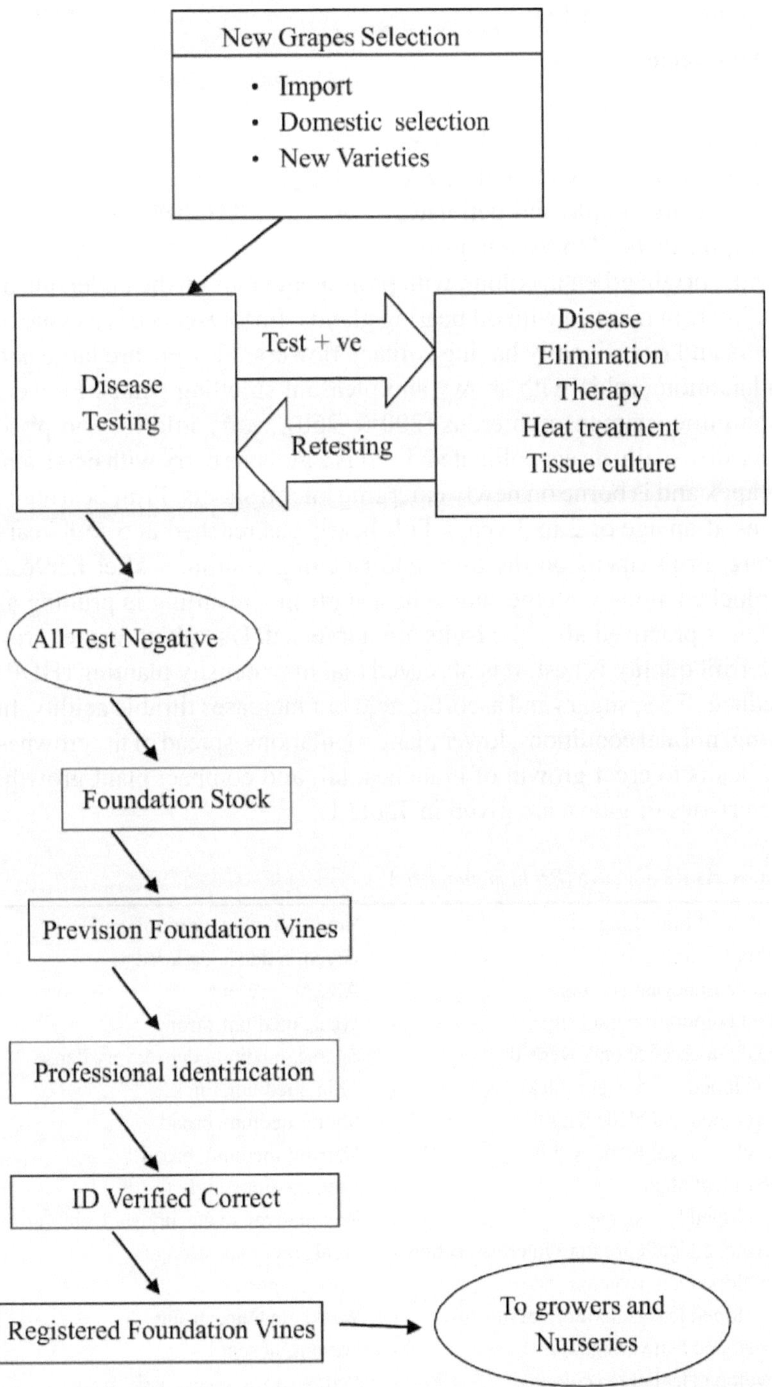

93. Guava

(*Psidium guajava* L.) (2n:18)

Family : Myrtaceae

Guava is an evergreen fruit tree with open and irregular crown. The bark of the stem is usually cream in colour with pink spots. Leaves are opposite simple and petiolate, oblong to elliptic shape, 3 to 6 cm long and 3 to 6 cm wide, bright green in colour, with prominent veins on the underside of leaves. Leaf margin is entire with oil bearing glands. Inflorescence is a cyme of 1 to 3 flowers and occasionally having solitary flowers. Flowers are large and stalked and actinomorphic with showy and pleasant smelling petals. Flowers are hermaphrodite, stamens numerous (200 to 250), ovary inferior and pistil one. Flowers are usually insect pollinated. Fruit is a globose berry with persistent calyx at its apex and is borne on newly emerging lateral shoots. Fruit bearing in guava begins at an age of 2 to 3 years. Full bearing is reached at 5 to 8 years after planting. Fruit ripens on the trees and ripening continues after harvest. Fruits are plucked along with the shoots bearing fruits, resulting in pruning as well. Pruning is practised after the fruits are harvested. Usually in winter crop harvest, the fruit quality is best. It is observed that high density planting (HDP) in guava reduces TSS, sugars and ascorbic acid but increases titrable acidity. In usual planting, normal conditions, lower plant populations, spread of the crown is more. HDP leads to erect growth of branches, tall and compact plant growth. The characteristics of guava are given in Table 1.

Table 1. Characteristics of guava (*Psidium guajava.* L.)

No.	Characteristic		Description
1.	Tree attitude of branching	-	Erect, spreading, drooping
2.	Young shoot colour	-	Green, reddish, dark red
3.	Young leaf anthocyanin colour	-	Absent, present
4.	Young leaf colouration intensity	-	Weak, medium, strong
5.	Young leaf pubescence on lower side	-	Sparse, medium, dense, very dense
6.	Fully developed shoot stem thickness	-	Thin, medium, thick
7.	Fully developed leaf blade length	-	Short, medium, broad
8.	Fully developed leaf blade width	-	Narrow, medium, broad
9.	L/B ratio of leaf blade	-	Low, medium, high
10.	Fully developed leaf shape	-	Round, oval, ovate, obovate, oblong
11.	Fully developed leaf curvature in cross section	-	Weak, medium, strong
12.	Fully developed leaf twisting	-	Absent, present
13.	Fully developed leaf - curvature of midrib	-	Weak, medium, strong
14	Fully developed leaf variegation	-	Present, absent
15.	Fully developed leaf green colour	-	Yellow green, green, dark green

16.	Fully developed leaf – colour of midrib on lower side	-	Cream, yellow
17.	Fully developed leaf – spacing of secondary veins	-	Close, medium, wide
18.	Fully developed leaf	-	Smooth, medium, wrinkled
19.	Fully developed leaf pubescence on lower side	-	Sparse, medium, dense
20.	Fully developed leaf – shape at base	-	Obtuse, rounded, cordate
21.	Fully developed leaf – undulations of margin	-	Weak, medium, strong
22.	Fully develop leaf – shape of tip	-	Attenuate, apiculate, acute, obtuse, rounded
23.	Inflorescence predominant numbers of flowers	-	One, one to three, three and more
24.	Flower size	-	Small, medium, large
25.	Flower – number of fully developed petals	-	Few, medium, many
26.	Flower – staminoid petals	-	Present, absent
27.	Flower – staminoid petals	-	Few, medium, many
28.	Fruit length	-	Short, medium, long
29.	Fruit width	-	Narrow, medium, branched
30.	Fruit skin colour	-	Green, pale, yellow, orange, orange green, dark green, red
31.	Fruit shape – stalk end	-	Broad, round, truncate
32.	Fruit width	-	Narrow, medium, broad
33.	Fruit surface	-	Smooth, rough, bumpy
34.	Fruit longitudinal ridges	-	Absent, present
35.	Fruit longitudinal ridges prominence	-	Weak, medium, strong
36.	Fruit longitudinal groove	-	Present, absent
37.	Size of sepals	-	Small, medium, large
38.	Fruit calyx cavity in relation to fruit diameter	-	Small, medium, large
39.	Fruit stalk length	-	Short, medium, long
40.	Fruit flesh colour	-	White, cream, pale pink, pink, dark pink
41.	Fruit flesh colour eveness	-	Even (uniform), mottled
42.	Fruit discolouration after cutting	-	Absent, present
43.	Fruit grittiness of outer flesh	-	Absent, present
44.	Fruit thickness of outer flesh in relation to core diameter	-	Very thin, thin, medium, thick, very thick
45.	Fruit puffiness	-	Absent, present
46.	Fruit degree of puffiness		Weak, medium, strong
47.	Fruit succulence juiciness	-	Dry, medium, juicy
48.	Fruit acidity	-	Very low, low, medium, high, very high
49.	Fruit sweetness	-	Low, medium, high
50.	Fruit meshiness	-	Absent, present
51.	Fruit – number of seeds	-	Very few, few, medium, many
52.	Fruit seed size	-	Small, medium, large
53.	Period from flowering to fruit maturity	-	Short, medium, long

(UPOV, 1987)

Nucleus stock production: Superior performing healthy true to type uniform trees of the concerned cultivar at full bearing capacity and around 8 to 10 years age are selected. Trees are verified for diseases and pests free nature and adequately fertilized with nutrients and micronutrients. From each of the selected superior tree, well developed healthy current season growth scion or bud sticks are identified and labelled after verifying the fruit characteristics. Healthy raised nursery bed is preferred and sown with seeds extracted from fruits collected from the selected healthy and vigorous trees with high productivity. The seedlings are utilized as root stock for budding or grafting. In guava, patch budding finds favour and high success is achieved if budding is done in the best favourable period in each location.

Layering and inarching are practised but the clonal multiplication is at low rate. The budded plants with buds from each selected plus tree are stocked separately constituting a clonal progeny.

In the nucleus seed production plot, each clonal progeny budded plants, are planted in two rows with 10 plants per row in pits of 1 m x 1m x 1m size in a spacing of 5 m x 5 m. This spacing could allow the plants to reach full bearing capacity in 5 to 8 years. Total number of clonal progenies depends on the identification and selection of nucleus germplasm source of the selected cultivar. Isolation distance of 5m. is provided. Recommended agronomic and horticultural practices, micronutrient application and ideal plant protection practices are adapted seasonally each year to establish a healthy productive plot of guava. Micronutrient sprays are applied at new flush, new shoots, flowering and fruit set. Plant training regulations at young age upto 2 years are followed. Past seasons terminal growth is pruned to a length 10 to 15 cm during the end of fruiting seasons to encourage more laterals.

In the progeny rows, individual plants are monitored from establishment to flowering and fruit bearing for trueness and uniformity for stem, bark, lateral branches, foliage, new flush growth and colour, flower colour and size, fruit setting, developing and ripening fruits and profuse bearing. This evaluation is made in each of the plant in each season and in each year. When the fruit in the guava plants bearing reaches full capacity, the performance and productivity over 2 to 3 seasons are assessed. Deviating trees showing non conformity are uprooted in the progeny rows. Among the retained trees in each progeny showing true to type, uniformity and stability, superior performing individual trees for productivity are identified and tagged. Fruits from the identified plus trees are evaluated in the laboratory for the characteristics of developing and ripe fruits such as shape, size, fruit surface, ridges and colour, nature of corolla still persisting at fruit trip. Cross section shape, flesh colour and consistency, extent of seediness or less seededness, aroma, flavour, taste, fruit pedicel length, fruit firmness, consumer appeal and shelf life are assessed. Estimation of T.S.S, sugars, ascorbic acid, titrable acidity and other nutritive contents is made in the laboratory. It is desirable to make above said evaluations in normal/usual space planted progenies.

Breeder Stock Production: In the next step, budded plants already produced by grafting on the buds from the selected 2 to 3 years old superior trees from the nucleus stock plot onto the seedling root stock are evaluated for true to type, uniformity and healthiness. Selected healthy budded plants are planted in breeder resource stock plot in well prepared pits. Isolation distance recommended in 5m. Recommended agronomical and horticultural practices, training and pruning regulations and healthcare steps are adapted to establish healthy breeder resource stock. Each year, growth and vigour and trueness to characteristics specific for growth phases are monitored and non conforming plants, if any are uprooted. Individual trees at first fruiting season are studied for 2 to 3 fruiting seasons to verify for trueness. This breeder stock is maintained in 20m. isolation and free of diseases and pests with utmost care to supply buds, scion shoots for layering and for other clonal multiplication methods to facilitate adequate supply of true to type clones of the variety.

94. Jackfruit

Artocarpus heterophyllus Lamk. (2n:56)

Family : Moraceae

Seeds are sown immediately after extraction. Softwodod grafting on 2 month old seedling, (cleft grafting) with scion from 4 month old shoots are prepared. Grafted plants (70 to 80% success) are planted in 6m.× 6m. spacing and start fruiting in 4 to 5 years after planting. Female flowers are hand pollinated with pollen collected from 5 to 10 cm. long male flowers spike that are borne in clusters. The grafted plants, are trained to a single stem/ trunk at 1.5 to 2.0 m height. Flower buds appear on trunk. Thimmaraju and Shivananda (1993) have presented details on floral biology pollination and fruit set in jack fruit. The characteristics are given in Table 1.

Active prolific and regular bearing elite plus trees of 8 years and above age are identified. At least 10 plus trees are selected on the basis of regular bearing, yield and fruit quality traits as described in the characteristics of the Jack fruit (Table 1). Systematic evaluation is done and at least ten grafts are prepared with scion from each selected tree. The grafted plants are planted in 6m × 6m spacing. Regular training is done to the trees. True to type nature is verified.

The fruit yield data are recorded from 4th year onwards upto 8 years. The trees performing above mean +SE are identified at 8 years age. Systematic efforts are made to clone the elite superior trees by epicotyl grafting and Nucleus Breeder/ Mother block is established in accordance with the practices of Mother Block establishment in perennial fruit trees. Minimum distance of 5m is ensured as isolation distance.

A guide book to Jackfruit is published by Priya *et al.* (2014).

Table 1. Characteristics in Jack fruit

1. Tree age
2. Tree vigour
3. Tree height
4. Tree circumference
5. Tree trunk surface- Smooth, rough, v. rough
6. Tree crown diameter
7. Tree crown shape pyramid, broadly pyramid, spherical, oblong, elliptical
8. Tree branching semi circular sparse, medium high density
9. Tree branching pattern- erect, open, vertical, horizontal, irregular
10. The leaf blade length
11. Leaf blade breadth
12. Leaf blade - shape
13. Leaf apex shape
14. Leaf blade colour- Light green, green, dark green
15. Leaf surface pubescence
16. Leaf base shape
17. Petiole- grooves
18. Flowering
19. Date of appearance of male inflorescence
20. Date of appearance of female inflorescence
21. Female inflorescence density- sparse
22. Female flower position the primary and secondary branches, intermediate, dense position
23. Male inflorescence
24. Number of years for first fruiting
25. Days from flowering to fruit
26. Fruit bearing position intermediate primary branch and secondary branch
27. Fruit clustering habit – solitary, clusters, others
28. Fruit shape- obloid, spherical elliptical oblong, clavate, irregular
29. Fruit stalk length
30. Fruit stalk attachment to the fruit – deep, flat inflated
31. Fruit length
32. Fruit diameter
33. Fruit weight
34. Fruit rind colour
35. Fruit rind thickness
36. Fruit rind surface- smooth, spiny
37. Fruit spine density- sparse, dense
38. Latex exudation- low, medium, high
39. Fruit quality combined aspects- poor, moderate, good, excellent

40. Number of flakes per kg (bulbs)
41. Weight of fresh flake with seed
42. Flake length
43. Flake breadth
44. Flake shape
45. Flake thickness
46. Flake fibre content
47. Flake texture
48. Flake nutrititive value
49. Pulp- taste and flavour
50. Pulp juiciness
51. Seed length
52. Seed width
53. Seed shape
54. 100 seed weight
55. Seed coat colour
56. Seed cooking quality

95. Mango

(*Mangifera indica* L.) (2n:40)

Family : Anacardiaceae

Trees are medium to large size, leaves alternate, simple, exstipulate, 15 cm. to 45 cm. long. Leaves are variable in shape, ovate lanceolate, obovate lanceolate, roundish oblong and leaf apex is rounded to acuminate. Leaf margin entire, sometimes wavy and undulating. Upper surface of leaf dark green and shiny and lower side light green and glabrous. Leaves are flaccid, appear in flushes and pendulous. Colour of young leaves varies from variety to variety like tan red, pink and yellowish brown. Matured leaves change to tan red to green and finally to dark green. Petiole swollen at base and 10 to 12 cm. long and grooved on the upper side. The characteristics are given in Table 1.

Table 1. Characteristics of mango (*Mangifera indica* L.)

1.	Attitude of main branch	-	Erect, spreading, drooping
2.	Young leaf – intensity of anthocyanin	-	Absent, very weak, weak, medium, strong, very strong
3.	Leaf blade length	-	Short, medium, long
4.	Leaf blade width	-	Narrow, medium, broad
5.	Leaf blade length / width ratio	-	Very small, small, medium, large, very large
6.	Leaf blade shape	-	Linear, oblong, elliptic, ovate, obovate
7.	Leaf blade colour	-	Yellow green, light green, medium, green, dark green
8.	Leaf blade twisting	-	Present, absent

9. Leaf blade shape in cross section - Straight, slightly concave, moderately concave, strongly concave
10. Leaf blade - spacing of secondary veins - Very close, close, medium, wide, very wide
11. Leaf blade margin modulation - Absent, very weak, medium, strong
12. Leaf blade shape at base - Acute, obtuse, rounded
13. Leaf blade shape at apex - Attenuate, acuminate, acute
14. Petiole attitude in relation to shoot - Erect, semierect, perpendicular, moderately recurved
15. Petiole length - Short, medium, long
16. Inflorescence length - Short, medium, long
17. Inflorescence diameter - Small, medium, large
18. Inflorescence length / diameter ratio - Small, medium, large
19. Inflorescence – number of primary branches - Few, medium, many
20. Inflorescence anthocyanin intensity of colouration of axis, branches - Absent, very weak, weak, medium, strong, very strong
21. Mature fruit length - Very short, short, medium, long, very long
22. Mature fruit width - Very narrow, narrow, medium, broad, very broad
23. Mature fruit – length / width - Very small, small, medium, large, very large
24. Mature fruit – CS shape - Narrow, elliptic, broad elliptic, circular
25. Mature fruit – colour of skin - Yellow, green, yellow green, green orange, green pink, green red, green purple
26. Mature fruit – density of lenticel - Sparse, medium, dense
27. Mature fruit colour contrast between lenticel and skin - Weak, medium, strong
28. Mature fruit - lenticel size - Small, medium, large
29. Mature fruit – surface roughening corkiness caused by lenticels - Absent, present
30. Mature fruit – stalk cavity - Absence or shallow, medium, deep
31. Mature fruit – presence of neck - Absent, present
32. Mature fruit – length of neck - Short, medium, strong
33. Mature fruit – shape of ventral shoulder - Rounded upward, rounded outward, rounded downward, sloping downward, falling abruptly
34. Mature fruit – shape of dorsal shoulder - Rounded upward, rounded outward, rounded downward, sloping downward, abruptly falling
35. Mature fruit - length of groove in the ventral shoulder - Absent or short, medium, long
36. Mature fruit – depth of groove in ventral shoulder - Absent or short, medium, long
37. Mature fruit – bulging of ventral shoulder - Absent, present
38. Mature fruit – presence of sinus - Absent, present
39. Mature fruit – depth of sinus - Shallow, medium, deep
40. Mature fruit – bulging of proximal of stylar scar - Absent, weak, medium, strong
41. Mature fruit-bulging at point at stylar scar - Absent or small, medium, large
42. Mature fruit-diameter of stalk attachment - Small, medium, large

43.	Ripe fruit – predominant colour of skin -	Green, yellow, green yellow, yellow orange, yellow red, red, orange purple, reddish purple, purple
44.	Ripe fruit – speckling of skin	Absent or very weak, weak, medium, strong
45.	Ripe fruit – skin thickness	Thin, medium, thick
46.	Ripe fruit – adherence of skin to the flesh	Weak, medium, strong
47.	Ripe fruit – main colour of flesh	Green, yellow, light yellow, medium yellow, light orange, medium orange, dark orange
48.	Ripe fruit – flesh firmness	Soft, medium, firm
49.	Ripe fruit juiciness	Dry, medium, juicy
50.	Ripe fruit – flesh texture	Fine, medium, coarse
51.	Ripe fruit – amount of fibre	Very low, low, medium, high, very high
52.	Ripe fruit – turpentine flavor	Absent, present
53.	Stone – relief of surface	Grooved, smooth, ridges
54.	Seed shape	Oblong, reniform,
55.	Seed embryony	Monoembryonic, polyembryonic
56.	Time of beginning of flowering	Early, medium, late
57.	Time of flower maturity	Very early, early, medium, late, very late

(PPV & FRA, 2009)

Inflorescences is pseudoterminal, narrow to broad, conical shape upto 45cm long. Panicle colour shows variation from yellowish green to light green with crimson colour flushes. Branching of inflorescence is tertiary, rarely quaternary and the ultimate branching is cymose. Panicle bears from 500 to 600 flowers of which 1% to 70% are hermaphrodite, remainder are male, depending upon the variety and temperature (Mukherjee, 1997). Mostly male flowers in a panicle are in large number. Flowers are subsessile, 6 to 8 mm in size borne on very short pedicels. Flowers pentamerous. Calyx five partite, lobes ovoid or oblong and concave. Corolla five, pale yellow and as long as calyx. Androecium five stamens of which 1 to 2 are fertile and others are staminodes. Fertile anthers and pistil are placed parallel or oblique to each other (Naik and Gangolly, 1950). Colour of anther is pink and turns to purple while shedding from flowers. Ovary sessile, one celled obliquely and slightly and laterally compressed. In Mango, hermaphrodite flowers are self fertile but benefits from cross pollination. Flowers open in the morning by 8 am and by noon time, the anthesis is completed. Flowers are receptive upto 72 hours. Mango is a cross pollinated crop (Naik and Rao, 1943). Flies, bees, thrips and other insects are pollinating agents. The fruit set is estimated to be around 1% in mango panicles and is independent of number of hermaphrodite flowers, pollination compatibilities of the varieties, pollen viability and environment conditions. Fruit is a fleshy drupe. Fruit size, shape, surface nature, colour, fibrous nature, flavour, projection at beak and sinus and other fruit characteristics show variation among the cultivars of mango. Fruit shape varies from round to ovate, oblong and oval. Length of fruit ranges from 2.5 cm. to 30 cm. in different varieties. Skin is gland dotted. At maturity, fruit colour may be in different mixtures of green, yellow and red shades. In mango, polyembrony is recesseively controlled by a single pair of genes (Anon *et al.*, 1998).

Ecogeographical dependence is observed in tree size, shape, canopy branching pattern and leaf characteristics in mango varieties in India. In mango, breeding and selections have culminated in many choice varieties and some of them are notably the result of open pollination arising as chance seedlings. Varieties like Mallika, Amrapali, Ratna, Sindhu and Arka Puneet are recent hybrids evolved from hybridization and selection (Chadha, 2001).

Mango is vegetatively propagated by epicotyl/stone grafting and also by inarching. Root stock nurseries are prepared and seeds are sown in the raised beds within 4 to 5 weeks after seed extraction. Alternatively, seeds are sown in polythene bags filled with ideal soil and compost mixture. The leaves of the germinating seedlings turn green in 2 to 4 weeks. If the bed system is followed, 2 to 4 week old seedlings are transplanted in polybag containers. After establishment, the seedlings are used for grafting at different ages following any one of the grafting methods.

Nucleus Stock Production: Minimum of 20 grafts are developed from each of the regular and heavy bearing mango trees for over years of the selected variety/hybrid. Healthy and disease free buds of scions of current season growth are selected for grafting onto the selected healthy and uniform root stock seedling derived from polyembryonic or monoembryonic origin. Well developed grafts which are the clonal progenies are planted in 1m. x 1m. x 1m. pits either at usual spacing or in high density planting. Isolation distance of 5m is provided. Training and pruning regulations are periodically practiced to space the branches at optimal density and to maintain the canopy in an ideal manner. Each progeny plot is of 2 rows and in each row ten grafts are planted.

One row of graft of pollinating tree is planted all around as border. Recommended agronomic and horticultural practices prescribed for mango crop are adapted properly and pest and disease free condition is ensured. Individual plants in each clonal progeny are critically examined for growth phase specific characteristics for stem, branch and leaf at preflowering phase and later on at post flowering phase for panicle and male and female flowers and developing and matured fruit characteristics. Besides, the health, vigor, growth, true to type, uniformity for flowering, regularity in fruit bearing and high productivity are looked into. Data on productivity are recorded for a minimum of 3 fruiting seasons in 3 different years and on the trees producing satisfactory crop. Each year, ripe mangoes are critically studied in the laboratory for various quality traits and absence of sponginess and for market and consumer acceptability. Processing qualities are determined. Post harvest stage storability and shelf life and transportability are scored on individual tree basis. The performance of individual clones grown in usual spacing and or in high density planting are initially assessed. At the closure of evaluation spread over minimum of three years fruiting seasons, trees showing non conformity for any one of the characteristics are uprooted. In the remaining trees that show conformity for all the trait characteristics, superior individual trees are selected and tagged. All such tagged trees for the true to type, uniform and stable nature constitute nucleus stock for the variety concerned.

Breeder Stock Production: In the next step, selected healthy scion shoots from superior performing trees which are in conformity, true to type and uniform are pooled together and buds are grafted onto the uniform healthy and fast growing appropriate root stock plants. Grafts are prepared and grown till they become suitable for field planting in usual spacing or in HDP. Isolation distance of 5m is adapted. Pollinating mango variety trees are planted in the same lot and tagged. All the trees are tagged indicating their identity and breeder stock status. Recommended pruning and training and agronomical and horticultural practices are followed. Individual trees are monitored for the productivity and conformity to the prescribed characteristics of branches, foliage, panicle, flower, young and matured fruits and later in the laboratory for ripe fruit and flesh characteristics followed by satisfying with the post harvest expression and suitability for specific product utilization. The intricate evaluation of mango trees for productivity and mango fruits is done spread over at least in two different fruiting seasons. The individual trees in the breeder stock resource plot are verified for regular and profuse, flowering, high fruit set, high pulp to stone ratio, high tolerance to diseases and pests, superior keeping quality and good for spoon eating. Rigorous steps and routine care are bestowed to maintain the health of the trees. The breeder stock mango orchard is an elite true to type, genetic/germplasm resource of that particular variety/hybrid conforming to the distinction, uniformity and stability and free from diseases and pests. Bud woods are collected from the breeder stock mango trees to raise foundation stock trees to establish Mango Mother Blocks of the specific mango variety.

96. Mulberry

(*Morus alba* L.) (2n:28)

Family : Moraceae

Mulberry is short to medium size tree with stem light green to dark green colour. Leaves simple, upto 3 lobed, stipulate, stipules lateral. Leaves cordate at base and acute to acuminate at tip. Leaf margin is serrated. Inflorescence axillary, pendulous, raceme like catkins. Flowers green coloured, 4 free imbricate petals. In male flowers, stamens 4, pistillode shaped. Female catkins 2 to 2.5 cm long, white to purple, pink and dark purple coloured. Female flowers borne on long or short spikes, ovary 1 to 2 loculed, ovule single and style bipartite. Fruit syncarp and many drupes embedded in fleshy perianth. Flowers are bisexual or unisexual on different branches of the same plant. Sometimes plant changes from one sex to another sex. Pollen disposal in mulberry is launched at a fast speed, half of the speed of sound (Taylor *et al.,* 2001). The characteristics are given in Table 1.

Table 1. Characteristics of mulberry

1. Growth habit
2. Number of leaves / branches
3. Internode colour and length (cm.)
4. Stem internode diameter (mm.)
5. Colour of matured internode
6. Surface of internode rind
7. Node bud shape / size
8. Leaf blade shape at tip and at base
9. Leaf blade size
10. Petiole length (cm.)
11. Leaf blade margin serration
12. Colour of male catkin
13. Colour of female / bisexual catkin
14. Fruit catkin colour at maturity
15. Leaf succulence, moisture level
16. Fibre status of leaf- low, medium, high
17. Withering of detached leaves-low, medium, fast
18. Withering of intact leaves on shoot-while shoot feeding, low, medium, fast
19. Nutrition qualities
20. Feed conversion ratio

Mulberry is propagated mostly by vegetative cuttings. Mulberry varieties are selected for fast growth, good quality leaf production, quick rooting and good field establishment and regeneration and quality leaf production. Mulberry varieties possessing resistance to high tolerance to leaf / foliar diseases and root rot infection are desired ones. In addition, mulberry varieties should possess superior leaf qualities and high to very high silkworm feed conversion efficiency.

It is ideal to have mulberry varieties like G2 suitable for chawki rearing and varieties like G4 for late age silkworm rearing.

Stem cuttings of the variety concerned, 15 to 20 cm. long with 3 to 4 healthy buds are prepared from matured plants from well maintained healthy disease and pest free gardens and planted in raised nursery beds. Spacing of 20 cm x 10 cm is provided. Green wood cuttings are also directly planted in poly bags and planted in field pits. Field planting may be done in single rows or paired rows. Plant population of 12345 per ha in pits of 30 cm x 30 cm x 30 cm size at 90 cm x 60 cm spacing to 13457 plants per ha in paired rows with 90 cm x 60 cm spacing and 120 cm between paired rows is maintained in the field. A population of 9677 per ha is accommodated in 3 m planting with 90 cm x 90 cm (two rows) intercepted by 120 cm spacing (CSRTI, Mysuru reports).

Nucleus Seed Production: Healthy growing uniform representative true to type individual plants are selected, examined / verified and labelled. Five hundred such true to type plants in the second year of field growing reflecting their survival, quick regeneration and uniform branches, coppicing and high quantity and quality foliage production are finally selected. From each plant, 2 to 3 matured shoots are harvested and 3 budded setts are prepared. Setts from single plants are planted in the small plot nursery for rooting and minimum of 10 rooted cuttings for each selected plant are field planted in single clonal progeny rows along with the check line of rooted cuttings of the same variety once in ten progenies rows. Two hundred clonal progenies are established with isolation of 3m. and studied for their rooting ability, rapid field establishment quick regeneration and in production of good foliage in terms of number of leaves, leaf area, luster, turgidity and medium leaf texture, low fiber content and durable moisture content retention. The field examination of the clonal progenies was done repeatedly in different stages of mulberry growth corresponding to the leaf harvest stages corresponding for respective instars of larval growth of mulberry silkworm in the rearing house.

Shoots are harvested at regular intervals to assess the suitability for leaf and shoot feeding of silk worms. After each cut is done, plants in clonal rows are examined for survival and coppicing and for regeneration with quick growth, leaf quality and for suitability for leaf and shoot feeding of silkworm and silkworm growth development and performance and cocoon yield. High tolerance to diseases infection and insect pests attack are also assessed along the shoot feeding of silkworm larvae. The above said procedure is followed for at least 3 to 4 cycles of cutting or more. In the second year beginning, the performance of the 200 clonal progenies are compared for their suitability as feed besides the true to type integrity and uniformity. Best performing 100 clones are selected and rooted cuttings from each selected clonal progeny are prepared. A nucleus clonal progeny nursery is raised comprising 50 stem cuttings per progeny.

Breeder Seed Production: This breeder seed production plot of 2000 sq.m is isolated by 3 meters from other mulberry growing fields. Good agronomic and horticultural practices, weeding and plant protection measures are followed. In the breeder seed production plot well established healthy individual plants alone are retained and poor health weak and disease infected plants are destroyed.

Critical examination is made from field establishment, to active shoot formation. In each growth phase, shoot growth and vigorous leaf development, leafiness, leaf size, succulence, sustained moisture content and excellent silkworm feeding quality, digestibility and high conversion efficiency are evaluated after establishing trueness or true to type and uniformity. Superior performing bushes over a period of 2 years are tagged and retained. Monitoring team visit is organized. Leaf samples are analyzed for nutritional and biochemical characteristics. Three budded setts are prepared and pretreated with fungicides and distributed as breeder seed. To meet the demands of mulberry seed clones, breeder seed stage can be repeated adapting the steps explained herein.

Nutritive content of tender and matured leaves are studied to decide the suitability for different stage of silkworm rearing. Feed conversion ratio is assessed.

97. Musk Melon

(*Cucumis melo* L.) (2n:24)

Family : Cucurbitaceae

Stem is glabrous or hairy, angled or striate. Leaves ovate to reniform. Leaves three to seven lobed. flowers are clustered. Pistillate flowers are solitary. Pedicels short and thick. In *C. melo* fruits are glabrous, or pubescent, smooth or netted, mostly with sutures. Stem is hairy and five angled. This species is highly polymorphic for leaf colour, shape and for the fruit shape and size, fruit skin marking-netting, flesh colour, crispness, flavor, sweetness, T.S.S. and seediness. Most of the cultivated varieties are monoecious or andromonoecious (hermaphrodite and staminate flowers). Gynoecious and monoecious plants exist but not commonly. Male flowers occur in clusters and pistillate flowers in solitary. Muskmelon is partly self pollinated and partly cross pollinated. Anthesis occurs from 5 am. to 6.30 am. True breeding monoecious lines are developed and maintained. Musk melon do not show inbreeding depression. The characteristics are given in Table 1.

Table 1. Characteristics of Muskmelon

Cucumis melo; *Cucumis melo* subs. *cantalupensis*; *Cucumis melo* subs. *reticulatus*

1.	Cotyledon length	Short, medium, long
2.	Cotyledon width	Narrow, medium, broad
3.	Leaf blade length	Short, medium, broad
4.	Leaf width	Narrow, medium, broad

5.	Leaf blade depth of terminal lobe	Weak, medium, strong
6.	Leaf blade length of terminal lobe	Short, medium, long
7.	Leaf blade dentation of margin	Weak, strong
8.	Leaf blade petiole length	Short, medium, long
9.	Sex expression at full flowering	Monoecious, andromonoecious
10.	Ovary length	Short, medium, long
11.	Ovary width	Narrow, medium, broad
12.	Ovary pubescence	Sparse, dense
13.	Fruit length	Short, medium, long
14.	Fruit diameter	Narrow, medium, broad
15.	Fruit shape (LS)	Oval, ovate, elongated globe, round, oblate, obovate, cylindrical
16.	Fruit rind colour	Creamy white, yellow, yellow green, orange
17.	Fruit patches	Absent, present
18.	Fruit peduncle at maturity	Slipable, nonslipable
19.	Fruit shape at peduncle end	Pointed, round, truncate
20.	Fruit shape at blossom end	Pointed, intermediate truncate
21.	Fruit diameter at blossom end	Small, medium, large
22.	Fruit surface	Smooth, grooved
23.	Fruit suture	Smooth, grooved
24.	Fruit suture colour	Absent, present
25.	Fruit surface netting	Absent, moderate, dense
26.	Fruit flesh thickness	Thin, medium, thick
27.	Fruit flesh colour	Cream white, grey orange, yellow with green, greenish orange
28.	Fruit flesh texture	Mealy, intermediate, crisp
29.	Fruit flavor	Mild, medium, strong
30.	Seed colour	Creamy white, yellowish

(*Source:* PPV & FRA)

Nucleus seed production: The nucleus and breeder seed are produced in isolated plots,1600m. from other melon crops. From the Breeder source seeds or stock seeds, 2000 seeds are selected and pregerminated seeds are sown and the melon crop is grown with good agronomical and horticultural produces practices. The spacing adapted is 1.5 m. x 1.0 m. and only single plant is retained in each hill. In the first season of crop, detailed examination is made from the seedling to juvenile to flowering phase for true to type character and the single plants that are in conformity to the true to type nature alone are retained before flowering stage. At flowering stage, 200 to 250 plants are selected and self fertilized by aided hand pollination of the protected female flowers with pollen from the protected male flowers in the same plant. At maturity, the self fertilized fruits are examined in the laboratory for all the prescribed characteristics (Table 1) of fruit, flesh, seediness and quality aspects such as TSS and selected.

The seeds of selected plants are sown in the following season in progenies rows in a well prepared field. The progeny are raised in two rows plots and in each row 20 plants are maintained. Two replications are maintained. Observations on individual plants are continuously made for true to type and deviating plants within progeny are uprooted along with the rest of the plants in that progeny. The progenies that are true to type breeding and showing superior performance are retained. At fruit maturity, the fruits are harvested from individual plants and labelled. In the laboratory, fruits are examined for the shape, size, fruit wall and flesh thickness, flesh consistency, texture, aroma, taste and T.S.S. Based on the vine, fruits and flesh qualities besides field tolerance to whitefly, fruitfly and other insects and diseases, the superior single plants within each progeny showing high performance over mean +2 SE are selected. Around 100 plants are self fertilized, seeds extracted and stored.

In the next generation, out of 100 plants selected, only 50% plants of superior performance, growth, vigor and healthiness are sown in 2 rows plots, each plot with 10 plants and replicated twice. The isolation provided is 1600m. The seeds of source variety are also sown once in 10 progenies for comparison. All descriptor data recordings and observations on true to type nature and uniformity and stability of the progenies are completed in respect of stem, foliage and fruit characteristics. Matured fruits are harvested from superior single plants from the selected progeny rows showing performance higher than mean+2SE. for yield and quality traits. Individual fruits are examined for the fruit, flesh and quality traits. Fruit market acceptability is assessed. Seeds are separated, dried and part of the seeds from each fruit is bulked together to constitute nucleus seed. Apart from this, the remainder of seeds of individual plants are stored for future cycle of nucleus breeder seed.

Breeder seed production: Seeds from the nucleus seed generation are sown in plots not previously grown to melon adopting the recommended spacing of 1.5 m x 1.0 m. Single plant per hill is maintained. Isolation distance of 1600 m. is maintained. All good agronomic and horticultural practices are adapted to raise healthy and clean crop. Multistage monitoring and evaluation for trueness is made in seedling, juvenile, preflowering, flowering and maturity phases. Off types and disease affected plants are removed. Plants showing deviations for the variety characteristics are uprooted. At midmaturity stage Monitoring Team visit is arranged. At fruit maturity, vine and fruit characteristics are examined and matured fruits are critically examined. Fruit quality features are evaluated. Seeds from selected plants are pooled and processed to constitute Breeder seed. Samples of breeder seed are tested for seed quality assurance and labelled with Breeder seed tags. GOT is arranged to arrive at genetic purity. Fruit quality and market acceptability are assessed.

98. Papaya

(*Carica papaya*) (2n:36)

Femily : Caricaceae

The papaya plant is referred to as tree but properly it is a large herb growing to 2 to 3 meters in the first year and reaching to 6 to 9 m in height. The stem is hollow green or deep-purple showing leaf scars. The leaves emerge directly from the upper part of the stem in a spiral on nearly horizontal petioles that are 30 cm to 105 cm long, hollow, succulent, green or more or less dark purple. The leaf blade is deeply divided into 5 to 9 main segments, each irregularly subdivided and has prominent yellowish ribs and veins. Both the stem and leaves contain copious white milky latex. The five petaled flowers are fleshy, waxy and slightly fragrant. The characteristics are given in Table 1.

Table 1. Characteristics of Papaya

1.	Number of nodes to first flower	Low (<30), intermediate (30 to 40), high (>40)
2.	Length of middle internode on tree (cm)	Short (<3), intermediate (3.45), long (>4.5)
3.	Colour of mature leaf petiole	Green, green with purple tinge, purple
4.	Sex type	Dioecious, gynodioecious
5.	Colour of inflorescence stalk	Green, purple/pink
6.	Height of first fruit (m) breast height	Low (<1.0), high (>1.0)
7.	Number of fruits on trunk at first harvest	Low (<30), intermediate 30 to 40, high (>40)
8.	Stem colour of adult plant	Greenish or greenish brown, green with shade of red, purple, pink
9.	Fruit shape	Pear shaped (Pyriform), lengthened cylindrical, oblong ellipsoid, globular high round, oblong, oval
10.	Fruit apex (Stylar tip)	Blunt, depressed, prominent
11.	Fruit pulp colour	Yellow, yellow orange, orange red
12.	Fruit shape of stalk end (Stylar base)	Depressed, flattened, inflated
13.	Ridging on fruit surface	Absent, present
14.	Fruit weight (kg)	Low (<0.5), intermediate (0.5-<1.00), high (>1.0)
15.	Fruit length (cm)	Short <15, intermediate (15-25), long (>25)
16.	Fruit diameter (cm)	Low (10<), intermediate (11 to 13), high (>13)
17.	Pulp thickness (cm)	Thin (<2.5), thick (>2.5)
18.	Central cavity diameters (cm)	Low (7<), intermediate (7 to 9), high >9
19.	Fruit cavity index (%)	Low (<25), high (>25), angular
20.	Shape of central cavity	Star shaped, circular or round
21.	TSS (Brix)	Low (<11), high (>11)
22.	Pulp firmness at edible stage (kg/cm^2)	Low (<3), intermediate (3 to 5), high >5

Sex Expression: In papaya, three types of flowers, such as male, female, hermaphrodite / gynodioecious are present. Inflorescence is a cyme. Male inflorescence is long, pendulous, freely branching, 25 to 100 cm long. Male flowers are sessile, calyx 5, cup shaped with tooth like rim, corolla 5, trumpet shaped, stamens 10 in 2 whorls and stamens alternating with petal lobes. In female tree, flowers are solitary or few and coloured. Clayx is cup shaped, corolla 5 twisted, ovary superior, ovules many and stigma lobed. Hermaphrodite flowers are either elongate types, short peduncle, 10 stamens in two series and elongated ovary. Male flowers are similar to female flowers but with 5 stamens. Intermediate forms occur in which some or all stamens are carpellate. Anthesis in papaya is between 5 am. and 6 am. Stigma receptivity is maximum on the day of anthesis. Fruits mature in 145 to 165 days after pollination.

The sex in papaya is regulated by a single gene with three alleles, namely, M^h, M and m. The hermaphrodite sex phenotype is regulated by the genotype M^hm, male sex by Mm and female sex by mm genotypes. MhMh genotype is not viable. In papaya, the plants are either male (Mhm) or female (mm) or gynodioecious (bisexual and female flowers). Male and hermaphrodite trees also undergo varying degrees of sex reversal depending on seasonal changes. The female form (mm) is most stable.

The inheritance for sex phenotypes in different crosses is presented below:

Sex inheritance in Papaya

S.No.	Cross	Genotype	Sex phenotype			Non viable combinations
			M.	F.	H.	
1.	F x M	mm x Mm	1	1	0	0
2.	F x H	mm x M^hm	0	1	1	0
3.	H x H (H self)	M^hm x M^hm	0	1	2	1 M^hM^h
4.	H x M	M^hm x Mm	1	1	1	1

M=Male (Mm), H=Hermaphrodite, (M^hm), Female (mm), M^hM^h is non viable (W.B. Storey 1953). (Ray Ming *et al.* (2007) Sex determination in papaya. Seminar in Cell and Development Biology : 18:401-408 Elsevier).

Nucleus Seed Production method is based on the sex expression of the concerned papaya variety. Isolation required is 2000 m from the fields growing papaya. Nursery is raised in polybags filled up with well decomposed mixture and in each bag 4 well developed seeds are dibbled. Seedlings are transplanted 60 DAS in well prepared pits of the size of 45 cm x 45 cm x 45 cm. In each pit, 4 seedlings are planted. After 4 to 5 months time, the plants will commence flowering. In each pit, only one well developed vigorous plant, either a male or a female plant is retained. In an overall manner, the retained male or female plant can be in alternate or adjacent pits to facilitate making paired crosses. At around one year growth stage, well developed, vigorous, true to type disease free and healthy female plants are selected and labelled. So also healthy vigorous disease free true to type male plants are labelled. At least 100 paired crosses are performed and the pollinated female flowers are labelled and protected with paper covers for another 15 days. While selecting individual female and male trees, extra care is bestowed to select only true to type, uniform plants with stable expression in all aspects of characteristics of stem, leaf, inflorescence, flowers and fruit and fruit qualities. After the crossed female flower started to develop, all other immature fruits in the same tree are thinned or removed to facilitate development of cross pollinated fruits. At maturity, seeds from fruits true to type from individual paired crosses are extracted, dried and labelled seriatum and stored.

In the next season, nursery is raised in polybags with seeds from paired crosses in plant progeny wise. In each polybag, 4 seeds are sown. Isolation distance of 2000m is provided. After planting them in pits , at time of flowering, only one plant either a male or female is retained in each pit in such way, in each of the paired cross progeny, 20 plants are maintained with equal number of male and female plants. In the progenies, individual plants, females as well as males are examined for true to type and uniformity in all its aspects. Progeny rows showing deviating plants are discarded. In the true breeding progenies, vigorous healthy male and female plants are selected and paired crosses are made within each progeny as described earlier. In this generation also, 100 paired crosses are made and seeds extracted and saved as individual paired crosses. While examining the paired cross progeny rows, attention is bestowed to examine the true to type nature of the plant type, canopy shape, leaf shape, size, leaf margin and colour, stem colour, thickness, internode length, petiole length, sturdiness and colour, flower size, perianth colour, size, shape, stigmatic lobes, extent of fruit set, fruit shape, unripe and ripe fruit colour, prominence of shoulders at the fruit top and fruit apex shape. Flesh colour in raw and ripe fruits, thickness of flesh, fleshiness, flavor, crispness, brix value, seed content, placental mass extent and fruit weight are further critically examined plantwise. The male plants are studied for the vigour, health, stem, leaf and the male spike, anther size and pollen colour to verify the true to type nature. Seeds collected from the paired crosses are examined in the laboratory for trueness to type and uniformity in development and seed traits. In case of dual purpose variety, in the sibling plants, latex production ability is assessed as per the prescribed standard practices. The latex composition is also evaluated in the laboratory. Progeny rows with deviating type of plants in male as well as female are rejected.

In the next year, following the procedure described earlier, 100 pairwise crosses are performed and 100 progenies are raised in 2000 m. isolation plots with each row having 20 to 30 plants per progeny with equal number of male and female plants and in two replications. Progenies showing deviations in any of the traits in male as well female are rejected. After assessment of the field performance, genetic uniformity and features of the fruit, progenies performing above mean + 2SE are identified. Utilizing selected elite male and female plants in the selected progenies, 100 paired crosses are made and fruits are collected. At maturity and ripe stage, each and every crossed fruit is examined and only from selected cross fertilized papaya fruits, seeds are extracted and saved in quality seed packets. Seeds are examined in the laboratory for seed traits and selected. Seeds from each packet of paired crosses are collected and pooled to constitute nucleus seed. The remnant seeds are pooled together to be used as stock seeds or seeds for next cycle of nucleus seed.

Breeder seed production: The breeder seed production plot is raised with minimum of 2000 m. isolation. Polybag nursery sown with 4 seeds per bag is raised and planted in pits. In each pit, only one plant either one male or one female plant is retained, preferably pit with one female plant is alternated with pit with one male plant in order to facilitate sibmating by paired crosses. Before paired crosses are initiated, all the male and female plants are examined periodically for true to type in multistages of development namely vegetative, preflowering, post flowering and fruit development. Deviant type of male and female plants are uprooted. One hundred paired crosses are made. Non crossed fruits from individuals are examined for fruit and fruit related characteristics and qualities and market acceptability and matched with prescribed characteristics. Male and female parents showing deviations are discarded along with the seeds from the respective paired crosses. Visit of Monitoring Team is arranged to examine the breeder seed production field. The seeds from crossed fruits alone are extracted and maintained in separate packets. Laboratory examination is made for the seed characteristics and seed packets are selected, bulked and Breeder seed is constituted. Samples are drawn from seed packets and seed quality parameters are tested. The quality assured seed is tagged with breeder seed label. GOT is conducted with seed samples drawn from breeder seed to test the genetic purity . Breeder seed II stage production may be continued depending on the requirement.

Nucleus seed production in Gynodioecious Papaya plants: In papaya, gynodioecious (GD) plants are with bisexual flowers (hermaphrodite) and female (pistillate) flowers. Self fertilizing the GD plant will give rise to 1 female to 2 bisexual plants.

Nursery is raised from source population stock seeds of Gynodioecious(GD) plants. Field planting is done and all good agronomic and horticultural practices are practised that are the same as in the previous section of Female x Male cross for breeder seed production. In the pits with four plants planted, only one bisexual plant is retained per pit and female plants are uprooted. All the plants in the population are grown in an isolation of 2000 m and are critically examined upto preflowering stage and deviating plants and off-types are uprooted. At flowering time, 100 plants each with bisexual flowers are identified and the bisexual flowers are self fertilized. The self fertilized seeds are then raised in the next generation as described earlier and in the plant progeny rows only 20 bisexual plants are retained in each row. Periodical examination of individual bisexual plants is made at seedling, juvenile, preflowering, flowering and fruit setting and fruit development stages. Open pollinated fruits from the self fertilized trees are examined for the fruit and fruit market related traits including keeping quality. Self ferilised seeds from the selected bisexual plants are stored in separate paper covers. This cycle will take 24 to 30 months.

In the next year, nursery in polybags is raised from the self fertilized seeds of GD and seedlings from single plants are planted in pits at the rate of 4 seedlings in each pit in well prepared field that is isolated from other papaya crop by 2000 m. distance. The progenies are planted in 2 replications. At flowering, female plants are uprooted retaining only one hermaphrodite plant. The progeny rows having deviant plants are uprooted. Self fertilization of bisexual flowers is performed in the individual plants in each selected progeny rows. The progeny rows with bisexual plants showing / exhibiting deviation for any of the characteristics are discarded as a whole. Field observations are continued on all the bisexual plants in each progeny row for the true to type stem, canopy, foliage, flower, pollen colour, stability in sex expression(GD), fruits setting, immature and mature fruits, cut fruits and flesh fruit quality traits and characteristics to check for the conformity to the prescribed characteristics. Very early and late flowering progenies are discarded to maintain uniformity in flowering. In each true to type progeny row, 15 to 20 plants are self fertilized. Evaluation for yield is made in the true to type progenies. Progeny rows showing performance above mean + 2SE are identified. From among the selected progeny rows, 100 self fertilized plants are selected conforming to the true to type nature and uniformity and seeds are extracted. After the laboratory examination, the seeds from individual plants are stored in separate paper covers. Seeds from each plant are bulked to constitute nucleus seed.

Breeder seed production: The breeder seed production in GD papaya is done in isolated field with 2000 m. distance. Polybag nursery with 4 seeds dibbled in each bag is raised and at 60 DAS, the seedling from each polybag is transplanted in pits of 45 cm x 45 cm x 45 cm with a spacing of 1.80 m on either side. All good agronomic and horticultural schedules with macro and micro nutrient inputs and plant protection measures were followed throughout the crop production upto 24 to 30 months. At flowering stage, the pistillate plants are uprooted leaving only one bisexual plant per pit. Periodical field monitoring is made to assess the true to type nature and deviating plants are uprooted. Diseases and pest affected plants are destroyed. Monitoring Team visit to the field is organized. At the time of flowering 100 to 200 bisexual plants are identified, self fertilized and seeds are extracted. The non self fertilised open pollinated fruits from each of the selfed plants are examined for conformity for trueness for the unripe, ripe and cut fruit details and for fruit and market related characters. Laboratory analysis for fruit quality is completed . Seeds from the finally selected self fertilized plants are pooled to constitute Breeder seed. Samples of breeder seed are drawn and seed quality parameters are tested. GOT is conducted for genetic purity. Breeder seed labels are tagged to assure the seed qualities. Depending on the quantity of seed required, the self fertilized seeds production can be planned.

Nutritive quality of papaya fruits are analysed and recorded. In case of papaya grown for latex/ papain the yield of papain and its quality are studied. Utility of green papaya after latex extraction should be explored. Dual purpose papaya varieties are assessed for fruit yield and quality and for latex yield.

Papaya

Nucleus and Breeder seed Production (Female and male parents)

Female (mm.)× Male (Mm.) crosses give raise to female (mm.) and male (Mm.) plants in equal proportion (1:1)

↓

Prepare 100 pits and sow 4 seeds in each pit. At flowering, one male plant is retained in each of the source the seeds. 50 pits and in the remaining 50 pits one female plant is retained.

↓

Examine the male and female plants for vigour, uniformity, true to type to the characteristics, stability in sex expression, free from virus, diseases and pests. At the end of 2nd year, female plants are scored for prolific and consistent fruiting and high fruit set. Fruit shape and size noted and selection of plants made. 45 to 50 paired crosses (female ×male) are performed. Crossed female flowers secured and protected with paper covers. 3 to 5 female flowers are crossed in each female tree. Seeds are collected from fruits set in paired crosses and labelled pairwise crosses

↓

In the next season, 50 paired cross progenies are raised with 10 plants in each progeny (5 female and 5 males). The plants in the paired cross progenies are individually examined for all aspects and for fruit and flesh as described in previous year crop. Individual male and female plants within each progeny row are labelled and paired crosses made. Seeds collected from paired crosses are examined in the laboratory.

↓

Seeds of selected paired crosses are raised in progeny rows and male and female plants are maintained in equal ratio in each paired cross row. After verifying the health, vigour and selection criteria, best performing progenies over and above mean value are selected and superior male and female plants are identified. Paired crosses are performed pairwise in each progeny. The ripe fruits and seed set in each cross is examined in the laboratory and seeds from the selected paired crosses are saved crosswise. Part of seeds from each paired crosses are bulked to constitute Nucleus seeds.

↓

In the following generation, breeder seed crop is raised and the plants are maintained in female to male plant in the ratio of 5:1. Very critical observations are made for the health, vigour, uniformity, stability of expression of characteristics of stem, leaf, canopy, male and female phases expression, free from pests, diseases and virus infections, flowering uniformity and high fruit setting and productivity. In the 2nd year, when the fruiting is at maximum, selected female trees are pollinated with vigorous male trees. The fruits and crossed seeds are examined in the laboratory for fruits, fruit flesh and seed characteristics. Part of the seeds from selected crosses are then bulked to constitute breeder seed. The breeder seeds are testing for the seed technological parameters for quality assurance GOT is conducted with seed samples drawn from the breeder seed. Nutritive and fruit qualities are assessed. In the meanwhile, fruits from the selected trees are examined for suitability for dual purpose (juice and papain extraction) or table purpose.
Based on the requirement, NSI is repeated to NSII;
Breeder seed Production is repeated.

Note:

1. The NS and BS production is repeated once in 3 years.
2. Crop rotation is followed to avoid successive papaya crops.
3. Isolation distance followed is 2000 m.
4. Clean field and clean cultivation should be practised.
5. Off-types and deviating plants are discarded before or at preflowering stage.
6. Vigorous, uniform and stability in sex expression and fruit traits are maintained.
7. Conformity to DUS characteristics.
8. Inbreeding depression duly managed.

Papaya

Nucleus and Breeder Seed Production in GYNODIOECIOUS Plants

Gynodioecious (GD) Plants upon selfing give raise 2 GD ($m^h m$) and 1 pistillate (mm) plants. (2:1 ratio for GD and Female plants)
Raise seedlings from source seeds in 100 pits. At flowering, examine the plants and cull out female plants. Retain only one GD plant per pit.

↓

Examine the GD plants for stable expression of GD nature and for true to type expression of the characteristics (DUS), vigour, disease and pest free nature.

↓

Identify GD plants with prolific fruit production, fruit set and select 50 GD plants and self fertilise.

↓

In the following period raise 50 progeny rows of GD plants. Examine for trueness, vigour and uniformity. Early and late flowering plants are removed. 25 progeny rows performing above mean value at the end of second year are selected and 5 highly productive GD plants in each selected progeny are identified and self fertilised. Seeds collected plantwise. Ripe fruits and seeds are examined in the laboratory for fruit flesh and seed characteristics, fruit /flesh quality and market acceptability. Seeds are saved plantwise.

↓

Part of seeds from selected individual GD plants are bulked to constitute Nucleus Seed of GD plants.

↓

Breeder Seed: 100 GD plants are selected after examination for trueness, fruit setting, pest, disease, virus free nature, non lodging nature. 3 to 5 flowers in each selected GD plants are pollinated by dusting or rubbing pollen from different GD plants (intercrossing). Pollinated flowers protected. Monitoring Team visit is arranged. Plantwise fruits and seeds are examined in the laboratory for trueness and fruit qualities. Seeds are saved selected plants wise. Quality assured seeds are labelled. Seeds from selected plants are bulked and tested for seed parameters and Breeder Seed labels attached.

↓

Fruit/flesh qualities and nutritive aspects are estimated.

↓

GOT is conducted to confirm genetic purity.

Note:

1. Once in 3 years the NS-BS cycle
2. Isolation distance 2000 m.
3. DUS characteristics
4. Crop rotation is followed.

99. Pineapple

(*Ananas comosus* L. Merr.) (2n:50)

Family : Bromeliacea

Pineapple is a herbaceous perennial producing 4 to 5 aerial suckers per year. Suckers arise from the leaf axils on the main stem and form roots. Slips are the suckers that arise immediately below the fruit. Sucker slip when planted reach bearing earlier than the crown slip suckers and produce larger fruits. Suckers should be planted 2 weeks after removing them from mother plant. Suckers from mother plants that are healthy and disease free alone should be selected.

The inflorescence is terminal consisting of 50 to 200 individual flowers borne spirally and capped by a crown made of approximately 150 short leaves on a short stem. Flowers are hermaphrodite. The individual flowers are typical of monocots, trimerous with three sepals and petals, six stamens (approximately half the length of petals) in two whorls and one pistil, which contains three carpels. Each flower is surrounded by a thick bract, which is covered by trichomes and pointed at its tip. The sepals are more or less triangular in shape and are comparable to the bracts in colour and texture. The flowers are arranged in a narrow compact tubular manner. The style is trifid and longer than the stamens. The ovary contains three trilocular carpels, with the ovules present in two single or double rows. Depending on the cultivar, the number of ovules can vary from 16 to 71. The stigma is made up of three lobes and is borne within the petals (Bartholomew *et al.,* 2003). Flowers are normally self sterile and fruit development is parthenocarpic. Uniformity in growth, flowering and fruiting is essential to facilitate the harvest at one time.

Propagation of this monocot is either by slips or suckers. Slips arise from fruiting stems and from the crown. Suckers originate from the underground part of the stem. Plants grown from suckers produce fruits in 18 months while the plants originating from slips and suckers from the disc cuttings take over 2 years time for fruiting. After fruits are harvested, the stalks are cut into pieces / discs and used for propagation. Leaf cuttings 10 to 15 from each crown can be utilized for planting but takes more time for fruiting. Planting material required for one ha is 43500 slips at a spacing of 90 cm x 60 cm. Suckers and slips weighing around 350 to 400 g are planted. The cropping consists of the main crop and 2 subsequent crops. Average duration is 18 to 24 months and the yield is 50 t./ha.

Nucleus Seed Production: Basic source stock is grown for maintenance in a disease and pest free and healthy conditions that have received good agronomic practices and proper manuring and micronutrient (Zn and Fe) application. At 35 to 40 leaf stage of growth, the crop should have received NAA or Ethrel and sodium carbonate spray to have uniform flowering and NAA spray after fruit formation in the recommended manner to increase the size of the fruit. From this source population planted in 90 cm x 60 cm, 2000 healthy, disease and pest free uniform and true to type single plants at flowering/harvest stage are selected. All the selected plants should conform to the varietal characteristics. After harvest, the sucker propagules are prepared from the individual harvested plants.

Suckers (weighing 400 to 500g) from the selected plants are planted plant wise with labelling in well prepared plots. All recommended good agronomic practices, fertilization and hormone application schedules are followed. Periodical observations are made to examine the conformity to the plant characteristics. Isolation distance of 3m. is followed. At maturity, fruit characteristics expression in the progeny rows is looked into. Progenies showing deviating type of plants are discarded. Progenies conforming to true to type nature and uniformity are harvested along with the suckers. In the next season, the suckers from selected true to type plants are grown plant wise (progenies). Based on the periodical observations progenies showing deviant plants are discarded and uprooted. The selected plant progenies are harvested and the suckers collected progeny wise are raised in progeny plots, each progeny in 2 rows. The original source variety is grown once in 10 progeny plots for comparison. Individual plants in each progeny rows are critically verified for the conformity to true to type, uniformity and stability for the characteristics. Progenies showing deviating type of plants are discarded. Progenies showing high productivity and fruit quality and fresh fruit market acceptability are identified at the time of harvest. Among the selected true to type progenies, highly productive and superior plants are tagged and fruits harvested and examined in the laboratory for fruit size, shape, fruit texture, aroma, flavor, taste and traits highly suitable for juice industry and other consumer preferences and market acceptability. Suckers are finally prepared plantwise from the selected plants and pooled together to constitute the nucleus seed stock The selected plants from the productive progenies are retained as ratoon crop for production of quality suckers and also as basic stock to develop next cycle of nucleus stock.

Breeder Seed Production: The suckers from nucleus stock are planted (Clones:Suckers/slips, the weight of the sucker should be 400 to 500g and slips 350-450g) in well prepared clean plots with spacing of 90 cm x 60 cm. All recommended good agronomic practices, fertilization, micronutrient and hormone application schedules are followed and 3m. isolation is adapted. Periodical observations are made to examine the true to type nature, uniformity and stability of the vegetative, floral and fruit characteristics from crop establishment to fruit maturity. At maturity, fruit characteristics expression in the individual plants is looked into. Deviating type of plants are discarded. Plants conforming to true to type nature, uniformity and stability are harvested .In the laboratory the harvested fruits from individual plants are examined for fruit size, shape, fruit texture, aroma, flavor, taste and traits highly suitable for juice industry and other consumer preferences and market acceptability. Plants are selected and suckers are collected plantwise from the selected plants in the breeder seed plot and bulked to constitute breeder seed stock The selected plants in the breeder seed plot are retained as ratoon crop with great care for production of quality suckers. The fruit qualities are estimated.

Micropropagation. Dormant axillary buds from the crowns are utilized for shoot induction. 1000 to 2000 plants can be developed in a year from single crown (Pine Apple Research Station KAU, Vazhakulam). Mealy bugs associated virus (PMWA) is to be eliminated. One lakh tissue cultured pineapple (TC plants) are produced in MD2, Amritha, Kew, Marritus and Nanas varieties. (Hort. Purdue). Vegetative production does not assure facsimile reproduction. Many mutations and distinct clones have occurred. (Hort. Purdue. Pineapple). The *in vitro* produced plants are tested for genetic fidelity, free from virus and hardened and planted as nucleus or breeder stock plants to supply planting material.

100. Pomegranate

(*Punica granatum* L.) (2n=16,18)

Family : Punicaceae

Pomegranate is a diploid with 2n : 16 and 18 chromosomes. It is often cross pollinated and is genotypically heterozygous. Best performing true to type and uniform plants in highly productive orchards are selected and 10 to 15 hard wood cuttings are prepared from each of the selected plant. Adopting mist propagation, 10 or more rooted cuttings per selected plus tree are prepared to develop the clones of the tree. Clonal orchard plots are then planted with 12 to 18 month old rooted cuttings, around 10 or more from each plus tree. The isolation distance of 3m. is followed. Detailed production practices are presented by Hiwala (2009) and Jalikop (2010).

Nucleus Stock: All recommended agrotechniques and horticulture techniques are followed to raise a good crop. At pre flowering stem and foliage characteristics are verified and true to type expressing clones bearing fruits alone are retained.

Fruits characteristics are documented on whole fruit basis and cut into pieces to evaluate the arils, colour, juiciness, taste and soft seed characteristics, fresh market quality & consumer appeal. Based on the productivity & quality of fruits and contents, top performing true to type clones are selected and maintained as nucleus stock. The characteristics of pomegranate is presented in the Table 1.

Table 1. Characteristics of pomegranate.

1.	Bush/ tree height (m)	Short (<1.5), medium (1.5-2.5), tall (>2.5)
2.	Growth habit	Upright, spreading drooping
3.	Precocity (years after planting)	Early (<2), medium (2-3), late (>3)
4.	Shoot thorniness (no. of thorns /m)	Absent, less (<5), medium (5-10), high (>10)
5.	Foliage density	Sparse, medium, dense
6.	Leaf blade length (cm)	Short (<2), medium (2-5) and long (>5)
7.	Leaf blade width (cm)	Narrow (<1), medium (1-2), broad (>2)
8.	Leaf blade shape	Acute, obtuse, rounded
9.	Petiole length	Low, medium, long
10.	Petiole anthocyanin (% part covered)	Low (<25), medium (25-50), high (>50)
11.	Calyx length (mm)	Short (<20), medium (20-40), long (>40)
12.	Calyx width (mm)	Narrow (<10), medium (10-15), long >15
13.	Calyx colour	Yellow, orange, red, others
14.	Corolla colour	White, orange, red, others
15.	Corolla type	Single, double
16.	Petal length (cm)	Short (<15), medium (15-25), long (>25)
17.	Petal width (mm)	Narrow, (<10),medium (10-20), broad (>20)
18.	Fruit length (cm)	Short (<6), medium (6-8), long (>8)
19.	Fruit dia (cm)	Small (<5), medium (5-7), large (>7)
20.	Fruit shape (length to lateral)	Round (1.0-1.1), oval (1.1-1.2), oval (1.2-1.3), elliptical (13>)
21.	Fruit colour	Yellow, red, deep red, others
22.	Rind thickness (mm)	Thin (<3), medium (3-5), thick (>5)
23.	Nipple (Fin)	Absent, present
24.	Crown length (mm)	Short (<15), medium (15-25), long (>25)
25.	Crown neck	Absent, present
26.	Aril colour	White, light yellow, high pink, pink, red, dark red, others
27.	Aril length (mm)	Short (<10), medium (10-15), broad (>15)
28.	Aril width (mm)	Narrow, (<5), medium (5-7.5), broad (>7.5)
29.	Seed hardiness	Soft, medium, hard
30.	Seed length (mm)	Short (<6), medium (6-10), long (>10)
31.	Seed width (mm)	Narrow (<2.5),medium (2.5-5.0), broad (>5.0)
32.	Fruit maturity days after anthesis	Medium (130-170), late (>170)
33.	T.S.S (°Brix)	Low (<12.5), medium (12.5-16.0), high (>16)
34.	Acidity (%)	Low (<0.5), medium (0.5-1.25), high (>1.25)
35.	Fruit juiciness (%)	Low (<50), medium (50-60), high (>60)

(*Source:* PPV & FRA)

Breeder Stock: Hardwood cuttings from the true to type plus trees of nucleus stock are prepared and planted in a well prepared field to raise breeder stock plants. True to type, uniformity, free from diseases are ensured. Monitoring Team visit is organized at the time of first fruit bearing. Cuttings from the breeder stock plants are distributed to the growers.

Shoot tip culture of best clones are developed by tissue culture. Micropropagated *in vitro* plants from each plus clone are hardened and rooted in the rooting soil mixture in mist chamber. When the shoots are well established with adequate root mass, at least 20 to 30 plants are established in the field for each plus clone in the variety. All usual recommended horticultural technologies in respect of training, pruning and plant protection measures are followed. True to type nature or genetic fidelity is established in the *in vitro* clones. Nucleus stock is raised from *in vitro* plants. Breeder stock is developed with rooted cuttings of nucleus stock.

101. Sapota

Manilkara achras (L.) P. Royen

Family : Sapotaceae

Sapota is an evergreen fruit tree with radial whorls of branches that bear flowers and fruits terminally at leaf axils. Flowers are small and solitary or in cymose clusters in leaf axis, produced on the past season growth of shoots. Fruit matures in 7 to 11 months after pollination. Sapota fruit is climacteric, to be artificially ripened.

Seedling raised sapota plants commence flowering and fruit setting in 7 to 8 years after planting while grafted plants starts fruiting in 2 to 3 years time. Grafted plants attain economic yields after 7 to 8 years from planting. Sapota flowers are protogynous with stigma growing above the petals and stamens two days before anthesis favouring and felicitating cross pollination. Self incompatibility in flowers is also reported in sapota leading to low fruit set in single tree planted in homesteads. Flowers have 6 sepals in two whorls of 3 in each in gamosepalous condition. Petals 6 arranged in bell shaped tubular manner. Stamens are 12 and of which 6 are petalloid staminodes and 6 are epipetalous, and fertile with short filaments. Pistil is syncarpous carpels, 4 to 5 fused; placentation axile, style is long and green with capitate style protruding above stamens two days before anthesis. Anthesis is at maximum from 8 to 9 hrs. Hand pollination gives maximum fruit set.

Sapota is clonally propogated by air layering, soft wood cutting by dipping in IBA 2000 ppm, grafting onto 2 years old seedlings of Rayan or Pala root stock (*Manilkhara hexandra* L.), inarching and softwood grafting. Grafts are planted in 10m. x 10m. spacing. Spacing of 5m. x 5m. is adopted in high density planting. It is reported that in 5m. x 5m. spacing, maximum fruiting occurs in 13 years after planting and then declines. Yield of fruit is 20 to 25 tons per ha. The trees require training and pruning to maintain effective plant frame and light penetration to produce maximum number of fruits. In the nursery and in the nucleus and breeder clone plots, it is necessary to maintain isolation distance of 10m. from sapota root stock or grafting nurseries. Nursery rotation should be done with other crops once in two times nursery raising of sapota.

Nucleus Stock Production: Healthy, disease free, regular bearing true to type trees are selected from the Institution or the breeder sources or mother stock. Bud sticks of suitable size bearing healthy and disease free buds are separated from each of the selected true to type mother trees and grafted onto the vigorous and healthy seedlings of two years grown from the seeds of Rayan or Pala (*Manilkhara hexandra* L.). When the grafts / budded clonal plants are ready for field planting, ten plants from each mother tree are planted in a single row in well prepared pits of 1m x 1m x 1m size at 10m x 10m spacing in usual practice or in 5m x 5m spacing in high density planting (HDP). 5m. isolation is adopted. Twenty clonal progenies are selected. Systematic and continuous monitoring of the nucleus clone plot is done for studying the conformity to trueness, uniformity and stability in expression of the characteristics. Progeny rows showing non conformity are uprooted / discarded at the earliest. Progeny rows showing trueness, uniformity and conforming to the characteristics alone are retained at juvenile / vegetative phase in a well prepared plot that is isolated by 10m. from other sapota plants. At the time of flowering and fruiting phases, individual clonal plants in each of the clonal progeny rows are examined for the stem, branch, foliage, flower and fruit characteristics, fruit set and fruiting ability. Conformity to the characteristics is verified. Clonal progeny showing non conformity or deviations is uprooted. The yield potential of the individual plants in each of clonal progeny showing conformity is recorded at least for 2 to 3 regular bearing seasons and superior performing (above mean) clonal plants are identified as nucleus clonal plants. The clonal plants will be planted to raise the breeder clone or mother germplasm resource garden.

Breeder Stock Garden: The grafts produced utilising the budwoods from nucleus clone are verified for health, trueness and for characteristics of stem, branch, foliage, flower and fruit. Any non conformity plant, if present, is uprooted. Genetic purity is thus assured. At fruiting stage in two to three regular bearing seasons, the productivity is assessed and superior plants alone are retained, tagged and well protected from pests, diseases and micronutrient disorders. Bud woods/scion buds collected from the appropriate growth stage serve as breeder stock buds source for grafting to produces mother blocks or foundation clones.

Visit of Monitoring Team is arranged to study the field standards and genetic uniformity and purity of the Breeder clone garden. The fruit quality aspects will be estimated. Meanwhile, quality assured seedlings of *Manilkara hexandra* of root stock will be raised in polybags in a dedicated root stock nursery for supplying root stock plants to the budding demands along with the true to type quality assured scion buds.

The specifications of the clonal plants are: 1. grafted plants/clones free of suckers; 2. diameter of the root stock is from 1.00 to 2.00 cm; 3. height of grafting is 15 to 25 cm; 4. diameter of the graft at 10 cm above the graft union should range from 1.0 to 2.5 cm and height 50 to 75 cm. In case of inarching and in air layering 30 to 60 cm 5. conformity of the clones for the specified size shall be 100% by number.

102. Tinda (Round Melon)

(*Praecitrullus fistulosus* (Stocks) Pangale) (2n:24)

Family : Cucurbitaceae

Tinda is also known as Round gourd, Apple gourd and Baby Pumpkin. Tinda is considered as a native species to India, Pakistan and Afghanistan. It is a monoecious species, annual, climbing or trailing habit with robust villous hairy stem. Tendrils 2 to 4 to 5 fids. Leaves slender, alternate, simple, stipules absent, petiole 6 mm. long hirsute. Lamina ovate, 15 cm long, cordate at base, hispid, hairy flowers solitary in leaf axils, 3 cm dia. regular pentamerous, pedicel short. Calyx campanulate, petals 5, connate, hairy. Male flowers with 3 stamens. Female flowers with inferior ovary and stigma 3. Fruit is a gooseberry, 6 to 12 cm diameter, hispid in immature fruit and smooth glabrous in matured fruit. Fruit pale green to dark green in exterior and creamy white to pale green in inner side. Seeds are many, ovate, oblong and compressed, 8 mm. long with ridged margin and smooth black in colour. Male flowers open one week before flowering of female flowers. Pollination is smoothly by bees. Fruits are harvested 13 to 15 weeks after sowing. Seed rate followed is 3 to 3.5 kg/ha; spacing recommended is 90 cm x 30 cm. and 2 seedlings, per pit are maintained. Yield expected is 10t/ha in 90 days. Four fruits per plant are obtained. Seeds of two released varieties are available for cultivation.

Nucleus seed production: A good crop of Tinda is raised in a 500 sq.m plot that is isolated by 1600 meters from other Tinda crops. The seeds are sourced from the institution or the breeder concerned. Recommended Tinda crop production practices are followed adapting spacing of 90 cm x 30 cm. to get good expression of plant traits. Border rows of 5 meter wide are raised all around the plot. At the time of flowering, very early and late flowering plants are removed. Uniform looking, true to type, healthy and vigorous true to type single plants are tagged 15 to 20 days after flowering. Based on the true to type nature of the developing immature fruits, 300 uniform and true to type plants are selected and labelled. In each selected 300 plants, fruits already set are removed and 2 to 3 flowers are self fertilized. The self fertilized flowers are tagged and indicator sticks are firmly fixed in the site. At maturity, 1 to 2 self fertilized fruits from individual plants are harvested and stored for 30 days to complete seed maturity. In the laboratory, the fruit characteristics and weight, flesh texture, colour and firmness, size, shape and glabrous nature at maturity are verified and seeds are extracted. Seed size, shape, colour, and development are recorded. Seeds from the finally selected fruits are stored plant wise (Fig. 1).

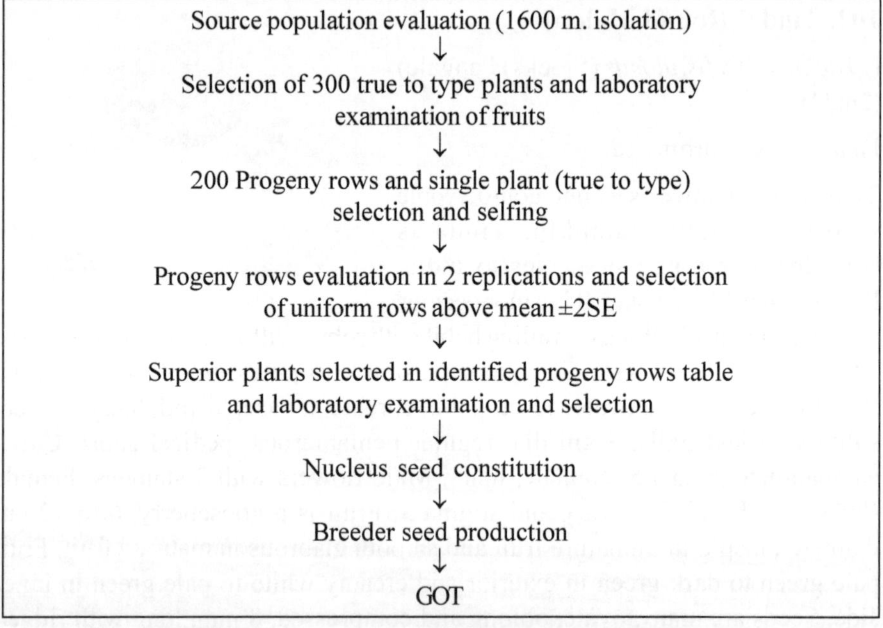

Fig. 1. Nucleus and Breeder seed production in Tinda

In the next favourable season, seeds from selected 200 individual plants are sown in 2 row progeny plots, each row 90 cm apart and 20 hills on the rows retaining single plant per hill and spaced 20 cm on the rows. Plant progenies are sown in two replications. Between progeny rows two meter space is provided.

Once in 10 progeny row plots, 2 rows of check plants are raised for comparison. Border rows, 5m wide are raised on all sides. The plot is isolated by 1600m from other tinda fields. All good agronomical and horticultural protocols are practised. Early and late flowering progeny rows are discarded. Progeny rows showing off-types and deviant plants, if noticed, are uprooted. Matured fruits are harvested and yield recorded in terms of number of fruits and weight. Based on the evaluation for productivity and fruit traits, progeny rows performing above mean +2SE are identified and seeds saved from superior plants in each selected progeny row. Seeds are examined in the laboratory and saved plantwise. Part of seeds from each plant is bulked together to constitute Nucleus seed.

Breeder seed production: In a well prepared field to fine tilth, not grown with tinda in the past 2 to 3 years, nucleus seeds of Tinda variety are sown in lines adapting 90 cm x 30 cm. spacing. The breeder seed production plot is provided with 1600 m. isolation distance. Border rows, 5 m. width are raised on all the sides of the plot. Best agronomical and horticultural practices are adapted. Only one healthy plant is retained per hill. Early and late flowering plants, off-types and deviant type of plants are uprooted. Deviant types, if noticed at flowering or after flowering are removed along with plants around them to avoid pollen contamination. Individual plants in the breeder seed production plot are examined for trueness, uniformity and stability of traits expression at juvenile, preflowering, flowering, fruit development and at maturity for stem, branches, foliar, flower and fruit characteristics. Plants showing good fruit s*et al.*one are retained. At mid maturity stage, visit of Monitoring Team is organized. Fruits are harvested from individual true to type plants, examined in the laboratory and selected for seed extraction. Seeds extracted are verified plantwise for trueness and saved plantwise. Part of the seeds are then bulked to constitute Breeder seed. Samples of breeder seed are tested in the Seed Testing Laboratory for various seed qualities. Quality assured seeds are labelled with Breeder Seed Tags. Samples of Breeder seed are tested in the Grow Out Tests to ascertain the genetic purity. Nutrient quality of fruits is estimated.

103. Watermelon

Citrullus lunatus (Thunb.) Matsum and Nakai (2n : 22)

Family : Cucurbitaceae

Watermelon is monoecious plant with angular stem and pinnately divided leaf lobes. Flowers occur singly, either pistillate or staminate occurring in the same plant.

The pistiliate flowers normally occur in every seventh leaf axil while the staminate flowers occur in the intervening axis. Corolla 5 lobed, rotate, yellow coloured. Flowers open shortly after sunrise and remain open for one day only. Usually anthers (three in number) dehisce before flower anthesis. The stigma is receptive throughout the day. The greater part of the fruit flesh is mostly derived from the placenta. Plant is naturally cross pollinated with considerable self pollination and sibling. The descriptor and characteristics are given in Table 1&2.

Table 1. Descriptor of watermelon

1. Ploidy	
2. Season of cultivation	Including suitability for intensive and/or extensive protected cropping
3. Plant vigour	Moderate, high
4. Foliage	leaf protection against sun-scald
5. Fruit	External shape : round, oval or long Relative size and weight at market maturity Rind color, bi-color, pattern, striped Internal : thickness or rind, flesh color, intensity of flesh color (especially towards the center of the fruit), absence of central cavity
6. Seed	External color when mature, striped or single color, relative size
7. Resistance to specific pests	Pathogens or other disorders, e.g. Fusarium sp., Didymella bryoniae (gummy stem blight) & Colletotrichum sp. (anthracnose or sun- scald).
8. Selection criteria in watermelon	i) Longer main stem ii) More number of primary branches iii) Early first female flower appearance iv) Desirable fruit shape. This can be done in seedling stage. Long narrow cotyledons indicate elongate fruit shape whereas round cotyledons show round fruits. v) Fruit length and diameter vi) Medium size fruit vii) Thick tough flexible rind- 1/16-1/8 inch thick and 3 inch long strip is bent into a circle. If it makes a complete circle, it is very tough. if it breaks early in this attempt, it is tender (explosive). viii) Firm flesh ix) Intense red flesh x) TSS > 10% xi) Resistance to Fusarium wilt, anthracnose, fruit fly.

Note: Descriptions are based on UPOV, PVP & FRA, Hari Har Ram (2005)

Table 2. Characteristics of watermelon *Citrullus lunatus* (Thunb) Mansf.

1. Cotyledon shape	Narrow elliptic, medium elliptic, broad elliptic
2. Plant length of internode (cm)	Short (<7), medium 7 to 8.5), long (>8.5)
3. Leaf blade length (cm.)	Short (<10), medium (10-14), long (>14)
4. Leaf blade width (cm.)	Narrow (<9), medium (9-12), broad (>12)

5. Leaf blade colour	Light green, green, dark green
6. Leaf blade degree of primary lobing	Weak, medium, strong
7. Leaf blade degree of secondary lobing	Weak, medium, strong
8. Petiole length (cm.)	Short (<7), medium (7-9), long (>9)
9. Appearance of first pistillate flower in 50% of plants (days)	Early (<45), medium (45 -50), late (>50)
10. Sex expression at full flowering	Monoecious, andromonoecious
11. Male sterility	Absent, present
12. Ovary length (cm.)	Short (<1), medium (1-2), long (>2)
13. Ovary width (cm.)	Narrow (<0.8), broad (>0.8)
14. Ovary pubescence	Sparse, dense
15. Ovary colour	Yellow, green, dark green
16. Fruit shape in LS	Round, flat, oval, cylindrical (oblong), elongated globe
17. Fruit colour of rind	Yellow, light green, medium green, dark green
18. Fruit grooves	Absent, present
19. Fruit stripes	Absent, weak, diffused clearly defined
20. Fruit length (cm.)	Short (<20), medium (20 to 30), long (>30)
21. Fruit diameter (Cm)	Narrow (<15), medium (15 to 25), broad (>25)
22. Fruit size (Kg.)	Small (<3), medium (3 to 6), large (>6)
23. Fruit flesh colour	White, yellow, orange reddish pink, light red, dark red
24. Seediness - number of seeds/ fruits	Absent rudimentary (<20), low (<150), medium (150-350), high (>350)
25. Seed length (cm.)	Short (<0.6), medium (0.6 to 1.0), long (>1.0)
26. Seed width(cm.)	Narrow (<0.4), medium (0.4 to 0.7), broad (>0.7)
27. Seed colour of coat	White, grey, red, brown, black, others

Source: PPV & FRA

Nucleus seed production: The seeds of the specified variety or the line from the source population in the maintenance breeding or stock seeds are used for sowing. In a well prepared field, long channels are formed 2.5m apart and on one side of the channel pre-germinated seeds are sown in pits 0.9m apart. After the germination of seeds, only one vigorous seedling per hill is retained. The field is isolated by 1600m from other fields grown with watermelon crop. Five hundred true to type similar looking single plants are finally selected in the field. The plants are checked for true to type for the descriptions, uniformity, vigour and free from diseases and pests at juvenile, pre flowering and at flowering phases. Off-types and deviating type of plants are removed. When the first female flower is ready for opening, the female buds that will open next day are protected by covering with butter paper bags to avoid cross pollination. Likewise male flowers in the same plant about to open next day are covered with paper bags. The female flowers that are protected from cross pollination when opened are pollinated by dusting pollen from the bagged male flowers in the same plant opening that day. At least 2 to 3 female flowers are self pollinated in each plant with pollen from the male flowers borne on the same plant. Open pollinated female flowers in the same plant are regularly removed to facilitate proper development of the self fertilized female flowers into fruits. The self fertilized flowers are labeled and stacks are fixed closely for easy locating them. Meanwhile, the plants are monitored from flowering to fruit maturity for true to type, uniformity, vigour, health and free from diseases and pests and deviating plants are uprooted. During the fruit development and maturity the size, shape, colour of fruits and stripe pattern of fruits in individual plants are examined and plants showing deviations in expression for fruit traits are uprooted / discarded. The fruits are retained in the plants till they mature. At the time of harvest, 100 superior plants are identified and the self fertilized fruits are labelled, with plant number to indicate the pedigree. The fruits are stored for a month to allow for seed maturity. In the laboratory, each fruit is now examined for shape, size, fruit, colour, fruit weight, rind thickness, flesh colour, consistency, flesh texture, taste ,aroma, uniformity in ripening, absence of pithiness or spongy nature and absence of central cavity. Degree of sweetness, TSS and extent of seediness, seed shape, size and thinness, sparse seed distribution pattern in longitudinal section besides resistance to heat, fruit cracking and diseases. Seeds extracted from the fruits of selected individual plants that fulfill all norms and trueness are processed and saved plantwise for raising individual plant progenies.

In the next generation, 50 to 60 progenies are raised in 2 replications. Each progeny is grown in 3 row plots, each row with 20 plants (single plant/hill). The source seeds of the variety are grown all round the plot in 3 meter breath border. Isolation distance of 1600 m. is maintained from fields grown with watermelon. In each of the progeny row over the 2 replications, superior performing plants are identified and labelled. At maturity self fertilized fruits from superior plants selected within each progeny and replication are labelled. Fruits are stored for a month for seed maturity. Fruits are examined for various traits of fruits, including flesh and seed characteristics and TSS as described earlier. Based on the fruit traits, seeds are separated from selected fruits that are from individual plants and processed and stored in separate packets. Part of the seeds plantwise from each seed packet is drawn and bulked to constitute nucleus seed. Remainder of the seeds are utilized for next nucleus seed-breeder seed cycle (Fig. 1).

Breeder seed production: In the next season, nucleus seeds are sown in field not previously grown to watermelon to raise breeder seed. Seeds are sown at a spacing of 2.5m x 0.90 m. Only one plant per hill is retained. Isolation distance of 1600 m. is provided. All good agronomical and horticultural practices are applied. Ethrel 250 ppm. (2.5 mg / l. of water) is applied as spray for 4 times at weekly intervals commencing from 15 days after sowing. Monitoring of individual plants is done at different stages of crop growth. Off-types and plants showing deviation from trueness and uniformity are uprooted. First formed female flowers in progeny rows performing above general mean +2SE are identified and self fertilized in all the selected plants. At time of fruit maturity, fruits are examined for trueness and harvested after the visit of Monitoring Team. The harvested fruits are stored for a month for seed maturity. Individual fruits are examined in the laboratory, for all the prescribed fruit characteristics, uniformity in ripening from fruit periphery to the inner fruit core, flesh crispness flavour, taste, TSS, sweetness, less seediness and thin seeds. From the selected fruits, seeds are separated, dried, processed and bulked to constitute breeder seed. The seed samples of selected watermelon plants are tested for seed quality norms and quality assurance and tagged with Breeder seed label. GOT is conducted as prescribed. Nutritive qualities of the fruit are estimated.

Grow source population of 800 to 1000 plants in 1600m. isolation plots. Discard weak and disease affected plants. Allow the crop to maturity and retain the fruits in the field. Survey the plant and the fruits for true to type. Select 100 plants based on the fruits production and trueness and uniformity. Conduct table and laboratory examination of fruit size, flesh, flavors, and market qualities and other specialty features. Collect seeds plant wise and save for the next season.
Raise 200 single plant wise progenies in the next season. In each plant self fertilize 2 to 3 flowers and at maturity of self fertilized fruits seeds are separated plant wise to raise progeny rows.

↓

Raise two rows progeny plots in 2 replications. Monitor / examine the 100 progeny plots from seedling to flowering to fruit for trueness and uniform maturity. Discard progenies showing variation and disease infections, Based on vigour, healthiness and product, and its performance above mean + 2SE, progenies are identified and in each identified progenies, 2 to 3 superior plants are selected and fruits are collected from individual plant wise and studied in the laboratory for fruits and seed features and also for fruit qualities. Seeds are then saved plant wise.

↓

Part of seeds from the selected plants are bulked to constitute Nucleus seed

↓

Breeder seed production and nutritional qualities estimated

↓

GOT is conducted

Fig. 1. Nucleus and Breeder seed production in Watermelon and Muskmelon

Spice and Plantation Crop

104. Ajwain

Carum copticum [L.] Benth.et Hook. (Syn. *Carum ajowan*, *Trachyspermum ammi*) (2n=18)

Family : Apiaceae

Other Names : Ajwain, carom, bishop's weed, omum

Ajawain looks like wild parsley (similar to caraway, celery and cumin seeds). It is a native of India and is grown throughout the country. The striped seeds are used as the spice. Ajowan is an erect, glabrous or minutely pubescent, branched annual that grows up to 90 cm. It is profusely branching and leafy. Stems are striate and leaves are distinct, pinnately divided and feathery. Flowers are small, ovoid 2 to 3 mm. long. Small white flowers are on terminal or seemingly lateral pedunculate, compound umbels. The fruits are small, ovoid, greyish brown, aromatic cremocarps with single seed. Flowers are self fertile, protandrous, with 5 sepals, 5 petals and 5 stames in each flower. Cross pollination by insects is prevalent. (Peter, 2004).

Ajawain grows on all kinds of soil but does well on loams or clayey loams, both as a dry crop and under irrigation. Seeds are sown from September to November. The plants flower in about two months and the fruits become ready for harvesting when the flower heads turn brown. The sensoric quality of ajowan is similar to thyme, but stronger and less subtle. The essential oil (2.5 to 5% in the dried fruits) is dominated by thymol (35 to 60%).

Nucleus Seed Production: Seeds of the variety concerned from the breeder or the Institution are sourced. In a well prepared field previously not grown with the same seed spice crop, the seeds are sown in rows 30 cm. apart and 10 cm. on the rows. After germination and establishment, seedlings are thinned to maintain 5 cm. spacing on the rows. Single plant is maintained per hill. An area of 500 sq m. is sown excluding border rows. Border rows 3 meters width around the plot are raised with the same source seed. This field is isolated by 400m. distance from the same seed spice crop on all sides.

The field is well maintained free from weeds, diseases infection and insect attack. Individual plants are examined continuously at different stages of growth till flowering. Off-types and deviating type of plants are removed. True to type plants for stem, branches and foliage characteristics alone are retained. Early and late flowering plants are removed to maintain flowering uniformity. Plants are examined at flowering phase, seed development and maturity stages for true to type nature. Plants showing variation are discarded. True to type plants are retained till harvest stage and 2000 to 2500 single true to type representative plants are labelled and matured umbels turning brown are collected plant wise in separate paper bags. The umbels are air dried under shade. Dried umbels are then examined individually for shape, size, number of umbels per plant. Extent of seed setting, number of seeds, size and shape of seed, aroma and related characteristics are recorded. Based on the laboratory examination of umbels, seeds and productivity, 200 to 250 plants are selected. Seeds from the selected plants are stored separately plantwise.

In the next favarourable season, single plant progenies are raised in two replications in the field with isolation distance of 400m. The recommended spacing and good agronomic production practices are adapted. Border rows of 3 meters width around the plot are raised with the same seed. The crop is maintained free from weeds, diseases and pests. After every 10 th progeny row, one row of check progeny is raised for comparison. Progenies that are poor in field establishment, weak health and poor vigour are uprooted. Progenies showing variations within progenies for stem, branches, foliage size, shape and colour are discarded before commencement of flowering. Progenies showing deviating plants at or after flowering are uprooted along with the progenies on either side in order to avoid pollen contamination. Progenies showing plants that are uniform, true to type and stability for the expression of stem, branches, leaves size, colour and aroma, umbel shape and size, flower and developing and matured seeds are labelled. The performance of labelled progenies for number of umbels and seed yield are recorded. In the laboratory, umbels and seeds are examined for the trueness and deviating type of progenies are discarded. Progenies showing true to type and performance above mean + 2 SE are identified and the seeds are saved. Part of the seed from each selected plant is pooled together to constitute nucleus seed. Sample from the nucleus seed is analysed for essential oil content, taste, aroma and flavor. Nucleus seed is labelled.

Breeder Seed Production: Nucleus seed is sown in well prepared field, not grown with the same seed spice crop in the past 2 to 3 years. The selected field is isolated by 400m. from other fields being grown with the same crop. The field sanitation is maintained. Recommended spacing of plants and single plant per hill is adapted. The crop is maintained free from diseases and insect pests from the establishment to harvest stage. Individual plants are verified for trueness, uniformity and stability of the characteristics expressed by the stem, branches, leaf, umbels, flowers and matured seed. Off-types and deviating plants noticed up to flowering stage are uprooted. Deviating plants occurring at flowering and post flowering stages are removed along with the plants around them. At mid maturity of seeds, visit of Monitoring Team is arranged to adjudge the field and genetic purity of the crop. At maturity, individual plants are harvested and in tact plants are examined in the laboratory for true to type in umbel and seed traits. Seeds from the selected plants are bulked to constitute breeder seed. Samples of breeder seed are tested for seed quality parameters and quality assured seeds are labelled with Breeder Seed Label. Grow out test is conducted with seed samples drawn from breeder seed. Laboratory estimation of essential oils and other qualities such as aroma, flavor, taste are evaluated.

105. Anise

(*Pimpinella anisum* Linn.) (2n=14)

Family : Apiaceae

Anise, *Pimpinella anisum* L., is a herbaceous annual. It is cultivated chiefly for its fruits, called aniseed. The plant grows to a height of about 75 cm. and requires a warm and long frost-free growing season of 120 days. It has long-stalked basal leaves and shorter, stalked stem leaves. Small and yellowish white flowers form loose umbels. The small white flowers bloom in midsummer. Flowers have each 5 sepals, 5 petals and 5 anthers. Seed maturation usually occurs one month after pollination, The fruit is nearly ovoid in shape, about 3.5 mm. long and has five longitudinal dorsal ridges. The fruit consists of two united carpels each containing an anise seed. The seed is small and curved, about 0.5 cm. long and greyish brown and usually contains hair-like protrusions from each end.

The fruiting umbels should be harvested when the seeds turn brown. The oil content in the dried fruit is about 2.5%. The fresh leaves possess a flavour similar to that of the seeds and may be used as needed during the season. The essential oil of anise is present at about 1.5 to 3.5% level. The major constituent in oil of anise is anethole. Aroma and flavor are well known in aniseed leaves.

Nucleus Seed Production: Seeds of the variety concerned from the breeder or the Institution are sourced. In a well prepared field previously not grown with the same spice crop, the seeds are sown in rows 30 cm x 30 cm on the rows. After germination and establishment , seedlings are thinned to maintain 5 cm spacing on the rows. Single plant is maintained per hill. An area of 500 sq. m. is sown excluding border rows. Border rows 3 meters around the plot are raised with the same seed. This field is isolated by 400 m. distance from the same seed spice crop on all sides.The field is well maintained free from diseases infection and insect attack.Individual plants are examined continuously at different stages of growth till flowering. Off-types and deviating type of plants are removed. True to type plants for stem, branches and foliage characteristics alone are retained. Early and late flowering plants are removed to maintain flowering uniformity. Plants are examined at flowering, seed development and maturity uniformity. Plants are examined at flowering, seed development and maturity stages for true to type nature. Plants showing variation are discarded. True to type plants are retained till harvest stage and 2000 to 2500 single true to type representative plants are labelled and matured umbels turning brown are collected plant wise in separate paper bags. The umbels are air dried under shade. Dried umbels are then examined individually for shape, size, number of umbels per plant. Extent of seed setting, number of seeds, size and shape of seed, aroma and related characteristics are recorded. Based on the laboratory examination of umbels, seeds and productivity, 200 to 250 plants are selected. Seeds from the selected plants are stored separately plant wise.

In the next season, single plant progenies are raised in two replications in the field with isolation distance of 400 m. adapting the recommended spacing and good agronomic production practices. Border rows 3 meters around the plot are raised with the same seed. The crop is maintained free from diseases and pests. After every 10th progeny row, one row of check entry is raised for comparison. Progenies that are poor in field establishment, weak health and poor vigour, are uprooted. Progenies showing variations within progenies for stem, branches, foliage size ,shape and colour are discarded before commencement of flowering. Progenies showing deviating plants at or after flowering are uprooted along with the progenies on either side in order to avoid pollen contamination. Progenies showing plants that are uniform, true to type and stability for the expression of stem, branches, leaves size, colour and aroma, umbel shape and size, flower and developing and matured seeds are labelled. The performance of labelled progenies for number of umbels and seed yield are recorded. Progenies showing performance above mean + 2 SE are identified and the seeds are saved. Part of the seed from each selected plant is pooled together to constitute the nucleus seed. Sample from the nucleus seed is analysed for essential oil content, taste, aroma and flavor. Nucleus seed is labelled with nucleus seed label.

Breeder Seed Production: Nucleus seed is sown in well prepared field, not grown with the same seed spice crop in the past 2 to 3 years. The field is isolated by 400 m. from other fields being grown with the same crop .Three meter wide border crop is raised with the same seed as in the breeder seed plot. The field sanitation is maintained. Recommended spacing of plants and single plant per hill are adapted. The crop is maintained free from diseases and insect pests from the establishment to harvest stage. Individual plants are verified for trueness, uniformity and stability of the characteristics expressed by the stem, branches, leaf, flowers, umbels, developing and matured seed. Off-types and deviating plants noticed up to flowering stage are uprooted. Deviating plants occurring in flowering and post flowering stages are removed along with the plants around them. At mid maturity of seeds, visit of Monitoring Team is arranged to adjudge the field and genetic purity of the crop. At maturity, individual plants are harvested and intact plants are examined in the laboratory for true to type in umbel and seed traits. Seeds from the selected plants are bulked to constitute breeder seed. Samples of breeder seed are tested for seed quality parameters and quality assured seeds are labelled with Breeder Seed Label. Grow out test is conducted with seed samples drawn from breeder seed. Laboratory estimation of essential oils and other qualities such as aroma, flavor, taste are made.

106. Arecanut

(*Areca catechu*) (2n:32)

Family : Arecaceae

India is number one in arecanut production with 57% of world's area and 53% production. Arecanut is a perennial palm with tropical adaptation. Trunk is solitary and crown is spherical to semispherical shape. Arecanut is monoecious. Inflorescence is a much branched spadix. Male flower is sessile and has 3 minute calyx, 3 corolla and 6 stamens arranged in 2 whorls. Male flowers are borne in the upper part of the rachis. Female flower is solitary, 3 sepals, 3 petals, stigma trifid, ovary monolocular. Fruit is an one seeded berry. Male phase duration in each spadix is from 25 to 50 days. Female flower commences opening after all the male flowers in the inflorescence are shed. Anthesis is from 5 am to 6 am spread over 3 to 10 days. Male and female phases in different spadix in the same palm tree overlap resulting in self pollination. Cross pollination by wind occurs when the two phases do not overlap. Pollen dispersal through wind may be upto 1 to 2 km. distance. From bloom to maturity, it takes 35 to 47 weeks for fruit maturity. Seed set is 12 % through open pollination. Spraying pollen mixed in sucrose solution results in 26 % fruit set. Characteristics of this crop are given in Table 1.

Table 1. Characteristics of arecanut

1.	Seedling girth (cm)	Low (‹2), medium (2-4), high (›4)
2.	Number of leaves per seedling	Few (‹4), medium (4-6), many (›6)
3.	Seedling height (cm)	Short (‹50), medium (50-110), tall (›110)
4.	Crown shape	Drooping, intermediate, upright
5.	Plant height (m)	Short (‹6), medium (6-8), tall(›8)
6.	Crown length (m)	Short (‹1.50), medium (1.5-2.0), long (>2)
7.	Internode length (cm)	Short (‹4), medium (4-10), long (›10)
8.	Leaf length (cm)	Short (‹120), medium (120-200), long (›200)
9.	Leaf bredth(cm)	Narrow (‹85), medium (85-100), broad (›100)
10.	Leaf sheath length (cm)	Short (‹50), medium (50-90), long (›90)
11.	Leaf sheath breadth (cm)	Narrow (‹25), medium (25-45), broad (›45)
12.	Initiation of flowering (months)	Early (‹30), medium (30-45), late (›45)

13.	Spadix length (cm)	Short (‹50), medium (50-70), long (›70)
14.	Spadix breadth (cm)	Narrow (‹12), medium (12-18), broad (›18)
15.	No of female flowers per inflorescence	Few (‹120), medium (120-180), many (›180)
16.	Orientation of the inflorescence	Upright, horizontal, drooping
17.	Color of ripe nuts	Pale yellow, yellow, deep yellow, pale orange,orange
18.	Shape of nuts	Round, oval, oblong
19.	Fresh fruit weight (g)	Low (‹24), medium (24-36), high (›36)
20.	Fruit length (cm)	Short (‹4.50), medium (4.50-5.50), long (›5.50)
21.	Fruit breadth(cm)	Narrow (‹3.50), medium (3.50-4.50), broad (›4.50)
22.	Dry fruit weight (g)	Low (‹9), medium (9-13), high (›13)
23.	Kernel length (mm)	Short (‹18), medium (18-26), long (›26)
24.	Kernel breadth (mm)	Narrow (‹16), medium (16-24), broad (›24)
25.	Kernel weight (dry)(g)	Low (‹6), medium (6-12), high (›12)
26.	Dry kernel weight/per palm(kg)	Low (‹1), medium (1-3), high (›3)
27.	Husk thickness (mm)	Low (‹4), medium (4-6), high (›6)
28.	Dry husk weight (g)	Low (‹3), medium (3-5), high (›5)
29.	Kernel recovery %	Low (‹15), medium (15-25), high(›25)
30.	Arecoline content mg/g dry nut weight	Low (‹0.32), medium (0.32-0.40), high (›0.40)
31.	Tannin (total polyphenols) mg/g dry nutweight	Low (‹160), medium (160-180), high (›180)

(*Source:* PPV & FRA, 2017)

Arecanut is a perennial palm which actively produces arecanut bunches year after year. Its prebearing age is commonly 5 to 6 years in many gardens. Being perennial, cross pollinated and heterozygous for many yield contributing, growth and agronomical traits, genetic and breeding intervention to enhance the performance requires long period of time. One cycle of improvement envisages selection of elite mother palms that are already 25 years or more age. Selection of good grade ripe bunches, followed by seed and seedling selection, establishing superior arecanut palm seed garden and generating quality seeds for nursery establishment and quality assured seedling selection will take 20 to 25 years. Establishing quality seedlings production nurseries and seedling distribution supply effectively facilitate the propagation of genetically superior arecanut palm seedling (Ahmat Bavappa and Nair, 1982).

Mother palm selection: Arecanut palm plantations with high number of palms that are well maintained are surveyed and plantations with high yielding palms are selected. While selecting the mother palm, care is taken to select mother palms that show earliness to first spadix formation indicative of earliness, semitall to dwarf stature based on short internodes, general vigour and good healthy leaf canopy and regular bearing, high fruit set and high fruit number. Trees on the border rows are not considered. High yielding plus trees from the middle area of the healthy productive plantations are selected based on visual judgment and in comparison with palms around. It is desirable to select mother palms that show 25 to 30% fresh fruit set, 300 to 400 nuts / palm / year and around 4 kg yield per palm per year.

Seednut selection: Ripe bunches from the identified palm trees are harvested and examined for the extent of fruit set, fruit size, shape and uniform development and the palms satisfying the conditions alone are selected. Ill developed, malformed and disease infected nuts and bunches are rejected.

Nursery selection : Nursery is raised with healthy, disease free planting the ripe fruits (each fruit around 350 g) from selected plus trees. Seeds are planted stalk end up in 6 cm x 6 cm spacing in the primary nursery. Germination is monitored. Ideal nursery and plant protection practices are followed. At 90 days, the seedlings that are vigorous, healthy and with fully emerged 2 to 3 leaves alone are selected and carried forward to the secondary nursery and the seedlings are planted with 30 cm x 30 cm spacing in the raised beds that are 150 cm. wide and long beds. Seedlings are again selected at 12 to 18 months in the secondary nursery. Seedling selection is based on shorter stem, stout girth at collar, more number of leaves, vigour, healthy, uniformity and stability. Selection index is formulated based on one year old seedlings growth.

Selection index = Multiply the leaf number by 40 and then deduct the seedling height in cm

SI = (leaf no x 40) – seedling height in cm.

Individual arecanut seedlings are selected on the high or very high SI values. Girth at collar one year after transplanting and number of nodes after 2 years after transplanting are highly correlated with arecanut yield. In the secondary nursery, at the end of second year, the seedling with poor collar growth and lesser number of nodes are discarded.

Nucleus Seed Garden: The arecanut fruits in conformity to the variety description alone are sown in the primary nursery. Seedlings are selected at 90 days after sowing with vigorous screening as described earlier. The selected true to type seedlings are transplanted in the secondary nursery. Seedlings are selected with rigid norms as described earlier. The quality assured standard seedlings of high genetic potential are supplied to the farmers. Five to ten per cent of the superior and elite seedlings are forwarded for isolation growing 1600m. for establishment of nucleus seed garden that will sustain the production of breeder quality seeds and delivery of standard planting materials. Systematic multistage selection in arecanut is envisaged from mother palm garden selection seednut to seedling selection in primary and secondary nurseries and at the end of first year and second year of seedling age (Fig. 1).

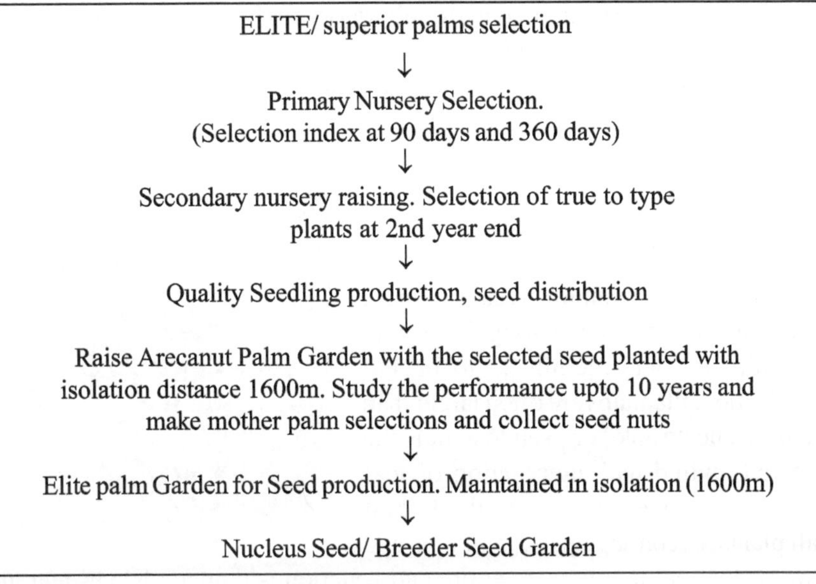

Fig. 1. Nucleus and Breeder Seed Production in Arecanut

Breeder Seed Garden: Seedlings, selected from secondary nursery raised with nucleus seed are planted in Breeder Seed garden with isolation distance of 1600m. Palms are verified for true to type nature and uniformity in development and expression of characteristics. Palms that are less vigorous and slow in growth and palms that are late flowering and shy flowering and low/poor fruit are uprooted. Highly productive plants showing high fruits *et al.* one are retained for collection of fruits/ seeds for distribution of Breeder seed. Source nurseries sown with breeder seeds are established and seedlings conforming to the standards are distributed. Elite seed gardens are arranged to be raised with superior seedlings selected from breeder source nurseries.

ELITE Mother Palm Seed Garden: In well prepared loamy soil field, the true to type seedlings selected from the secondary nursery. Breeder seed nuts are planted in well prepared 60 cm x 60 cm x 60 cm pits that are 2.7 m x 2.7 m in spacement and 1600m isolation. The garden is continuously monitored. Palms that are showing poor growth, vigour and development, poor in collar girth and tall with long internodes are discarded. Prebearing period is scored and the palms that are late to produce spadices are discarded. Unproductive and poor performing palms and palms showing poor fruit set, low number of fruit and only few spadices bearing are removed periodically upto 10 year period. Disease affected palms are uprooted. Ideally the mother palm garden is planted in compact areas with isolation distance of 1600m. from other arecenut plams. The bunches and fruits selected in the elite palms from the mother palm garden at 10 year age or later years are examined in the laboratory for healthiness, high fruit set in each bunch, fruit size, uniformity and shape besides the nut quality. Palms bearing small sized and ill formed seed nuts are discarded.

107. Banana

(*Musa* species) (2n=22,33)

Family : Musaceae

Both banana and plantain trees are large, herbaceous monocots, reaching 7 to 8m. in some cultivars, but generally 2.5 to 5m. tall in cultivation. Plantains are often larger than bananas. The "trunk" or pseudostem is the clustered, cylindrical aggregation of leaf stalk bases. Leaves are among the largest of all plants, becoming up to 3.5m. long and 0.6m. wide. Leaf margins are entire and venation is pinnate; leaves tear along the veins in windy conditions, giving a feathered or tattered look. There are 5 to 15 leaves on each plant, with 10 considered the minimum for properly maturing a bunch of fruit. The perennial portion of the plant is the rhizome and is known as corm. It produces suckers or vegetative shoots, which are thinned to 2 per plant-one "parent" sucker for fruiting and one "follower" to take the place of the parent after it fruits and dies back. Banana produces roots and that serve as a storage organ for the plant. The characteristics of this plant are given in Table 1.

Table 1. Characteristics of Banana (*Musa* sp)

1.	Pseudostem length (m)	Very short (<2), short (2-2.5), medium (26-30), long (3-4),very long (>4)1
2.	Pseudostem colour	Green, yellow green, red, others
3.	Purple blotches on young leaf	With blotches, without blotches
4.	Colour of cigar leaf (lower side)	Green, red, purple, others
5.	Leaf orientation	Upright, spreading, drooping
6.	Petiole canal	Open, with margin, spreading, wide with erect margin, straight with erect margin, margin curved inwards, margin overlapping
7.	Petiole length (cm)	Short (<30), medium (30-69), long (>70)
8.	Leaf blade shape at base	Both sides rounded, one side rounded, one side acute, both sides acute
9.	Peduncle length (cm)	Short (30-40), medium (41-70), long >70
10.	Peduncle colour	Light green, green, dark green, red, pinkish purple

11.	Peduncle pubescence	Present, absent
12.	Bunch shape	Cylindrical, irregular, conical
13.	Bunch position	Hanging vertical, hanging at an angle, horizontal
14.	Bunch compactness	Loose, lax, medium, compact
15.	Rachis orientation of male phase	Hanging, inclined, curved with vertical end, horizontal
16.	Rachis appearance	Base bract abtuse, middle, presence of withered bract on the whole stalk, rachis absent
17.	Rachis prominence of bract scar	Weak, moderate, strong
18.	Male bud	Absent, degenerated present
19.	Male bud colour	Yellow, green, purple, red, others
20.	Male bud shape	Lanceolate, ovate, round
21.	Male flower colour	White, orange, yellow, pink, others
22.	Stigma colour	Creamy white, green, orange, others
23.	Style shape	Straight, curved, Curved under the base
24.	Bract curling	Revolute, not revolute
25.	Persistence of male bud	Absent, weak, strong
26.	Fruit orientation	Perpendicular to the axis, curved upwards, curved towards stalk
27.	Fruit length (cm)	Short (<6), medium (6-15), long >15
28.	Fruit shape	Straight, slightly curved, curved at distal part
29.	Fruit transverse section	Rounded, slight ridge prominent ridges
30.	Fruit apex	Pointed, blunt, tip, bottleneck, truncate
31.	Persistence of floral organs	Absent, present
32.	Fruit pedicel attachment	Weak, medium, strong
33.	Pedicel surface	Glabrous, pubescent
34.	Pedicel length (cm)	Very short (<0.6), short (0.6-1.0), medium (1.1-1.5) long (>1.5)
35.	Pedicel colour	Pale green, green, dark green, red, purple
36.	Peel adherence	Weak, medium, strong
37.	Fruit waxiness	Present, absent
38.	Peel colour at fruit ripe stage	Pale yellow, golden yellow, yellow, green, reddish orange
39.	Ripe fruit pulp colour	White, cream, yellow, orange, yellow
40.	Number of hands per bunch	Few (<6), medium (7 to 8), many (>8)
41.	Number of fruits per hand	Few (<9), medium (9 to 13), many (>13)

Source : PPV & FRA (2012)

The inflorescence is a spike. Initially, it appears above the last leaf in an upright position, and consists only of a large, purple, tapered bud. As the bud opens, the narrow, white, tubular, toothed flowers are revealed, clustered in whorled double rows along the stalk; each cluster covered by a thick, purple bract. The flower stalk begins to droop down under its own weight after opening. The flowers are negatively geotropic and turn upright during growth. Bananas are male sterile and those of the Cavendish group are female sterile as well. Fruit is set parthenocarpically. Fruit is an epigynous berry. Fruits are borne in "hands" of up to 20 fruit, with 5 to 20 hands per spike. Fruits reach harvest maturity in 90 to 120 days after flower opening. The terminal bud on the stalk may be removed if fruit set is high, to allow more complete filling of fruits.

Banana varieties widely differ in their genomic constitution, plant vigour, leaf petiole groove and suckering or ratooning ability, bunch and fruit shape, size and resistance or tolerance to virus and fungal diseases.

Nucleus /Breeder Seed / Clonal Seed Production: As a first step, the individual banana palnts in the source population or in the maintenance breeding stock are critically examined for descriptor traits at different stages of growth. The selected banana plants should be free from virus, wilt and other foliar diseases. A total of 1500 disease free, uniform and true to type plants for stem, foliar, fruit bunch and fruit characteristics are ear marked at the harvest stage. The bunch, fruits and hands and foliage and pseudostem characteristics are critically examined and 1000 diseases free plants are finally identified and labelled. Suckers are removed from each plant and graded. Only sword suckers, phsiologically well developed and weighing around 1.5 to 2.0 kg are selected and numbered as plant progeny maintaining the identity of the selected single plants. The selected suckers are well trimmed of the roots and 20 cm. pseudostem above the corm alone is retained. The corm surface is then cleaned of the soil and pralinage was done with 40 g of carbofuron 3G granules per sucker to control nematodes. To avoid wilt disease, the infected portions of the corm may be pared and dipped for 5 minutes in carbendazim (0.1%). (TNAU, 2013). The suckers thus prepared are planted with variety specific spacing in well prepared field that is not grown with banana in last 3 years and isolated from other banana fields by 5 m to avoid mixing of clones at the time of harvesting, separation of suckers and preparing selected suckers for planting in the next clonal generation.

In the first clonal crop, continuous observations are made to find out physiological vigour, plant establishment and expression, pseudostem and leaf features, colour, time of flower emergence, flowering and fruit bunch and fruit characteristics. Uniformity in flowering is achieved by culling out early and very late flowering progenies. Based on the critical observations, true to type plants are identified. Off-types and plants showing deviations are uprooted. At maturity of the first clonal generation, harvest of bunches is made and fruit characteristics are recorded before and after ripening. Fruit taste and quality parameters and other related aspects including laboratory assessment are carried out. The suckers from the selected plants are then removed from mother plants, graded, selected and prepared and numbered for the second clonal (progeny) generation planting as described earlier.

In the second clonal generation planting, suckers of all the selected plants are planted in individual progeny rows with space between two progenies for easy monitoring. Careful and continuous observations are recorded in multistages from field establishment to fruit maturity and disease and pest free nature. Progenies showing plants with deviations form true to type nature are uprooted and discarded as whole progeny. The selected progenies are finally evaluated for physiological vigour, growth rate, health and vigour of stem and foliage and fruit characters for genetic uniformity and trueness. The ripe furit traits are examined in the labraotry and market quality and taste, flavor and nutritive aspects assessed. After thorough put evaluation, clonal plant progneis are selected. Suckers were separated from mother plants in the selected individual progenies, labelled progenywise. Corms are treated with plant protection chemicals and planted as described earlier as clonal plant progeny in the third clonal generation. All observations and evaluations were followed as in the first and second clonal generations. At the end of the third clonal generatin, based on the productivity, superior performing pedigreed clones are selected and tagged. The suckers are removed from the selected superior performing clones that are virus and disease free, vigorous, highly productive and high fruit quality clones. The suckers selected are trimmed to allow 20 cm pseudostem and prepared by removing the roots and thorough cleaning and pretreating with fungicide and nematicide. The suckers from different clonal progeny plants are bulked to constitute nucleus clonal stock.

Breeder Seed Clonal Stock: The suckers selected from the nucleus stock plants should be free from virus, wilt and other foliar diseases. The suckers are planted in well prepared clean field not gown to banana crop in the past 2 years and this field is isolated from other banana growing fields by 5 m. distance. The plants are individually verified for the quick field establishment, vigour, leaf development, uniform and trueness, uniformity and stability for stem, foliar , fruit bunch and fruit characteristics and disease and pest free conditions till the maturity stage. The bunch, fruits and hands and foliage and pseudostem characteristics in diseases free plants are critically examined again at the harvest time of the fruit bunches. In the laboratory bunches are allowed to ripe. The bunch shape and size and number of hands per bunch and fruits per hand besides the fruit shape, size and colour and the peeled fruit taste, aroma, sweetness, non seededness, keeping qualities and market acceptability traits are recorded. Based on the data, the single plants are finally selected and tagged. Visit of Monitoring Team to the Breeder stock is organized at the time of harvest of branch at midmaturity. The suckers collected from the selected plants alone are bulked to constitute breeder seed stock. The selected plants are maintained in the field as breeder stock for ratooning with extereme sanitation and clean cultivation adopting good agronomic production protocol.

Suckers are removed from each plant and graded. Only sword suckers, phsiologically well developed and weighing around 1.5 to 2.0 kg. are selected from the selected single plants. The selected suckers are well trimmed of the roots and retaining 20 cm. pseudo stem above the corm. Suckers are supplied from this breeder stock plants. It is desirable to continue the nucleus - breeder stock development cycle once in 3 years considering the vegetative multiplication and need to clean the propagules free of pests and disease infections.

Micropropagation: Suckers selected from superior true to type plants from breeder seed lot and that are virus and disease free are utilized for plant tissue culture preparing ex plants for in *in vitro* multiplication adopting optimum protocols that ensure healthy, physiologically and uniformly developed paintlets that are true to type ,absolute similarity in genetic fidelity. Well developed *in vitro* plants are hardened and treated for disease and pest free nature. Tissue culture banana plants with 5 to 6 leaves are delivered for raising productive banana plantations. Good agronomic and horticultural practices are adopted to raise a very healthy banana crop. The plant tissue culture protocols for banana are shown in Fig. 1.

It is advisable to grow at least 1500 to 2000 healthy disease free *in vitro* developed plants or suckers in Breeder seed stock clones production plots with the spacing of 2m x 2m for monitoring true to type and stability for plant and fruit characteristics. If deviating plants are noticed, they need to be uprooted. This breeder seed clonal plot is isolated with 5m. distance from other banana crops and is maintained in highly clean and well drained condition and free of insect pests, nematode and fungal and virus infections. This plot will become source for plant resources for further selection of suckers and meristems for micropropagation.

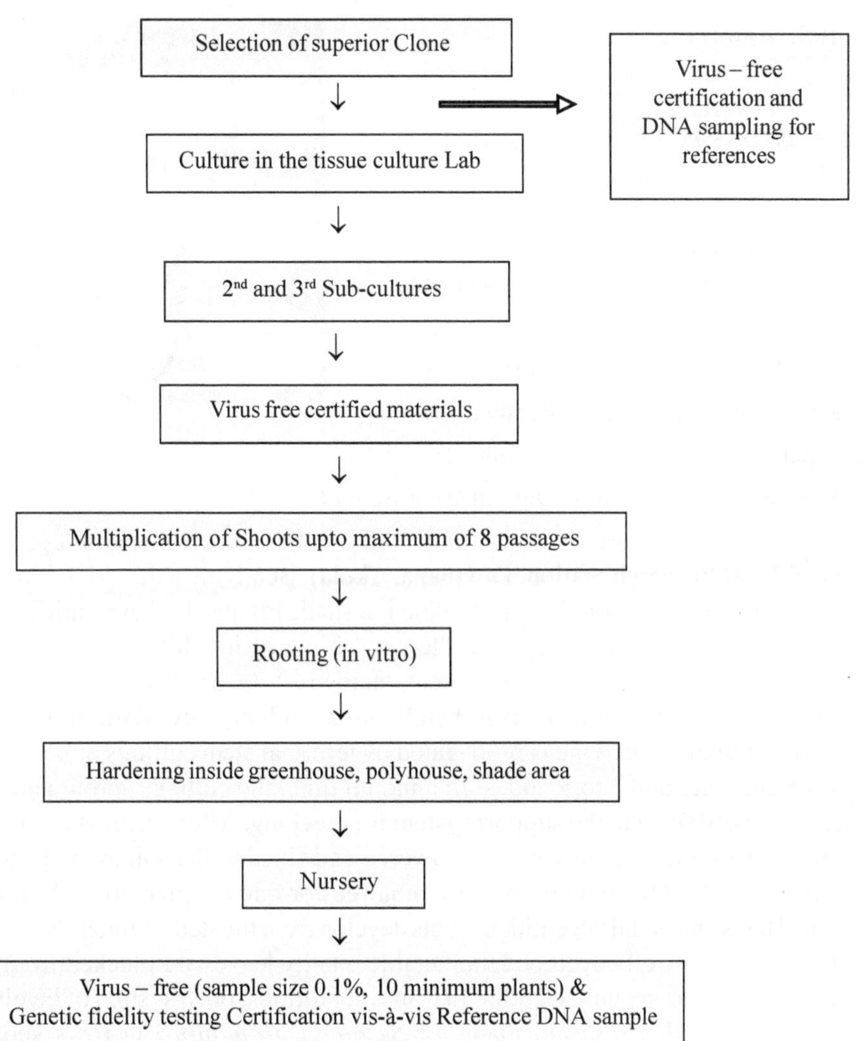

Fig. 1. Procedure and Standard Parameters for Banana Tissue Culture
Source: Department of Biotechnology, New Delhi

Somaclonal variants, if any, are identified in the breeder plot and also in multiplication field and discarded.

Monitoring Team visit is organized two times in the first year of the crop. The quality assured clone is tagged with Breeder seed label. Grow out test is conducted as per the specifications of GOT.

Fruit nutritional qualities and market acceptailities of the fruit are assessed.

108. Betelvine

(*Piper betle* L.) (2n=26)

Family : Piperaceae

More than 40 betelvine varieties are cultivated in India. They are developed by clonal selection. The cultivated betelvine varieties belong to the following groups:

Desawari - North India

Bangla - North India

Karpssri Kaniyur kodi - Peninsular India

Sanchi - Peninsualr India

Meetha - Only in West Bengal

Twenty types/CVs are grown in Tamil Nadu. Betelvine is a dioecious evergreen vine (Petel Research Station, Diwthana, Akola). Betelvine cultivation is almost on organic farming practices. Betelvine is a shade loving climber requiring high humidity. Leaves are glossy and alternate. Flowers in white catkins; flowers actionomorphic, calyx, corolla absent, stamens 2. Ovary superior. Dioecious nature or sexual dimorpsrism in betelvine is explained by Maiti and Biswas (1991). Usually betelvine is propagated by terminal stem cuttings or setts of 30 to 45 cm. long and 3 to 5 nodes. In a month time, the cuttings sprout and start creeping and fixed to the support system for creeping. After reaching to a height of 2.5 to 3.0 meters, the vines are lowered and fixed to the soil at the bends at ground level. One lakh cuttings are required for one ha. plantation. Vigorous apical buds and nodal adventitious roots develop from the stem cuttings. Normally, 75 to 100 lakh well developed marketable quality leaves are plucked from one ha area per year in peninsul ndia. In tropic conditions, open system of betelvine cultivation with *Erythrina* plants or *Sesbania grandiflora* in rows support system for betelvine plants is practicised. Betelvine is also cultivated under protected conditions as in North India during cold weather. It is mostly supported by splash irrigation system. Betelvine grows to harvest stages in 140 to 160 days after planting and the vines remain healthier and productive stock upto 24 to 30 months. The characteristics are given in Table 1.

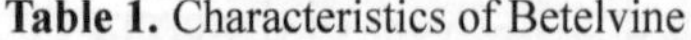

Table 1. Characteristics of Betelvine

No.	Character	Description
1.	Plant adventious root production	Few (<5), medium (5 to 10), many (>10)
2.	Plant stem colour (orthotropic)	Lightgreen, green, moderately green, dark green
3.	Plant – stem internode length (cm)	Very short (<4), short (4-6), medium (6-8), long (>8)
4.	Leaf lamina colour	Light green, green, dark green

5. Leaf lamina length (cm)	Very short (<7), short (7 to 11.5), medium (115-14), long (>14)
6. Leaf breadth (cm)	Narrow (<905), medium (9.5 to 12.50), broad (>12.5)
7. Leaf petiole length (cm)	Short (<6), medium (6 to 8), long (>8)
8. Leaf thickness	Thin, medium, thick
9. Leaf- depth of sinus (cm)	Shallow (<0.6), medium (0.6 to 1.25), deep (>1.20)
10. Leaf width of leaf lobe (cm)	Short (<3.5), medium (3.5 to 5.0), long (>5)
11. Leaf lamina orientation	Flat, V shaped
12. Leaf apex shape	Acuminate, acute
13. Leaf texture	Coriaceous, membranous
14. Leaf-proximity of basal lobes S.No 1 to 14 relate the orthotropic stem.	Overlapped, close separate
15. Number of plagiotropic shoots per meter	Low (<3), medium (3-6), high (>6)
16. Stem colour	Light green, yellow green, green
17. Leaf lamina colour	Light green, green, dark green
18. Leaf lamina shape	Elliptic, ovate oblong, ovate
19. Leaf apex shape	Acuminate, acute
20. Leaf texture (S.No 15 to 20 relate to plagiotropic stem)	Coriaceous, membranous
21. Sex of the plant	Female, male, hermaphrodite
22. Flowerring habit	Shy, moderate, profuse
23. Female inflorescence colour	Beige, yellow
24. Female inflorescence length (cm)	Short (<2.5), medium (2.5 to 4), long (>4)
25. Male inflorescence length (cm)	Short (<7), medium (7 to 10), long (>10)
26. Number of inflorescence per plagiotropic branch	Low(<2), medium (2 to 4), high (>4)
27. Leaf taste	Sweet, low pungent, moderate pungent, highly pungent
28. Eugenol content (%)	Absent (0), low (0 to 10), moderate (10 to 20), high (20 to 40), Very high (>40)

Source: PPV & FRA

Natural population of betelvine shows high degree of variation in respect of vigour, rooting ability, leaf shape, colour size and leaf bitterness. The aroma and taste of leaves may be pungent to non pungent and even sweet as in Meetha type.

The crop improvement in betelvine is by clonal selection of superior productive vine type bearing high quality and superior texture leaves. Clonal multiplication can be achieved in developing a well planned clonal chain of Basic stock – Nucleus stock - Breeder stock - Clonal stock. Maintaining the vines in a disease and pest free and healthy manner is utmost important. Phytosanitary meansures against anthracnose, powdery mildew, phytopthora leaf spot, nematode and mealy bugs, mites and scales are utmost important to safeguard the delicate vines and in ensuring clean and healthy clonal seed multiplication.

In vitro micropropagation: Short nodal segment culture, rooting, hardering and selection of proper size *in vitro* plants is recommended for fast multiplication of disease free clonal plants.

Nucleus stock development: Highly productive betelvine gardens growing the same variety type of vine are surveyed and from them, 1000 cuttings from disease free individual plants are prepared. The vine cuttings are pretreated in insecticide and furgicide and planted in the well prepared land with disease free and nematode free soil and not grown to betelvine in the previous 2 to 3 years. The betelvine crop is raised in two rows in 2m. wide beds and maintained with ideal production practices for betel vine with support plants. Live/support agathi (*Sesbania grandiflora*) in long rows. Operation of sowing of training of the vines and lowering of vines are carried out.

The vines are examined for quick rooting, health, vigour, root growth and trained to creeping on the supporting plants. Leaf characters, from shot tip to matured stem, young developing to matured leaves are studied for the healthy, smoothness, glossiness, bitterness and aroma. Five hundred true to type, uniform plants for vine and leaf characteristics are retained and non conforming vines are discarded. The vines are lowered after they grow to 3m. growth and fixed at the bottom of the support and the tillering/branching capacity is asseseed. The quality leaf productivity and the performance of the vines and tillering are adjudged on the individual vine basis and healthy superior vines are identified and labelled. In the next generation, ten stem cuttings from each selected true to type vine are prepared and planted in clonal progeny rows adaping the ideal crop production practices and providing live support system. When vines grow upto 0.5 to 2.0 meters, the vines are lowered. Individual vines are studied from the establishment upto 18 months for the vigour, productivity and healthiness. True to type nature, uniformity and stability of the stem and leaf characteristics, leaf quality, overall tillering ability, freedom from pests and diseases are assessed. Based on the assessment, 250 to 300 true to type superior plants are selected and labelled and from each such plant, from the middle of the stem, vine cuttings are prepared and planted as nucleus stock.

Nucleus stock betelvine is raised following the crop production practices described earlier. After 18 to 24 months, the true to type and uniform and highly productive vines that produce high quality and market accepted betel leaves are identified and retained. The rest of the vines are uprooted. The identified and retained vines are the Nucleus Stock.

Breeder Stock of betelvine: Stem cuttings, 2000 or more obtained from the nucleus stock are planted in a well prepared field plot that contains diseases and nematode infection free soil and not grown to betelvine earlier. All good betelvine production practices are adapted. Individual vines are studied for good field establishment vigour, fast growth and leaf development, good tillering/branching, quality leaf and high productivity, true to type nature, uniformity and stability of individual vines are studied. Vines showing low vigour and growth deviating type of plants, and disease affected plants are discarded. Leaf plucking operations are done. Around 18 months after planting, visit of monitoring team is arranged. Well developed vine cutting are prepared and Breeder stock is constituted. The breeder stock field is maintained for 2 to 2 ½ years and continuous supply of planting material is arranged. Quality assured cuttings are supplied with breeder seed label. Grow out test is conducted with vine cuttings by growing upto 6 to 8 months to assess the genetic purity. Isolation distance of 5m is maintaine in the nucleus and breeder seed stock.

Betel vines are affected at different growth stages by diseases such as Phytophthora, bacterial leaf spot, blight and bacterial stem rot, anthracrose, powdery mildew and affected vines should be removed away from the garden and burnt. The basins from where disease affected plants are removed and basins adjoining plants should be drenched with 0.25% Bordeaux mixture. The disease prevention/management practices need to be taken in basic vine stock, nucleus and breeder stock and clonal stock production fields.

Two year old disease free, healthy betelvine population of the concerned variety is selected after assessing the regularity in productivity and leaf quality characteristics, healthiness, and vigour and free from diseases and pests. Stem cuttings from 1000 superior plants are prepared and planted in well prepared field with ideal soil conditions and addition of Organic / Vermicompost.

↓

Individual plants are monitored for quick rooting and field establishment, fast and vigorous growth of stem and foliage / leaves. Periodical betelvine cultivation steps of providing live support system, raising the active growing shoots on the support are done True to type uniformity, healthy nature and stability of different betelvine characteristics are recorded in individual vines. Number of leaves and yield recorded are compiled. Superior individual lines that do not show decline in health and freedom from pests and diseases are labelled and 500 plants are selected. Vine cutting of 15 to 20 cm. are prepared from each selected plant.

↓

In the next step, 100 clonal progeny rows are raised, from the selected plants. The betelvine growing is done as per the recommended practices. Among the progenies, the healthy and fast growing vines, vigorous root clasping on the support plants, free from diseases and pests, high productivity progenies with healthy marketable leaves are tagged. Progenies showing trueness, uniformity and stability of stem and leaf characteristics are identified. Healthy vine cuttings are prepared from selected superior plants in the true to type clonal progeny and bulked to constitute nucleus clonal stock.

↓

Clonal stock vine cuttings from nucleus stock source are planted adapting the recommended practices for betelvine cultivation including raising of the live support system plant. Systematic and regular examination is made on individual plants for fast growth and rooting and vine thickness, number of leaves per harvest, leaves per plant and leaf characteristics

↓

At the end of second year, quality of marketable leaves, and market acceptance are assessed Vines showing deviating characteristics and disease and pest infestations are discarded. True to type, uniform and stable expression are evaluated. Monitoring Team Visit is organized. Vine cuttings from the selected plants are bulked to constitute the breeder stock. The betelvine plantation of selected plants in the Breeder Stock nursery is maintained for effecting continuous supply of quality assured and disease free seed vines to the growers. GOT is organized.

Fig. 1. Nucleus and Breeder Seed Production in Betelvine

109. Black Pepper

Piper nigrum L. (2n=52)

Family : Piperaceae

Black pepper, the king of spices, is a perennial woody climber. The orthotropic stem clings to the support. Plagiotropic branches bear the inflorescence and fruit spikes. Leaves are simple, alternative and stipulate, leaf blade is entire, ovate and coriaceous. Inflorescence is a catkin like spike having 50 to 120 flowers. The flowers are either hermaphrodite or unisexual. Hermaphrodite flowers have two stamens on either side of the ovary. Style is very short or absent. Stigma has 3 to 5 fleshy branches. Fruit is a sessile globe drupe. The dried fruit is black pepper and the white pepper is the seed of pepper. Bisexual flower is protogynous but experimental evidences indicate that pepper is a self pollinated plant.

Rooted cuttings developed from the runner shoots of high yielding, healthy vines are planted in 2.7 m× 2.7m. spacing. As the vines grows, they are tied and guided to the standard or supports. The vines grow and reach fruit bearing and harvest 3 years after planting. Vines yield regularly upto 20 years age. Berries reach maturity from 180 to 200 days after flowering.

Nucleus Seed Stock: The methods for rapid multiplication of pepper is standardised. The blackpepper plantations that are in active and regular bearing are examined/studied for the presence of vigorous, healthy high yielding and true type plants of the variety concerned and twenty productive vines are selected and from each of the selected vines ten rooted cuttings are prepared. Healthy rooted cuttings are planted in a well prepared clean disease free field in a spacing of 2.7 m × 2.7m in a single row. Each progeny row of 10 plants are planted in a single row and the recommended production practices are adapted. The support by planting standards to facilitate the vines to cling and climb is provided from the time of growth of the vines. The training and pruning operations are done to the support the pepper vines. The flowering and berry setting are critically observed from the third year after planting. When the berries start turning red in the spikes, the whole berry is hand picked. The yield will be 2 to 3 kg per vine per year three years after planting.

The well established vines in each of the progeny row are examined continuously for fast and vigorous growth and health. At regular intervals, the true to type expression of the characteristics and stability and uniformity, for stem, leaf, flower, berry characteristics upto the end of 7th or 8th year after planting. The characteristics of pepper are given in detail in the Table 1. The performance of the twenty progeny rows is analysed. Progeny rows performing above mean +SE are selected. Individual plants with each of the selected progeny rows are identified. Rooted cuttings are prepared from the laterals of each of the identified plant. The rooted cuttings/constitute Nucleus Stock.

Breeder Seed Stock: In a well prepared clean field with disease free soil, breeder stock garden is raised with the healthy vigorous rooted cuttings selected from lateral branches of the nucleus stock. Isolation distance of 20 m is provided. The standard recommended pepper production practices are applied with the support system. Disease, pest and other kinds of stresses are efficiently managed to maintain an ideal plantation of black pepper. Plant to plant uniformity and true to type or trueness are assured and ideally maintained. At the end of 5 years after planting, visit of the monitoring Team is arranged. Rooted cuttings are prepared and regular supply of quality assured planting material are arranged to the growers and nurseries.

Table 1. Characteristics of Black Pepper

1.	Plant-shoot tip colour	Light green, light purple, dark purple
2.	Leaf length (cm)	Short (<10), medium (10 to 16) long (>16)
3.	Leaf width (cm)	Narrow (<7), medium (7-10), long >10
4.	Leaf petiole length (cm)	Short (<2), medium (2 to 3), long (>3)
5.	Leaf lamina shape	Ovate, ovate lanceolate, ovate elliptic
6.	Leaf base shape	Round, corolate, acute even, wavy
7.	Leaf margin	Even, wavy
8.	Lateral branch length (cm)	Short (<30), medium (30-40), long (>40)
9.	Number of nodes per lateral branch	Few (<20), medium (20-40), many (>40)
10.	Lateral branch pattern	Semierect, horizontal, hanging
11.	Spike colour	Light brown, green yellow
12.	Spike peduncle length (cm)	Short (<1), medium (1 to 2), long (>15)
13.	Spike length (cm)	Short (<10), medium (10 to 15), long (>15)
14.	Spike twisted	Absent, present
15.	Spike proliferation	Absent, present
16.	Number of berries per spike	Few (<25), medium (25-30), more (>50)
17.	Spike setting	Loose, compact
18.	Number of spikes per lateral branch	Few (<4), medium (4-7), more (>7)
19.	Berry shape	Round, oval
20.	Berry size (mm)	Small (<3.01), medium (3.01 to 4.26), bold (>4.26)
21.	Time of harvest maturity (months)	Early (<7), medium (7 to 8), more (>8)
22.	Bulk density (g/litre) (g)	Low (500), medium (500-600), high (>600)

Spike Composition

Number of fruits/ spike - 84-1000

Fruit volume cc - 108

1000 fruit weight (g.) - 107

Bisexual flowers - 95.5%

Female flowers - 4.0%

Male flowers - 0.0%

Fruit sets - 82%

Dry fruit recovery - 34%

Yield of green pepper per vine - 5 kg.

Composition

Piperin - 4.7%

Oleoresin - 12.5%

Essential oil - 3.4%

Spike length - 11.5cm

High oborecin in the twisted spike.

110. Caraway

(*Carum carvi* Linn.) (2n:14)

Family : Umbelliferae; Apiaceae

Other Names : Carvies; Wild cumin; Roman cumin; Persian cumin

The seed or fruit of caraway (*Carum carvi*) is obtained from a biennial or occasionally an annual form of umbelliferous plant cultivated in many parts of the world. It is cultivated in a limited scale in Kashmir, Kumaon, Garhwar and Chamba area in India. Caraway is one of the world's oldest culinary spices.

Caraway is a perennial or biennial herb, grows to about 60 cm and has feathery, compound leaves. Flowers are small with 5 sepals, petals and 5 stamens. The roots are tubers and thick. The flowers are small and white borne on umbels. It blooms in every two years to produce large creamy flowers. The seeds are mericarps as each seed is a half of the fruit. Each single seed or carpel is about 0.5 cm long, tan to brown, and curved with five lighter coloured ridges along the length of the seed.

Caraway seeds contain about 3.5 to 5.0% volatile oil. The main component (50 to 60%) of the essential oil is carvone. It also contains limonene, dihydrocarvone., dihydrocarveol, carveol, acetaldeyde, methyl alcohol, furfural and diacetyl.

Nucleus Seed Production and Breeder Seed Production: Procedure as being followed in the Nucleus and Breeder Seed production in Ajawain could be followed.

111. Cardamom

(*Elettaria cardamomum* Maton) (2n:52)

Family : Zingiberaceae

Cardamum plant is 2 to 4m, tall, herbaceous perennial with branched subterranean rhizome from which several leafy shoots arise. The first year is mainly vegetative growth, second year for reproductive growth of flowers and fruits and third year a senescence stage. New buds and shoots develop from base of old shoots; flowers are borne on erect or semi erect inflorscences. Leaves are 25 to 90 cm long, 5 to 15 cm broad. Inflorescence arise from the base of the leafy shoot, 45 to 120 cm long.

In cardamom, flowers are borne in recemes. Flowers are bisexual, zygomorphic (4 cm x 1.5 cm) and calyx is tubular, green, 3 toothed, persistent. Corolla tubular, long green and lobed. Stamens are modified into labellum, petalloid and only one stamen is functional with long anther. Ovary inferior, tricarpellary. Style slender rests on the top of the anther. Flowering takes 20 to 30 days. Fruits mature in 110 to 140 days after pollination. Fruit is a trilocular capsule with 15 to 30 seeds. Pollination is self compatible but cross pollination is more common. Flowers remain in bloom for 15 to 18 hours. Pollen viability and stigma receptivity are maximum from 8 am to 10 am.

Nucleus Seed: Highly productive healthy 6 to 8 years old cardamom plantations are surveyed for elite plus performing individual plants. Five hundred true to types elite plus clones of 5 years and above are selected. Besides the true to type nature, the elite plants are examined for uniform looking and producing uniform size raceme and flower features and mature bold green pods are selected. Seeds collected from pods from the selected uniform plus plants are sown in single plant nurseries on individual plus plant basis. Seeds are allowed to germinate in the primary nursery and develop into healthy seedlings before planting in the secondary nursery. Seedling selection is made in the primary nursery for vigour, growth, uniformity and disease free nature. Selected seedlings (60 to 75 cm height, 8 month age, 2 to 3 tillers, upto 5 leaves, vigour and healthy) from each of the 500 plant progenies alone are planted in the secondary nursery and at 18 month old age (including the period in primary nursery), the seedlings are verified and selected based on the true to type vigour, growth, leaf and other traits and planted in the main field with isolation distance of 400m. provided. Borders rows are planted with plants of the same variety. All recommended field production practices are followed. Deviating plants are discarded at the age 18 months after planting. Based on the productivity from 3rd year onwards upto the 5th year, plus performing cardamom plants that are uniform flowering are identified and selected. The racemes and pods are verified for trueness. Seeds are collected from selected individual single plants. Based on the disease free nature, health, vigour and pod setting, pod and seed characteristics such as colour, aroma, and other quality virtues, 200 to 250 individual plants are finally selected in the 5th year. The seeds from the selected plants are bulked and the nucleus seed is constituted.

Breeder Seed Production: Seed to seed: Breeder seed production is done by sowing the nucles seed in the primary nursery. Selected seedlings are planted in the secondary nursery. The selected seedlings from the secondary nursery are planted in the mainfield isolated by 400m. The production practices prescribed for cardmom are followed to raise a good and healthy crop. Off-types and deviating plants are discarded. True to type plants are retained. Performance of individual plants from 3rd to 5th year is recorded for capsule yield. Seeds from plus plants are saved. At the end of each cycle of breeder seed production the seeds form the selected plus plants will be constituted as breeders seed. In the view of the 5 to 6 years time required to identify plus performing plants, this cycles of nucleus and breeder seed may be repeated.

Breeder Seed: Rhizome to seed: Rhizomes of plus performing seedling raised plants are selected from productive and healthy plantation and planted in 1.8m. x 0.6m. well filled trenches. Isolation distance provided is 400m. Periodical observations are made on their disease free nature, suckering and growth in the different clonal progenies for vigour, true to type, disease and nematode free nature. From 3rd year onwards till end of 5th year, the pod production performance is recorded and seeds collected from individual plus clones are examined in the laboratory. Seeds from the selected plus elite clones are bulked to constitute the breeder seed.

Micropropagation of the superior elite clones of cardamom is successfully adopted by R&D companies for multiplying the superior cardamom stock plants yielding large, green and bold capsules borne in the spikes that show the absence of deformed/shriveled capsules.

112. Cashew

(*Anacardium occidentale* L.) (24=42)

Family : Anacardiaceae

The Cashew tree is a bushy, low-branched tree growing up to 10 to to 12 m in height and width. The simple leaves have prominent pinnate venation, smooth margins and are oval to elliptical in shape. In South India, flowers appear from December to January. The tiny, clustered flowers begin as green and turn yellowish pink to reddish pink, sometimes with stripes. Often male flowers and bisexual flowers are present in the same inflorescence. There is high degree of variation in the proportion of male, female and bisexual flowers due to high influence of weather and location factors. The flowers, each has five sepals and five petals. In female flowers, the ovary is single and in male flowers, there are 7 to 15 stamens, often 10 stamens, of which one is larger and nine are smaller. Stigma in the bisexual flower is longer than the stamens. Anthesis starts from 6 am. and continues till 4 pm. The bisexual flowers open mostly between 9 am. and 11 am. The stigma is receptive for only one day (Madhava Rao and Vazir Hassan, 1957). The flower produces abundance of nectar which is highly attractive to flies, bees, ants, and other insects. Madhava Rao and Vazir Hassan (1957) reported that the panicles consist predominantly of male flowers (~60%) and some bisexual (perfect) flowers There are rarely any female flowers. About 5 percent of the hermaphrodite flowers produce fruits. In general, the fewer the hermaphrodite flowers, the lower the fruit percent set. The characteristics of this crop are given in Table 1. Detailed pollination studies have been made by Jeyalakshmi and John (2005).

Fruit is borne singly or in small clusters and matures in 60 to 90 days. The true fruit of the tree is the cashew nut resembling the letter C. An interesting feature of the cashew is that the nut develops first and when it is full-grown but not yet ripe, its peduncle (receptacle) fills out, becomes plump, fleshy, pear-shaped or rhomboid-to-ovate with waxy, yellow, red, or red-and-yellow skin and spongy, fibrous, very juicy, astringent, acid to subacid, yellow pulp which is conspicuous so-called cashew apple.

Table 1. Characteristics of Cashewnut

1.	Plant height (m)	Dwarf (‹2.5), semitall (2.5-4.0), tall (›4.0)
2.	Plant branching pattern	Extensive, intensive
3.	Plant – internodal length (cm)	Short (‹1.0), medium (2.0-3.0), long (›3.0)
4.	Leaf – colour of young leaves	Red, yellow red, green yellow, purple, green, darkgreen
5.	Leaf shape	Elliptic, obovate, oval
6.	Leaf apex shape	Pointed, round, slightly notched
7.	Leaf area (sq.cm)	Small(<60), intermediate (60-120), large (>120)
8.	Flower: compactness of inflorescence	Loose, compact
9.	Shape of inflorescence	Conical, deltoic, broadly pyramidal
10.	Pseudo fruit colour of peduncle of tender fruit	Green, grey, purple
11.	Fruit – color of tender nuts	Green, slightly purple, purple
12.	Pseudo fruit – cashew apple shape	Cylindrical, conical to obovate, round pyriform
13.	Pseudo fruit- matured cashew apple colour	Yellow, red, yellow red, red purple

14. Pseudo fruit – weight of cashew apple (g)	Low (‹30), intermediate (30-50), high (›50)
15. Fruit apple to nut ratio	Low (<6), intermediate (6-12), high (>12)
16. Fruit – nut shape	Kidney, oblong-ellipsoid
17. Fruit nut weight (g)	Low (<5), intermediate (5-7), high (7-9), veryhigh (>9)
18. Fruit shape thickness (mm)	Thin (<2.5), intermediate (2.5 – 4.0), thick (>4.0)
19. Fruit – presence of cashew nut shell liquid	Present, absent
20. Fruit – shelling %	Low (<18), intermediate (18 - 28), high (>28)
21. Kernel weight (g)	Low (<1.2), medium (1.2 – 2.5), high (>2.5)
22. Seed- Attachment of peel to kernel	Loose, tight
23. Astringent or tannin content in cashew applejuice (mg/ml)	Low (<1), intermediate (1-4), high (>4)
24. Sugars in cashew apple juice (clarified) (mg/ml)	Low (<50), intermediate(50-200), high (>200)
25. Ascorbic acid in clarified juice (mg/ml)	Low (<1), intermediate (1.0 – 4.0), high (>4)

(*Source:* PPV & FRA, 2017)

Propagation of cashew is by seed, grafting, layering and cuttings. Seed sown crop commences fruit bearing in 4 to 5 years. By vegetative propagation the fruiting can be obtained in 2 to 3 years from planting. Self-pollination and cross-pollination play an important role in the formation of cashew seed. Seedlings therefore show great variation and true to type progeny is not obtained. Selected trees should preferably be multiplied by grafting or air layering because vegetative propagation will ensure the best productivity, true to type and quality nuts . Trees that are precocious and bear high number of fruits and grow vigorously are selected. Field establishment of air layers have been found to be poor. Hence softwood grafts, which give a high rate of establishment and early flowering are recommended for planting.

Softwood grafting and Raising Rootstocks: Among the vegetative propagation methods, softwood grafting is found to be the best for multiplication of cashew. Quality seed nuts collected from superior, proven, regular and high bearing healthy mother trees are sourced and seeds are selected by immersing in water or 10% saline solution. Seeds, which sink in water are selected for sowing in poly bags, Medium sized nuts (7 to 9 g) are selected to get vigorously growing seedlings. Fresh seed nuts are used for raising rootstock. Pre soaked seed nuts are planted in properly prepared polythene bags. Nuts are sown at weekly intervals to get continuous supply of rootstocks. The seedlings will be ready for grafting in 50 to 60 days after germination and 50 to 60 day old healthy seedlings having single main stem grown in the centre of the polythene bag are selected as rootstock.

Selection of mother trees: Mother trees of 15 to 25 years of age and having the characteristics like good health, vigorous growth and intensive branching habit with panicles having high percentage of hermaphrodite flowers, regular bearing of nuts of medium size and weight (5 to 8 g/nut), 7 to 8 nuts per panicle and recording an average yield of 15 kg nuts per annum per tree are identified and selected. Mature, medium sized nuts, which sink in water are selected as seeds after drying in sun for two to three days. Pre soaked seeds are sown in poly bags of size 20 cm x 15 cm and seedlings are raised. Seeds germinate in seven to ten days. Plus mother trees are the source for scion.

Selection of scions: High yielding plus trees of the variety of cashew already selected are the source mother plants to collect adequate number of scions. The scions should be 3 to 5 month old non-flowering lateral shoots of current season's growth and 10 to 12 cm long, straight, uniformly round and pencil thick with brown colour having dormant plumpy terminal bud. The top 4 to 5 leaves should be dark green in colour indicating proper maturity of the scion. The selected quality scions are then precured by clipping off three fourth portion of leaf blades. The scions will be ready for grafting in 7 to 10 days after the removal of leaf. The scions are collected before the terminal buds sprout.

Grafting is done by inserting the wedge of the scion into the cleft of the rootstock, taking care to ensure that the cambium layers of stock and scion are in perfect contact with each other. The grafted plants are to be kept under shade for 10 to 15 days to enable sprouting of the terminal buds. The grafts will be ready for planting in 5 to 6 months after grafting. The softwood grafts are planted in pits of size 50 cm x 50 cm x 50 cm during June-July in a spacing of 7 to 8 m on both sides. Initial training/shape pruning is continued. The plants should be allowed to grow by maintaining a single stem up to 0.75 to 1.00 m from ground level. This can be achieved by removing the side shoots or side branches gradually as the plants start growing from the second year of planting. Weak and criss-cross branches can also be removed. The flower panicles emerging from the grafts during the first and second year of planting should also be removed in order to allow the plant to put up good vegetative growth. The plants are allowed to flower and fruit only from the third year onwards.

In older cashew plantations, removal of dried or dead wood, criss-cross branches, water shoots etc. should be attended to at least once in 2 to 3 years. This allows proper growth of the canopy and receipt of adequate sunlight on all the branches. Pruning of cashew plants should be done during May/June.

High density planting: This technique involves planting more number of grafts per unit area and thinning at later stages, 312 to 625 grafts will be planted in one ha.

Nucleus seed source: Individual clonal plants from each selected superior plus trees are raised in isolated plots replicated twice, with very good care and management of training and pruning. The individual trees are regularly verified and monitored for the expression of vegetative and reproductive and fruit characteristics. The regularity in bearing and productivity and the fruit and nut traits are studied each year plant wise, within each clone. Based on the progressive performance and period for reaching economic productivity, the clones performing above mean + 2 SE are identified. Systematic efforts are bestowed to rejuvenate the selected individual uniform plus plants from true to type clones that are in complete conformity to the characteristics to facilitate collection of highly healthy scion shoots for collection of buds for grafting. The selected elite plants are maintained in a separate mother block with ideal attention and management, training and pruning regulations and free from pests and diseases. The scion bud woods constitute nucleus source capital.

Breeder Stock Source: Meanwhile a dedicated breeder stock clonal bank with the buds sourced from the selected nucleus superior plus trees of the clonal orchard is established separately and plants maintained in a highly healthy and hygienic conditions with suitable pruning and training regulations. This breeder plant service could be supported by modern protected nursery and propagation facilities and freeing from pests and diseases and improved phytosanitary measures. As an essential adjunct, a dedicated root stock source plantation will be also developed and maintained to source compatible root stock specific for the scion source for better productivity. The quality of the nuts are studied and recorded.

113. Cocoa

(*Theobroma cacao*) (2n:20)

Family : Sterculiaceae

Cocoa plants are self incompatible with homomorphic sporophytically controlled multiallelic gene system (S1 to S6). Flowers are perfect, pentandrous, calyx 5, corolla 5, stamens 5, staminodes 5 and ovary 5 carpelled developing into pods. Cocoa flowers open in the early morning. Aided pollination is made by insect vectors. Flowers are borne in cauliflorous tissues that are protruding directly from woody branches and trunk. Upon pollination, pollen tubes grow in the style but fertilization is not taking place. Only 1 to 5% of the flowers are successfully pollinated and form pods. Pods ripen 5 to 6 months after pollination. Pods are attached to the cushion by fruit stalk. Seeds have no dormancy. Seedling growth is orthotropic. After one to 3 years, terminal bud is divided to form 3 to 5 branches. Cocoa bears fruit at 4 to 5 years age. Cocoa pods are yellowish or orange in colour, 25 cm. long, 400 to 500 g in weight and with 30 to 40 seeds per pod.

Three botanical cultivars are recognized in cocoa 1. Criollo, 2. Forestero and 3. Trinitorio. The share of Forestero CVs is about 80% in chocholate production on global basis while Trinitorio contributes to 10 to 15% and Criollo, which is highly prized and more aromatic contributes to 5 to 10% of chocholate production. The Trinitorio varieties are considered to be the hybrids of Criollo x Forestero. Among the cocao varieties Forestero are found to be good varieties in adaptability and high yielding and recommended for commercial cultivation in India.

Nucleus and Breeder Seed: Propagation is by 1. cloning of self compatible superior and selected plants and 2. by seed production in monoclonal and biclonal orchards. Propagation by seedlings is not being followed owing the genetic variations found in seedlings. Vegetative propagation or cloning of selected, healthy, vigorous seedlings or mother trees selected from hybrid derivation is the practice. Selection of parents and production of hybrids commenced in 1990's. Varieties and Hybrids have been developed and released by the Kerala Agricultural University (KAU). Clones CCRP 1 to CCRP 7 and hybrids CCRP 8 to CCRP 10 are released for cultivation and all of them are self compatible. Seeds are produced in biclonal or polyclonal seed orchards planted with, superior plants that are self compatible in CCRP farm in Kerala Agricultural University, KAU and Kidur farm of the Central Plantation Crop Research Institute (CPCRI). Vegetative propagation of superior plants is done by budding (budded plants). For large scale cultivation, vegetative propagation is preferred to plant uniform material. Bud woods/scions from chupons alone are utilized. Bud woods from fans may develop into bushy plant with spreading habit. Root stock of 6 to 12 months age are utilized. Isolation distance is 10m.

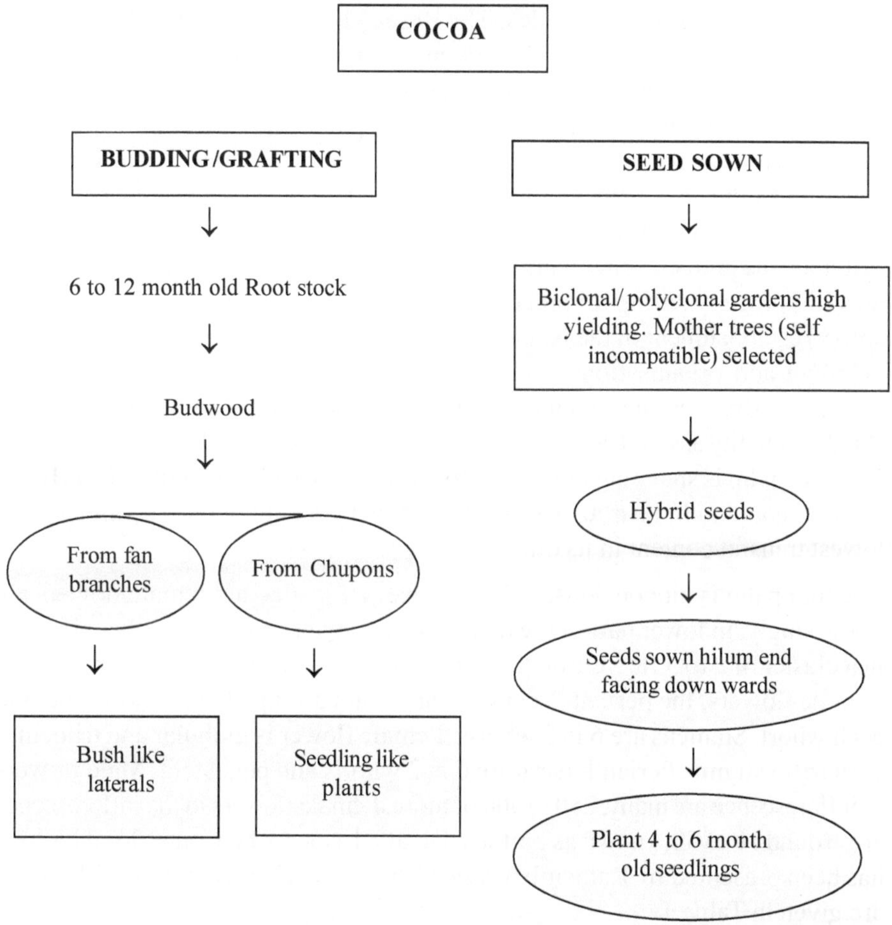

COCOA

BUDDING /GRAFTING → 6 to 12 month old Root stock → Budwood → From fan branches → Bush like laterals; From Chupons → Seedling like plants

SEED SOWN → Biclonal/ polyclonal gardens high yielding. Mother trees (self incompatible) selected → Hybrid seeds → Seeds sown hilum end facing down wards → Plant 4 to 6 month old seedlings

Micro budding technology (K.L. Chadha, 2001) is budding onto very young plants less than 15 days old with green bud of standard size from the scion. The green bud is inserted inside the pad of the bark and tied. The tap is removed after a week.

Fig. 1. Grafted plant and seedling production in cocoa

114. Coconut

(*Cocos nucifera* Linn.) (2n:32)

Family : Araceae

Coconut is the only commercial cultivated crop extensively grown in 90 countries. Coconut is a monotypicus genus. It is a perennial tropical palm, economically important and multipurpose utility. Improvement in coconut in genetic upgradation and breeding performance is taken up in the past eight decades. The Research and Development progress in this direction is considerably slow owing to the long time taken between two successive generations to assess the genetic and breeding performance and gains. The reliability on the superior mother palm selection can be arrived at after 20 years at the earliest to adjudge the ability of mother palms to yield consistently, besides early and regular bearing nature. Nearly 20 years or more period is reqiured starting with the high probability of selecting elite mother palm to initiate the protocols involving (i) identifying elite mother palm (ii) selection of nuts (iii) progeny testing for prepotency (iv) selection of quality assured seedling and (v) establishment of the isolated mother palm garden. These are the genetic, breeding and organisational initiatives for establishing systematic steps for Nucleus - Breeder seed chain for the production of coconut seed nuts and standard quality assured coconut seedlings to acclerate the yield gains in coconut. Coconut palm is spanning upto 60 years in active production of nuts. Nearly 5% of the coconut in India is also utilized for tender coconut juice. Coconut has the lowest transfat content in its oil.

Coconut palm is monoecious, inflorescence is a spadix and female flowers are borne singly on lower part of the rachis. Male flowers are borne either single or in a cluster of 2 to 3 flowers on the upper part of the rachis. Flowers are sessile. In male flowers, the perianth segments are 6 and arranged in 2 whorls and 3 in each whorl. Stamens are 6 in 2 whorls. Female flower is globular and trilocular with trifid stigma. Perianth parts are 6 in 2 whorls and persistent. Male flowers in inflorescence are many, 250 to 300 or more. Female flowers in the inflorescence are around 10 to 50, known as buttons. Detailed aspect of coconut floral biology has been presented by Ratnambal *et al.* (2003). The characteristics of this crop are given in Table 1.

Table 1. Characteristics of Coconut

1.	Plant petiole colour	Green, yellow, reddish yellow, brown, grey
2.	Collar girth (cm)	Low (<8), medium (8-12), high (>12)
3.	Plant - number of split leaves	Absent, present
4.	Plant height (cm)	Short (<130), medium (130-160), tall (>160)
5.	Number of leaves	Few (<6), medium (6-10), many (>10)
6.	Leaflet length (cm)	Short (120<), medium (120-200), long (>200)
7.	Leaflet breadth (cm)	Narrow (<2), medium (2-4), broad (>4)
8.	Leaf tip	Straight, intermediate drooping
	Note: 1 t o 8 observations in one year old plant	
9.	Inflorescence emergence in months	Early (<30), medium (30-60), late (>60)
10.	Inflorescence length (cm)	Short (<70), medium (70-90), long (>90)
11.	Inflorescence girth (cm)	Thin (<8), medium (8-10), thick (>10)
12.	Number of spikelets per inflorescence	Sparse (<5), few (5-15), medium (6-36), many (>36)

13.	Number of female flowers per inflorescence	Few (<20), medium (20-30), many (>30)
14.	Duaration of female phase (days)	Short (<5), medium (5-8), long (>8)
15.	Interspadix overlapping of male and female phase	Absent, present
16.	Fruit quality of tender nut, quantity of water (ml)	Low (<200), medium (200-300), high (301-400), very high (>400)
17.	Fruit colour	Yellow, yellow red, reddish yellow, reddishgreen, green golden yellow and brown.
18.	Fruit shape (polar view)	Oval, round, oblong
19.	Fruit shape (equatorial view)	Flat, round, angular
20.	Fruit length (cm)	Short (<15), medium (15-25), long (25-35), very long (>35)
21.	Fruit breadth (cm)	Narrow (<10), medium (10-15), broad (>15)
22.	Fruit husk thickness (cm)	Thin (<2), medium (2-3), thick (>3)
23.	Husked fruit shape	Oval, round, oblong
24.	Husked fruit length (cm)	Short (<8), medium (8-12), long (>13)
25.	Husked fruit breadth (cm)	Short (<8), medium (8-13), broad (>13)
26.	Shell thickness (mm)	Thin (<3), medium(3-5), thick(>14)
27.	Fresh endosperm thickness (mm)	Very low (<400), low (400-700), medium (700-1200), high (1200-1900), very high (>2000)
28.	Fruit weight (g)	Very low (<250), low (251-450), medium (451-850), high (851-1500), veryhigh (>1000)
29.	Husked fruit weight (g.)	Very low (<150), low (150-300), medium (300-600), high (>600)
30.	Husk weight (g.)	Very low (<80), low (80-120), medium (121-150), high (>150)
31.	Endosperm dry weight (g.) (6% moisture)	Low (<100), medium (100-200), high (>200)

(*Source:* PPV & FRA, 2001)

Anthesis in coconut is between 5 am. and 6 am. Female flowers become receptive after all the male flowers in the same spike shed their pollen. Inflorescence maturation, anthesis and pollen dehiscence continues for 18 to 20 days. The female flowers begin to open after 21 days and the female phase lasts for 4 to 7 days with the stigmatic receptivity for 1 to 3 days. Cross pollination is customary by insect and wind pollination. Flower to fruit set is 25 to 30%. Dehusked coconut fruit exhibit 3 eyes of which one is functional with the presence of an embryo while the 2 others are non functional. Endosperm is liquid natured. Button shedding in coconut tree is prevalent upto 55%. Spraying of 2, 4-D can prevent button shedding to the extent of 50 to 60%.

Coconut palms are broadly three types: Tall type, Nana (Dwarf) and Javanica (intermediate form). Tall palms are in general allogamons (cross pollinated), dwarf palms are autogamous (self pollinated) and the intermediate type is self and cross pollinated. In the intermediate types, male and female phases of two successive inflorescence overlap and this characteristic is also influenced by G x E. interaction.

In coconut, leaf primordial is differentiated 30 months before emergence of leaf and the life of leaf is 2.5 to 3 years. In the adult coconut palm, there are 8 to 10 leaves with bunches in leaf axil, 10 to 12 leaves with nuts of varying size in development and 10 to 12 leaves with spadix in different stage of development and a central apical bud. The inflorescence emergence is 32 months after inflorescence primodia begins to differentiate. Inbreeding depression of embryo and endosperm weight in relation to open pollinated nuts is reported.

Mother Palm Selection: Palms credited with (i) high record of performance in 25 to 30 years age range (ii) high proportion of regular bearing palm with high number of leaves in crown (iii) palms representing best of 10% of the population (iv) visible features associated with high yield like closely placed leaf scars, full spherical to semispherical canopy crown, short strong bunch stalk (v) 30 to 40 fully opened leaves (vi) 10 to 12 spadi per crown (vii) consistent and regular high yielding (regular bearing and high yielders) (viii) 80 to 100 nuts yield per tree per year / palm (ix) ripe nuts weight 650 g. and more and (x) high copra content (150 to180 g. per nut) are considered.

Prepotency: A prepotent mother palm is one which transmits its desirable characteristics to its progenies under random mating which can be evaluated by progeny testing. Superiority of progenies is tested at nursery stage from certain characteristics like early and fast growth, collar girth, faster rate of leaf production and early flowering. Prepotent parents transmit high yield of copra to the progenies in spite of the fact they have been derived from randomly pollinated nuts of seed parents. Prepotency testing is also useful for selecting best combiners. The prepotency (of an animal) is to impress upon its off springs to such an extent that they resemble that parent to each other more closely than usual (Lush, 1946). Harland (1957) used this term in coconut, a palm that is able to transmit the character high copra yield to the offspring in spite of random pollination. Liyanage (1961) attributed the prepotency in coconut due to high degree of additive genetic value and or the general combining ability besides degree of dominance and also on linkage and epistasis.

Selection of Mother Palm Gardens: Well maintained, disease and pest free plantations having high proportion of regular bearing and high yielders are identified by survey of the coconut belts / plantations in compact areas. Palms in the borders and those palms not surrounded by other palms are not considered. High performing plantations of 25 to 50 years old may be good source for nuts. Nuts are selected on the basis of size, shape, uniformity and disease free nature.

The ideal indicators are: ripe seed nut weighing on dry basis 670 to 675 g of whole nut, 340 to 345 g husked nut, husk weight 330 g, shell weight 155 g, endosperm 180 to 185 g, nut water 6 to 8 g and copra 180 to 190 g (6% moisture basis). As the coconut palm is monoecious, prepotency may be either on female or male side and the term female transmitters or male transmitters may be used. It is not known that the female transmitter will balance equally efficiently as male transmitter. Harland (1957) considered that just as the progeny of a single mother palm may be superior, whatever nature of the male parent, it is evident that reverse situation also holds good. If once female transmitters are identified, the pollen of the same palm may be used on phenotypically high yielding mother palm, thus making a very large quantity/number of good planting material available at a very short time. Large quantity of seedlings need to be produced by hand pollination with identified prepotent palms.

Seed gardens that are spatially isolated are established in 1955 in Sri Lanka. (Liyanage, 1961).

Nursery Raising: Seed nuts are planted in flat or horizontal planting in 60 cm x 60 cm spacing in well prepared nursery beds. Usually selection of seed nuts is made from February to August and December to May. Seed nuts are sown 1½ to 2 months after harvest in case of Tall variety and after 15 days in case of Dwarf palms. In an area of 120 sq.m, 1000 nuts are planted. Nuts from each palm is grown in individual progeny rows. Germination is completed in 4 to 5 months time. Vigorous seedlings put forth 2 to 3 leaves in 3 month and 6 to 8 leaves in one year. Early splitting of leaves, fast growth, thick collar development, short but thick leaf stalks and growth of collar to 10 to 12 cm at one year age are some of the visual indicators for quality and heathy seedlings. Within each progeny row, weakly developed seedlings, abnormal seedlings and other deviant types with poor stand are rejected. Around 60 to 65% of seedlings will be of standard quality that will show good field establishment, ideal recovery of growth and expression of plant type. Progeny rows are studied for high percentage of quality seedlings and high expression of prepotency.

Nucleus Seed Production in Coconut Palm: Productive palm gardens are selected and superior coconut palms based on yield and visual characters are identified in the middle of the garden and their progenies are tested in well laid coconut seedling nurseries. Superior palms showing high prepotency in seedlings are selected. Seed nuts from the superior palms are collected and used as Nucleus seed.

Breeder seed: Breeder seed Mother Palm garden: Selection of superior mother palms is done in mother palm garden planted with prepotency tested palms. The gardens are grown with 1600m isolation Seed nuts are harvested from superior elite adult palms. The seed nuts are then selected for size, shape and for trueness and uniformity. Seed nuts from superior palm origin are then planted in progeny rows. Selection is made in the nursery grown with progeny rows for quality seedlings conforming to the prescribed growth parameters. Selected quality seedlings are distributed for raising commercial gardens and parents for hybrid seed production

Elite mother palms are established by planting 5 to 10% of the superior quality seedlings, selected from families / progenies showing high degree of prepotency.

Isolated Seed Garden (ISG): Isolated Seed Garden (ISG) was first established in Sri Lanka in 1955 (Liyanage 1961). ISG is a plantation garden planted with seedlings of known parents in isolation plots of 1200 m. Male transmitters within the Isolated seed garden (ISG) are identified (Fig. 1). Index is computed and low value palms are removed. Establishing a coconut garden and identifying male transmitters will take about 20 years. High productive palms are used as females. Known male transmitters are used as pollen parents. Seeds from ISG are used for raising nurseries and for planting coconut plantations.

Isolated Seed Garden (ISG) (Isolated Spatial 1600m)

↓

Plant with seedling of known parents

↓

Identify male transmitters by selection index and reject those with low index

↓

Emasculate the palms other than male transmitters

↓

Raise Progenies and observe for 20 years period

↓

Select Palms of High Breeding Value

↓

Nursery Selection and raise or establish seedling plantations

(Based on Liyanage, 1961)

Fig. 1. Coconut-Selection of High Breeding Value Palms

Selection index (SI) for seedlings developed by Liyanage (1961) is arrived at as follows:

$SI = X_1 - 14.70 X_2 - 4.47 X_3$

where

SI is selection index

X_1 = Number of nuts per palm per year

X_2 = Weight of husked nuts in lb.

X_3 = Flowering period of the palm in months

It is estimated that 10 to 12 year old palms produce 6200 nuts / year /ha. In 16 to 18 years, the palms reach optimum production. The optimum production will be 10000 nuts / ha /year. Production of palms will continue upto 60 years. Coconut plantations can be replanted with quality seedlings .Replacement of old sinile trees, with productive palms and the rejuvenation can be done periodically with high quality assured seedlings.

Isolated seed gardens (ISG) are suggested in view of difficulties in doing large scale controlled pollination(Liyanage,1961). In ISG, palms with known parentage are planted. Within the garden, palms with high breeding value are used as females and the pollen source is limited to known male transmitters thereby ensuring that the seed to be collected is the result of crossing between palms, one of which is at least a proven parent.

Nair (1991, 1993) has presented methods of breeding for higher yield in coconut and exploitation of coconut seedlings for the genetic progress. In the coconut nursery, seedlings have been scored for the ideal growth and balanced development from germination (first 2 to 3 months after sowing), fifth month and at the end of twelve months. It is found that 18 to 20% of the seedlings are defective and need to be discarded at the age of 5 moths. The rest of the seedlings exhibit variation for vigour and growth in terms of number of leaves and split leaves produced. Sree Rangasamy and Sridharan (1993) observed that around 15 to 18 % of seedlings raised from selected mother palms are defective and are to be discarded. In the rest of the seedling, selection need to be excercised at different ages of seedling necessitating multistage selection in the nursery for quality seedling planting material. The selection criteria of hybrid coconut seedlings have been outlined by Luke Rahtna Kumar *et al.* (1993) and Luke Rathnakumar and Sridharan (1993).

115. Coffee

Coffea arabica L. (Arabica) and *Coffea canephora* Pierre Ex Frenchna

(Robusta) (2n=44)

Family : Rubiaceae

Coffee flowers are axillary and develop in clusters. Flowers are white in colour, calyx 5 lobed, corolla tubular, stamens 5 and style is bifid. Arabica coffee is autogamous while robusta coffee is cross pollinated.Coffee crop is seed propagated. Two types of coffee varieties are known-Arabica and Robusta. Propagation by seed is the preferred practice for new cultivars in most countries. The descriptors for this crop is given in Table 1.

Table 1. Descriptor table of Coffee (*Coffea arabica, Coffea canephora and C. arabica × C. canephora hybrid*)

1.	Plant shape	Conical, ellipsoidal, cylindrical
2.	Plant height	Very short, short, medium, tall, tall, very tall.
3.	Plant canopy (diameter)	Very small, small, medium, large, very large
4.	Plagiotropic primary branch length of inter node	Short, medium long
5.	Plagiotropic primary branching intensity of ramification	Very weak, weak, medium, strong, very strong
6.	Leaf length	Very short, short, medium long, very long
7.	Leaf width	Very narrow, narrow, medium broad, very broad
8.	Leaf shape	Lanceolate, ovate, elliptic
9.	Young leaf anthocyanin	Absent, very weak, strong
10.	Leaf undulation of margin	Very weak, weak, medium strong
11.	Leaf depth of secondary vein	Absent, present
12.	Leaf diameter	Shallow, medium, deep
13.	Inflorescence number of flowers	Few, medium, many
14.	Fruit size	Very small, small medium, large, very large
15.	Fruit shape	Elliptic, circular, oblong
16.	Fruit colour	Yellow, orange, light red
17.	Fruit adherence to the branch	Weak, medium, strong
18.	Fruit dry weight (100 fruits)	Low, medium, high
19.	Seed length	Very short, short, medium, long, very long
20.	Seed width	Narrow, medium, broad
21.	Seed length/width ratio	Small, medium, large
22.	Seed thickness	Thick, medium, thin
23.	Seed weight (100nos.)	Very low, low, medium, high, very high
24.	Period between flowering and harvest	Very short, short, medium, long, very long
25.	Seed caffeine content	Low, medium, high
26.	Seed sucrose content	Low, medium, high
27.	Seed total chlorogenic acid content	Low, medium, high

Arabica:	**Low**	**Medium**	**High**
1. Caffeine constent	<0.9	0.9-1.0	>1.2
2. Sucrose content	<7.0	7.0-9.0	>9.0
3. Total chlorogenic acid	<4.5	4.5-6.0	>6.5
Robusta and canephora:			
1. Caffeine content	1.8	1.8-2.5	>2.5
2. Sucrose content	<4.5	4.5-7.0	>7.0
3. Total chlorogenic acid	<7.0	7.0-8.0	>8.0

(*Source:* UPOV TG249/2008)

In Arabica coffee, in which the output is pure line, spatially isolated gardens are established for seed multiplication and distribution in India (Srinivasan, 1996). In Columbia, it is synthetic mix of seeds from selected lines (Moreno, 1944). Hybrid seeds are produced by hand pollination of previously emasculated and bagged flowers. Male sterility, governed by a single recessive gene in Arabica has been deducted for quick development of female fertile and male sterile coffee plant.

In Robusta coffee, seeds are produced in isolated gardens planted with selected seedling population. Synthetic hybrids are produced in seed gardens planted with two clones of known combining ability (Montagnan *et al.,* 1998). Self incompatibility ensures cross pollination and such clonal hybrids are not very uniform quality. Uniform hybrids are produced with parents developed by doubled haploid method (Lashermes *et al.,* 1994). Clonal propagation in coffee is ten times more expensive than seed propagation. Robusta Clones find limited application only.

Clonal hybrids of Arabica is more demanding due to slow rate of multiplication and hardening problems. Zamarip *et al.* (1991) report that somatic embryoids appear to be much promising for efficient mass propagation. In promising Arabica hybrids, this method is being adopted as an alternative to hybrid seed production (Etenne *et al.,* 2018). Rooted cuttings of coffee tend to be shallow rooted and consequently less tolerant to prolonged dry spells.

Arabica coffee is planted with a spacing of 1.5 to 2.0 m. either way. Robusta coffee is planted 2.5 m. apart on either side. Bearing of berries commences in 5th year, or little later.

In 2013-2014, Coffee Board of India has distributed 6978.5 kg of Arabica seed and 2011.5 kg of seeds of Robusta in traditional coffee growing areas. In the non traditional areas, 4410 kg of seeds in Arabica and 0.5 kg seeds in Robusta have been distributed (Annual Report, Coffee Board, 2013-2014). Rooted clones are also being prepared for distribution by the Coffee Board of India.

Vegetative propagation: In coffee vegetative propagation is adopted for preserving unique characteristics. Five to six month old orthotropic branches are selected from the highly productive plants. The horizontal plagiotropic branches are not selected. Single node cuttings (10 cm length) from the orthotropic branches retaining a pair of leaves at each node are selected. The leaves are cut to half the size. The cut ends of the shoots are dipped in IBA (500 mg IBA + 50 ml alcohol in 100 ml water) for facilitating early rooting. In about 3 months, cuttings will begin rooting and this is followed by 3 months duration hardening before field planting.

Seed Propagation: The Central Coffee Research Institute of India, Chikmagalur provides Robusta and C x R seed coffee for preparing nursery. After selection of plants, blossom should be forced on these plants selected by irrigating well before the general blossom shower. This step will avoid any cross pollination from the undesiable plants. After harvest of seeds, segregate the unripe and floats. Graded seeds are used for sowing (Ref. Coffee Cultivation Guide, 2008

Central Coffee Research Institute, Chikmagalur-577117 Karnataka state) .Triangular, elephant beans and cut seeds are removed. Seeds are pretreated before sowing. The seedlings are transferred to polybags and hardened 5000 seedlings in bags occupy 12 m x 8 m area of nursery.

Seedling Production: Seeds of coffee are collected from repeatedly well performing plants. Cherries are allowed to ripe well, harvested, graded or selected. Hand squeezing is done to pulp the selected fruits followed by fermenting and washing the seeds. Seeds are dried upto 25% moisture content and graded. Parchment with two leaves are selected and sown in flatbeds at 1 cm. depth and 2 cm apart with flat side of the bean downwards. Seedlings are retransferred to polybags of 17 cm dia. 25 cm depth size when first pair of leaf appears. Seedlings with poor growth, vigour, deviating types and abnormal development of shoot and root are discarded. The nursery is provided with shade net to regulate temperature, light and humidity. Seedlings are examined for shoot and root development. Seedlings with 5 to 7 pairs of leaves and 25 to 30 cm. height are transplanted in the field. Isolation distance of 400m is provided. Top grafting is made on 4 to 6 month old seedlings, whenever there is a need. Coffee flowers in 3 to 4 years after planting.

Nucleus Seed Production: The seedlins are grown to flowering and fruiting in isolated field and plus trees are selected and cherries berries collected and seeds extracted and seedlings raised. In the seedling planted crop provided with isolation of 1600m the seedlings are selected mother plantwise and planted in progeny rows of 20 plants in each progeny replicated twice. In each progeny, uniformity in flowering, true to type nature, and field establishment and fast growth, profuse flowering and high fruit/berry set are observed in individual plants. Inferior and deviating and poor performing plants are discarded. At the end of fourth year, superior performing plants in each progeny are tagged and berries collected and seeds extracted plantwise. In the laboratory seeds are examined for the size, uniformity, development and colour and selected. The selected seeds are bulked to make the nucleus seeds.

Breeder Seed Production: Healthy, quality seedlings are raised from the nucleus seed by polybag method and hardenced. Weak, poor and deviating type of seedlings are discarded. The selected seedlings are planted in isolated garden by 400m and good coffee crop production practices are followed. At different phases of crop growth, examination of single plants are made and weak, poor, very late, very early, flowering plants and deviating plants are uprooted. Selected healthy profuse flowering and berry setting plants are retained as breeder seed crop. Seeds from breeder crop are distributed for planting after ascertaining the genetic purity quality of the seeds. The seeds are tested for caffeine content and other commercial qualities.

Survey the productive coffee plantations and select elite high yielding plants. Collect ripe berries, extract the seeds and sow the graded and hand picked seeds in poly bag nurseries as single plant progenies.

↓

Evaluate the seedlings for fast growth, vigour, health, and true to type nature. Seedlings selected at six pairs of leaves are field planted in progeny rows 10 plants per row in 2 rows plots, replicated twice at 1.5m×1.5m spacing. Isolation distance of 1600m is maintained. Recommended crop production practices followed.

↓

Individual plants in the progeny rows are examined for fast growth, vigour, uniformity and true to type nature. Characteristics are studied to confirm descriptor (DUS) (Table 1) Off-types and deviating type of plants are uprooted. At first flowering/blossom, 3rd year after planting, plants are studied and performance recorded on single plant basis for flowering, uniformity, true to type nature of flowers and berries, fruit setting and yield and free from diseases and pests.

↓

Based on the performance of progeny rows for 2 to 3 harvest seasons for berry number and seed yield, the progeny rows recording above mean +2SE are identified. Superior plants in each identified progenies are labelled and berries harvested and seeds saved plant wise. Seeds from different selected plants are exammed in the laboratory and bulked and nucleus seed constituted.

↓

Seeds from nucleus seeds are raised in polybag nurseries and selected seedlings alone are planted in 1.5m×1.5m spacing with 400m. isolation distance in Breeder Seed Plot . Individual plants are now monitored for vigour, fast growth, true to type nature and uniformity from seedlings stage to flowering and berry setting. Plants with high berry setting and yield at 3rd , 4th, 5th year after planting are retained. Poor performing plants and deviating type of plants are uprooted. Monitoring Team visit is arranged at berry maturity.

↓

Berries are harvested from selected individual plants and seeds saved plantwise. Seeds are examined in the laboratory for true to type and sound bean development. Caffeine content in seeds from each plant is estimated and plants are selected. Seeds from the selected plants are bulked to constitute breeder seed. The seeds are tested in seed testing laboratory for seed qualities. Quality assured seeds are then tagged with breeder seed label.

↓

Molecular Grow Out test performed to ascertain the genetic purity.

Fig. 1. Nucleus and Breeder Seed Production in Coffee

116. Cumin

Cuminum cyminum L. Syn. *Cuminum odorum* Salisb. (2n:14)

Family : Apiaceae

Cumin, a small, annual herbaceous plant grows to a height of about 25 cm. and small white or pink flowers are borne on small compound umbels. The small, boat-shaped seed has nine ridges and it is brown-yellow in colour.

The reported life zone of cumin is 9 to 26 degrees centigrade with moderate annual precipitation and a soil pH of 4.5 to 8.3. Cumin thrives on rich, well-drained sandy loam soil. The plant, which needs mild temperatures during a three to four month growing season, is intolerant to long periods of dry heat. Cumin flower is small, sepals 5, petals 5, stames are 5 in number. It flourishes best in sunny places with low rainfall. The seeds are sown in a sunny location. The fruit or seed is greenish tan, long and narrow with ridges down its length. The flavour is warm and bitter. The three major types of cumin seed on the market, Iranian, Indian and Middle Eastern differ in seed color, quantity of essential oil and flavor. The odour and flavour of cumin is derived largely from the essential oil which contains cumaldehyde or cuminic aldehyde as the main constituent. Synthetic cuminaldehyde is an adulterant to cumin oil and is very difficult to detect chemically. The dried seed of cumin has 2.5 to 5 percent essential oil on a dry weight basis and is obtained by steam distillation.

Nucleus Seed Production: Seeds of the variety concerned from the breeder or the Institution are sourced. In a well prepared field previously not grown with the same seed spice crop, the seeds are sown in rows with 60 cm x 25 cm spacing. After germination and establishment , seedlings are thinned to maintain 10 cm spacing on the rows. Single plant is maintained per hill. An area of 500 sq m. is sown excluding border rows. Border rows 3 meters around the plot are raised with the same seed. This field is isolated by 400 m distance from the same seed spice crop on all sides. The field is well maintained free from weed, diseases infection and insect attack.Individual plants are examined continuously at different stages of growth till flowering. Off-types and deviating type of plants are removed. True to type plants for stem, branches and foliage characteristics alone are retained. Early and late flowering plants are removed to maintain flowering uniformity. Plants are examined at flowering phase, seed development and maturity stages for true to type nature. Plants showing variation are discarded. True to type plants are retained till harvest stage and 2000 to

2500 single true to type representative plants are labelled and matured umbels turning brown are collected plant wise in separate paper bags. The umbels are air dried under shade. Dried umbels are then examined individually for shape, size, number of umbels per plant. Extent of seed setting, number of seeds, size and shape of seed, aroma and related characteristics are recorded. Based on the laboratory examination of umbels, the selected plants are stored separately plant wise.

Single plant progenies are raised in the next season, in two replications in the field with isolation distance of 400 m and adapting the recommended spacing and good agronomic production practices. Border rows 3 meters around the plot are raised with the same seed. The crop is maintained free of weed, diseases and pests. After every 10th progeny row, one row of check progeny is sown for comparison. Progenies that are poor in field establishment, weak health, poor vigour are uprooted. Progenies showing variations within progenies for stem, branches, foliage size, shape and colour are discarded before commencement of flowering. Progenies showing deviating plants at or after flowering are uprooted along with the progenies on either side in order to avoid pollen contamination. Progenies showing plants that are uniform, true to type and stability for the expression of stem, branches, leaves size, colour and aroma, umbel shape and size, flower and developing and matured seeds are labelled. The performance of labelled progenies for number of umbels and seed yield are recorded. In the laboratory, umbels and seeds are examined for the trueness and deviating type of progenies are discarded. The progenies showing performance above mean + 2 SE are identified and the seeds are saved progeny wise. Part of the seed from each selected plant is pooled together to constitute nucleus seed. Sample from the nucleus seed is analysed for the essential oil content, taste, aroma and flavor of seed.

Breeder Seed Production: Nucleus seed is sown in well prepared field, not grown with the same seed spice crop in the past 2 to 3 years. The selected field is isolated by 400 m from other fields being grown with the same crop. Three meter wide border crop is raised with the same seed as in the breeder seed plot. The field sanitation is maintained. Recommended spacing of plants and single plant per hill is adopted. The crop is maintained free from weeds, diseases and insect pests from the establishment to harvest stage. Individual plants are verified for trueness, uniformity and stability of the characteristics expressed by the stem, branches, leaf, flowers, umbels and developing and matured seed. Off-types and deviating plants noticed up to flowering stage are uprooted. Deviating plants occurring at flowering and post flowering stages are removed along with the plants around them. At mid maturity of seeds, visit of Monitoring Team is arranged to adjudge the crop. At maturity, individual plants are harvested in tact and plants are examined in the laboratory for true to type in umbel and seed traits. Seeds from the selected plants are bulked to constitute breeder seed. Samples of breeder seed are tested for seed quality parameters and quality assured seeds are labelled with Breeder Seed Label. Grow out test is conducted with seed samples drawn from breeder seed. Laboratory estimation of essential oils is done. Qualities such as aroma, flavor, taste are verified.

117. Dill Seed Spice

(*Anethum graveolens* L.) (2n:22)

Family : Apiaceae

Dill is a fennel like herb, annual, growing upto 90 to 120 cm. Leaves are compound, each leaf having 2 to 3 or more discrete leaflets. Leaf margin is smooth. Leaves are 12 to 35 cm. long and alternately arranged. Flowers are borne in umbels and symmetrical and are pentamerous with 5 sepals, 5 petals and 5 stamens. Petal colour is yellow, ovary inferior, styles 2 to 3. Fruit is 3 to 7 mm. long, dry, flattened and not splitting. Seeds are schizocarp. Dill is heterogamous in the umbels and highly cross pollinated. Dill seed contains 2 to 5% volatile oil. (Indian Institute of Spice Research, Calicut). The characteristics of this crop is given in Table 1.

Nucleus Seed Production: Source seeds of dill are obtained from the breeder concerned or the Institution. Seeds are sown in a well prepared isolated plot of 500 sq. m. area. Seeds are sown in rows 45 cm apart and with a spacing of 30 cm. After 15 days, seedlings are thinned and only a single vigorous, healthy plant is retained per hill. Border rows, 5 meter wide, are raised around the field plot with the same source seeds. Isolation distance provided is 400m. Off-types and deviant plants are uprooted. Early and late flowering plants are discarded to maintain flowering uniformity. Plants showing deviations at flowering are removed/uprooted along with the surrounding plants. At maturity, 1000 true to type, healthy, uniform vigorous well matured plants are tagged from the middle of the plot and harvested with shoot and leaf in tact. In the laboratory, the selected and harvested plants are examined for trueness to the leaf type, leaf aroma, umbels shape, seed set and matured seed shape, size, colour and aroma. Three hundred true to type plants are finally selected and seeds separated plant wise.

Table 1. Characteristics of Dill

1.	Young plant anthocyanim colouration	-	Absent, present
2.	Attitude of leaves 3 to 5 leaf stage	-	Erect, semi erect, dense
3.	Plant density foliage	-	Large, medium, dense
4.	No. of primary branches	-	Few, medium, many
5.	Plant height	-	Short, medium, tall
6.	Length of main stem	-	Short, medium, long
7.	Stem diameter at middle third	-	Small, medium, large
8.	Stem blue hue	-	Absent, present
9.	Stem – intensity of green colour	-	Light, medium, dark
10.	Stem waxiness	-	Absent, very weak, weak, medium strong, very strong
11.	Leaf shape	-	Triangular, rhomboidal, oblong
12.	Leaf density of feathering	-	Loose, medium, dense
13.	Leaf width of segment	-	Narrow, medium, dense
14.	Leaf length	-	Short, medium, long
15.	Leaf width	-	Narrow, medium, broad
16.	Leaf blue hue	-	Absent, present
17.	Leaf intensity of green colour	-	Light, medium, dark
18.	Leaf waxiness	-	Weak, medium, strong
19.	Main umbel diameter	-	Small, medium, large
20.	Main umbel no. of peduncles	-	Few, medium, many
21.	Time of appearance of main umbel	-	Early, medium, late
22.	Time of beginning of flowering	-	Early, medium, late

(*Source:* UPOV, 1999)

In the next season, plant progeny rows are raised in 2 row plots 5 m in length with a spacing of 45 cm x 30 cm. Between plots, 60 cm wide spacing is provided to facilitate examination of individual plants. The plot is isolated by minimum of 400 meters distance from other dill growing fields. Around the progeny rows plots, 5 meter wide border rows with the same source seeds of dill are raised. A check plot of 2 rows of dill variety same as the source variety is raised once in every 10 plots for comparison. Single healthy plant only is retained in each hill in the progeny rows. Systematic observations are made on individual plants in the progeny rows from seedling stage to maturity. Off-types are uprooted as and when they appear and within and between progenies, true to type and uniformity are verified. Progeny rows showing deviant plants observed at flowering, are discarded along with the progeny rows on either side. Progeny rows are assessed for plant type, stem and foliage, flower, umbel features and seed characteristics. The productivity and performance of the progeny rows are evaluated. Progeny rows showing performance above mean +SE are identified. Individual plants are harvested and seeds saved. Seeds of superior plants within each selected progeny are again verified in the laboratory for trueness. Part of the seeds from each plant are bulked/pooled to constitute the nucleus seed and remaining seeds saved for future use in developing nucleus seed. The sample of the nucleus seed is studied for various aromatic characteristics.

Breeder Seed Production: In the next favourable season, seeds from nucleus seed are sown in a well prepared breeder seed production plot adopting a spacing of 45 cm x 30 cm. All around the plot, 5 meter wide border rows are raised with the same source seeds. Only one healthy single plant is retained per hill. The breeder seed production plot isolated by 400m from other dill growing field. Observations at multistages are made for vigour, plant type, leaf orientation, leaf lamina colour, umbel size and shape, flower colour, size and shape and furrows on matured seeds. Non conforming plants have already been uprooted before flowering and the deviating plants occurring after flowering are removed along with the surrounding plants. At mid maturity stage, Monitoring Team visit is arranged to adjudge the breeder seed crop of dill. At maturity, plants are harvested and examined in the laboratory for trueness, uniformity and conformity of leaf, umbel flower and matured seeds. Seeds from selected plants are separated plant wise and saved. Seeds from different selected plants are bulked together to constitute Breeder seed. Sample of the breeder seed is tested for various seed quality parameters and the quality assured seeds are labelled with Breeder seed label. Samples drawn from the Breeder seed is evaluated in GOT to ascertain the genetic purity. Aromatic oil and other qualities of the seed spice are estimated.

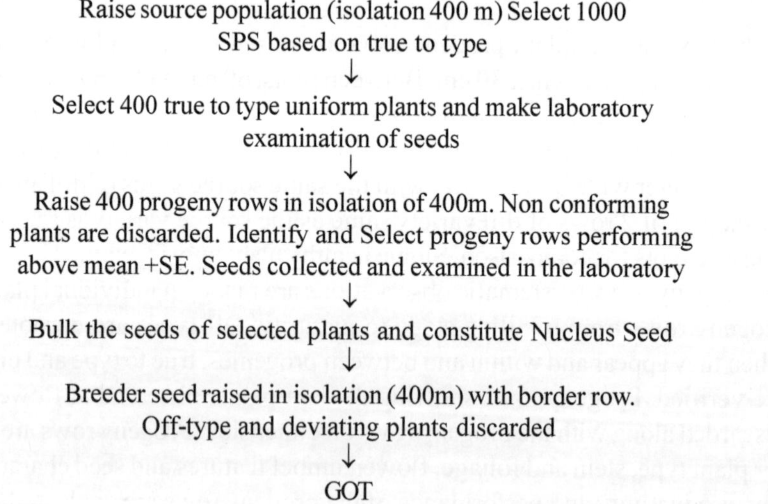

Fig. 1. Nucleus and Breeder seed Production in Dill (Anethum)

118. Fennel

(*Foeniculum vulgare* Mill.) (2n:22)

Family : Apiaceae

Other Names : Common fennel; sweet fennel; Florence fennel; spigel

Fennel is an erect growing perennial herb native to southern Europe and the Mediterranean area and is commonly grown for flavoring purposes. The mature seeds or dried fruits of this herb are used commercially and the young tender shoots and leaves of the sweet fennel are used in foods in European countries. This plant has a thickened base of leaves which can be blanched or boiled and eaten as a vegetable. The seeds should be planted early in spring to a depth of one cm. and in a spacing of 90 cm x 20 cm and thinned to one seedling per hill. The descriptors are given in Table 1.

Fennel is a greyish-green, strong-smelling herbaceous perennial, with slim stems, bearing soft lacy, dark green leaves with thread-like lobes and swollen bases. Creeping rootstock gradually extends plant into a sparse clump grumolo. It grows to 150 cm. height. The plant has small mustard-yellow flowers on a compound umbel. Flowers are yellow and small borne in umbels. Calyx five, petals 5 and stamens are 5. The fruit splits into two seeds which are oval in shape with five ridges. The seed is light green to grey, about 0.75 cm. long, and curved. Ordinarily the plants do not flower in the first season but a full crop of seed will be produced for several seasons thereafter. Although a perennial, fennel is generally grown as an annual or biennial crop. Yields of the dried fruit, commonly thought of as fennel seed, are low in the first year but increase in the second year. The leaves loose most of their sweet aromatic flavor on drying. Approximately 60 percent of the essential oil is located in the fruit, with the rest in the rays of the umbel and other green plant parts.

Table 1. Descriptor characteristics of fennel

1.	Length of cotyleden	-	Short, medium, long
2.	Length of first leaf petiole	-	Short, medium, long
3.	Plant height at harvest	-	Short, medium, long
4.	Foliage attitude	-	Erect, semi erect, horizontal
5.	Foliage density	-	Sparse, medium, dense
6.	Foliage green colour intensity	-	Very light, light, medium, dark, very dark
7.	Main stem height at flowering	-	Short, medium, tall
8.	Leaf length	-	Short, medium, long
9.	Leaf tip curvature	-	Absent, weak expression, strong expression

10.	Plant grumolo formation	-	Absent, present
11.	Grumolo growth height	-	Short, medium, tall
12.	Grumolo width	-	Narrow, medium, broad
13.	Grumolo thickness	-	Thin, medium, thick
14.	Grumolo shape (cs)	-	Round, elliptic
15.	Grumolo colour	-	Whitish, greenish
16.	Grumolo - intensity of green colour	-	Light, medium, dark
17.	Grumolo sheath ribbing	-	Weak, medium, strong
18.	Grumolo sheath overlapping	-	Weak, medium, strong
19.	Time of grumolo maturity	-	Early, medium, late

(*Source:* UPOV, 2001)

Fennel contains 1 to 3% of a volatile oil which is composed of about 50 to 60% anethole and about 20% d-fenchone. Oil of fennel is commonly available. An oleoresin is also available with a volatile oil content of only about 3 to 6%. Bitter fennel oil is thought to contain a bitter mixture with a camphor-like odour and flavour and less anethole than sweet fennel oil. Sweet fennel oil is of a superior quality with a more pleasing aroma and flavour.

Nucleus Seed Production: Seeds of the variety concerned from the breeder or the Institution are sourced and sown in a well prepared field previously not grown with the same seed spice crop. The seeds are sown in rows 90 cm apart and 5 cm on the rows. After germination and establishment, seedlings are thinned to maintain 5 cm spacing on the rows. Single plant is maintained per hill. An area of 500 sq m is sown excluding border rows. Border rows 3 meters around the plot are raised with the same seed. This field is isolated by 400 m distance from the same seed spice crop on all sides. The field is well maintained free from weeds, diseases infection and insect pests. Individual plants are examined continuously at different stages of growth till flowering. Off-types and deviating type of plants are removed. True to type plants for stem, branches and foliage characteristics alone are retained. Early and late flowering plants are removed to maintain uniformity in flowering. Plants are examined at flowering phase, seed development and maturity stages for true to type nature. Plants showing variation are discarded. True to type plants are retained till harvest stage and 2000 to 2500 single true to type representative plants are labelled and matured umbels turning brown are collected plant wise in separate paper bags. The umbels are air dried under shade. Dried umbels are then examined individually for shape, size and number of umbels per plant. Extent of seed setting, number of seeds, size and shape of seed, aroma and related characteristics are recorded. Based on the laboratory examination of umbels, seeds and productivity, 200 to 250 plants are selected. Seeds from the selected plants are stored separately plant wise.

Single plant progenies in two rows plants are raised in the next season, in two replications in the field with isolation distance of 400 m adapting the recommended spacing and good agronomic production practices. Border rows 3 meters around the plot are raised with the same seed. The crop is maintained free from weeds, diseases and pests After every 10th progeny row, one row of check progeny is raised for comparison. Progenies that are poor in field establishment, weak health and poor vigour, are uprooted. Progenies showing variations within progenies for stem, branches, foliage size, shape and colour are discarded before commencement of flowering. Progenies showing deviating plants at or after flowering are uprooted along with the progenies on either side in order to avoid pollen contamination. Progenies showing plants that are uniform, true to type and stable for the expression of stem, branches, leaf size, colour and aroma, umbel shape and size, flower and developing and matured seeds are labelled. The performance of labelled progenies for number of umbels and seed yield are recorded. In the laboratory, umbels and seeds are examined for the trueness and deviating type of progenies are discarded. Progenies showing performance above mean + 2 SE are identified and the seeds are saved progeny wise.

Part of the seed from each selected plant is pooled together to constitute nucleus seed. Sample from the nucleus seed is estimated for essential oil content, taste, aroma and flavor.

Breeder Seed Production: Nucleus seed is sown in well prepared field, not grown with the Fennel seed spice crop in the past 2 to 3 years .The selected field is isolated by 400 m. from other fields being grown with the fennel crop. Three meter wide border crop is raised with the same seed as in the breeder seed plot. The field sanitation is maintained. Recommended spacing of plants and single plant per hill are followed. The crop is maintained free of weeds, diseases and insect pests from the establishment to harvest stage. Individual plants are verified for trueness, uniformity and stability of the characteristics expressed by the stem, branches, leaf, flowers and umbels developing stage. Deviating type of plants occurring in flowering and post flowering stages are removed along with the plants around them. Visit of Monitoring Team is arranged. At maturity, individual plants are harvested and in tact plants are examined in the laboratory for true to type in umbel and seed traits. Seeds from the selected plants are bulked to constitute breeder seed. Samples of breeder seed are tested for seed quality parameters and quality assured seeds are labelled with Breeder Seed Label. Grow out test is conducted with seed samples drawn from breeder seed. Laboratory estimation of essential oils and other qualities such as aroma, flavor, taste are made.

119. Fenugreek

(*Trigonella foenum graceum* L.)

Family : Fabaceae

It is an annual herb crop used as green leaf vegetable condiment and grain legume. The main stem is erect. Leaves, green to light green, pinnately trifoliate, Inflorescence is axillary in cymes. Flowers are papilionaceous. Anthers 10 in 9 plus 1 arrangement. Anthesis occurs from 9 am. to 6 pm. Anthers dehisce between 11 am. and 12.30 noon. Stigma becomes receptive 12 hours before flower opening and remains receptive till 12 hours after flower opening. Flowering duration is from 7 to 18 days. Fenugreek is self pollinated and cleistogamous crop. Crop duration and harvest are 20 to 25 days for green leaves and 90 to 100 days for grains. The characteristics of this crop are given in Table 1.

Table 1. Characteristics in Fenugreek

1.	Basal shape of first leaf blade	Acute, oblate, rounded
2.	Apex shape of first leaf blade	Obtuse, rounded
3.	Size of leaflet on first primary axis for length and width	Small length and width, small length and wider blade, long length and narrow blade, larger and wider
4.	Basal shape of leaf blade on first primary branch axis	Acute, obtuse
5.	Apex shape of leaf blade on first primary branch axis	Acute, obtuse, rounded
6.	Basal shape of leaf blade on first pod axis	Acute, obtuse
7.	Apex shape of leaf blade on first pod axis	Acute, obtuse, rounded
8.	Size of leaf on full grown terminal leaf	Small in length and width, small in length and breadth, width, long length but narrow blade
9.	Apex shape of leaf blade on full grown terminal leaf	Acute, obtuse
10.	Number of primary branches	Low (<6), more (>6)
11.	Plant growth pattern	V type, U type
12.	Plant grown habit	Determinate, indeterminate
13.	Plant height (cm)	Short (<45), Tall (>45)
14.	Pods per plants	Low (<50), high (>50)
15.	Pod length (cm)	Short (<11), medium (11-12), long (>12)
16.	Pod curvature	Moderately curved, strongly curved
17.	1000 seed weight (g)	Low (<16), medium (16-18), high (>18)

(*Source:* PPV & FRA)

Nucleus Seed Production: Fenugreek crop is raised with wide spacing in an area of 400 sq m with seeds accessed from authentic breeder source. Isolation distance of 20 m is provided. Individual plants are critically examined for conforming to the characteristics, vigour, health and growth at vegetative, green leaf stage (20 to 25 days after sowing) and at pod maturity around 100 days. One thousand single plants are identified, labelled and pods collected. In the laboratory, pods and seeds are examined for true to type nature and selected. In the next season, in well prepared field, 1000 progeny rows are sown in 2 rows each and all the recommended good agronomic production practices are followed. Progenies are critically studied for trueness, uniformity and for green leaf yield at 20 to 25 days and at maturity for grain yield. Isolation distance of 20 m is maintained. Off-types and deviant plants are removed along with the rest of the plants in the progeny row. The progenies that show greater uniformity and true to type breeding and stability alone are retained and the rest discarded. In the uniform looking progenies, 600 to 700 single plants are labelled and harvested and seeds separated. Pods and seeds are examined in the laboratory and deviating types are discarded. In the next generation, 500 plant progenies are raised in 2 row plots and studied for uniformity, true to type breeding and stability for the varietal characteristics. Out of the 2 rows in each progeny, one row is utilized at appropriate age for green leaf characteristics and quality at 25 to 30 days after sowing. Laboratory evaluation is made for the green leaf biomass, its tenderness and cooking quality and for pod and seed traits at maturity. Progenies showing deviant plants and disease affected plants are discarded as a whole. The progenies are evaluated for grain yield. Uniform looking progenies performing above mean +SE are identified and from them one hundred plants that are superior are then tagged and harvested separately. Seeds from the superior selected plants are examined in the laboratory for trueness and uniformity and bulked to constitute nucleus seed.

Breeder Seed Production: Samples of the nucleus seed are utilised for production of breeder seed adopting 20 m isolation distance Single plant per hill is maintained. Field sanitation and all recommended package of production are practised. Frequent monitoring for the genetic uniformity, true to type breeding and stability is done. Off-types and deviant plants are removed along with adjoining plants to avoid pollen contamination. Monitoring Team visit is organized at mid maturity stage. At maturity, pods are harvested, seeds separated, dried and processed. Laboratory tests are made for seed quality and Breeder Seed label is affixed to the quality assured breeder seed. Seeds samples from the Breeder seed lot is used for conducting grow out test to verify the genetic purity. Nutritional qualities of green leaf and seeds are studied/assessed of this specialty fenugreek.

120. Garlic

(*Allium sativum* L.) (2n:16)

Family : Alliaceae

Garlic is a vegetative propagated bulbous crop. This crop plant is highly sterile and no seed set is obtained. Plants grow up to 150 cm. tall; real stem is very short, formed at the base of the plant in the form of a disk, with adventitious roots at base. Bulb is solitary, depressed globose to ovoid, upto 7 cm in diameter, whitish to purplish, composed of 7 to 15 sessile cloves, that are ovoid to ellipsoid-oblong and borne in the axil of the 2, 3 or 4 to 5 last leaves. Each clove consist of a protective sheath, a single thickened storage leaf sheath and a small central bud. Leaves 4 to 10, distichously alternate, glabrous, with tubular sheath; blade linear-oblong, up to 50 cm × 2.5 cm, nearly flat or V-shaped in cross-section, acute at apex, smooth or crenulate at margins. Inflorescence is a spherical umbel up to 2.5 cm in diameter, on a solid scape up to 150 cm long, initially curved as a crook, then coiled and eventually straight; umbel composed of flowers and bulbils or only bulbils, initially surrounded by a membranous spathe splitting on one side. Flowers usually poorly developed or absent; pedicel slender, up to 1.5 cm long; tepals 6, in 2 whorls, free, lanceolate, up to 3 mm long, greenish white or pale pink; stamens 6, usually rudimentary; ovary superior, 3-celled, style shorter than tepals. Fruit abortive, seedless. The characteristics of this crop are given in Table 1.

Table 1. Characteristics of Garlic

1.	Plant density of leaves	Sparse, medium, dense
2.	No. of leaves per pseudo stem	Few (<9), medium (9-12), many (>12)
3.	Foliage attitude	Erect, semierect, drooping
4.	Leaves- intensity of green colour	Light, medium, dark green
5.	Leaf waxiness	Absent, present
6.	Leaf length (cm)	Short (<25), medium (25-35), long (>35)
7.	Leaf width (widest) (Cm)	Narrow (<1.5), medium (1.5-2.5), broad (>2.5)
8.	Leaf shape (cs)	Flat, slightly concave, concave
9.	Pseudostem length (cm)	Small (<5), medium (5-10), long (>10)
10.	Pseudostem width(cm)	Narrow (<1.0), medium (1.0-1.5), broad (>1.5)
11.	Pseudostem-intensiy of athocyanin at base	Absent, present
12.	Flower stem	Absent, present
13.	Flower stem curvature	Absent, present
14.	Flowering stem length (Spl character for the temperate condition) (cm)	Short (<70), medium (70-80), long (>80)
15.	Flowering stem bulbils	Absent, present

16. Days to maturity	Early (<130), medium (130-160), late (>1 to-)
17. Bulbs size (dia) cm.	Small (<2.5), medium (2.5-3.5), large (3.5-5.0), verylarge (>5.0)
18. Bulb shape (L.S)	Elliptic, ovate, circular
19. Bulb shape(CS)	Elliptic, circular
20. Bulb position of root disc	Inserted, flat exserted
21. Bulb compactness of clones	Loose, medium, compact
22. Bulb shape of base	Recessed, flat, rounded
23. Bulb groud colour of dry externals scale	White, yellow, white reddish, pink
24. Bulb anthocyanin stripes on dry external scale	Absent, present
25. Number of cloves in a bulb	Few (<10), medium(11-20), many (>20)
26. Bulb distribution of cloves	Radial, non radial
27. Bulb skin adherence of dry external scales	Absent, present
28. Clove size diameter (cm)	Small (<1), medium (1-2), large (>2)
29. Clove colour of scale	White, cream, pink, purple
30. Clove-colour of flesh	White, yellow

It is usually propagated by bulb to bulb cycle by seed bulbs known as cloves. Seed rate followed is 500 to 600 kg/ha. The cloves are sown in a spacing of 10 cm on the furrows that are shallow and width 15 cm between furrows. Important precaution is avoiding excess N use to minimize rubberisation of cloves. N should not be applied as urea. Ammonium sulphate is recommended as N source. Spraying 1500 ppm of CCC or MH on 30th day of planting is beneficial.

Nucleus Seed Production: Healthy, vigorous, disease and nematode infestation free plants that are true to type and uniform and representative are selected from the source nursery or source population / maintenance breeding population. After harvest, the selected garlic plants in sufficient number are examined for true to type nature ,uniformity free of bulb rot, root and nematode infestation. Cloves are excised from the selected plants. Laboratory selection of garlic mother bulb <2.5 cm. and above is made based on leaf colour and size and the size, colour, shape and number of uniformly developed cloves per clump. The cloves separated from selected mother clump are raised in the following season. The field is isolated by 5m. from other garlic crops. The individual plants from the cloves are examined for vigour, fast growth, truness, uniformity in clump development and maturity free from nematode infestation Off-types and deviating plant types are uprooted. Selected true to type and healthy plants are harvested and clumps are verified in the laboratory for the uniformity in development and truness. Selected clumps are shade dried and cloves separated and graded and pooled to constitute the Nucleus seed cloves.Based on the requirements, the cycle of developing mother bulb clumps are repeated to produce required quantity of nucleus seed in Garlic (Fig. 1).

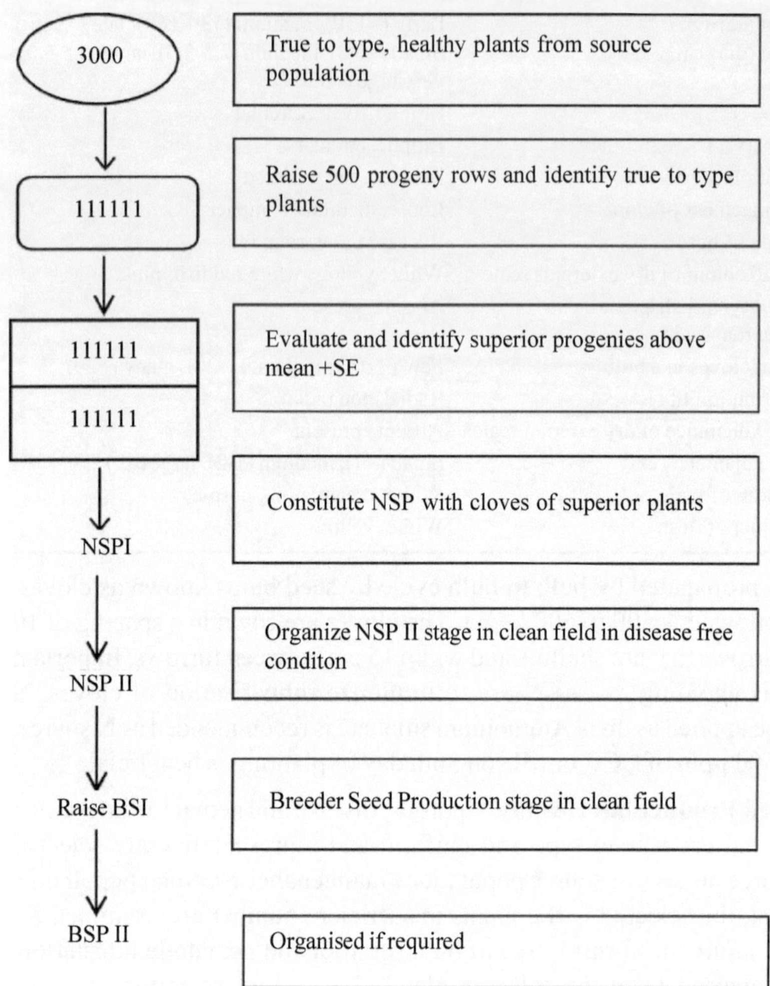

Fig. 1. Nucleus and Breeder Seed Production in Garlic

Breeder Seed Production: The breeder seed of cloves of garlic is produced in disease free field adopting the techniques of quality assured bulb to bulb (clove to clove) production in garlic. Minimum distance of 5m isolation is necessary. Sound and healthy cloves are dibbled in well prepared field not grown to garlic in the previous 2 to 3 seasons and that will be free from nematode and diseases inoculum. All good agronomic and horticultural production practices are adopted. Single plant per hill is retained. Excess application of N fertliser is avoided and the frequency of irrigation is reduced to minimize rubberisation of cloves. Spraying of 0f 1500 ppm of CCC or MH on 30th day of planting is recommended. Application of boron (0.2%) and sodium molybdate (0.1%) on 30th,60th and 90th day after planting is desirable (TNAU Horticulture Production

Guide, 2013). Individual plants are critically studied for trueness, uniformity and health. Fast growing plants are retained and weak plants are removed. Off-types and deviating plants are discarded. At maturity stage Monitoring Team visit is organized.

At proper stage, matured bulbs are harvested and uniform size true to type bulbs are separated from the stem. In the laboratory, uniform bulbs alone are examined for the clove development and maturity and for the prescribed (true to type) characteristics of the garlic variety. Cloves are shade dried, graded and tested for clove seed qualities. Quality assured clove seed is labelled with breeder seed label. Grow out test is conducted as per the norms.

Breeder seed BSP I and BSP II stages are then organized in clean field with disease free nucleus seed.

121. Nigella (Black Cumin)

(*Nigella sativa* L.) (2n=12) Karunjeeragam

Family : Ranunculaceae

It is a cool season crop. Black cumin is a small herb grows up to 60 to 70 cm height. Leaves are linear, lanceolate segments, 2.5 to 3 cm. long. Flower hermaphrodite, solitary on the main axis; size 3 cm. length and breadth ; colour pigeon blue; flower without any involucres of bracts, peduncle long erect; petalloid sepals broad, ovate in a single whorl, 4 to 6, mostly 5 and characterized by the presence of nectaries; corolla absent; Stamens in 3 to 4 whorls, numerous (32 to 66). Anthesis is from 7 am. to 10 am. and the filament bent outward during male phase and pollination is effected on synchronization of both male and female phases; gynoecium 5,completely united follicles and composed of variable number of multiovule carpel developing into a follicle after pollination. Fruit is a capsule. Flowers are protandrous, cross pollinated by insects. Fertility on an average is 89%.

Nucleus Seed Production: Seeds of the variety concerned accessed from the breeder or the Institution/from AVT are sown in a well prepared field previously not grown with black cumin crop. The seeds are sown in rows 30 cm apart and 30 cm on the rows. Single plant is maintained per hill. An area of 600 sq m is sown excluding border rows. Border rows 3 meters around the plot is raised with the same seed. This field is isolated by 400 m distance from the same seed spice crop on all sides. The field is well maintained free from weeds, diseases infection and insect attack. Individual plants are examined continuously at different stages of growth till flowering. Off-types and deviating type of plants are removed. True to type plants for stem, branches and foliage characteristics alone are retained. Early and late flowering plants are removed to maintain flowering uniformity. Plants are examined at flowering phase, seed development and maturity stages for true to type nature. Plants showing variation are discarded. True to type plants are retained till harvest stage and 2000 to 2500 single true to type representative plants are labelled and matured capsules turning brown are collected plant wise in separate paper bags. The matured flowers are air dried under shade. Dried flowers/heads are then examined individually for shape, size and number of heads per plant. Extent of seed setting, number of seeds, size and shape of seed, aroma and related characteristics are recorded. Based on the laboratory examination of flowers, seeds and productivity, 200 to 250 plants are selected. Seeds from the selected plants are stored separately plant wise.

In the next season, single plant progenies in 2 rows plots are raised in two replications in the field with isolation distance of 400 m adapting the recommended spacing and good agronomic production practices. Border rows 3 meters around the plot are raised with the same seed. The crop is maintained free from diseases and pests. After every 10th progeny row, one row of the original variety is raised for comparison. Progenies that are poor in field establishment, weak health and poor in vigour are uprooted. Progenies showing variation for stem, branches, foliage size, shape and colour are discarded before commencement of flowering. Progenies showing deviating plants at or after flowering are uprooted along with the progenies on either side in order to avoid pollen contamination. Progenies showing plants that are uniform, true to type and stability for the expression of stem, branches, leaves size, colour and aroma, flower shape and size, flower and developing and mature capsules and seeds are labelled. The performance of labelled progenies for number of flowers and seed yield are recorded. In the laboratory, flowers and seeds are examined for the trueness and deviating type of progenies are discarded. Progenies showing performance above mean + 2 SE are identified and the seeds are saved progeny wise.

Part of the seed from each selected plant is pooled together to constitute the nucleus seed. Sample from the nucleus seed is analysed for essential oil content, taste, aroma and flavor of seed.

Breeder Seed Production: Nucleus seed is sown in a well prepared field, not grown with black cumin spice crop in the past 2 to 3 years. The selected field is isolated by 400m from other fields being grown with the same crop. Three meter wide border crop is raised with the same seed as in the breeder seed plot. The field sanitation is maintained. Recommended spacing of plants is followed and single plant per hill is maintained. The crop is maintained free from diseases and insect pests from the establishment to harvest stage. Individual plants are verified for trueness, uniformity and stability of the characteristics expressed by the stem, branches, leaf, flowers, developing and matured capsules. Off-types and deviating plants noticed upto flowering stage are uprooted. Deviating plants present at flowering and post flowering stages are removed along with the plants around them. At midmaturity of seeds, visit of Monitoring Team is arranged to adjudge the purity of the crop. At maturity, individual plants are harvested and in tact plants are examined in the laboratory for true to type in flower and seed traits. Seeds from the selected plants are bulked to constitute breeder seed. Samples of breeder seed are tested for seed quality parameters and quality assured seeds are labelled with Breeder Seed Label. Grow out test is conducted with seed samples drawn from breeder seed. Laboratory estimation of essential oils and other qualities such as aroma, flavor, taste are carried out.

122. Nutmeg

(*Myristica fragrams* Hutt.) (2n:40)

Family : Myristicaceae

Flowers are fragrant and produce nectar. It is a dioecious plant and female trees are cross pollinated by pollen from male trees. Female and male trees occur almost in equal proportion in the seedling raised plants along with a very few bisexual plants. Fruit is a fleshy drupe pale yellow and grooves in the center of the fruit wall. There are no named varieties. Seeds are sown in polybags and sprouting occurs 20 days after sowing. Epicotyle grafting developed by NRCS, Calicut is followed for clonal multiplication of the elite plus true to type trees. Grafted plants with orthrotropic scions flower 6 years after planting and peak production is after 20 years of age. In the planted field 10% of the trees are male plants. Fruits develop 9 to 12 months after planting. Pericarp is used for food preparations. Nutmeg oil is 4 to 16% in nutmeg mace. Fruits weight is 60g, 7g. of seed and 3 to 4 g. of mace. The yield is estimated as 1000 kg of nutmeg and 80 kg of mace per ha. Eartliness and high yielding tree are the ideal objectives in Nutmeg breeding and selection. The characteritics of nutmeg are given in Table 1.

Table 1. Characteristics of Nutmeg

1.	Leaf shape	Eliiptical, oblong, obovate
2.	Leaf surface texture	Smooth, shiny, pumby; smooth shiny not pumby, smooth dull and bumby, smooth dull and bumby smooth, dull and bumpy smoothdull and not bumpy
3.	Leaf colour	Green, dark green, yellowish green
4.	Leaf size	Small, medium, large, very large

5. Leaf tip sharpness	Slightly pointed, pointed, short
6. Shape of petiole base	Flat, looped
7. Plant height (m)	Short (<25), medium (25 to 35), tall (>35)
8. Seeds with shell (dia)	2.75 to 2.75 cm
9. Seed- mace thickness	1.00 to 1.80 mm
10. Seed-compactness of mace on seeds	Very spasse, spasse, dense, very dense
11. Fruit shape	Round, short round, oval, round spherical
12. Fruit diameter/length	Low, medium, high
13. Fruit skin colour	Yellow, light yellow, yellowish green
14. Fruit flesh thickness	0.9 to 1.02 cm
15. Fruit base shape	Straight, bended forward, bended forward with a bump
16. Fruit grooves (medium)	Flat, deep, rather deep

Flowers are borne in axillary umbellate cymes. Pedicel is 1 to 1.5 cm long. Flowers are 1 cm long, creamish yellow, waxy, fleshy and fragrant. Petals are absent. Female flowers pubescent, sessile and one celled ovary. Stigma is two lipped, male flowers bear 8 to 12 stamens. Fruit is one seeded drupe. Nutmeg crop is systematically removed at an early age of 4 to 15 years (Joseph, 1980). Morphological diversity of nutmeg mother trees and seedling variation have been estimated and presented by Krishnaoorthy *et al.,* 1996 and Helen *et al.,* 2015. Recent reviw on Nutmeg is by Thangaselvibai *et al.* (2011).

Nucleus and Breeder Seed Production: The grafted plants from selected elite nutmeg trees that are highly productive and regular bearing as adjudged at age of six years or more after planting are selected. They form the basic material. Clonal progenies from 20 selected elite trees, at least five in number per plant developed by grafting are planted in single row plots and grown with the male plants and recommended production practices. True to type nature and uniformity among the clonal progenies are verified. So also the characteristics shown in Table 1. are examined. Based on yield data of fruits and fruit qualities, true to type clonal progeny performing above mean values are identified. Within each clonal progeny elite and plus plants are selected. The selected plants constitute Nucleus/Breeder Stock and the selected true to type plants are raised in a mother Block with isolation distance of 10m. in view of long generation time and perrenial nature. The unselected progenies in nucleus/ breeder stock are uprooted retaining the elite plus trees only and maintained as mother Block. Grafted male plant (10% of grafts) are planted in the Mother Block.

123. Palmyra Palm

(*Borassus flabellifer* L.) (2n=36)

Family : Arecaceae

The Palmyra palm is distributed in India, Sri Lanka and South East Asia and Africa. Genome of palms from Indonesia, India and Africa have been compared by Arunachalam, (2012). Palmyrah palm possess 2n:36 cromosomes. Sree Rangasamy and Devasahayam (1972) studied the sex chromosomes and showed that male palms are of XY and female palms are of XX constitution in sex chromosome pairs. Marker assested sex detection in Palmyrah has been developed by George and Karun (2015).

In Tamil Nadu, Palmyrah Palm, *Borassus flabellifer* is considered in high honour and is designated as the State Tree in 1978.

Stem is tall, unbranched and showing leaf scars prominently. Palmyra palm grows to a height of 25 to 30 m and stem diameter ranges from 60 to 90 cm. Stem cortex is narrow. Root system is adventitious. Matured roots have a sclerotic exodermis (Peter, 2001). Palmyra Palm crown is spherical consisting of 25 to 35 erect fan shaped green leaves with nearly 100 cm long petiole which is strong by grooved adaxial upper surface. The petiole margins are adorned by chitinase saw like teeth providing strength to the petiole to bear the leaves. Leaves of palmyrah show 80 to 90 compound segments per leaf, each segment in duplicate. The leaf segments are dried and processed after leaf maturity and used in wide range of varieties of palmyra leaf handicrafts and decorations. More than that, the long leaf segments possessing veins parallel to each other and mid rib are used as palm leaf papers to write scripts, poems, text, literature, calender, records etc using metal stylets for writing that are either straight or semi curved. Dating back to 3000 to 3500 BC, the palm leaves are valued sources for hand carved or written manuscripts and treasured.

Palmyrah palms, male as well as female, passes through long juvenile or pre flowering period, usually 10 to 12 years or more to putforth male spadix or female spadix respectively. The male spadix is 5 to 10 branched. Each spike is cylindrical, 30 to 40 cm long and 2.5 to 4.0 cm. wide. Number of spikelets per spike ranges from 800 to 1000 and each spikelet shows upto 10 sessile small florets. Male flowers are sessile with three sepals, three petals and six connate filamentous anthers. It takes 16 to 25 days for flowers to open after the spathes open. Pollen shedding is simultaneous with flower opening.

Female spadix has 3 to 5 branches and female flowers range from 30 to 75 in number in each spadix. Female flowers are globose, large, showing 6 lobed perianth, that are persistent and fleshy. Flowers show six staminodes. Ovary is globose, 3 to 4 celled, pistils 3 to 4, syncarpous, stigma three, recurved. Ovules are basal and erect. Fruit is a fleshy drupe. Epicarp color varies from dark brown, light brown, greenish yellow to yellowish orange at the early development stage and becomes dark brownish/ black and yellowish pink at maturity. Mesocarp is yellow, fleshy, fibrous and upon fruit ripening it is tasty and pleasantly smelling. The pyrenes 1 to 3 are bilobed. Endocarp medium hard. Testa is brownish and the endosperm is jelly like and gelatinous. The individual cell cavity is filled up in 60 to 75 days time after fertilization. As the fruit matures, the gelatenous endosperm hardens. In Tamil Nadu, flowering is from March to July, with peak period in third week of May. Male flowers open mostly between 8 am. and 11 am. Cross pollination occurs through insects. Time taken in female trees from flowering to fruit maturity is from 120 to 130 days.

Scientists are selecting for dwarf to semi dwarf palms and precocious in bearing habit. Palmyra palms both male and female trees, yield sap upon tapping. The sap or juice is utilized for making sugar, sugar candy and sugar syrup preparations, Neera and fermented sap known as toddy. The tree trunk is used in low cost housing, thatched sheds with palmyrah leaves and split open. Sized palm trunks for shed and other utilities are well known and common.

It is cross pollinated due to dioecious nature. There is no well planned planting geometry. Palmyrah is solely propagated by pyrenes or seeds. That are planted at random and mostly on field bunds or in self sown manner in natural population. In either case, the male and female palms are found growing in equal proportion. Till date, no organized quality seed production or quality seedlings nursery and distribution are in vogue. The breeding improvement is aimed at early maturity and dwarf or semi dwarf palms and stepped up systematic efforts to increase the sap yield per se in male and female trees and in enhanced fruit/jelly yields in female trees Peter (2007) of Quality seedlings need to be produced for distribution. (Ratnam, 2001).

Organized production of nucleus/ breeder seed is planned as a part of the breeding activities. The plan for establishing isolated palmyrah mother palm garden is presented in Fig. 1 and Fig. 2.

Survey and identify highly productive female palms and high number of male spathe bearing male palms in a well established 30 to 40 years age palm garden. Select uniform true to type female palms and nearby male palms and also ascertain the regular and prolific bearing female trees as source trees.

↓

1. Identify 250 to 300 high yielding female palms trees and collect fruit bunches (1 to 3 from each palm) at fruit ripening stage from each female palm. Allow fruit bunches to fully ripe and dry the true to type ripe fruits for 30 days in shade.

↓

2. Collect 3 celled fruits, that are uniform true to type fruits from each palm ; examine, plantwise fruit shape, size, skin color, measophyll texture, mesophyll taste, flavour and the pyrene (matured seed) size and shape.

↓

3. After 30 days, separate the seeds from each 3 celled fruit palm wise. Raise palm wise 50 to 100 plant progenies. The well developed seeds are retained with seed coat fibres still adhering to the seed testa.

↓

4. Plant single seed each in a polybag(12cm. x 8cm.) nursery in, palm wise rows and water daily. Seeds germinate in 22 to 25 days.

↓

5. Plant the seeds in polybag and keep inside a humid chamber (2m. x 1m. x 0.5m. pits) and close with polythene sheet and water daily. Allow for 45 to 50 days within the humid chamber. Advantage: (i) 90 to 98% germination (versus 56 to 65% in conventional) (ii) less extent of mortality and (iii) good field establishment.

↓

6. Roots of seedlings are cut and treated with 2% DAP, 2% MOP and 25 ppm NAA(3hrs.) and planted in polybags. (45cm. x 20cm.)

↓

7. Select fast growing healthy, uniform true to type seedlings alone and transfer to large seedling containers and maintain for 12 months to 24 months.

↓

8. Selected healthy, quality, true to type and vigorous seedlings are distributed for field planting.

↓

9. When the palms reach flowering in 10 to 12 years after planting, the male and female palms will be in equal proportion. Both male and female palms, 10 to 12 years are more or less equally productive in terms of juice or sap; In case of female trees tapping for sap and fruit harvests are obtained. In practice fruiting is preferred in lieu of sap tapping.

↓

10. In a naturally existing populations of palmyrah, it was found that 68.4% population is poor in performance while 26.40% population is moderate and 5.26% population only is high in yield (Kurian and Peter 2007). It is recommended to select the palms for vigour and productivity.

Fig. 1. Proposed protocol for production of superior quality planting material in palmyrah palm.

Tamil Nadu state has established palmyrah development mission, 2021 to establish palmyrah palms by supplying 7.6 million seed nut and 0.1 million quality seedlings. It is planned to supply 100 million seeds and seedling during the mission period.

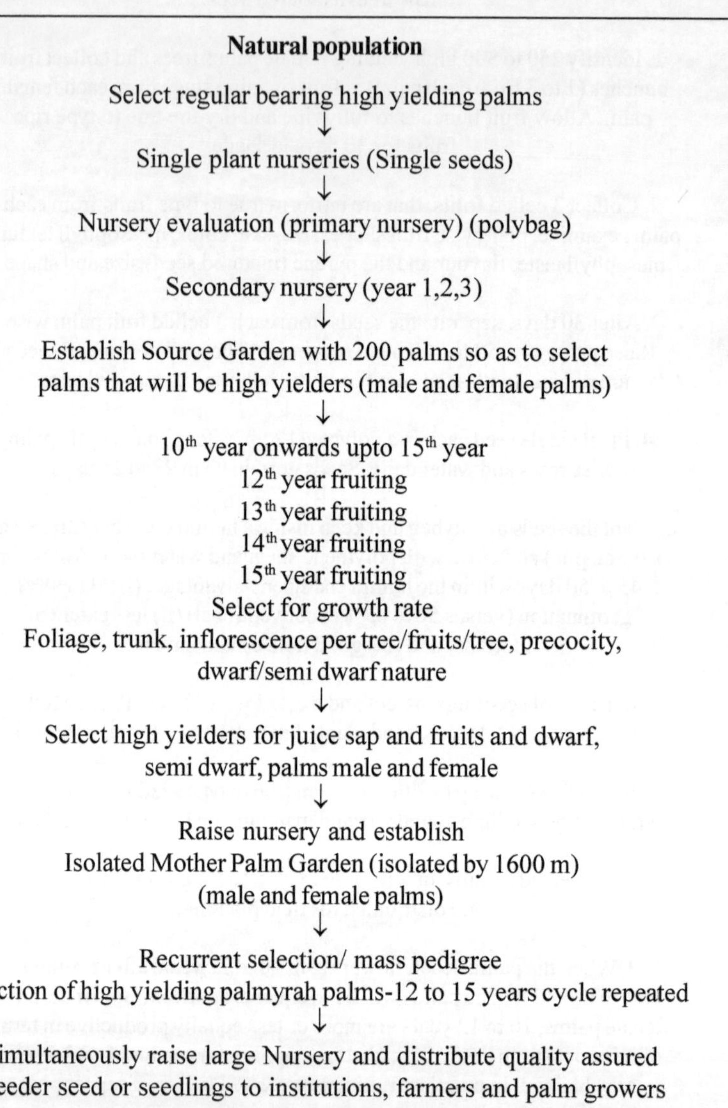

Fig. 2. Establishing ISOLATED Palmyrah Mother Palm Garden

124. Periwinkle or Sadhaphuli

Catharanthus roseus Linn. (2n:34) (syn. *Vinca rosea* L.)

Family : Apocynaceae

It is an evergreen, erect, bushy perennial herb and native of Madagascar (syn. *Vinca rosea* L). It has the unique distinctions of ornamental exterior versus therapeutic interior and the two faces of versatile herb (Nejat *et al.,* 2015). Periwinkle grows upto one meter height with plant canopy spread of 100 cm x 100 cm. Leaf orientation is opposite and leaves are 2.0 to 9.0 cm long x 1.0 to 3.0 cm breadth and glossy green on the upper surface. Inflorescence is a raceme. Flowers are soft pinkish or red tinged and pentamerous, actinomorphic and with 5 corolla. Corolla throat is narrow and cylindrical, 2 to 3 cm long with corolla lobes attached. Stamens are five, inserted below the corolla tube mouth with short whitish filaments attached to the subsessile anthers. Pistil 1.5 to 2.5 cm long. Carpels two, stigma glabrous, ovary accommodates 2 seriate ovules. Each fruit contains 2 free narrow cylindrical follicles, 2 to 4.5 cm long, 3 mm wide enclosing 10 to 20 oblong minute seeds that are 2 to 3 mm long and black colour. Cotyledon is black in colour. Periwinkle varieties are either white or pink rose or white with rose purple spots in the flowers. In India, periwinkle is cultivated in Tamil Nadu, Karnataka and little area in Gujarat and Maharashtra. It is suitable for organized organic cultivation farms and for ornamental plant culture. Periwinkle is almost resistance to pests and diseases and adapted to drought and saline soils. It is very well known as a naturally occurring anticancer drug. The characteristics are given in Table 1.

Table 1. Characteristics of periwinkle (*Catharanthus roseus*) L.G. Don

1.	Plant growth habit	-	Erect, semi erect, spreading
2.	Plant height (cm.)	-	Short (<50), medium (50-80), tall (>80)
3.	Plant width (cm.)	-	Narrow (<40), medium (40-80), broad (>80)
4.	Stem anthocyanin pigmentation	-	Absent, present
5.	Number of primary branches	-	Low (<10), medium (10-20), high (>20)
6.	Number of secondary branches	-	Low (<30), medium (30-40), high (>40)
7.	Internode length (cm.)	-	Short (<1), medium (1-2), long (>2)
8.	Leaf shape	-	Linear, oblong, elliptic
9.	Leaf blade length (cm.)	-	Short (<5), medium (5-6), long (>6)
10.	Leaf blade width (cm.)	-	Narrow (<2), medium (2-3), broad (>3)
11.	Leaf lamina margin	-	Entire, wavy
12.	Diameter of flower (cm.)	-	Small (<3), medium (3-4), large (>4)
13.	Flower petal colour	-	White, pink, purple

14.	Flower colour – eye zone	-	Cream, pink, red
15.	Flower tube length (cm)	-	Short (<2), long (>2)
16.	Siliqua length (cm)	-	Short (<2), long (>2)
17.	Number of seeds per siliqua	-	Low (<2), high (>2)
18.	1000 seeds weight (g)	-	Low (<1), high (>1)
19.	Pod dry weight (g)	-	Low (<5), high (>5)
20.	Main root length (cm)	-	Low (<10), high (>10)
21.	Alkaloid content - total leaf alkaloid content in shade dried leaf (%)	-	Low (<0.8), high (>0.8)
22.	Total root alkaloid content in shade driedroot (%)	-	Low (<1.5), medium (1.5-2.0), high (>2.0)

Note : 1. Characteristics 1 to 11 and 16 to 20 are observed / recorded 120 to 150 days after sowing 2. Characteristics 21 and 22 recorded at 150 DAS 3. Characteristics 12 to 15 recorded at 70 to 80 DAS (*Source:* PPV & FRA, 2007)

Seeds are sown in nursery beds (500 g in 200 sq.m. raised beds). Germination is completed in 10 to 12 days. Sixty days old seedlings are transplanted in 45 cm x 30 cm spacing. Seed rate followed is 2 to 3 kg/ha. Periwinkle is self compatible and intraflower self pollination occurs. Often, it behaves like allogamous cross pollinating species. The pollinators are mostly seasonal moths and butterflies. Since both kinds of pollination, self and cross pollination occur, the progenies show genetic variation. Plant population of upto 75000 plants per ha is maintained to get high leaf yield. Close planting on the rows (5 plants in a meter) is advised. 15 to 20t/ha of FYM and 40 kg N and 20 kg P per ha. are applied to the crop while transplanting. Dry matter yield of root, stem and leaves is 1.8 t, 2.2 t and 3.0 t. respectively per ha. under irrigated conditions. The crop needs to be protected from attack of pests like aphids, mites, thrips, scale and mealy bug and diseases such as *Alternaria, Colletotrichum, Botrytus, Fusarium* and *Rhizoctonia.* Periwinkle is a proven trap crop for root knot nematodes. Periwinkle can also be raised by rooted cuttings: Cattle avoid grazing the periwinkle. Plants can be maintained by pruning. It regrows from brown woody stem. For 2 to 3 years, the crop production can be maintained.

All parts of the plants are of medicinal use and important alkaloids are concentrated in root bark. (Vincristine alkaloid) (40g of vincristine is obtained from one ton of periwinkle leaves) (NCL, Pune).

In periwinkle, leaf harvests are made thrice 120 and 150 days after planting, and 200 days at time of uprooting. Root growth is high between 50 and 200 days when alkaloids in root are at maximum.

Nucleus seed production: Periwinkle crop grown in isolation of 400 meters from other periwinkle crops is selected and at least 1500 healthy disease free, true to type and uniform looking plants are tagged at flowering commencement. The tagged plants are selected from the middle portion of the field showing flowering uniformity and flower colour trueness. The tagged plants are monitored for flower traits, profuse flower production and good pod setting and well developed stem and root system. At maturity, 1000 tagged plants are selected and gently uprooted. In the laboratory, individual plants are studied for the trueness of characteristics of leaf, stem and root biomass. After thorough examination in the laboratory, 300 to 500 plants showing trueness, uniformity, vigour and health are selected, seeds collected and saved plant wise.

In the following favourable season, seeds from the selected 200 to 250 plants are sown in 2 row progeny plots with a spacing of 45 cm x 30 cm in a 500 sq.m. plot with 5 m border rows on all sides. Isolation distance of 400 m is maintained. One row plot of check is raised after every 10th plot for comparison. Only one vigorous plants is allowed per hill. Recommended good agronomic practices are made. Off-types are removed. Progeny rows showing the deviating plants are discarded as a whole progeny. When deviant plants are noticed in progeny rows/plots at the time of flowering or after flowering, the progeny rows concerned and the adjoining rows on either side are removed. Individual plants in the progeny rows are critically verified for trueness and uniformity and for the conformity to the stem, leaf, flower and fruit characteristics. At maturity, plants are uprooted intact and leaves, stem, branches, matured fruits and root portions are separated and dry weight recorded. Progeny rows showing higher performance above mean +SE are identified. Seeds from superior plants in each of the identified progenies are dried and stored plant wise. Alkaloids are estimated in bulked samples of leaf, stem and roots from identified progenies. Part of the seeds saved from the individual plants are bulked and Nucleus seed is constituted.

Breeder seed production: In the next ideal season, under irrigated scheduling, well developed healthy seedlings raised from nucleus seed are transplanted in 60 cm x 30 cm spacing in a well prepared field that is not grown with periwinkle in the previous years. The seed production plot is raised in 2000 sq.m. area provided with 400 m isolation and grown with 5 meter wide border rows on all the four sides. Single healthy plant per hill is maintained and weak plants are uprooted. Periodically critical examination is made to verify for the uniformity, trueness and conformity to the vigour and habit and characteristics of leaf colour, size and attitude, main stem and branches, flower size, petal and throat colour, pod size, seed colour and matured stem base and root size. Off-types are rogued out. Deviant plants occurring, if any, before flowering are uprooted. Deviating plants noticed after flowering are discarded along with surrounding plants to avoid pollen contamination. Detailed verifications for trueness and conformity are made in different stages of crop growth and development. Monitoring Team visit is organized at mid maturity stage. At maturity, individual plants are uprooted with roots intact and examined in the laboratory for plant type, stem and branches, root size, foliar and pod and seed. Alkaloid contents are estimated in leaf, stem and root samples. Seeds from selected plants are separated and stored plantwise. After verification for seed shape and colour, major part of seeds from each plant are bulked together to constitute Breeder seed. Samples of breeder seeds are tested for prescribed seed quality parameters. Quality assured seeds lot is tagged with breeder seed label. Samples drawn from breeder seed lot is tested in Grow Out Test for genetic purity.

125. Rubber

Hevea brasiliensis Muell (2n=36)

Family : Euphorbiaceae

Hevea brasiliensis is monoecious and flowers are arranged in a pyramidal shaped panicles which is many branched, pubescent and bearing flowers of male and female sex. Large size female flowers are borne at the end of central axis of the panicle and the main branches of the panicle. Small and numerous male flowers are borne on the other parts of the panicle. Flowers are short stalked. Calyx five, triangular, and petals are absent. Staminal column consists of 10 stamens that are small and arranged in 2 rings of 5, one above the other. Ovary is 3 celled and conical styles are three in number, short sticky and sessile. Rubber is naturally cross pollinated to the extent of 64%. Inbreeding coefficient is estimated to be 26%. Tapping is commenced at seventh year after planting and peak yield will be at 14th year. High yielding budded plants yield 800 to 1000 kg/ha of rubber. (Priyadharshan *et al.,* 2007 and 2009). The characteristics of this plant are given in Table 1.

Table 1. Characteristics of Rubber

1.	Leaf shape	Acute, obtuse, round, flattened
2.	Leaf central lobe shape compared to lateral	Same or slightly different, intermediate very different
3.	Leaf intensity of green colour on upper side	Light, medium, dark
4.	Leaf glossiness of upper surface	Absent, weak, medium
5.	Leaf upper surface texture	Smooth, slightly rough, medium rough, very rough
6.	Leaf pubescence-vein on lower side	Absent, present
7.	Leaf blade attitude in relation to petiole	Semierect, horizontal, semidrooping
8.	Leaf blade length	Short, medium, long

9. Leaf blade postion of broadest part	Towards base, at middle, toward apex
10. Leaf blade undulation of margin	Absent, weak, medium, strong
11. Leaf blade shape at base	Acute, obtuse, rounded
12. Leaf blade apex shape	Acute, obtuse, rounded
13. Petiole attitude	Semierect, horizontal, semi drooping
14. Trunk curvature of axis	Straight, slightly curved, medium curved, strongly curved
15. Trunk diameter	Small, medium, large
16. Trunk main colour of bark	Reddish brown, brown, grey
17. Trunk texture of bark	Smooth, slightly rough, moderate rough, very rough
18. Tree shape	Triangular, ovate, circular oblate
19. Tree density of foliage	Sparse, medium, dense
20. Coagulum-colour of surface	White, light yellow, medium yellow, dark yellow
21. Tree winter leaf shedding	Absent, partial full
22. Seed length	Short, medium, long
23. Seed width	Narrow, medium, broad
24. Seed thickness	Thin, medium, thick
25. Seed shape in dorsal view	Elliptical, cordate oblong, obovate

(UPOV. TG 254-1/2009)

Rubber seed weighs 3.5 to 6.0 g. Germination is hypogeal. Germination occurs within a week after sowing. Leaves are glabrous with long petioles. Seedlings of 3 months to one year age are used as root stock plants for grafting buds from elite clones. Green and brown budding are in vogue. In brown budding vigorously growing seedlings of one year age are used as root stock plants with brown buds collected from matured shoots of the same age from the elite scion parent. In green budding, 3 to 5 months old green vigorous seedling are used as root stock and tender green colour buds are used as scion. The stock-scion union is completed in 3 weeks time. Budded grafted plants are planted in 3 m × 2m spacing. Pinching of stem tip is done at height of 2 to 2.5m to induce branching and enhancing trunk girthing.

Flowering occurs 4 year after planting. Latex harvest is initiated by 7th year at a height of 150 cm from the bud union when the tree attains a girth of 50 cm. Latex is harvested by tapping by controlled excision of thin bark strip. (MOEF 2016). The juvenile phase in rubber is 5 to 8 years period before start of flowering phase is the limitation to shorten the breeding cycle. Early and vigorous clones show early performance at 4 1/2 years and recording 23 to 40% yield increase over existing clones during first year of tapping (Rubber Board Released clones) (Mydin, 2014).

The limitations are: 1. long juvenile period; 2. season dependent bud dormancy; 3. need for alternating vegetative and reproductive phase; 4. narrow genetic base; 5. heterozygous nature of rubber crop; 6. seasonal and non synchronous flowering; 7. low fruit set, long breeding and selection cycle 8. lack of full and reliable and early selection parameters of latex production. Rubber breeding also aims at evolving and developing rubber into a dual purpose tree for latex and wood. (Rubber Board India). Relationship between ortet and ramet correlation is worked out by Bombanatio and Concalves (2015). It is followed that more productive ortet selected at nursery stage produces more productive ramets. Progeny mean heritability is high and that selection on progeny level is highly realised. Genetic variance for dry mater production, bark thickness, number of lateral vessels rings remained higher when compared with values of residual variance even after selection. The superiority of GCV resulted in high relative CV for all traits studied indicating favourable situation to select in the clonal population (Priyadharshan *et al.,* 2017).

Nucleus and Breeder Stock development: Ten to fifteen years old healthy plantation in active production of latex is selected and the individual plants are examined. The high performing plants/trees of vegetative graft origin with high degree of stock-scion union and compatibility are selected. Based on the performance of latex yield, plant structure and health and vigour, 25 elite superior trees of clonal origin released variety are identified. Twenty five grafts are prepared from each of the selected 25 trees by grafting on to rootstock plant of proven high degree of compatibility in graft union. It is assured that the root stock plant seedlings are raised from seeds of the same mother tree. Variation among bud grafted populations is significant influence of stock-scion interaction and need to be taken note of and budgrafts are selected. Ten selected grafted plants from each selected high yielding and healthy tree are planted. Plants are raised in progeny rows with 3m on the row and 3m. between rows. Plantation is raised with isolation space of 20m and as per recommended production practices and guidelines. Tapping for latex yield commenced in 8th year after planting. Year wise yield of latext is recorded per tree upto 10th to 12th year after planting to arrive at latex yield average of the progeny rows. In the meanwhile, the progeny rows are examined for health, uniformity, free from diseases and pests. The true to type nature is assessed on individual trees in conformity with the DUS characteristics (Table 1). Trees showing deviations or variations in the characteristics are discarded and the rest of the trees in the particular progeny row are also discarded. True to type nature for the tree appearance, trunk characters, leaf and flower characteristics and latex appearence and yield and other prescribed characteristics are again verified for conformity. True to type progeny rows performing above mean +SE values for latex yield are selected and elite superior trees within each selected clonal progeny are now tagged or numbered. The selected trees constitute Nucleus stock.

Systematically prepared grafted plants with scion buds from the selected trees in the nucleus stock are now planted in Breeder Stock field adopting all recommended production practices. All around the plot, 20m space is provided as isolation distance.

Juvenile stage selection, marker assisted selection and other latest breeding approaches are suggested to develop a high yielding clone in 17 years instead of the normal cycle of 35 to 40 years. Selection within plants per genotype and large scale testing with 128 plants per genotype are suggested (Chandrasekar *et al.,* 2007 and Priyadharshan, 2017).

126. Stevia: The Sweet Leaf Herb

(*Stevia rebaudiana* Berton) (2n : 22)

Family : Asteraceae

Stevia is known by many names like Sweet leaf, Honey leaf, Candey leaf etc. Stevia plant produces sweet steviol glycosides discovered in 1931 and are responsible for the sweetness in the plant. Stevia is a perennial, semi shrub, growing upto 30 cm or little more. Stem is weak and pubescent at bottom. Leaves are 3 to 4 cm long, sessile, elongated lanceolate or spatulate shape; lamina tip blunt and margin serrate from middle to tip; upper surface of the leaf is slightly glandular pubescent. Rhizome has slightly branching roots. Flowers composite of involucres of epicalyx. Capitulum is loose, irregular, sympodial cyme. Flowers are pentamerous, light purple. Fruit is an achene and five ridged spindle shaped. Height of plant, number of leaves, growth per day, total leaf biomass and stevioside content are the important contributing factors. The plant density recommended is 83000 to 111000 per ha. High density planting can be attempted. Flowering occurs in 90 to 100 days after planting or still earlier. Long days increase the stevioside content in the leaves (Vasanthi *et al.,* 2011). Harvesting is done prior to flowering when stevioside glycoside is maximum. Stevia yields upto 8 kg seeds per ha. One thousand seed weight is from 0.15 to 0.30 g. Seeds are better usable when stored at 0°C. Stevia seeds are small and quick to loose viability. (Bharathi 2017). An account of cultivation of stevia is given by Ramesh *et al.* (2006) and by Srikant *et al.,* 2017, Vasanthi *et al.* (2011).

Stevia has potential as an alternate to cane sugar sweetening source. It is being used in many parts of the world as non calorie sweetener. Stevia leaf powder is highly safe to use with no processing, calorie free and around 30 times sweeter than cane sugar (Bharathi, 2017). Stevia has antioxidant activity. The taste of stevia is pleasant and inhibits fat absorption and lower the blood pressure (Swathi *et al.,* 2010).

Land is prepared to fine tilth and 15 cm high raised beds are formed and nursery is sown. Seedling are planted in 30 cm x 15 cm spacing. Sprinkling of water daily in summer and once in 3 days is non summer period is advised. Protective measures are needed against root grubs. Pinching of main stem to produce side branches is done. Good agronomic and horticultural practices are followed to sustain the stevia - a perennial plant. After harvest of leaves before flowering, stem may be cut above 15 cm from base. The stevia crop can be maintained for 3 to 4 years period (Bharathi 2017, CIMAP). It is a highly heterozygous plant and seedlings show variation. Constant selection and reselection are needed to maintain mother seed plots.

Nucleus Seed Production by stem cuttings: Vigorous, uniform, healthy and high leaf biomass plants are tagged at preflowering phase from the crop grown in an isolated place (1600m). One thousand such plants are selected. The plants selected should have reduced potency for flowering. Leaves should have high rebaudioside - A and low in stevioside associated with bitter taste. Stem cuttings 10 to 15 cm long with at least 2 to 3 buds arising from leaf axils are prepared from highly leaf biomass producing true to type, healthy and vigorous individual plants. The stem cuttings are treated with IBA 500 ppm. for inducting rooting. Stevia requires partial shade during hot long summer months (24 to 35°C). Soil moisture supply and wetting are to be done. With good moisture, plants with 2 to 3 shoots / branches can be developed. Fertilizer application of 30t/ha organic matter, 70:35:45 kg /ha of N:P:K respectively is done. Leaves are harvested after 4 months (first harvest) and subsequently in 4 months interval. In a year, 3 to 4 harvests are made. Leaves are dried at 50° hot air for 6 hours.

Nucleus seed production by seeds: One thousand true to type plants from the middle of the plot that are already identified at seed maturity are removed intact and examined in the laboratory for stem, foliage, flower, head and seed trait characteristics and out of this 500 true to type plants are identified and seeds are saved plantwise.

In the next favourable season, well prepared raised beds of 90 cm wide and 5 m. long are prepared. Seeds of single plants are sown. Only single plant is maintained per hill. Isolation distance of 1600m is provided. At flowering stage, progenies showing low leaf biomass and high extent and profuse flowering are rejected. Progenies that are vigorous and produce more number of healthy leaves and high leaf biomass and lesser extent of flower head production are selected and from each of the selected progenies, uniform and superior plants are labelled. In each plant 2 to 3 cuttings or harvests of leaf are made. At the end of the first year, selected superior plants are allowed to flower and matured heads are harvested and seeds stored. The leaf stevioside glycoside is estimated in the leaf biomass, preferably from each of the selected progeny. Based on the consideration of conformity of different plant characteristics and alkaloid content, seeds are separated from each of the superior plant or line. Part of the seeds from each plant / line is bulked to constitute nucleus seed. The remainder of the seeds are stored and utilised in future for nucleus seed development.

Breeder seed production: Nucleus seeds are sown in well prepared long and narrow nursery bed of 90 cm x 5 meter long beds. Healthy, vigorous seedlings alone are transplanted. Single healthy vigorous plant is retained in each hill. Border rows of 5 plant rows are planted on all the sides of the breeder seed plot. The breeder seed field is isolated by 1600m from other Stevia fields. Breeder seed plot is critically examined before the Stevia plants reach flowering. Weak plants, off-types and deviant plants are uprooted. At flowering stage, plants showing deviations and non conformity are uprooted. Deviant plants noticed after flowering phase are uprooted along with adjoining ones to prevent pollen contamination.

First harvest is done by removing the top portion of the uniform, true to type plants from 15 cm. above the ground level. Plants showing slow, poor regeneration of branches and leaf biomass are culled. The stevia leaves are estimated for the specific alkaloid content. This procedure is practiced at 2nd and 3rd cutting of harvest and removal of leaf above 15cm from ground level in each plant. Nearly one year after planting, the plants are allowed to flower and set seeds. Monitoring Team visit is arranged. The seeds of individual plants are examined along with leaves for trueness and selected. Leaf alkaloid content is estimated. After final verification, seeds are bulked, processed and examined in the laboratory for breeder seed quality parameters. Quality assured seed is labelled with breeder seed label. Grow Out Test is conducted with seed samples from breeder seed lot.

127. Tamarind

(*Tamarindus indica* L.) (2n:24)

Family : Fabaceae

Tamarind is a multipurpose monotypic condiment tree species. It is highly adapted to long duration dry regions. Tamarind tree is drought resistant and an evergreen tree. Leaves are alternately arranged on branches and have 20 pairs of leaflets in each leaf. Flowers are few to many are borne on lax racemes. Lateral flowers are drooping. Flowers are protogynous, endomophilous, largely cross pollinated. Self pollination also occurs. Flowers are hermaphrodite, irregular and pedicellate. Calyx 5. Petals 5 of which 2 are anterior and three are upper ones. Stamens 3 to 5. Ovary superior with ovules few to upto 18. Nectar is produced at the base of the ovary. Fruit bearing commences 7 years after planting the seeds, whereas vegetatively propagated trees bear in 3 to 4 years. Fruit is an indehiscent pod. Pod length varies from 8 to 20 cm. The fruit pericarp is thin to start with and becomes brittle shell at fruit maturity. As the pods develop and matures, the pods are filled with juicy pulp combined with development of fibrous mesocarp that is acidified and brown to reddish brown in colour. Pod matures in approximately 10 months after flowering. Pods will remain upto next year on the tree unless harvested. The characteristics of this tree are given in Table 1.

Table 1. Characteristics of Tamarind

1.	Growth habit of tree	Spreading, semi spreading, upright
2.	Tree foliage type	Sparse, dense
3.	New flush colour	Reddish green, reddish brown
4.	Inflorescence- paniclelength (cm)	Low (<10), medium (10-20), high (>12)
5.	Matured fruit colour	Brown, reddish brown, grey
6.	Matured pod shape	Curved, straight
7.	Matured pod width (cm)	Low (<1.5), medium (1.5-2.5), high (>2.5)
8.	Matured pod length (cm)	Short (<10), medium (10-15), long (>15)
9.	Matured fruit pulp colour	Brown, reddish brown
10.	Pulp TSS at ripe stage (°B)	Low (<60), medium (60-70), high (> 70)
11.	Titrable acidity (%)	Low (<8), medium (8-10), high (>10)
12.	Seed weight per pod (g)	Low(<5), medium (5-6) high (>6)
13.	Ripening period (days)	Early (250), medium (270), late (280)

(*Source:* PPV & FRA)

Vegetative reproduction is the fast method of multiplication of exceptional plus trees with prolific bearing and excellent fruit qualities. Rooted cuttings, stem cuttings and airlayering are well known methods. Stem cuttings using terminal soft wood cuttings are easy to follow (F.R.S. Maddimadagu, A.P) (Sri Vasuki *et al.,* 1990). This method advocates dipping the cuttings in 1000 ppm IBA and 50% isopropyl alcohol for 10 seconds and placing in mist propagation chambers (70 to 80% RH) for 10 to 15 days. The success in rooting is 90% and above. Fungal infection in terminal cuttings is very low. In soft wood cuttings or semiwood cuttings, 15 to 20 cm long cuttings from one to two years old branches can be rooted after dipping in 1000 ppm IBA and placing in sand bed mist chambers. New buds develop after 20 days and new leaves appear 45 days after planting.

Distinct variations are noticed in tamarind in respect of cluster bearing nature, fruit size and shape, seediness, pulp texture, taste, pulp colour and overall productivity in terms of pod yield and pulp yield. Seeds also show variation in size and glossiness in some tamarind types. The inner covering of the seed is very distinctly white in a few trees. Planting season is from June to December. Seedlings population in tamarind show variation due to the heterozygous nature of the trees and cross pollination. In natural populations variation for almost all traits are documented.

Improved varieties of plus trees have been developed by cloning by Tamilnadu Agricultural university (TNAU). They are Tamarind PKM 1 and Urigam variety. Softwood grafts are usually produced during. March-April and the softwood grafts are the main propagation material. Air layering by treating the shoots with 4000 ppm IBA is also prepared. Varieties: PKM 1, Urigam, Hassanur, Tumkur, DTS 1, Yogeshwari and Pradhistan.

Clone Development: Elite and uniform looking regular bearning highly productive trees known for fruit qualities are identified and selected. Approach grafts are made with seedling root stocks. The seed sources for the seedling root stocks are selected previously from disease free trees. Utmost care to promote superior ideal synergistic root stock-scion compatibility for life long period and harmony and sustaining productivity is bestowed. Usually 2 years old well united uniform size grafts are planted in 8 m x 8 m spacing or in high density planting spacing of 6 m x 6 m. The grafts reach pod bearing stage in 4 to 5 years after planting as against 10 years period for seedling plantings. Five year old graft planted tree yields 50 kg pods per tree and the 10 years old tree yields 250 kg pods (80 kg pulp). Approach grafting results in 95% success with root stock of proper size and development. Approach grafting is presently adapted in the clonal multiplication of improved tamarind variety PKM 1 with very high success. PKM 1 is a cluster bearing clonal selection and has yielded 263 kg of pods per tree which is 59% more than unselected cultivar. Upto 500 kg pods per tree yield potential has been recorded by Jambulingam and Fernandes (1956). Bud sports are also identified in Tamarind. Sweet pulp tamarind trees are the result of budsports identification and clone development.

Clone Selections: Natural population/provenances of tamarind are surveyed and studied for tree stature, flowering, pod set and pod maturity, pods per cluster, matured pod and pulp qualities and yield of pods and pulp per tree. Trees around 6 to 9 years age will form the meaningful agronomic bases for evaluation and selection. Regionwise distribution and scouting for plus individual trees are advocated for selection. Selected elite trees are cloned by approach grafting, terminal shoot tip cuttings and softwood cuttings and are established with space planting of 8 m x 8 m or high density planting in compact area. The elite clonal plantation is very well maintained with high phytosanitary standards and appropriate agronomic schedules, training and pruning practices. Yearswise data on phenology, flowering, pod set, pods per cluster and earliness indices and pulp quality are generated.

Nucleus and Breeder Stock: In the clonal collections, based on yearwise observations upto 6 to 9 years and recordings, trees that are similar, true to type and uniform looking, stable and regular bearing are retained. The base clonal orchard stock is constituted. This will provide clonal stock plants for sourcing hard wood and semiwood cuttings and stem tip cuttings, or for approach grafting. Approach grafts 20 or more made from each of the elite superior trees are planted in the Clonal Stock Orchard. The growth rate and development and earliness to flower and fruit set and fruit qualities are recorded. At the end of 6 to 9 years, the best performing clones are identified and the elite trees selected within such superior clones are identified as superior Nucleus mother trees for propagation by any one of the suitable method of vegetative propagation. Breeder Stock Mother trees are developed from the identified superior nucleus stock trees by adopting any one of the clonal propagation methods. Simultaneously, specific *in vitro* micropropagation protocols for mass propagation is also standardised along with the appropriate rooting, hardening and quality standard evaluation methods. A dedicated *in vitro* and *ex vitro* facility need to be designed and established for maintenance of superior nucleus clonal stock orchard and distribution /supply of quality planting Breeder stocks to the nursery and farmers. Biochemical properties/qualities of tamarind are adjudged.

128. Tea

Camellia sinensis (L) O. Kuntz (2n:2x=30)

Family : Theaceae

Tea flower is borne in solitary or in clusters in axils of scaly leaves. Flowers are fragrant, pedicel is short, calyx persistent, sepals 5 to 7, petals 5 to 7, stamens numerous with 2 celled yellow anthers; ovary superior with 4 to 6 ovules per carpel, style short, stigmatic lobes 3 to 5. Fruit is a capsule, thick walled, 3 lobed, 3 celled, dehiscing by apical splitting. The characteristics of tea are given in Table 1.

Tea is self sterile out breeder and predominantly cross pollinated plant. Genetic upgradation is by hybridization and selection. Clone to seed cycle is being followed as an important method for continued improvement of tea plant. Tea is known for its large plant size, long duration and long generation cycle of 5 to 10 years. The breeding objectives are height, large stem diameter, root-shoot ratio, high field performance, quick rapid regeneration, quality leaf, high yield, highly tolerant to drought, fairly resistant to pests and diseases, besides excellent cup characters of made tea.

Selfing and seed set were estimated by Rajkumar *et al.* (1997). Seed set in tea is three times high on cross pollination as compared to the seed set on self pollination. Selfing results in small sized seeds with reduced or no germination at all. Seedling populations are heterozygous and heterogenous and variation exists. Clones are multiplied by single leaf internode cuttings. Identification and release of tested clones for commercial cultivation needs 7 to 10 years time. Polyclonal seeds are produced by growing selected clones in isolated field. The performance of polyclonal seed is unpredictable.

Biclonal seeds stocks are produced by growing two high yielding and high combining ability clones in isolation of 1600m. Biclonal seeds germinate early in the nursery and establishes well in the field and are suitable for infilling, replanting and extension planting and for combining good yield with drought tolerance. Seed Bari block seed orchard is established and maintained. (Mondal, 2014; Burman, 2011).

Elite Mother Bushes known for desirable attributes and high leaf yield are identified and repeatedly studied and selected for vegetative propagation. Clones are recognized as high yield clone, estate clones, quality clones and moderate yield and drought tolerant clones. Tissue cultured plants are produced and tested for uniformity and performance. DNA marker applications with SCAR and SSR markers are successfully adapted for selection of clones resistance to drought and diseases (Devarumath *et al.,* 2002).

Bari: Seed Bari Blocks for Biclonal seed. Tea is propagated by seed and vegetative cuttings. Seeds are produced in seed Bari (Seed orchard). Isolation distance or pollen dilution zone of 20 m. width is practised. Tea seed is recalcitrant and has low viability. Seedlings of 12 to 18 months age are field planted. Single leaf cuttings show root induction in 45 to 60 days and are planted in polybags and grown for 8 to 12 months before planting. In tea, from pollination to seed ripening takes 12 months time. In each capsule 1 to 3 seeds are set. Pollen dilution is recorded in Tea (Muoki *et al.,* 2007).

Organized seed production in seed Bari was done as early as 1960 while vegetative propagation started in still earlier year 1940. Seed progenies or seed jats are the major sources for planting material. Tea growing areas in NE region is still existing with inferior plant material like old jats. Use of at least 5 improved varieties in tea estates and replanting with suitable clones and biclonal seeds are recommended. No clone should occupy more than 10% of total area of the Tea Estate. Planting ratio of clone to seed population in 1:1 is considered ideal. First Tea clone in Tea in the world was developed in 1949 in TRA, Tocklai. Organized seed Bari and seed Jats have been developed by TRA, Tocklai and later in UPASI, Tata Tea Ltd., and in Tea Board (Refer Upasi.org.tocklai.ent.teaboard.gov.in). Biclonol seed stock development is explained by Singh (1992) and Vinod (2016).

Nucleus and Breeder Seed Production: Mother Bushes are carefully selected for vigour and fast growth of shoots and leaf pubescence which is positively correlated to the tea quality, high leaf yield, tolerant to resistant to pests and diseases, high quality of cup tea and other specific selection criteria. Application of 0.5% $AlSO_4$ and 1% $MgSO_4$ as foliar spray followed after a week time by application of 2% $ZnSO_4$. After another one week, 1% urea as foliar spray is applied. This is followed by tea fertiliser mixture of N and K. Usually cuttings are taken from identified elite mother bushes in April-May and August-September (Crop Production Horticulture crops TNAU-2014). Semi hardwood cuttings are prepared with one full leaf and one internode with a slanting cut at bottom of the cutting. In 10 to 12 weeks, rooting is completed. Hardening of the rooted cutting and planting is done for 4 to 6 months. Single hedge planting is done with a spacing of 1.20 m x 0.75 m (10800 plants per ha). In double hedge planting 13200 plants per ha are planted with spacing of 1.35 m x 0.75 m x 0.75 m. Shade providing plants are planted in the field. The tea bushes are properly trained by centering, tipping and pruning. The different stepwise protocols for Nucleus and Breeder Stock production are presented in Fig. 1.

Micropropagation protocols are developed by DBT and could be adopted.

Table 1. Characteristic of Tea

Camelia sinensis and *C. assamica*

1.	Growth habit	Erect, semi erect, horizontal
2.	Young leaf colour	Yellow green, dark green, purple green
3.	Maturing leaf colour	Dark green, yellow green
4.	Petiole anthocyanic pigmentation	Absent, present
5.	Leaf shape	Lanceolate, elliptic, ovate
6.	Leaf size l/b ratio	Small (<2.75), medium (2.75-3.25) large (>3.25)
7.	Leaf upper surface	Absent, present
8.	Leaf apex	Acute, acuminate
9.	Leaf base	Attenuate, obtuse
10.	Leaf margin	Serrate, biserrate, blint
11.	Leaf blade attitude	Erect, semierect, horizontal
12.	Leaf pubescence density	Sparce, medium, dense
13.	Position of stigma in relation to stamen	Introse, coplanner, extrose

14.	Position of style splitting	Geniculate, ascending united
15.	Flower colour of inner petals	White, pale yellow, light pink
16.	Flower diameter	Small, medium, large
17.	Flushing behavior	Early, medium late, prolong
18.	Leaf blade shape (cross section)	Folded, upward, flat, recurved
19.	Branch zig sagging	Absent, present
20.	Leaf blade, undulation of leaf edges	Absent, weak, medium, strong

(*Source:* PPV & FRA)

1. Caffeine content in tea measured with two leaves and a bud sample harvested in the first flush of the year (method ISO 10727:1995)

Absent or very low	< 0.5%
Low	0.6 to 2.0%
Medium	2.1 to 3.5%
High	3.6 to 5.0%
Very high	> 5.0%

2. Tea polyphenol content estimated by method ISO 14502:2015 (**Ref** : TG/ 238 Tea 2008. UPOV)

Survey elite productive Tea plantation and select elite mother Bushes with repeatable performance for yield. Mother bushes treated with tea mixture AlSo4 (0.5%), Mg So4(2%), Zn So4 (2%) and urea 1%. Prepare semi hard wood cuttings with one full leaf and an internode (cloning) from each selected mother Bush. Six month old cuttings (seedlings) are sorted out for vigour, growth and leaf development. Twenty such healthy cuttings are selected in each bush and progeny row with 20 cuttings in each bush are field planted. All recommended production practices are adopted: Isolation distance of 20m is provided. Raise clonal progeny rows with 20 cuttings planted for rooting and hardening in polybags in each selected bush in isolation block (20m).

↓

Evaluate the clonal progenies for quick field establishment, fast growth, vigour and uniformity. Examine for DUS characteristics, true to type nature, uniformity. Record leaf yield (2 leaves one bud) and quality of cup tea in each individual plants in each clonal progeny . Evaluate for 2 to 3 harvests and study rejunevation after the first harvest of leaves 3 years afer planting.

↓

Select the clonal progenies performing above mean +SE for leaf yield. Prepare semiwood cuttings from the true to type superior plants in each selected progeny row. The cuttings from the selected plants are bulked and constitute nucleus stock.

↓

Raise Breeder Seed/Stock plot by planting the Nucleus Stock cuttings (20m isolation). Evaluated the plants growth for good field establishment vigour, fast growth, true to type and uniformity.

↓

Record leaf yield (2 leaves and bud) in individual plants and assess cup tea qualities. Plants not upto the cup tea quality level and leaf yield and poor regeneration and deviating type of plants are discarded.

↓

Uniformity of the bushes for the stem and quick and fast regeneration after each leaf plucking are observed. At proper age stage of crop growth and bush establishment visit of Monitoring Team is organised. The clones (semi hard wood cuttings) from the selected bushes are prepared, bulked and breeder stock is constituted. Leaf quality parameters are estimated.

Fig. 1. Nucleus and Breeder Seed Production in Tea

129. Turmeric

Curcuma longa (2n=63)

Family : Zingiberaceae

Turmeric crop flowers in 110 to 155 days, usually in July-October. The well developed inflorescence blooms in 8 to 12 days after emergence. Flower opens in 8 to 10 days. No seed set is noticed although pollen germinates and reaches the ovary/ovule. In general fruit and seed set has not been noticed in turmeric. The characteristics of turmeric are given in Table 1.

Genetic Resources: Improved varieties that are available for cultivation are: CO.I, BSR.I, BSR 2, Roma, Swarna Sudharsana, Suguna Sughandham,. Ranga, Rashmi, Rajendra, Sonia, Krishna, Suroma and Allepy Finger Turmeric (AFT). Seed rate : 1500 to 2000 kg fresh rhizomes/ha. Crop duration is 9 months. Yield potential: 25 to 30 t/ha of fresh rhizome and 5 to 6 t/ha of cured rhizome.

Table 1. Characteristics of turmeric

1.	Plant-Pseudostem habit	Compact, open
2.	Plant height (cm)	Short (<85), medium (85-100), tall (>100)
3.	Plant number of shoots	Few (<3), medium (3 to 5), many (>5)
4.	Plant number of leaves on main shoot	Few (<5), intermediate (5 to 10), many (>10)
5.	Plant leaf disposition	Erect, semierect, horizontal
6.	Leaf petiole length (cm)	Short (<15), intermediate (15.1 to 25), long (25)
7.	Leaf lamina length (cm)	Short (<30), medium (30 to 40), long (>40)
8.	Leaf lamina width (cm)	Narrow (<10), medium (10-15), broad (>15)
9.	Leaf colour on dorsal side	Light green, green, dark green
10.	Leaf colour on ventral side	Green, dark green
11.	Leaf venation pattern	Close, distant
12.	Leaf margin	Even, wavy
13.	Pseudostem anthocyanin colouration	Absent, present
14.	Bract colour	White, coloured
15.	Bract tip colour	White, rose, purple, green
16.	Rhizome habit	Compact, intermediate, loose
17.	Rhizome shape	Straight, curved

18.	Rhizome length (primary) (cm)	Short (<5), medium (5 to 10), long (>10)
19.	Rhizome-number of mother rhizomes	1, 2 to 3, >3
20.	Rhizome internode pattern (cm)	Close (<1), distant (>1)
21.	Rhizome-status of tertiary rhizome	Absent, present
22.	Rhizome inner core colour	Orange, lemon yellow, reddish yellow
23.	Duration (number of days)	Shrot (<180), medium (181 to 200), long (>200)
24.	Dry recovery (%)	Low(<15), intermediate (15-20), high (>20)

Source: PPV & FRA (2007)

Propagation is done by mother rhizomes – whole or split. In Andhra Pradesh primary finger is used as seed material. In the mother rhizome planting, 5 to 6 plants from each rhizome will be developing. Mother rhizome exhibits 2 months dormancy. Tissue culture is an alternative propagation method. Quality seeds (rhizomes) contribute to 20% increased yield in turmeric.

Rhizomes of pedigreed varieties or improved authenticated stock seed rhizomes of varieties from the Breeders or Institutions / Organizations are sourced. Source seed rhizomes are examined for the rhizomes characters, uniformity and for diseases and pest free nature. Rhizomes are pretreated with carbandazim (1g/lit) and monochrotophos 36 WSC for controlling rhizome rots and scale insects respectively.

Nucleus Seed: Rhizome bits with active buds are planted in well prepared clean field previously not grown with turmeric. Uniform mother rhizome bits or primary fingers of selected plants are planted in 1000 square meter area in 40 cm × 20 cm spacing with 5m isolation distance. All good agronomic practices and fertilizer and micronutuient application are made in timely manner. The crop is well protected from pest and disease. Periodical observations at bimonthly intervels are made on individual plants on their vigour, fast growth, tillering ability, healthiness and expression of foliage, rhizome development and pest and disease free nature. All deviant type of plants are uprooted along with the rhizomes. At maturity, when plants starts lodging in about 9 months, with leaves yellowing and drying, the rhizomes of individual plants are collected in tact. Individual plants along with rhizomes are examined for leaf size, shape and colour of fingers and 2500 to 3000 disease free single plants are selected that are conforming to the variety or clone descriptor in respect of plant type, foliage and rhizomes (Fig. 1).

In the next season, mother rhizome and or primary finger of 2000 selected single plants are cut into 6 to 7 bits, each bit with a well developed bud, and sown. The bits of rhizome are planted in the well prepared clean field as single plant progenies. Isolation distance of 5m is followed. Critical observations are recorded on single plant progeny basis in a periodical manner. Only such of those plant progenies that are uniform among themselves and conforming to the varietal description are selected and labelled. Progeny rows showing differences among the progenies are rejected. All good agronomical practices and fertilisation, micronutrient application and plant protection management are adapted.

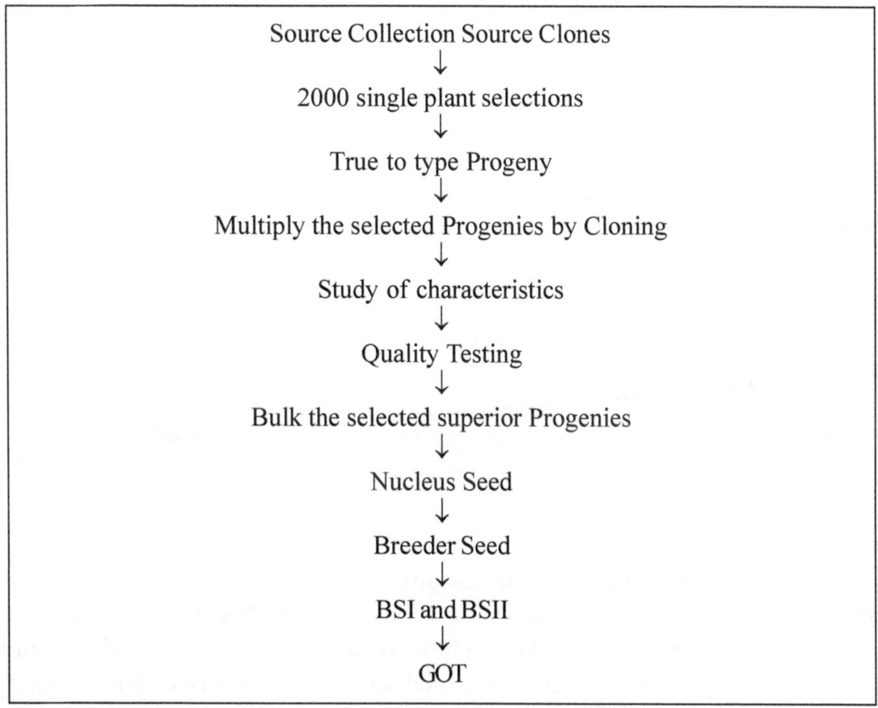

Fig. 1. Nucleus and Breeder Seed Production in Ginger and Turmeric

The mother rhizomes of the single plant progenies that are selected stringently are pooled together as one selection. Like that all the other selections are constituted after discarding off-types and deviants in growth and shoot and rhizome traits. In a well prepared clean field, not grown with turmeric in the previous years, rhizome bits from individual rhizomes in each selection are planted rhizome wise in a continuous manner maintaining 100 cm space vacant between rhizome progenies. All periodical observations on characteristics are recorded on progeny or selectionwise. Selected selections alone are harvested and bulked selectionwise. In each selection, 30 to 35 mother rhizomes will be available. In the next year, selections are raised in 2 replications in comparison with the source seed or the variety or variety bulk. Selections that are uniform, true to type and stable for rhizome traits showing performance above mean +2 SE are finally selected and the rhizomes of the identified selections are bulked to constitute nucleus stock seed.

Breeder Seed Production: Breeder seed stock I and BS II stages are to be followed if the requirement of Breeder seed is high. Nucleus to breeder seed is to be repeated every three years by clean, careful and dedicated growing following the protocols. The scale of BS & FS production is suggested as follows.

	Area (ha.)	Production (q.)
Breeder seed	10	200
Foundation seed I	50	1000
Foundation seed II	250	5000

It is ideal to introduce Monitoring Team visit and GOT to assess the field stand, crop health, quality of rhizomes and crop vigour.

130. Zinger

(*Zingiber officinale* Rosc.) (2n:22)

Family : Zingiberaceae

In zinger, flowering is from October to November. Flower buds 20 to 25 numbers develop from bud initials and flowering is completed in 23 to 28 days in an inflorescence. Anthesis is between 1.30 pm. and 3.30 pm. Flower opening and anther dehiscence is almost simultaneous and stigma is receptive at anthesis. Pollen fertility ranges from 0.9 to 40.0%. Natural seed set in zinger has not been observed so far. It is reported that the spiny stigmatic surface prevents pollen adherence and germination of pollen. The pollen germination is slow and the pollen tube becomes coiled structure in the advance stage. Flowers abscise within 12 hours. Self incompatibility is reported due to pin and drupe type incompatibility. Heterostyly is a contributing factor for sterility. Bud pollination and other pollination manipulations did not succeed (Nybe 2001). No seed is obtained either on selfing or cross pollination. Zinger is universally cultivated by rhizome (Dev Sharma and Vipin, 2014). The characteristics of zinger are given in Table 1.

Table 1. Characteristics of Zinger

1.	Plant growth habit	Erect, semi spreading, spreading
2.	Plant height (cm)	Short (<100), medium (100-200), tall (>120)
3.	Plant – number of shoots	Few (<10), medium (10-15), many (>15)
4.	Plant height of shoot (cm)	Short (<75), medium (75-90), tall (>90)
5.	Shoot diameter (cm)	Narrow (<3), medium (3-5), broad (>5)
6.	Shoot intensity of green colour	Light green, green, dark green

7.	Number of leaves on the stem	Few (<25), medium (25-35), many (>35)
8.	Leaf length (cm)	Short(<25), medium(25-30), long (>30)
9.	Leaf width (cm)	Narrow (<2.5), medium (2.5-3.5), broad (>3.5)
10.	Leaf intensity of green colour	Light green, green, dark green
11.	Leaf petiole length (cm)	Short (<0.5), medium (0.5-0.7), long (>0.7)
12.	Spike length (cm)	Short (<25), medium (25-35), long (>35)
13.	Colour of bract tip	Crimson, yellowish white tip
14.	Rhizome thickness (m)	Thin (<2), medium (2-3), bold (>3)
15.	Rhizome shape	Straight, curved, zig zagged
16.	Crop duration (days)	Shot (<200), medium (200-210), long (>210)
17.	Dry recovery (%)	Low (<16), medium (16-18), high (>18)

The source population that is genetically improved clonal selection and released varieties and also pre release cultures form the basis for developing breeder seed. The stepwise presentation is given below:

1. Source population is genetically improved selections, released varieties and clonal stock seeds.
2. Individual zinger entries are raised in wide space planting in 1000 sq.m. by planting disease free, well developed and physiologically uniform rhizomes.
3. The source crop is established in the clean field which is not grown with ginger in the previous 2 to 3 years. This field is isolated from other zinger growing fields by at least 20 m. Periodical bimonthly observations are made and each time the deviant types are uprooted along with the rhizomes. Uniform and true to type plants alone are retained and from among them, 3000 plants are selected. At maturity, single plants rhizomes intact with aerial parts are harvested and examined in the laboratory for the plant health, leaf characters, rhizome characteristics true to type and uniformity and for rhizome development and colour. Based on the observations, 2000 plants are finally selected and rhizomes are separated and stored in moist sand.

In the following season, in well prepared fields. rhizome bits of single plants, each bit with a well formed bud are dibbled plants wise in continuous rows. Good agronomic practices and application of fertilizers and micronutrients and plant protection methods are followed as prescribed for ginger crop. Periodical observations are made on the plant type and expression, health, vigour, foliage traits and growth rate. Single plant progenies showing off-types and deviant plants are uprooted along with the rhizomes. At maturity, all selected plant progenies are harvested with rhizomes in tact and examined in the laboratory for true to type, uniformity and stability and to reject the deviant types. If variation is noticed among the rhizomes, all plants in such progenies are rejected. Rhizomes of the selected progenies are stored progeny wise. In the next favourable season, selected individual progeny rows are raised comprising all the plants in each progeny. For comparison, rhizomes of the source selction is raised once in 10 progeny rows plot. This is the third year of the progeny growing. The same periodical procedure in field observation is followed till maturity. Progenies showing deviations for plant expression are uprooted. Progenies showing performance for rhizome yield above mean value are selected and the rhizomes are examined progenywise and selected. The rhizome qualities are estimated and rhizomes are selected.

In the third year, rhizomes of the selected progenies are pooled together and raised in isolation of 20m from other ginger crops following the production practices, field sanitation and periodical field examinations and removal of off-types and deviant types, if any. After the maturity, uniform and true to type are harvested and the rhizomes are table examined before pooling to constitute the nucleus seed of the ginger variety.

Breeder Seed Production: Breeder seed production is made in clean fertile once in 3 years. Breeder seed is freshly introduced in the seed chain. BSI and BSII stages are arranged to produce sufficient quantity of quality assured seed material.

Medicinal and Aromatic Plants

131. Aloe Vera

(*A. barbadensis*) Linn Burmf (2n:14)

Family : Asphodelaceae

Aloe is a shallow rooted perennial. Stem is short 30 to 60 cm in height. Roots are in multiples and shallow. Aloe vera does not have a true stem and presents rosette shape bearing densely packed fleshy thick leaves, 60 to 160 cm long, 10 to 14 cm broad and 15 to 20 cm thick. Leaves are highly cuticular with spiny margins. Agave produces showy flowers that are regular and actinomorphic and are borne in axillary spikes. Flowers have 2 whorls of perianth, three perianths in each row. Stamens 6 and present in two whorls. Plants produce only a very few viable seeds. The characteristics of this crop are given in Table 1.

Three types of Aloe are described

1. Small leaf *Aloe vera* (upto 20 cm)
2. Medium leaf *Aloe vera* (21 to 40 cm) and
3. Large leaf *Aloe vera* (above 40 cm)

Aloe vera leaves are more frequently green coloured. The leaf orientation shows variations such as spiral, parallel and nearly parallel. Leaf spine colour is mostly green. Red colour spines are noticed in less than 20% of varieties. Aloe vera leaves contain 99.0 to 99.5% water and in the remaining solid content, 45 different ingredients such as vitamins, minerals, enzymes, amino acids, salicylic acid, sugars, lignins and others are present. Aloe plant juice or gel contains proteins, lipids, carbohydrates and minerals. The gel shows TSS of 0.5 to 0.7%.

Aloe vera is propagated by suckers, root suckers and by rhizome cuttings (TNAU Agriportal Ramsingh, 2011). Rhizomes are cut to 5 to 6 cm. long bits with 2 to 3 nodes and planted. Otherwise the rhizome bits are rooted in sand medium or sand beds and sprouted bits are planted at 45 cm x 40 cm to 60 cm x 30 cm. The field is applied with 20 tons of farm yard manure and 10:50:50 kg/ha respectively of N, P and K. At the time of planting, 36500 rooted plants are maintained per ha and the crop is adequately irrigated. Pests and diseases management are timely fulfilled. Aloe vera crop improvement is reported by Kaur (2011).

Harvest of fleshy well formed leaves is initiated from second year onwards. Three times in a year, the harvest is scheduled. Plants can be maintained upto 5 years for good harvest. In each plant, 3 to 4 leaves are harvested with proper and gentle cut with sharp knives. Side suckers are used as planting material. An estimated yield of 45 to 50 tons/ ha of fresh leaves is realized. Suckers from 50 to 60% plants can be used as planting material. *Aloe vera* crop is maintained weed free by periodical spading or earthing up.

Nucleus Stock Production: Healthy vigorous and true to type uniform plants for leaf and raceme inflorescence and flower characteristics are tagged in three to four years old crop. As many as 800 to 1000 uniform true to type plants are selected. Suckers and side shoots are carefully separated and labelled with serial numbers plant wise. After laboratory examination for different characteristics, uniform suckers from individual plants are planted in progeny rows with 60 cm x 60 cm or 60 cm x 40 cm spacing and well cared for in the field. The field is isolated by 5m. Single plant is maintained per hill. The field is kept pests, disease and weed free. Periodical observations are recorded for true to type nature and uniformity. Progeny rows showing off-types and deviant plants are removed. At the end of first year, plants in each progeny are gently uprooted and suckers are separated and examined in the laboratory. In the next step, progeny rows are planted with single plant derived suckers by adopting 60 cm x 30 cm or 60 cm x 40 cm spacing. Only healthy and uniform suckers from single plants in each progeny are selected and carried forward for progeny rows evaluation with 5m. isolation. As many as 500 progenies are selected and carried forward as nucleus stock.

Individual plants of sucker origin in each progeny row are examined for conformity, trueness and uniformity for maturity and leaf development, leaf characteristics, leaf succulence and also for flowering uniformity and inflorescence and floral traits. Field evaluation at the end of first year is completed and the identified progeny rows that are true to type uniform and similar looking are selected. In each selected progeny rows, superior plants are labelled and uprooted along with suckers. In the laboratory, the suckers are examined. So also the matured leaves in the selected plants are evaluated for biomass, thickness, colour, leaf spots, juiciness and other scorable features. The suckers of the selected superior plants are pooled to constitute the nucleus stock after laboratory examination.

Meanwhile root tubers of individual plants of the previous crop are uprooted in each progeny, examined for true to type, healthy and free from diseases and 15 cm. long cuttings or minisetts are prepared. The minisetts from each progeny are planted in sand bed for rooting. Rooted sprouts or cuttings are planted in a separate plot in progeny rows and observed periodically. Plants from true to type progeny are selected to make nucleus stock.

Breeder Stock Production: In the next favourable season, the uniform and disease free suckers from nucleus stock are planted in 60 cm x 30 cm spacing in a well prepared field to fine tilth not grown with Aloe in the past years and isolated by 5m. from other Aloe vera crop. Good agronomic and horticultural practices are adapted to raise a disease free uniform crop. Early and late flowering plants, poor and weak plants are culled out. At maturity, at the end of first year, leaf harvests are made to adjudge the marketable and technical features. Monitoring Team visit is organized. Shoot suckers are gently removed and rooted and pooled together to constitute Breeder stock. One more Breeder Stock stage is performed to produce adequate Breeder Stock stage II.

Table 1. Characteristics of *Aloe vera* L.

No.	Characteristic	Description
1.	Plant height	- Very short, short, medium, tall, very tall
2.	Plant width	- Very narrow, narrow, medium, broad, very broad
3.	Plant - number of inflorescences	- Very few, few, medium, many
4.	Leaf length	- Very short, medium, long, very long
5.	Leaf width	- Very narrow, narrow, medium, broad, very broad
6.	Leaf breadth	- Low, medium, high
7.	Leaf thickness	- Thin, medium, thick
8.	Leaf curvature	- Strongly incurved, moderately incurved, straight, moderately recurved, strongly recurved
9.	Leaf width at apex	- Narrow, medium, broad
10.	Leaf colour of upper side	- Yellow green, light green, dark green, blue green, blue grey
11.	Leaf spots on upper side	- Present, absent
12.	Leaf stripes on upper side	- Present, absent
13.	Leaf marginal teeth	- Small, medium, large, very large
14.	Inflorescence branch	- Absent, primary, secondary, tertiary
15.	Inflorescence length	- Very short, short, medium, long, very long
16.	Peduncle length	- Very short, short, medium, long, very long
17.	Colour of peduncle	- Green, green reddish, reddish brown
18.	Terminal raceme - length of flowering part	- Very short, short, medium, long, very long
19.	Terminal raceme - density of flowers	- Sparse, medium, dense
20.	Mature bud - colour of pedicel	- Greenish, yellowish, orange, reddish,brownish
21.	Mature bud main colour	- RHS colour chart
22.	Pedicel - main colour	- Greenish, yellowish, orange, reddish,brownish
23.	Flower basal swelling	- Weak, medium, strong
24.	Perianth length	- Very short, short, medium, long, very long
25.	Perianth diameter	- Small, medium, large
26.	Outer perianth segment recurving of apex	- Absent, weak, medium, strong
27.	Inner perianth main colour of apex ofinnerside	- White, green, yellowish red, purple, brown
28.	Stamens protrusion	- Absent, weak, medium, strong
29.	Filament anthocyanin	- Absent, weak, medium, strong
30.	Beginning of flowering	- Early, medium, late
31.	Terminal raceme shape	- Capitate, corymb, capitate to conical,conical, conico cylindrical, cylindrical

(*Source :* UPOV 2015)

132. Avaram

Avaram Senna : *Cassia auriculata.* L. (2n=14)

(*Senna auriculata* (L.) Roxb)

Family : Fabaceae

Shrub with erect solid stem that is much branched; stem is dark brown and smooth. Branches close to the stem: leaves are pale green to green, stipulate, paripinnate, compound usually crowded at the top of plant. Leaf rachis 10 cm long, much pubescent, furrowed leaflets, moderate size, 16 to 24 pairs per leaf, glabrous. Flowers are conspicuous, showy, bulged, borne on racemose inflorescence. Many flowers prominently displayed on the top of the plant. Each raceme consists of 8 to 16 flowers; matured ones are loosely showy with bright yellow petals. Flowers are irregular, bisexual, pedicels glabrous, sepals 5 distinct, imbricate, unequal with two outer ones much larger than inner ones. Petals five, free imbricate, bright yellow with orange coloured veins. Anthers 10, upper three barren. Ovary superior, unicarpellary, axile placentation. Matured fruit pod is flattened, 7 to 10 cm long and 1.5 cm broad. Seeds are 10 to 18 per pod.

Senna is propagated by seeds and by stem cuttings. Seeds are scarified before sowing. Stem cuttings are planted at 30 cm apart on rows. Space of 60 cm is maintained between rows. In case of seedlings, thinning is done 20 days after sowing and single plant is maintained per hill. At the end of 3rd year from sowing or planting, the stems can be harvested and stripped for twig bark and an estimated yield of 1500 kg/ha of bark is obtained. Subsequently, each year, shoots are harvested from coppiced plants. Jose *et al.* (2012) has reported the medicinal value of Avaram.

Nucleus Seed Production: Naturally established self sown origin *Cassia auriculata* (Avaram Senna) populations are identified in four to five nearby field locations. In each location, based on number of field visit examinations, 500 well grown healthy true to type, representative plants of 3 years growth that grow side by side but not in exceeding 60 cm spacing between plants are selected. In all, a total of 2000 matured healthy plants are thus identified and tagged and examined for uniformity, healthiness and trueness for stem, bark texture, leaf attitude and colour, plant canopy shape, size and shape of raceme inflorescence, flower blooming and shape and colour, of young and matured pods and seed shape and colour. From among the tagged plants, after laboratory examination, 250 true to type representative plants are identified and seeds are saved plantwise.

In the next favourable season, scarified seeds of each plant are sown in well prepared field in two rows progeny plots adopting 60 cm x 30 cm spacing in

2 replications. Check seeds are sown in 2 rows plot for comparison once in every tenth progeny plot. Isolation distance of 100 meters is ensured. All good agronomic production practices are followed to raise a healthy and uniform crop. A thorough verification is made at each stage of crop development namely seedling, juvenile, branching, pre flowering, flowering and pod setting stage for different characteristics. Progeny rows showing deviating plants at pre flowering are discarded. Off types are uprooted. Progeny rows are examined for deviant plants at flowering and post flowering and if found are uprooted along with the adjoining progeny rows. Uniform looking, true to type progenies are finally retained. At the end of first and second years, yield of fresh flowers and their dry mass are compiled based on periodical harvest of flower heads. Periodical nipping off stem tips is done to maintain good branching. Stems are harvested from individual plants in each progeny and bark separated and bark weight recorded. Plants are coppiced at the third year end and allowed to regenerate shoots. There afterwards evaluation is made at end of each year for stem and bark and 3 to 4 times for flower mass.

The performance and productivity are compiled for the progenies and better performing true to type progenies are identified with higher performance above mean +SE. Pods and seeds from superior plants in each identified high performing progenies are harvested and seeds separated and examined for trueness, uniformity and freedom from pathogens and insect pests. Major part of the seeds from each superior plant are bulked to constitute nucleus seeds. Medicinal properties are analysed from leaf, flower and bark samples drawn from plants contributing to the nucleus seed. The remainder of seeds from individual plants are saved for developing nucleus seed stock in future.

Breeder Seed Production: Nucleus seeds are scarified and sown in well prepared field not grown to Avaram Senna in the previous years. Isolation distance of 100 meters is maintained, All good agronomic practices are applied. Spacing of 60 cm x 30 cm was followed and single plant per hill is retained. As described in the nucleus seed production, all stepwise protocols on observing characteristics at different growth phases till maturity are carried out. Tipping the plants 90 days after sowing was done. Off-types and deviating plants and plants around deviant plants are uprooted. Yield of fresh flower biomass is recorded in 2 to 3 times and stem and bark yield are assessed at the end of each year. At mid maturity stage, Monitoring Team visit is arranged. At maturity, at the end of first year, matured pods are collected from individual true to type plants and examined in the laboratory. Seeds from selected plants are bulked to constitute breeder seed. Samples of breeder seed are tested in the seed testing laboratory for seed quality parameters and quality assured seeds are tagged with breeder seed label. Grow out test is arranged with sample seeds drawn from labelled breeder seed.

133. Cassia Senna

Cassia angustifolia Vahl

Family : Fabaceae

Annual, herbaceous, stem erect, growing upto 90 cm tall. Pale green, long spreading branches; leaves alternate, leaflets 4 to 5 pairs, obovate, unequally oblique at base and exudes foetid smell when crushed. Flowers yellow in racemes arising axillary or sub terminally. Flowers are in racemes, 14 to 16 flowers per raceme. Flowers are bright yellow coloured, petals loosely held, flowers are bisexual. Sepals 5, petals 5, anthers 10 of which 3 are barren. Ovary superior. Matured pods are flat with 6 to 10 seeds. Plants flower in 60 to 70 days after planting. Pods 5 cm x 2 cm size, maturing in 90 to 95 days after sowing. Seeds 5 to 8 per pod, yellowish colour and flat. Duration of the crop is 110 to 130 days (Prota.org). It is a self pollinated plant but often out crossing could be high to very high (>20%) due to the visit of beetles (Al fredo, 2002) and Balashankar *et al.* (2013).

Senna is non nodulating plant, sown in 45 cm x 30 cm spacing with a seed rate of 20 kg/ha and plants are thinned to single plant per hill. February-March and September-October sowing is done depending on the growing situations. Plant population of 60000 to 70000 per ha is maintained. Fertilizer application of 15 t/ ha of compost/ FYM, and 40 and 50 kg/ha respectively of N and P at sowing time is suitable. Leaves and pods are hand pick harvested at 50, 90 and 130 days after sowing. Before first harvest of leaves, stem is tipped off to induce branching. An estimated yield of 1.2 to 1.5 t/ha of dried leaves and 0.35 to 0.40 t/ha of dried pod are realizable under irrigation. The total sennosides content is 3.5%. Presence of dianthrone glycoside (1.5 to 2.0%) and main sennosides A and B along with minor quantities of sennosides C and D are obtained from dried leaves and pods of senna plants. Senna is affected by root rot, leaf eating caterpillar, aphids and leaf spot (Ambrose *et al.,* 2016).

Nucleus Seed Production: A healthy senna crop is raised in an area of 500 sq.m. by sowing the seeds in 45 cm x 35 cm spacing in the ideal season in a well prepared field to fine tilth. The seeds are sourced from the Institution or breeder concerned. Isolation distance of 100m. is maintained from other senna crops. Only single plant is maintained per hill. Good agronomic production practices are followed. Observations are made on individual plants to judge healthiness, uniformity and true to type nature. Off-types and deviating plants are discarded and uprooted. Plants are examined for the trueness in stem, leaf shape and size, branching at preflowering stage and deviating plants are uprooted. Deviating plants spotted at flowering and post flowering are uprooted along with the surrounding plants to avoid pollen contamination. At maturity, 1000 true to type uniform plants for plant type, branching, leaf, flower and pod characteristics are tagged in the middle of the field. Single panicles are harvested and saved plant wise. The dried pods are examined in the laboratory for trueness and seeds extracted. The seeds are verified for true to type nature and uniformity and seeds of 200 superior individual plants are selected and saved.

In the next ideal season, in a well prepared field isolated by 100m and not grown to senna in the previous years, 2 rows progeny plots in 2 replications are sown with the seeds from 200 selected superior plants. The seeds are raised in progeny plots in four blocks with 50 progenies in each block. Once in 10 progeny plots, one 2 row plot of check seed is raised for comparison. Single plant per hill is retained. Off-types and deviating plants occurring at pre flowering are uprooted. Early and late flowering progeny rows are discarded to maintain flowering uniformity. Individual plants in each progeny plots are examined and verified for stem, branches, leaf, inflorescence, flowers and pod characteristics. Progeny rows showing deviating plants noticed at or after flowering are discarded along with the adjoining progeny rows. Based on the leaf productivity at 50th, 80th and 130th day and the pod yield, superior progenies performing above mean +2SE are identified. In each of the identified progeny rows, seeds from 5 to 10 superior plus plants are selected. The pods and seeds are verified in the laboratory and bulked to constitute nucleus seed. The seeds from selected remaining single plants are saved to develop nucleus seed in future.

Breeder Seed Production: In a well prepared field plot of 800 sq.m. isolated by 100m distance from other senna crops, seeds of nucleus seed lot are sown in 60 cm wide rows and 40 cm on the rows. All recommended good agronomic practices to raise a healthy crop are adapted. Five meter wide border rows are raised with the same source seeds. Only a single healthy plant is retained per hill. A thorough examination is made to verify single plants from germination, juvenile, adult, flowering and pod maturity stages. Off-types and deviating types at preflowering are removed and deviating plants noticed at flowering and pods flowering stages are uprooted along with the adjoining and surrounding plants to prevent pollen contamination. Visit of Monitoring Team at midmaturity stage is organized to assess the breeder seed production plot. At maturity stage, healthy and vigorous plants with high leaf biomass and pod setting are tagged and individually harvested. In the laboratory single plants are verified for trueness for pods and seeds selected. Seeds from selected plants are bulked to constitute breeder seed. Sample of breeder seed is tested in the seed testing laboratory for seed quality parameters. Quality assured seed is labelled with breeder seed label. Grow Out Test is conducted with sample seeds from breeder seed packs.

134. Glory Lily

(*Gloriosa superba* Linn.) (2n:14)

Family : Colchicaceae

Gloriosa superba is accorded the status of the State Flower of Tamil Nadu, India

Much efforts have not been focussed in selection and improvement of Gloriosa lily containing high percentage of colchicine in the tubers. Released variety is not available. Plant selection that is being cultivated possesses around 0.39% colchicine content in tubers. Colchicine content in seed, stem and leaf is respectively 0.69%, 0.05% and 0.16% (Seema *et al.,* 2017). Planting material required is 1000 kg of tubers per ha. Tubers are treated with 0.1% carbendazim for 30 minutes and then planted in a spacing of 60 cm. between rows and 30 cm on the rows. Tubers have dormancy and care should be taken to plant the tubers after the dormancy period is over. Seed sown plants take 3 1/2 to 4 years to bloom whereas tuber raised plants put forth 3 to 4 stems and flower after 6 to 8 weeks after planting and then continue to flower. (le Roux and Robert, 1994).

Bloom interval is two weeks. Anthesis occurs one day later than stigma emergence. The stigma remains receptive for 4 days time. Anthers dehisce a day after anthesis. In Gloriosa, anther dehiscence is from 8.30 to 10.00 am. Stigma becomes receptive next day after anthesis and remain receptive for 30 to 32 hours and viability is 14 hours after dehiscence. Gloriosa is both self compatible and self incompatible.

Tendrils are formed at the tip of leaves. Flowers are very showy. Petals are reflexed prominently, petal colour ranges from greenish yellow to greenish red and rarely deep purplish red. Stamens are six long and persistent, hypogynous extrose anthers. Ovary superior 3 celled, ovules numerous, vary from 25 to 30 in each cell; style is deflected at base and projecting less horizontally. Fruit is a loculicidal capsule. *Gloriosa superba* is a scardent plant. Perianth segments are accrescent, during anthesis and become reflexed and attractive coloured. Flower initiation takes place after 30 to 40 leaves formation depending upon the light and temperature and other environment conditions (Rajamani 2017, Chithra and Rajamani 2009). Gloriosa sets almost 2 times seed yield (680 kg/ha) on self or aided pollination as compared to 319/kg ha seed yield in natural pollinated condition. Flowers mature into capsules in 60 to 65 days after pollination. Mostly the number of seeds per pod is from 60 to 65. Spraying of GA3 at 50ppm recorded high seed yield performance. Gloriosa crop can be maintained for 3 to 5 years to complete economic life. (Rajamani, 2017). Detail of cultivation of gloriosa (Saravanan and Bhubaneswar, 2003) and quantification of colchicine is different parts of gloriosa (Seema *et al.,* 2017) are available.

Artificial or aided pollination between 8 and 11 am results in almost 90% and above increased seed set and higher seed yield. Branching is correlated with tuber weight. Pods are harvested 160 to 180 days after tuber sprouting. Tuber yield is 300 to 350 kg/ha.

Although there are several axillary buds in the tubers, only two tubers are produced per plant. Bulblets are obtained by dividing the L or V shaped matured tubers in every third year. Bud sprouting in the tuber is irregular and in 30 days time around 60% of buds sprout. Small tubers of 50 to 60 g. size show high success in crop establishment. Each tuber can be divided into 2 to 3 pieces each with one bud. Technology utilizing single budded tubers is not being attempted for raising commercial crop of Glory Lily.

The best alternative for planting material is micropropagation. Tubers are used as explants. Auxin enriched medium produces higher number of plantlets. A growth medium containing thiamine hydrochloride yield large number of mature microtubers in 3 to 4 months. Periodic dark treatment accelerates the microtuber formation.

Nucleus Seed Production: Tuber to Tuber method

1. Select 1000 superior glory lily plants that are regular flowering and high fruit and seed set from the clonal population. The criteria for selection are healthy, disease free, true to type vigorous growing plants with strong stem, profuseness in flowering and good pod and seed set. The tubers of selected plants are usually L or V shaped and each rhizome branch is cut into 3 to 4 pieces , each piece with atleast one well matured bud. It is possible by this approach to have 3 to 4 tuber buds growing into plants in each selected clonal plant progeny for field evaluation. Polybag nurseries can be also introduced for single bud planting.
2. Clonal progenies are raised in fine tilth field in rows by planting single budded tubers with demarcation between clonal progenies. Support system is provided. Continued critical observations are made for field establishment, fast growth, vigour, days to flowering, fruiting, seed set and seed yield. Matured seeds from each true to type clonal progeny rows are collected and examined in the laboratory for uniformity and development. Isolation distance of 5m is provided. Tubers in each selected progenies are planted in the next clonal progeny generation. The performance of clonal progenies are again assessed. Vigorous highly productive clonal progenies that are true to type nature are selected. The tubers separated from true to type plants are bulked and carried forward to constitute nucleus seed. Colchicine content is estimated in clonal progenies.

Breeder Seed Production: Tubers from nucleus stock bulbs collected from true to type, very healthy, fast growing and productive plants are washed well, cut into required number of clonal tubers each with one well developed bud. The planting material is treated with fungicide and planted in rows in well prepared field not grown with gloriosa lily in the past 3 to 4 years. Support system is provided.

Healthy and true to type glory lily plants are examined and at maturity true to type matured plants are selected. Pods and rhizomes are collected plant wise. Seed are separated and saved on plant basis. Based on the seed yield performance, 200 high performing plants are identified for the next cycle and rhizomes of high performing 200 plants are then closely examined for true to type and uniformity. The L and Y shaped rhizomes of the individual plants are cut into 3 to 4 pieces, each piece with a well developed bud and clonal progeny of single plant is established in a well prepared plot which is isolated by 5 meters from other fields grown with glory lily.

The spacing adopted is 60 cm x 40 cm. Tubers are planted after the dormancy is over. Isolation distance of 5m. is provided to avoid mixing of tubers from nearby field of glory lily. All suitable crop production protocols are followed. The trailing plants are provided with stakes length wise connected by criss cross tying of sticks to enable the plants to spread and blossom. Continuous monitoring /evaluation is made for growth ,vigour, trueness, stability and uniformity for days to flowering and fruiting. Aided pollination is made to improve the seed set. Visit of Monitoring Team is arranged. Matured pods are harvested and examined in the laboratory for conformity of the characteristics of the pod and seeds. The tubers of the selected plants for vegetative, floral, pod and seed characteristics and high seed yield are uprooted, gently washed and stored for raising another cycle of breeder seed of tubers. Colchicine content in seeds and tubers is estimated.

135. Mentha Mint

(*Mentha arvensis* L.) (2n=72)

Family : Labiatae

Mentha is an aromatic herb growing up to 60 to 100 cm. It is a hardy perennial, branching freely and vigorously. Stem is quadrangular and leaves are broad or long narrow. Inflorescence is a spike arising from leaf axil. Spikes bear lilac flowers. Leaves produce high menthol oil content (75 to 80%).

The plant produce suckers and stolons facilitating clonal propagation. Flowers are borne in spikes that arise from axils of leaves. Flower color is lilac. Calyx are 4 to 5, campanulate, 5 veined. Corolla 4 to 5, tubular, purple coloured; corolla are 2 lipped with four unequal lobes, upper lobe being larger. Stamens are 4; fruits are lacking or if present, with four nutlets. The plants show shallow root system. Clonal breeding in mint is explained by Fejer *et al.* (2014).

The propagation is by planting rooted stem cuttings, shoot tips or planting the tubers. Spacing adopted is 40 cm x 20 cm. Stems develop rapidly and five months after planting, stem cuttings can be taken for field planting or for nursery raising and thereafter cuttings can be removed at 3 months interval. The crop production can be continued upto 4 years with good agronomic and plant production packages/management.

Herbage yield: 15 to 20 t/ha/year

Mint oil yield: 50 to 100 kg/ha/year

Different Mint types are cultivated. They are;

1. Mint - *Mentha labiatae* L.
2. Japanese Menthol – *Mentha arvensis* L.
3. Pepper Mint – *Mentha piperita* L.
4. Spear Mint – *Mentha spicata* L.
5. Bengamot Mint – *Mentha citrate* L.

   (National Horticultural Board)

The characteristics of the Mentha - Mint are presented in Table 1.

Nucleus & Breeder Seed Production stepwise flow chart is presented in Fig 1.

Table 1. Characteristics of Mentha Mint

S.No.	Characters	State
1.	Plant growth habit	Erect, semi erect, spreading
2.	Plant height (cm)	Short (50), medium (51-70), tall (›70)
3.	Plant: no of branches on main shoot	Low (15), medium (16-20), high (›20)
4.	Stem: anthocyanin pigmentation	Light, medium, dark
5.	Leaf: Stem fresh weight ratio	Narrow (‹1.0), medium (1.1-1.5), high (›1.5)
6.	Leaf blade length (cm)	Short (‹4.0), long (›4.0)
7.	Leaf blade width (cm)	Narrow (‹2.5), broad (›2.5)
8.	Leaf area (sq.cm)	Low (‹10), medium (10-12), high(›12)
9.	Leaf hairiness (upper side)	Absent, present
10.	Leaf: intensity of hairiness (lower side)	Weak, strong
11.	Leaf: intensity of green colour	Light, medium, dark
12.	Leaf: anthocyanin colouration of vein lower side	Weak, medium, strong

13.	Leaf: anthocyanin colouration of margins	Absent, present
14.	Leaf lamina margins	Dentate, serrate, sinuate
15.	Leaf: shape of apex	Acute, obtuse, rounded
16.	Crop duration: days to maturity	Short (‹110 days), medium (120 days), long (›120 days)
17.	Essential oil content (%)	Low (‹0.5), medium (0.5-0.8), high (›0.8)
18.	Essential oil (menthol content in essential oil %)	Low (‹70), medium (70-75), high (›75)

Source: PPV & FRA, 2011

Source clonal population nursery raised in clean field planted with disease free clean 1000 to 1500 SPS that are pretreated with fungicides

↓

Select 500 disease free healthy vigorous true to type single plants that are true to type and uniform in flowering and aroma.
Table and laboratory Examination. Select 100 SPS

↓

Prepare clones shoot tips/root suckers. Plant clonal progenies with check entry once in 10 rows. Evaluate clonal progenies of shoot origin and root sucker origin separately for health vigorous trueness/uniformity/stability, free from disease.

↓

Identify clonal progeny rows performing above mean.
Select superior plants and constitute Nucleus seed.

↓

Breeder Seed

↓

GOT

Fig. 1. Nucleus and Breeder Seed Production in Mentha Mint

Note: 1. Isolation distance 5m; 2. Clean field growing; 3. Essential oil content, menthol content and aroma of leaves and other quality traits are assessed in Nucleus and Breeder Seed stage.

136. Noni

Morinda citrifolia L. (2n: 22,44)

Family : Rubiaceae

Noni juice is well known for restorative properties. Noni is an evergreen tree or shrub, growing to 3 to 6 meters height. Plant forms are variable in size, leaf and fruit characteristics and palatability. Noni flowers throughout year. Leaves are opposite in arrangement, glossy, elliptic to ovate shape, somewhat large size 20 to 45 cm long, and 10 to 20 cm width. Stipules are connate, flowers are perfect, hermaphrodite, found in clusters of ovoid global heads. Calyx truncated rim, corolla whorl tubular, 4 to 6 lobed, stamens 5, style

12 to 15 mm. long exserted above stamens, ovary two celled, flowers are adapted for self or cross pollination. Fruit is a fleshy syncarp, smelly, fetid, medium to large size. The seeds have distinct air chamber. A productive crop yields 80t/ha. Flowering commences third year after planting the stem cuttings or graft and from 5th year onward regular extent of fruiting can be expected. Root system is deep, well developed and spreading. Wood of noni is yellow. Generally noni adapts to 20°C to 35°C (Rajamani, 2017; Nelson, 2003).

Noni is propagated by seed and stem cuttings and also by root cuttings and by planting leaf petiole. Commonly vertically planted stem cuttings 20 to 40 cm long is being followed. Plants established in 6 to 9 weeks. Plants flowers in first year but the regular bearing of fruits is from the third year onwards.

Noni exhibits high degree of variation in foliage and fruit characteristics. The seeds are unique in possessing air cavities inside the seeds. The different characteristics of noni are given in Table.

Table 1. Characteristics of Noni (*Morinda citrifolia*)

1.	Tree canopy – cylindrical, conical, elliptical, spherical, bushy
2.	Tree height (cm) – small, medium, tall (> 400 cm)
3.	Tree diameter (cm) - < 20, 20 – 30, > 30.
4.	Tree bark surface – smooth, rough
5.	Internode length (cm) <8, 8 – 10, > 10
6.	Leaf shape – elliptic, lanceolate, ovate
7.	Leaf apex - acute, accuminate, obtuse, cuspidate
8.	Leaf lamina colour – light green, green, dark green
9.	Leaf length (cm): very short (15-20), short (20-25), medium (25-30), long (30-35)
10.	Leaf width – narrow, medium, broad
11.	Leaf petiole colour – light green, green, whitish green, reddish green
12.	Flower colour – white, yellowish white, yellow
13.	Position of stigma in relation to anther – above anther, at the same level, position below
14.	Presence of pistillate flower – absent, present
15.	Flower length (cm): <1, 1-2, >2
16.	Fruit colour – whitish green, yellowish green, green
17.	Fruit surface texture – smooth, rough
18.	Fruit shape – obovate elongate, obovate wide, round, irregular
19.	Fruit length (cm) –very short (<2), short (2-4), short to long (4-6) long (6-8)
20.	Fruit width – narrow, medium, broad, very broad (>8cm.)
21.	Fruit weight – very light, light, medium, heavy
22.	Fruit bunching – absent, present
23.	Fruit shape at base: cordate, tapering, depressed.
24.	Colour of floral eye ring – green, brown
25.	Fruit Length – sessile, pedicellate

26. Days to harvest from flowering to fruit ripening (days) – early (100), medium (100-120), late (120-130)
27. Fruit flesh colour – white, cream, pink
28. Average number of seeds per fruit – less (<50), medium (50-100), high (100-150)
29. Seed length (cm) – short (<0.5), medium (0.5-0.8), long (0.8)
30. Seed width (cm) – Narrow (<0.5), medium (0.5-0.8), broad (> 0.7)
31. 100 seed weight (g) – light (<2), medium (5), heavy (>7)
32. Seed colour – silver brown, reddish brown, dark brown, black
33. Seed wing – absent, present

Source: PPV & FRA

Nucleus and Breeder Seed Production: Noni plantation gardens of the same variety is critically surveyed and studied for the regular bearing throughout the year. The fruit quality is also monitored. Based on the survey as many as 150 to 200 true to type plants regular in fruit bearing are tagged and from each plant at least 20 to 25 stem cuttings each with 10 to 12 nodes are prepared to have atleast 1 to 2 cuttings from each fruit bearing stems. The cuttings are planted plant wise/progeny wise in separate rows and maintained in the polybags adapting good agronomic cultural practices for nursery. The established rooted cuttings are then planted progeny wise rows with the recommended production practices and with 5m. isolation. Continuous monitoring is made for good rooting ability, quick field establishment, shoot / branch formation and for early flowering in the first year of planting. In each clonal progeny row, plants showing deviations, poor health and disease and pest infections are uprooted. Continuous observation is made in the clonal progeny rows in first, second and third year of the crop and continued upto 6th year.

True to type nature, uniformity and stability of the crop are regularly observed and deviating plants are culled out. Stability of the flowering and fruiting throughout the year and the fruit number and fruit yield are recorded plant wise from 3rd to 6th year to assess the regularity in bearing and fruit yield potential. The data are compiled and the best performing progeny row above mean + sd are identified and superior plants in each identified progeny are tagged. Mature stems are selected and stem cuttings of the same size are prepared from each such selected plants. The stem cuttings are pooled together to constitute the nucleus seed.

In the well prepared disease free soil, previously not grown to noni, breeder seed crop is raised with the well developed stem cuttings and with 5m. isolation. All recommended agronomic and horticultural production practices are adapted. Plants with slow and poor growth, poor rooting and late flowering and showing deviations in leaf, flowers and fruit size and shape are discarded in the first year of crop. Like wise critical monitoring is bestowed during second, third and fourth year of the crop. By the time true to type nature, conformity of the crop to the varietal characteristics, fruit yield and stability are verified. Monitoring Team visit is organized to adjudge field purity of the clonal breeder seed crop. Fruit and fruit juice qualities are estimated.

Stem cuttings, are supplied to the breeder / foundation crop growers. Based on the demand breeder seed stage II can be also grown.

14

Long-term Conservation of Crop Varieties

Conservation of released varieties requires a sound scientific and technical basis. There are two main strategies for conserving agricultural biodiversity, namely *ex-situ* and *in-situ* conservation, both of which are equally important. ***In-situ*** conservation refers to the maintenance of a species in its natural habitat. ***Ex-situ*** conservation is the conservation of components of biodiversity outside their natural habitats. It is conserved in the form of seeds (*seed gene bank*), live plants (*field gene bank*), tissues (*tissue gene bank*), pollens (*pollen gene bank*) and DNA materials (*DNA bank*). However, the most effective *ex-situ* conservation measures are the conservation of living plants in botanical gardens, field germplasm repositories and seed gene banks. Recently, *cryo gene bank* is gaining importance as germplasm (seed/embryo/pollen or any explant) can be preserved at ultra low temperature of -196°C in liquid nitrogen for indefinite period. Thus, these conservation measures help to protect plant species facing extinction.

Ramiah Gene Bank at TNAU

Realizing the need for agro biodiversity conservation, TNAU in the year 2010 established a seed gene bank facility at Department of Plant Genetic Resources, Centre for Plant Breeding and Genetics, Coimbatore. It is named after the legendary rice breeder Dr. K. Ramiah, with the capacity to conserve more than 1,00,000 germplasm accessions both under medium (> 20 years) and long term (> 100 years) cold storage units. This seed bank is the first of its kind among State Agricultural Universities in the country.

Facilities at Ramiah Gene Bank

The facilities available at Ramiah Gene bank include a medium-term cold storage unit capable of storing seeds of 50,000 germplasm accessions at 5°C and a long-term cold storage unit for conservation of 50,000 entries at -20°C. The cold storage units are equipped with stainless steel cabinets to store the seed packets. The cold storage units are supported by a 200 KVA power generator for uninterrupted power supply. In addition, the gene bank has a walk- in precision-controlled, programmed, de-humidifier unit for drying seeds; a documentation system for effective storage and retrieval of passport and characterization data; a vacuum packing and sealing unit and an electronic seed moisture meter. The gene bank is attached with a molecular laboratory with facilities for molecular and biochemical characterization of germplasm entries.

Seeds are collected either through exploration or received from plant breeders/ scientists of TNAU / farmers. Once the seeds are received, they are transferred to temporary storage unit where the temperature is maintained at 15°C and relative humidity at 45%. The mandatory requirements for submission of germplasm seed samples in RGB (Ramiah Gene Bank) are:

a) Seeds should not be treated with any chemicals like fungicides and pesticides as they may act as mutagens during long term storage.
b) A minimum of 500 seeds in self-fertilized crops and 1000 seeds in cross-fertilized species will be used for storage. Lesser number of seeds are also accepted for conservation in case of rare and precious germplasm where seeds are normally meager.
c) Good quality seeds preferably freshly harvested, with a minimum viability of 80% or above will be accepted.
d) Seeds should be submitted along with passport information.

Seed processing procedures at Ramiah Gene Bank

Once the seeds submitted for deposition satisfy the above mentioned criteria, seeds are processed as detailed below:

1. Initial seed viability and physical purity testing

The seeds of germplasm accession intended for storage are checked for its physical purity and viability. Physical purity of the germplasm accessions will be done by using physical purity work board. Other crop seeds, weed seeds, immature seeds, insect damaged and diseased seeds and inert matter will be removed. Germination test is the standard method used to assess viability of *ex- situ* conserved seeds· From each accession, a random sample of 10% of total seeds received will be subjected to germination test 11 in replications (50 seeds of eight replications or 100 seeds of four replications) at viability monitoring unit. For small, medium and big size seeds, germination test will be conducted by using petriplate, roll towel and sand medium method, respectively (Fig. 1). For dormant seeds, tetrazolium test will be conducted to check the viability percentage. Seed samples that possess a minimum viability of 80% alone will be subjected to the registration process in the documentation unit.

2. Registration and documentation of released varieties

The registration is done through an in house database TNGRID (Tamil Nadu Genetic Resources Integrated Database) using a high end computer server provided with bar-coding facilities. There are three steps in the registration process *viz*., assigning a genebank ID, passport data entry and assigning barcodes.

Step 1: Assigning a genebank ID

The genebank ID is assigned to each accession in the form of a thirteen character alpha-numeric string. The first four characters are alphabets 'TNAU' which stands for Institute name, Tamil Nadu Agricultural University. Fifth character is an alphabet to indicate the crop code, which may be either 'F' for a Field crop or 'H' for a Horticulture crop, or 'T' for a tree crop and 'M' for miscellaneous category. The next three characters will be numerals to accommodate '999' crop species under each crop code. The remaining five characters will also be numerals to indicate the accession ID of each entry. The entire process of assigning gene bank ID will be guided by a menu driven procedure of TNGRID.

Step 2: Passport data entry

The second step in registration process is to feed the passport data of the accessions received. This is also accomplished by a user friendly menu driven data entry form in the front end of TNGRID.

Step 3: Assigning barcodes

The third step of registration process involves the assignment of a unique barcode for each accession through a software 'barcode tender'. The documentation system is integrated with a barcode scanner and printer to facilitate this process. The barcode labels pertaining to each accession are printed out in duplicates from which one will be pasted on the aluminum pouch assigned for storage and the second label will be pasted on a printed out document that holds the information on the corresponding accession ID and the vernacular ID of an accession . This bar-coding and documentation procedure will help quick retrieval of the seed packs physically from the gene bank as well as retrieval of data records from the database.

3. Storage temperature and seed moisture content control

The maintenance of seed viability over long periods of time in seed genebanks is a key element in conservation of plant genetic resources. Standards aimed at maintaining seed viability have been developed and applied in genebank management procedures, including drying and storage under low seed moisture content and temperature. The storage temperature control is easily accomplished through the fully automated refrigeration systems of the gene bank. Seed moisture content will be assessed directly by using hot air oven and indirectly using electronic seed moisture meter. However, in RGB, moisture meter is used as it is easy, rapid, requires few seeds and the error will be to the tune of ±0.1% only when compared to hot air oven method of moisture estimation.

For seed drying to the desired moisture content, a simple and cost effective silica gel based rapid desiccation system is followed. The germplasm seeds with high moisture content are then transferred along with labels into cloth bags, which are then piled up in a plastic box containing a layer of approximately one inch thick desiccated silica gel. The plastic box is then closed with an air tight lid and the seeds are allowed to dry for specified period that depends on the drying properties of the seeds. Seed moisture content will be estimated periodically till it reaches the desired level of 4 to 6% moisture content for long term cold storage at -20°C and at 6 to 8% for medium term cold storage at 5°C. The greatest advantage of the silica gel desiccation system is that silica is nontoxic and hydrated silica gel can be recycled indefinitely through microwave heating.

4. Vacuum packing and sealing of released varieties

After drying to the required level of moisture content, seeds are transferred from the desiccation unit to the vacuum packing unit. In this unit, the seeds from the cloth bags are transferred to tri-laminated aluminium pouches. The aluminium pouches containing seeds are then subjected to vacuum sealing and packing process. This process is accomplished through an 'automated vacuum sealing and packing machine (Audion Vac™). It involves two steps and in the first step, the aluminium pouches are filled with a specified quantity of soft filtered air to maintain a uniform pressure throughout the pouches. In the second step, the pouches are vaccuumized to Zero Bar pressure and simultaneously subjected to a bi-layered sealing. The entire process of vacuumization and bi-layered sealing is guided by preset programmes which vary with seed properties like size and density.

5. Deposition of released varieties in gene bank

The vacuum packed and sealed seeds in the aluminium pouches will be stacked as per genebank inside a stainless steel basket. The stainless steel baskets with the stacked pouches will be placed in a specified stainless steel cabinet drawer placed inside the cold storage facility to aid the process of easy location of the stored pouches. We assign a physical position number which is alpha-numeric string of six digits. For example, the physical position number C1-D1-B1, indicates that a particular pouch is placed in cabinet 1, drawer 1 and basket 1 in the genebank. This physical position number will be fed in the TNGRID as well. This help in easy location and speedy retrieval of the germplasm seed materials.

6. Monitoring of seed viability

Seeds in long-term storage will eventually lose their viability; therefore assessment of seed deterioration over time is required. At Ramiah Gene Bank, subsequent monitoring of viability will be done once in every five and ten years in seeds stored at medium and long term cold storage units, respectively. When the percentage of germination goes below 60%, the seed packs will be sent back to the depositor for rejuvenation and re-deposition at RGB for long term conservation.

7. Exchange and utilization

Plant breeders, researchers and students are the major users of varieties conserved in genebanks. Germplasm accessions/varieties will be supplied to the researchers of TNAU on request with the approval of Institutional Germplasm Identification and Exchange Committee (IGIEC) in Form III after getting the consent from the original depositor. Germplasm exchange to other institutions for research will be done after getting approval from both depositor and receiver in the Standard Material Transfer Agreement Form (SMTA) Form IV. Exchange is facilitated with the condition that the same quantity of genetically pure multiplied seeds should be submitted to the gene bank for further maintenance.

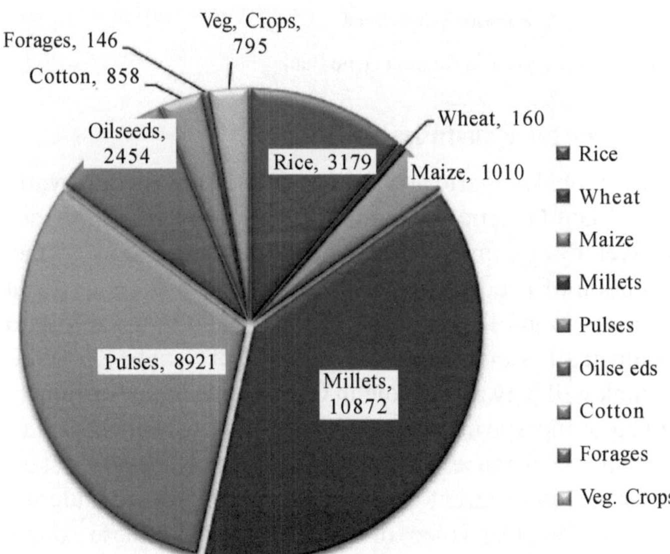

Fig. 1: Current status of germplasms conserved in various crops at RGB

Germplasm collection at Ramiah Gene Bank includes modern and obsolete varieties, land races, mutants, breeding lines and wild species of various crops. Presently a total of 28,646 accessions (Fig. 1) of more than 51 species are conserved in the gene bank. Most of the germplasm so far maintained at field gene banks of TNAU colleges, research stations and Krishi Vigyan Kendras are now conserved in the form of seeds at RGB and further depositions are being made through regular explorations. These resources will form an important reserve for sustainable economic and social development in future. Status of released varieties conserved under this gene bank is given in Fig. 2.

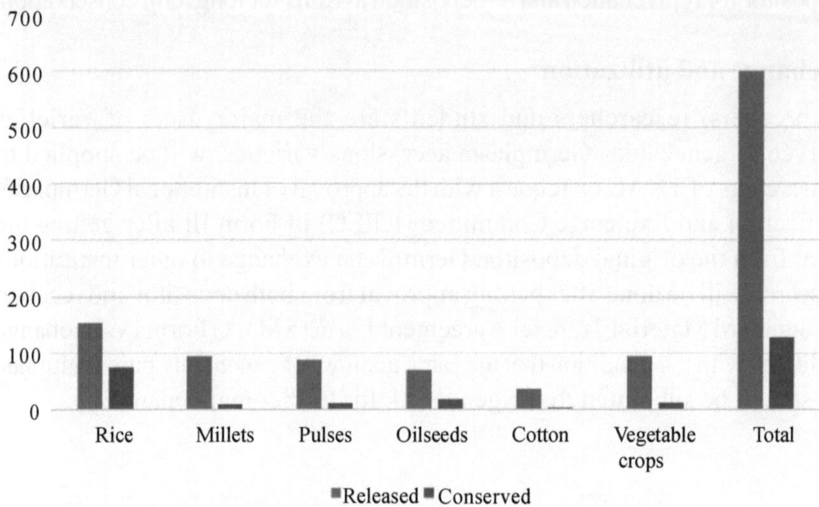

Fig. 2: Conservation of Released varieties in Ramiah Gene Bank

Plant Genetic Resources for a sustainable future

Considering the current global scenario and policies related to the conservation and sustainable use of Plant Genetic Resources, the initiative taken by Tamil Nadu Agricultural University by commissioning the Ramiah Gene Bank is timely in the lines of Convention of Biodiversity. This gene bank will act as a state repository for the conservation of germplasm that will help in the efforts on characterization, documentation and sustainable use of Plant Genetic Resources. Thus, Ramiah gene bank will play a vital role in Climate Resilient Sustainable Agriculture by providing germplasm materials to develop climate resilient varieties for drought tolerance, salinity tolerance, early maturity, pest and disease resistance and submergence tolerance in different crops. Intensive efforts are underway for collection and conservation of land races of native crops, so that these valuable genetic resources will be available for the future generations in climate change scenarios.

15

Protection of Released Varieties

In order to provide for the establishment of an effective system for protection of plant varieties, the rights of farmers and plant breeders and to encourage the development of new varieties of plants it has been considered necessary to recognize and protect the rights of the farmers in respect of their contribution made at any time in conserving, improving and making available plant genetic resources for the development of the new plant varieties. Moreover to accelerate agricultural development, it is necessary to protect plants breeders' rights to stimulate investment for research and development for the development of new plant varieties.

Such protection is likely to facilitate the growth of the seed industry which will ensure the availability of high quality seeds and planting material to the farmers. India having ratified the Agreement on Trade Related Aspects of the Intellectual Property Rights (IPR) has to make provision for giving effect to Agreement. To give effect to the aforesaid objectives the **Protection of Plant Varieties and Farmers' Rights Act, 2001** has been enacted in India.

For the purposes of this Act, Protection of Plant Varieties and Farmers' Rights Authority has been established and is located at: NASC Complex, DPS Marg, Opp-Todapur, New Delhi-110 012.

In order to develop a suitable IPR system for protection of plant varieties, Government of India developed a special act for protection of plant varieties while balancing right of farmers and breeders. This Act provides a way to establish proprietary claim on the crop genetic resources of the country. In essence, it provides a plant breeder right over the variety he has developed by giving exclusive right over the material for a fixed period of time. On the other hand, it allows farmers to save and resow seeds of a particular variety, unless, of course the farmer exploits the material on a commercial mode. Even then, the farmer has the scope to defend himself under certain situations.

The broad objectives of the Act are as follows

i) To provide for the establishment of an effective system for protection of plant varieties.

ii) To provide for the rights of farmers and plant breeders.

iii) To stimulate investment for research and development and to facilitate growth of the seed industry.

iv) To ensure availability of high quality seeds and planting materials of improved varieties to farmers.

The process of development an IP system for plant variety protection was initiated in India in late 1980s and the first draft bill was prepared in 1993. The basic structure of the Act is based on UPOV plant protection systems. However, it differs from the UPOV 1991 guideline in some major aspects, of which farmers' right is a major issue. It also differs from the UPOV framework by extending protection over extant varieties and farmers's varieties, which are not protected in the UPOV system. This makes the Act an effective plant genetic resources protection system, instead of serving as only an IPR system for breeders.

The essential features of the Act

	Salient features of the PPV&FR Act, 2001
Authority	Protection of Plant Varieties and Farmers' Rights Authority. The authority maintains a National Register of Plant Varieties and is the supreme body ensuring the Act.
Range of protection	Extant variety, Farmer's variety, New variety, Essentially Derived variety of any plant species
Eligibility of criteria for protection	1. Novelty: The material under registration should be new and should not disposed or sold in public domain before one year of registration. 2. Distinctness: Clearly distinguishable from other varieties by at least one essential character. 3. Uniformity: The expression of the essential character should be uniform in the population. 4. Stability: The progeny of the variety should retain the essential characteristics.
Exclusion from protection	1. Natural biodiversity cannot be protected. However, any traditionally cultivated plant variety which has undergone the process of domestication/improvement through human interventions can be registered and protected. 2. A variety using genetic use restriction technology (GURT, commonly known as Terminator technology) cannot be protected. 3. Any variety having technology injurious to human, animal or plant health or in any way affecting public morality cannot be protected.

Who can apply for protection	1. A breeder of the variety 2. Successor of the breeder 3. Any person assigned by breeder 4. A farmer or farmers group or community who have bred a variety 5. Authorized persons on behalf of farmers or farming community 6. University or Public Institution breeding the variety

In series of definitions the Act clarifies the different forms of plant varieties that come under the protection. These are:

a) *New variety*: a variety that fulfills the criteria of novelty, distinctness, uniformity and stability is eligible for protection. The protection period is initially nine years and renewable up to eighteen years in case of tree and vines. For other plants protection period is initially six years, and renewable up to fifteen years.

b) *Extant variety* : a variety that has been notified by the Seeds Act, 1966, or a farmers' variety, a variety of which there is common knowledge or any other variety in the public domain. The protection period is 15 years from the date of notification.

c) *Farmers' variety* : a variety that has been traditionally cultivated and evolved by the farmers, or any landrace or wild relative of a variety of which farmers have common knowledge. The range of protection is not well specified in the Act.

d) *Essentially derived variety*: a variety that has been developed from an initial variety by retaining the essential characteristics of the initial variety, but is clearly distinguishable from the initial variety by some characteristics. This includes the genetically modified plant varieties that are developed by inserting a foreign gene. Protection period is same as new variety. However, the authorization of breeder of initial variety is required for protection.

Registration Fees

Type of variety	Fees for Registration
Essentially Derived Varieties	Individual Rs. 7000/- Educational Rs. 10000/- Commercial Rs. 50000/-
Extant variety notified under section 5 of Seeds Act, 1966	Rs. 2000/-
New Variety	Individual Rs. 7000/-Educational Rs. 10000/- Commercial Rs. 50000/
Extant variety about which there is common knowledge	Individual Rs. 7000/- Educational Rs. 10000/- Commercial Rs. 50000/-
Farmers' Variety	No fee

Source: https://plantauthority.gov.in/fees-details (2022)

The certificate of registration issued will be valid for nine years in case of trees and vines and six years in case of other crops. It may be reviewed and renewed for the remaining period on payment of renewal fees subject to the condition that total period of validity shall not exceed eighteen years in case of trees and vines from the date registration of the variety, fifteen years from the date of notification of variety under Seeds Act, 1966 and in other cases fifteen years from the date registration of the variety.

References

Adhikari, S., Bandoupadyay, T., Ghosh P. (2014). Assessment of genetic stability of *Cucumis sativus* regenerated from encapsulated shoot tip. Sci. Horti. 170: 115-122.

Archana C.P., Deepu, V. Geetha S.P. and Indira B. (2013). RAPD assessment for identification of clonal fidelity of micro rhizome induced plants in Turmeric (*Curcuma longa*) CVs. Internat. Food Res J: 20: 3325-3328.

Ahmad Bavappa, K.V. and Nair, M.K. (1982). Arecanut. Publ. Coconut Board.

Ahrat, D.K. and Cariness C.E. (1994). Natural cross pollination in Soybean cultivars. Crop Sci. 34: 76-378.

Alfredo, A.C.A. (2002). Cassava Botany and Physiology in Cassava production and utilisation. ed. R.J. Hillocles and A.C. Belloti. CAB. International.

Alizadeh, M., Krishna, H., Eftekhar, M. and Modareskia, M. (2015). Assessment of clonal fidelity in micropropagated horticultural plants. Journal of Chemical and Pharmaceutical Research, 7: 977-990.

Al Snafi, A.E. (2015). The chemical constituents and pharamacological effects of *Chenapodium album*. L An overview. International J. Pharma Screening methods. 5: 10-17.

Alice Kurien and K.V. Peter (2007). Commercial Crop Technologies. NIPA New Delhi.

Allard. R.W., Jain S.K. and Workman, P.L. (1968). The genetics of inbreeding populations. Advances Genetics. 14: 55-13.

Ambrose D.C.P. Manickavasagam and Ravindra Naik (2016). Leafy medicinal herbs (botany, chemistry, post harvest Tech and uses) CABI. UK.

Anonymous (2010). Protection of Plant Variety and Farmers Right Authority (PPV & FRA) www. Plant Authority.gov.in crop New Delhi.

Arasakesary, S.J., S. Manonmani, R. Pushpam and S. Robin (2015). New Temperature Sensitive Genic Male sterile lines with better out crossing for production of Two line Hybrid Rice. Rice Science 22: 45-52.

Arpita P., Lakshmana Reddy, D.C., Ramesh, S. and Chenna Reddy A. (2018). Comparision of traditional Grow out test and DNA based PCR assay to estimate F1 hybrid purity in cauliflower. Currrent Science 115: 2095.

Arunachalam, V. (2012). Genomics of cultivated Palms. Elsevier Publishers.

AVRDC (2006). Amaranthus. Fact sheets: Saving seeds of amaranth (1000 m. isolation)

Ahmed, W. (2013). Somatic embryogenesis in coffee (*Coffea arabica*) hybrid by using leaf explants. J. Hort. Sci. and Biotech 88: 469-475.

Ahlawatt, I.P.S. (2012). Lentil – Rabi crop IARI 2012.

Anon, Y., Czosneck, H., S. Gojit and Degani, C. (1998). Polyembryony in mango is controlled by single dominant gene. Hort. Sci. 1241-1242.

Bakang K., Kgasudi, G. Kranthi Rekha, K. Uma Jyothi and K. Sasikala (2020). Performance of cluster bean [*Cyamopsis tetragonoloba* (L.) Taub.] genotypes for growth, yield and quality characters. Electronic Journal of Plant Breeding. 11: 373-381.

Balaguru P., V. A. Sathiyamurthy, G.J. Janavi and S. Santha (2020). Variability in perennial moringa (*Moringa oleifera* Lam.) genotypes for quantitative and qualitative traits. Electronic Journal of Plant Breeding, 11: 515-520.

Balashankar D., Vennilarasu, K., Selva Preetha, P. Rajeswari, S., Umadevi M. Dhebjit, B. (2013). Senna. A medical miracle plant. J. Med. Plants Studies 1: 41-47.

Balinder Singh, Jaspreet Singh, Amritpal Kaur (2014). Agro production, processing and utilization of *Stevia rebaudiana* as Natural Sweetener J. Agri. Engineering and Production Tech. 11: 28-31.

Bandupriya *et al.*. (2019). Genetic fidelity testing using SSR marker array confirm trueness to type of micropropagated coconut (*Cocos nucifera* L.) Plantlets derived from unfertilised ovaries. The Open Plant Science J. Vol. 11. 2019 ISSN 1874/2947.

Bateman A.J. (1947a). Contamination in seed crops. I. Insect Pollination. J. Genetics 48: 257-275.

Bateman. A.J. (1947b). Contamination in seed crops. II wind pollination. Heredity 1: 225-241.

Bateman A.J. (1947c). Contamination in seed crops. III Relation with isolation distance. Heredity 1: 303-336.

Bavappa, K.V.A. and P.R. Ramachander (1967). Improvement in *Areca catechu*. Indian J. Gene., 27: 93-100.

Bharathi, N. (2017). Stevia. The calorie free natural sweetener. Growmore Biotech Hosur.

Bhat K.M., F.A. Bawday, M.A. Mirza. Rather and G. Hussain (2013). *In vitro* grafting in Apple, Karnataka J. Agric. Sci. 26: 399-402.

Besirbi G. *et al.* (2017). Organic seed production of lettuce in Turkey. ISHS.Act.Hort. 1164: 47-51.

Bhatnagar, S.K., I.S. Khairwal and S. Pareak (2003). Pearlmillet in Nuclues and Breeder seed production Manual (Ed) R.K. Chowdhury and S.K. Lal 2003. NSP. IARI. New Delhi, p. 36-43.

Barman T.S. (2011). Tea cultivation for N.E. India. Science & Culture 77(9-10): 391-395.

Beevi, S.M. , M.S. Sahoo and H.L. Sharma (1984). Estimation of natural cross pollinations in Egyptian Clover. Euphytica: 34: 147-15.

Bombanatio, A.L. and P.S. Goncalves (2015). Rubber tree orpet-ramut genetic correlation and early selection efficiency to reduce rubber tree breeding cycle. Industrial crops and products. 77: 855-860.

Biswas M.K., Dutt, M., Roy, U.K. Islam (2009). Development and evaluation of *in vitro* somaclonal variation in strawberry. Sci. Hort. 122: 977-990.

Bisognin D.A. (2011). Breeding vegetatively propagated horticultural crops. Crop Breeding and Applied Biotechnology. 11: 35-43.

Chadha K.L. (2006). Report of the Working Group on Horticulture, Plantation Crops, and Organic Farming. XI Five Year Plan (2007-2012). Planning Commission. New Delhi.

Chadha. K.L. (2001). Hand Book of Horticulture. ICAR, New Delhi.

Chakrabarthi. S.K. (2018). Training on seed potato production ICAR – CPRI Shimla pp. 46.

Chandra, K. Panday, A., Kumar P. (2018). Synthetic seeds- a future prospects in crop improvement. Int.J.Agric innovation and Res. 6: 120-125.

Chandrasekhar T.R. *et al.* (2007). Age of yield stabilisation and its implications for optimising selection and shortening breeding cycle in rubber. Euphytica. 156: 67-75

Chaudhary, R., Kohli, P. Srimas K., Mohapatra, T. and R.P. Sharma (2001). Identification and characterization of aromatic rice based on DNA fingerprinting. Euphytica. 118: 243-251.

Chowdhury R.K. and S.K. Lal (2003). General guidelines. Chapter in Nucleus and Breeder Seed Manual. NSP IARI, New Delhi.

Chinnappareddy, L.R.D., Khandagale, K., Srinivas Reddy, S.H., Kanupriya, C., Chennareddy, A., and Singh, T.H. (2012). SSR-based DNA barcodes as a tool for identification of eggplant genotypes. International Journal of Vegetable Science, 18(3): 260-271.

Chithra R. and Rajamani K. (2009). Genetic variability in Gloriosa lily – Trop. Agric. 2009. A7: 77-79. and Plant Sci. 32: 112-115.

Choubey R.N. and A.K. Roy 2003 Oats P. 176 – 180 in Chowdhury R.K. and S. K. Lal (Ed) 2003. Nucleus and Breeder Seed Production Manual. National Seed Project IARI, New Delhi.

Chumi Lal, P. Manivel, A.L. Rathnakumar and Ranvir Singh (2003). Groundnut P104 – 110 in Chowdhury R.K. and S.K. Lal (Ed) Nucleus and Breeder Seed Production Manual. National Seed Project. IARI, New Delhi.

CIMAP – Madhu of Stevia (2011). Variety Release (CIMAP, Lucknow).

Currence. T.M. and J.M. Jenkins (1942). Natural cross pollination in tomato as related to distance. Proc. Amer. Soc. Hort. Sci. 41: 272-276.

Cohen H.H. *et al.* (2012). Marker development for onion genetic purity using SSR markers Korea. Br. Science. 44: 421-432.

Cowan, R.S., Chase, M.W., Kress, W.J., and Savolainen, V. (2006). 3,00,000 species to identify: problems, progress, and prospects in DNA barcoding of land plants. Taxon, 55(3): 611-616.

Crane M.B. and K. Mather (1943). Natural cross pollination of Crop plants with particular reference to Radish. Ann. Appl. Biology 30: 301-308.

Chand, S. Chandra, K., Indu. K. Champa Lal (2020). Variety Release, Notification and denotification system in India in Plant Breeding current and Future views (Ed). Ibrokim. INTECH OPEN/ DOI. 10.5772/94212/Oct 2020.

Choudhury, B. and A. Choumsai (1970). Indian J. Agric. Sci. 40: 805-812.

Debnath S.C. (2012). Morphological and molecular analysis in micropropagated berry plants acclimatized *ex vitro*. Canadian J. Plant. Sci 2012: 92: 1065-1073.

Debnath, S.C. (2017). Molecular markers and clonal fidelity and antioxidant properties of micropropagated horticultural crops. In VI International Symposium on Production and Establishment of Micropropagated Plants 1155: 261-268.

De-Lucas, A.I., Santana, J.C., Recio, P., and Hidalgo, E. (2008). SSR-based tool for identification and certification of commercial Populus clones in Spain. Annals of Forest Science, 65(1): 1-7.

De Hanns H. (1953). Maintaining varities of self pollinated crops. Euphytica 2: 37-45.

Deosingh K., T.G.K. Moorthy, K. Sakala A.I. Narayanan and J.A.V. Prasad Rao (2003). Tobaco in R. K. Chowdhury and S. K. Lal (ed) 2003. Nucleus and Breeder Seed Production manual NSP, IARI, New Delhi.

De Silva E.F., da Silva. L.M. and Ricardo, M. (2005). Crossing rate and distance in upland rice. Bragantia 64(2): ISSN 1678-4499.

Devarumath R.M., Nandy S. Rani.V. Marimuthu S., Muraleedharan. N and Raina S.N. (2002). RAPD, SSR and AFLP fingerprinting as useful markers to evaluate the genetic integrity of micro propagated plants of diploid and triploid elite tea clones of *Camellia sinensis* (china type) and *C. assamica* (Assam- India). Plant cell Reports 21: 166-173.

Dev Sharma, H. and Vipin Sharma (2014). Recent approaches for breeding and production of Ginger –Capacity Building Program. ICAR. Iasri. Res. in. e. book.

DBT (2006). National Certification system for Tissue culture raised plants NCS – TCP, Department of Biotechnology. GOI, New Delhi Oct 2006.

Diaz C., Hossain, M. Merca S. and T. Mew (1998). Seed quality and effect on rice yield: findings from farmer participatory experiments in Central Luzon, Philippines J. Crop Sci. 23: 111-119.

Dixit P.G. and D.P. Tripathi (2003). Mung, urd, Lentil, Lathyrus, Rajmash and Pea (MULLARP) P 70 to 79 in Choudhury. R. K. and S.K. Lal. Nucleus and Breeder Seed Production Manual. NSP IARI, New Delhi.

Dongre, A.B., Raut, M.P., Bhandarkar, M.R., and Meshram, K.J. (2011). Identification and genetic purity testing of cotton F 1 hybrid using molecular markers. Indian J. Biotech 10: 301-306.

Doyle Mackay *et al.* (2010). Evolutionary ecology of clonaly propagated plants, New Phytologist 186: 318-332.

Dua R.P. and Shiv Sewak (2003). Chickpea P. 58-63 in R.K. Chowdhury and S.K. Lal (Ed) (2003). Nucleus and Breeder Seed Production Manual. NSP, IARI, New Delhi.

Duhoon S.S. (2003). Niger p. 129-130 in R. K. Chowdhury and S.K. Lal (Ed) 2003. Nucleus and Breeder Seed Production Manual NSP-IARI, New Delhi.

Duhoon S.S. (2003). Sesame P. 136-144. in R.K. Chowdhury and S.K. Lal (Ed) 2003. Nucleus and Breeder Seed Production Manual NSP, IARI, New Delhi.

Dutt, M., Roy, U.K., Islam, R., and Hossain, M. (2009). Development and evaluation of *in vitro* somaclonal variation in strawberry for improved horticultural traits. Scientia horticulturae, 122(3): 409-416.

Dorrenbos, J. (1953). Rejuvenation of *Hedera helix* in graft combination Preb. 115. The Netherlands. 98-102.

El-Awady, M.A., El-Tarras, A.A., and Hassan, M.M. (2012). Genetic diversity and DNA fingerprint study in tomato (*Solanum lycopersicum* L) cultivars grown in Egypt using simple sequence repeats (SSR) markers. African Journal of Biotechnology, 11(96): 16233-16240.

EL Kassabya *et al.* (2007). Breeding without breeding. PLOS one 6:e257.

El siddique *et al.* (2006). Tamarind Southampton center of under utilised crops. Sussex England.

Errea, P., Garay, L. and Marin J.A. (2001). Early detection of graft incompatability in apricot using *in vitro* techniques. Physiol. Plant. 112: 135-141.

Etienne H. *et al.* (2018). Somatic embryogenesis in coffee Frontier in Plant sciences 9: 1-21. (doi.10.3389/01630)

Faisal, M. and Alatar, A. (2019). (eds.) Synthetic Seeds. Springer. Cham.

Fejer, J.D. Grulova, I. Salmon (2014). Clonal breeding of Peppermint with high content of menthol ISHS Acta Horticulture 1023-25. Internat Symp. on MAP.

Feistritzer, W.P. (1975). Role of seed technology for agricultural development. Seed Sci and Technology 3: 415-429.

Forneck A. (2005). Plant Breeding: Clonality - A concept for stability and variability during vegetative propagation. Progress in Botany 66:164-183.

FAO, Statistic (2006).

Ferreira, M.A. *et al.* (2002). Sexual expression and mating system in watermelon Crop Breeding and Appl. Biotech 2: 39-48.

Galbács, Z.S., Molnar, S., Halász, G., Kozma, P., Hoffmann, S., Kovacs, L., and Kiss, E. (2015). Identification of grapevine cultivars using microsatellite-based DNA barcodes. VITIS-Journal of Grapevine Research, 48(1): 17.

George R.A.T. (1990). Vegetable production Guide. CABI Publ. pp. 328.

George, J. and A. Karun (2011). Marker Assisted detection of Seed sex ratio in palmyrah palm. Curr. Sci 100 (6): 922-925.

Gibbs, T. (1984). Insect cross pollination in Okra. Seed Saver Exchange. 28-32a.

Giriraj, K. and A. Viswanathan (2003). Sunflower. P. 86 – In: R.K. Chowdhury and S.K. Lal (Ed) Nucleus and Breeder Seed Production Manual. NSP IARI, New Delhi.

Ghosh, S.P., Ramanujam T., Jos. J.S. Moorthy, S. and R.G. Nair (1988). Tuber crops. Oxford 1BH Publ. Co. New Delhi.

Ghosh, B., S.M. Hacque (2019). Synthetic seeds- an alternate approach for clonal propagation. In Faisal. M and Alatar. A (Eds). Synthetic Seeds. Springer – p. 77-112.

Godschichmid (2014). Plant grafting new mechanisms. Evolutionary implications. Front Plant Sci. 5: 727.

Gopalakrishnan, T.R. (2007). Vegetable Crops. New India Publishing Agency, Pitampura, New Delhi.

Gorak Singh (2012). Protocols and standard for vegetative propagation of fruit crops. ICAR, New Delhi, pp. 64.

GOI, MOA (2013). Indian Minimum Seed Certification Standards, 2013. Central Seed Certification Board. MOA. GOI. New Delhi.

Govinden-Soulange, J., Somanah, D., Ranghoo-Sanmukhiya, M., Boodia, N. and Rajkomar, B. (2010). Detection of somaclonal variation in micropropagated *Hibiscus sabdariffa* L. using RAPD markers. University of Mauritius Research Journal, 16(1): 435-447.

Goyal, A.K., Pradhan, S., Basistha, B.C., and Sen, A. (2015). Micropropagation and assessment of genetic fidelity of *Dendrocalamus strictus* (Roxb.) using RAPD and ISSR markers. Biotech, 5(4): 473-482.

Georget, F. *et al.* (2017). Somatic embryogenesis derived coffee plants can be effectively propagated by rooted minicuttings Scientia Hort, 216: 177-185.

George R.A.T. (2013). Vegetable Seed Production CABI Publ. pp. 338.

Grumberg, K. *et al.* (2006). The effect of genotypes and culture medium on somatic embryogenesis and plant regeneration from mature embryos of fourteen apomictic CV. of *Cenchrus ciliaris* grass. Forage Sci. 6: 2-8.

Harland S.C. (1957). The improvement of the Coconut palm through breeding and sellection. Bull. No. 15. Coc. Res. Inst. Ceylon.

Hari Har Ram (2005). Vegetable. Breeding Principles and Practices. Kalyani Puhshers, New Delhi.

Harrington J.B. (1952). Cereal Breeding procedure FAO Development Paper No28. Rome.

Hartmann, H.T., Kester, D.E., Paires, F.T. and Geneve, R.L. (2004). Plant Propagation Principles and Practices. Prentice Hall of India Pvt. Ltd., New Delhi.

Helen, H., Simmon H.T., Rahalo, K. and M.L. Hehame (2015). Morphological diversity of nutmeg mother trees and seedlings. Proc. 1st internate seminar of Basic Sciences FMIPA Unpatti Univ. Indonesia.

Hiwala S.P. (2009). Pomegranate. New India Publ. House, New Delhi.

Hussain G Wani M.S. Mir. M.A.J.A. Rather and K.M. Bhat (2014). Micrografting for fruit crop improvement. African J. Biotechnology 13: 2474-2483.

Hussain, S.M., P.G Karmarkar and Sushi Pandey (2003). Soybean in R. K. Chowdhury and S. K. Lal (Ed) 2003. Nucleus and Breeder Seed Production Manual. NSP. IARI, New Delhi. P. 80-85.

Holmes, F.O. (1956). Elimination of sperm virus from the nightingale *Chrysanthemum*. Phytopath 46: 599-600.

ICAR (2017). Hand Book of Seed and Planting Material and Testing Manual for Horticultural Crops. Division of Horticulture, ICAR PP263 ed. B.M.C. Reddy.

ICAR (2018). Technical Bulletin No 75 Production of quality seed and planting material. Central Institute for Arid Horticulture.

IFOAM (2011). International Federation of Organic Agriculture Movement. Position Paper 2011 on the use of organic seed. Plant propagation material in organic Agriculture.

Jagadeesh Kanth, R. Sambaranarayan and V. Suresh (2017). Effect of HDP on quality of curry leaf *Murraya koeinigi*. Int .Curr. microbio app. Sci. 2017. 6(12): 2861-2864.

Jagadesh Selvan, N. Muthukumar, H. Rahman, Senthil, N. (2015). Development and validation of SSR markers in fingermillet. Internatl. J. Tropical Agri 33.

Jalikop (2010). Pomegranats Breeding. Global Science Book. P. 26-32.

Janick, J., Sherry L. Kitto and Y. Kim (1989). Production of Synthetic seed by dessication and encapsulation. *In vitro* culture and Devp. Biology 25: 1167-1172.

Jenkins. D.A. *et al.* (2015). Attraction of pollination to atemoya using commercial lures and food attractants. J. Economic Entomology 108: 1023-1029.

Jeyalakshmi V.G. and John P.S. (2005). Pollination studies in Cashew. Indian Cashew 19: 20-27.

J.W. Seeds Company USA. (2018). Lettuce Seed Production.

Jothi, P.V. Athwin J.B.C.S. Reddy, (1990). Pollen ecology of Moringa. Proc Indian Acad Sci. (Plant Science) 100: 33-42.

Joseph S. and KV., Peter (1985). Curry leaf (*Murraya koenigi*) perennial nutritious leaf vegetable Economic Bot: 39: 68-73.

Jose V. *et al.* (2012). Medicinal values of Avaram *Cassia auricularia*- A review- Internatl. J. Current Pharma Research 4: 1-3.

Joseph, J. (1980). Nutmeg, its botany, agronomy, production and uses. J. Plantation Crop. 19: 69-72.

Justina, E.V., Fonseca, M.E.N. Ferresh *et al.* (2018). Estimate of natural cross pollination in *Capsium annum* using a codominant marker associated with fruit pungency Genet. Mol. Biol. 17. DOI. 104228

Jalaja, N.C., Neelamathi D. and Sreenivasam T.V. (2008). Micropropagation for quality seed production in sugarcane in Asia and Pacific. Rome.

Jambulingam, R. and E.C.M. Fernandes (1956). Tamarind – multipurpose tree in TN. Agro-Forestry System 4: 17-32.

Kadam, B.S. (1942). Deterioration of varieties of crops and the task of Plants Breeder Indian J. Genet Pl. Br. 2: 159

Kanna Babu N. and N. Seetharama (2003). Sorghum p. 44 to 53. In R. k. Chowdhury and S. K. Lal (Ed) 2003. Nucleus and Breeder Seed Production Manual. NSP IARI, New Delhi.

Kanthasamy V. (2006). Studies on Phenology and Floral biology in *Moringa oleifera.* International J. Agric. Sci. 2: 341-343.

Kanupriya, Latha, P.M., Aswath, C., Reddy, L., Padmakar, B., Vasugi, C., and Dinesh, M.R. (2011). Cultivar identification and genetic fingerprinting of guava (*Psidium guajava*) using microsatellite markers. International Fruit science, 11(2): 184-196.

Kaur, R. (2011). Evaluation and improvement of *Aloe vera* from North India cited in Shodhganga-inflib.net.ac.in.

Kgakudi B.K., Kanthi R. Umajothi K. and K. Sasikala (2020). Studies on genetic diversity in clusterbean Elec. J. Pl. Br. 11: 322-327.

Khatoon. U., Khushawn. J.K. Sharma L. Pande Y. and Kumar, R. (2017). Synthetic seed technique in vegetables. A Review. Ecol. Envt. Conservn 23: 127-138.

Khar, A., Lawande, K.E., and Negi, K.S. (2011). Microsatellite marker based analysis of genetic diversity in short day tropical Indian onion and cross amplification in related *Allium* spp. Genetic Resources and Crop Evolution, 58(5): 741-752.

Kherki, S.A., R.L. Srivastva and S.J. Rubey (2003). Linseed p. 120-128 in R.K. Chowdhury and S.K. Lal (Ed) Nucleus and Breeder Seed Production Manual. NSP. IARI, New Delhi.

Keerthi C.M., Byregowda. M., Girish. G. Ramesh S. Mahadev, P. (2015). Descriptor of *Dolichos lab lab purpureus*, L.J. Food legumes. 28: 203-214.

Krishnamoorthy, B., Sasikumar. B. and D.J. Rames (1996). Genetics variability and segregation in nutmeg (*Myristica fragrans*) J. Plantation Crops. 21: 468-472.

Kumar M.B.A., Venkateswaran, V., Krishnasamy, V. (2005). Enhancement of synthetic seed conversion to seedlings in hybrid rice. Plant Cell Tissue Organ Culture. 81: 97-100.

Kumar D., A. Saha and D. Gupta (2003). Jute and Mesta p. 155-168. Chowdhury R.K. and S.K. Lal (Ed) 2003. Nucleus and Breeder Seed Production Manual. NSP. IARI, New Delhi.

Kumar, R. and J.C. Wasiner (2011). Natural crossing in watermelon- a review Cucurbit. Genetics coop. Report, 33: 42-43.

Kundan Kishore, A.K.Shukla, Naresh Babu (2012). Pollination biology of *Annona squamosa* L. Evidences of pollination syndrome. Scientia Horticulture 144: 212-217.

Kurian Alice and K.V. Peter (2007). Commercial Crops Technology, NIPA, New Delhi. PP. 475.

Kanwar, J.S. and Gills, B.S. (2000). Natural cross pollination and pollen dispersal in brinjal Veg. Sci., 27: 145-147.

Kasa. M.A. (2016). Role of Biotechnology on Coffee propagation. J. Biol agriculture and Health care 6: 13-19.

Khadke *et al.* (2012). Development of SCAR markers for sex determination in dioecious betelvine (*Piper betle* L) curr Sci. 103: 712-716.

Kanakadurga (2012). Variability and divergence in horsegram (*Dolichos uniflorus*). J. Arid land, 4: 71-78.

Lakshmanan, V., Reddampalli Venkataramareddy, S., and Neelwarne, B. (2007). Molecular analysis of genetic stability in long-term micropropagated shoots of banana using RAPD and ISSR markers. Electronic Journal of Biotechnology, 10(1): 106-113.

Lakshmana Reddy, D. Chinnappareddy, Kiran, K.C. Kanupriya, Srikanth, H.S.R., Aswath, C. and T.H. Singh (2012). SSR based DNA barcode as a tool for identification of egg plants genotype. J. Veg. Sci.18: 260-271.

Lammerts J. Van Bueren and James R. Myers (Eds) (2012). Organic Crop Breeding 2012 Wiley-Blackwell.

Larkin, P.J., and Scowcroft, W.R. (1981). Somaclonal variation—a novel source of variability from cell cultures for plant improvement. Theoretical and Applied genetics, 60(4): 197-214.

Laverack, G.K. (1991). Management of breeders seed production. Seed Science and Technology 22: 551-563.

Le, Roux, L.K. Robbert P.J. (1994). Tuber Ontogeny morphology and vegetative reproduction in *Gloriosa*. South Aftican J. Bot to : 321-324.

Lekha, K.M., Kathirvel, M., Rao, G.V. and Nagaraju, J. (2001). DNA profiling of disputed chilli samples (*Capsicum annum*) using ISSR-PCR and FISSR-PCR marker assays. Forensic Sci. Int, 116: 63-68.

Liu L.W., Hou .X.L., Hou Hong T.Q., Zhang Y.L. (2004). Application of molecular marker in variety identification and purity testing in vegetable crops. Molecular Plant Breeding, 2(4): 563-568.

Liyanage, D.V. (1961). The use of isolated seed gardens for coconut seed production. Ceylon Coconut Q. 12: 245-252.

Luke Rahtnakumar A., Henry Louis I.S.R. Sree Rangasamy and M. Murugesan (1993). Selection criteria for Hybrid coconut seedling. Advances in Coconut R & D 105-106.

Luke Rathnakumar A., Sridharan. C.S. (1993). Genetic improvement of Copra outturn in Coconut Palms – South Indian Horticulture 41: 33-39.

Madhava Rao V.N. and Vazir Hussain (1957). Preliminary studies on the floral biology of cashew (*Anacardium occidentale*) Indian J. Agric. Sci. 27: 277-288.

Maiti S., S.R. Biwas (1991). Sexual dimorphison in Betelvine – Indian J. Agric Sci 61: 656-658.

Malhotra S.E. 2016. (Ed) Advances in Planting Material Production Technology in spices National Seminar April 2016 Kozhilode.

Malaviyaa D.R., A.K. Roy and P. Kaushal (2003). Berseem (P 163-168). In R.K. Chowdhury and S.K. Lal (Ed) 2003. Nucleus and Breeder and Seed Manual NSP, IARI, New Delhi.

Manickam S.K. Rathnavel and K.N. Gururajan Cotton (P-145-154). In: R.K. chowdhury and S. K. Lal Ed (2003) Nucleus and Breeder Seed Production Manual NSP – IARI, New Delhi.

Manonmani S. Pushpam R and S. Robin (2016). Stability of TGMS Lines under different temperature regimes for pollen sterility. Journal of Rice Research, 9(1):17-20.

Marathi, B. and K.K. Jena (2015). Floral traits to enhance outcrossing for high hybrid seed production in rice. Present status and future. Euphytica, 201: 1-14.

Masheer Mohammed *et al.* (2017). Testing of genetic homogeneity of elite Eucalyptus clones. Tropical Plant Res. 4: 391-395.

Mather, K. (1941). Variation and selection of polygene characters J. Genet. 47: 159-163.

Mather K. (1949). The genetic theory of continuous variation. Heredity 5: 159-193.

MC Cormick J.H. (2004). Isolation distance – principles and practices of isolation distance for seed crops. Caroline Farm Steward ship Association NC 2731.

Misra S.R. and R.K. Singh (2003). Sugarcane. (P 181-190). In. R.K. Chowdhury and S.K. Lal (eds) 2003.

Nucleus and Breeder Seed Manual. NSP. IARI New Delhi.

MOEF & C.C. and RRI Kottayam (2016). Biology of Hervea breeding (Rubber) Series of Crop specific document. icar.gov.in.

Mondal T.K. (2014). Breeding and Biotechnology of Tea and its wild sp. Springer Palm.

Morgante, M., Hanafey, M., and Powell, W. (2002). Microsatellites are preferentially associated with non repetitive DNA in plant genomes. Nature Genetics, 30(2): 194-200.

Mulvi GM. Sprent Jr. Ode. D. and W. Powell (2004). Estimate of out crossing rates in *Moringa oleifera* using amplified of length polymorphism (AFLP) Fragment. African J. Biotech 3: 146 -151.

Muoki R.C., Wachira F.N. Pathak R. and S.M. Kamunga (2007). Potential male gametophyte competition in isolated Biclonal seed orchards in Tea. African Crop Sci. J.15: 59-66.

Murasighe T. (1977). Plant cell and organ culture as horticultural practices Acta. Hort. 78: 17.

Murthy, N.S.R. and Murthy, B.S. (1962). Natural cross pollination in chilli. Andhra Agric. J. 9: 162-170.

Mydin, K.K. (2014). Genetic improvement in *Hevea brasiliensis*. Sixty years of breeding efforts in India. Rubber Sci. 27(2): 153-181.

Muehlbauerind, Mc and Rhee K.E. (2005). Lentil (*Lens culinaris*). Genetic resources in Grain legumes Ed. R.J. Singh and P. Jauhar P. 219-228 CRC. Press.

Mukherjee, S.K. (1953). The mango its Botonay. Econ. Bot. 7: 130-162.

Nair M.K. (1991-1992). Annual Report CPCRI. Production of nucleus and superior planting material at Kidur. CPCRI.

Nair M.K. (1993). (Ed.) Advances in Coconut R&D Oxford and 1BH publication PP 702.

Nair, R.R. Gupta S.D. (2007). *In vitro* plant regeneration from encapsulated somatic embryos of Black pepper (*Piper nigrum*). J. plant Sci. 2: 283-292.

Naik G.S. and J.K. Karihalbo (2003). Micropropagation in potato seed production. Potato J. 30: 267-273.

NBPGR (2001). National Bureau of Plant Genetic Resources, ICAR. Minimal Descriptor of Agri – Horticultural Crops. NBPGR New Delhi: Part I (2000) Cereal and other crops, Part II (2000) Vegetable Crops, Part III (2002) Fruits Crops and Part IV (2002) Medicinal and Aromatic Plants.

Nandakumar, N., Singh, A.K., Sharma, R.K., Mohapatra, T., Prabhu, K.V., and Zaman, F.U. (2004). Molecular fingerprinting of hybrids and assessment of genetic purity of hybrid seeds in rice using microsatellite markers. Euphytica, 136(3): 257-264.

Nelson, C. (2003). *Morinda citrifolia* Rubiaceae in USA. Permanent Agri. Resource vision 2013.

Nejal, N., Valdiarin D. Cachill, Y.H. Tan M., Maziah M. and R. Abiri (2015). Ornamental exterior versus therapeutic interior of Madagascar Periwinkle *Catharanthus roseus.* L) the two faces of the vessatile herb. Scientific world J. 2015. Article 982414 P.18.

Nirmalaruban R.N. Meenakishiganesan, P. Anantharaj (2020). Breeding system study in Sunnhemp (*Crotalaria juncea* L) A stepping stone in crop improvement. Electronic J. Pl. Br. 11: 204-209.

Nookaraju Akil *et al.* (2003). Propagation of grapevine root stock by tissue culture Natl. Symp on Biodiversity. Univ. of Pune 2003.

NRC National Research Center Citrus. Nagpur (Citrus Improvement Nrccitrus.nic.in)

Naik, K.C. and Gangully S.R. (1950). A. Monograph on south Indian Mangoes Govt Press. Madras. PP 311.

Naik, K.C. and Rao, M. (1943). Studies on blossom biology and pollination in mango. Indian J. Hort. 1: 107-119.

Nathan, J.P., Raghuram, H. Elangovan V. and K. Thangavel (2005). Bat pollination of Kapok tree (*Ceiba pentandra*). Curr Sci. 88: 1679-1681.

Neelamathi D. and Shobakumari (2007). *In vitro* degree and regeneration of encapsulated micro propagules in sugarcane. Phytomirphology 57: 103-107.

Neelamathi D. (2013). *In vitro* clonal propagation in sugarcane, TNAU Agri Tech Portal.

Ning, R., Li, S., Li, X. and C.Li. (2014). New maize hybrid has larger and deeper post silking root than old ones. Field crop res., 166: 66-71.

OECD Seed Scheme (2014). Hand Book on OECD varietal certification in India 2014 National Designated Authority OECD Seed Scheme MOA GOI September 2014.

Parlevliet J.E. (2007). How to maintain cultivars. Euphytica 153: 353-362.

Parthasarathy (2010). An overview of genetic resources of tree spices. Proc. Natl. Seminar Tree spices. HRS. Kanyakumari.

Pathak H. and Dhawan V. (2010). Molecular analysis of micropropagated applerootstock using ISSR markers for ascertaining clonal fidelity Acta Hort 865: 73 -80.

Pathak R.K. and R.A. Ram (2013). Approaches for green food production in Biodynamic Agricultural. In Crop Production Techniques of Horticultural Crops, TNAU, Coimbatore. Pages. 277-288.

Paroda, R.S. (2020). The changing face of Indian Seed Sector. www. Seedsector.org.

Parthipan, K.T. (2017). Miniclonal Technology for tree crops. In Forest Technologies a complete value chain approach (Eds.) K.T. Parthipan and R. Sreenivasan. Scientific Publ. Jodhpur PP. 103-110.

Plant Authority- Protection of Plant varieties and Farmers rights. Authority Ministry of Agriculture GOI, New Delhi.

Purusothaman.K and Narasimha Rao S.B.S. (1990). Propagation of tamarind by Veneer and softwood grafting. South Indian Hort. 38: 225.

Parvathaneni, R.K., Natesan, S., Devaraj, A.A., Muthuraja, R., Venkatachalam, R., Subramani, A.P., and Laxmanan, P. (2011). Fingerprinting in cucumber and melon (*Cucumis spp.*) genotypes using morphological and ISSR markers. J. Crop Science and Biotechnology, 14(1): 39-43.

Pathivana R. (1994). Natural cross pollination in Sesamum- Plant Breeding 112: 167-170.

Peng *et al.*, G.S.Khush (2010). The importance of Maintenance Breeding. A case study of the first miracle rice var. IR.8.Field crops. Res: 200: 343-347.

Prakash, D.P., Narayanaswamy, P., and Sondur, S.N. (2002). Analysis of molecular diversity in guava using RAPD markers. J. Horticultural Science and Biotechnology, 77(3): 287-293.

Peter K.V. (2002). Plantation Crops. National Book Trust. New Delhi.

Pitough A.N. Sykur, M. *et al.* (2018). Assessment of NCP in chilli pepper. Earth and Environment science 196: 012008.

Priya Devi, S *et al.* (2014). A guide to Jackfruits Tech Bull 41. ICAR Res complex Goa ICAR–CPRI.

Prota.org *Senna auriculata*. Plant use, Netherlands.

Priyadharsan, P.M., P.S. de Concalves and K.O. Omocafe (2009). Breeding Hevea rubber in Breeding plantation Tree Crops. Ch. 13.P. 469-522.

Priyadarshan, P.M. (2017). Refinements to hevea breeding. Tree genetics and genomies 13:20 (2017). doi.org.10.1007

Pusadee T. Jamjod, S. Change Y. Perkaran B. and B. Schaif (2009). Gene structure and selection distance by distance in land race of Thai Rice. P.N.A.S. 2009 106: 13880-13825.

Radenbaugh, K. (1990). Application of artificial seeds to Tropical crops. Horticulture. Sci. 25: 251-255.

Ramesh K. Singh, V. and N.W. Mageji (2006). Stevia cultivation. Adv. Agronomy 89: 137-177.

Radhamani, A., Nichodemis, A. Nagarajen, B. Mandel A.K. and K.N. Subramanian (1993). Reproductive Biology and breeding system studies on *Tamarindus indica* in Pollination in Tropics. International union for social insects. Indian chapter P. 33-35.

Raghu, D., Senthil, N., Saraswathi, T., Raveendran, M., Gnanam, R., Venkatachalam, R. and Mohan, C. (2007). Morphological and simple sequence repeats (SSR) based finger printing of south Indian Cassava germplasm. Int. J. Integr. Biol, 1(2): 141-148.

Rajasekharan, P.E., V. Ramanathan Rao (2019). (Eds). Conservation of Horticultural Genetic Resources Springer. Lmk.

Rajender Singh, Deilbagh Singh and J.S. Kanwar (2012). Studies on the extent of genetic contamination in seed production in exserted tomato seed prodiction. Hort. Flora. Res. Spl. 301-311.

Rajkumar, S.S. Bisht, J. Shoron and D.S. Chauhan (2003). Wheat and Barley, P. 11-17. In R.K. Chowdhury and S. K. Lal (Ed) 2003. Nucleus and Breeder Seed Production Manual, N.S.P., I.A.R.I. New Delhi P. 11-17.

Rajkumar *et al.* (1997). Selfing and seed set in Tea UPASI Tea. upasi org.

Rajendra Prasad, S. Charhan J.K. Sripathy (2017). National and international quality assurance system and strategies for emerging seed production chain of field crops in india. India J. Agric. Sci. 87: 287-300.

Rajamani, K. (2017). Medicinal plants tnau.ac.in (2017).

Rajamani, K. (2017). Noni. Agri Tech. TNAU.

Rajamani (2020). Makoi (Personal communication).

Rahman, M S., Sokag M.K.H. and Rahman, L. (2009). Microsatelite markers based DNA finger printing of 28 local rice varieties of Bangladesh. J. Bangladesh Agri cultural Univ 8: 7-17.

Rai M.K., D. Pooja, S.K. Singh, K.S., Jaiswal (2009). The encapsulation technology in fruit plants. Biotech Advances 27: 671-679.

Rani, A., Kumar V., Shukla S., Tha P. Rawal R. (2016). DNA barcoding of Indian Soybean Varieties constructed through SSR markers. Seed Sci. and Tech. 44: 367-396.

Ram Chandra (2012). Guidelines for establishing and managing Mother Blocks in fruit crops. P.29 in Singh. H.P. *et al.* (eds) 2012. Quality Seed and Planting Material in Horticultural Crops I.I.H.R, Bengaluru.

Rani K.V., S. Narayanaswamy S. Siddar Ramegowda (2013). Characterization of Horsegram *Macrotyloma uniflorum* L. Mysore J. Agri. Sciences 47: 530-546.

Ravishankar, K.V., Anand, L. and Dinesh, M.R. (2000). Assessment of genetic relatedness among mango cultivars of India using RAPD markers. Journal of Horticultural Science and Biotechnology, 75: 198-201.

Ratnam. C. (2001). Palmyrah palm in Chadha. K.L. (2001). Hanbok of Horticulture ICAR Publ. p. 666-670.

Robin, S., Jeyaprakash, P., Amudha, K., Pushpam, R., Rajeswari, S., Manonmani, S. and Ganesamurthy, K. (2019). Rice CR1009 Sub1 (IET 22187)- A new flood tolerant rice vaiety. Electronic Journal of Plant Breeding, 10: 995-1004.

Ray Ming, Q.Yu, P.H. Moore (2007). Sex determination in Papaya. Seminar in Cell and Development Biology 18: 401-408.

Rick C.M. (1949). Rate of natural cross pollination in tomato varieties in male sterile lines. Proc Amer Soc Hort Sci. 5H-237-252.

Ramsingh, J. (2011). *Aloe vera* P. 854-875. Genetic Resources, Chromosome Engineering and Crop Improvements CRC. Press.

Rihan. H.Z., Kareem F. El Mahrouk and M.P. Fulgev (2019). Artificial seeds. Principles, aspects, application. Agronomy 7:71. Doi 10.3390.

Ravi, D. and P. Anand (2012). Production and application of artificial seeds. A Review. Internatl. Res. J. Biol. Sci. 1: 74-79.

Ratnambal *et al.* (2003). Floral biology of coconut accessions. J. Plantation crop 31: 14-22.

Rasime Kalkon (2020). Molecular cytogenetics Technique. INTECH.

Ray ming *et al.* (2007). Sex determination in papaya. Cell and development Biology 18: 401-408.

Ritlank, K. (1990). Asercs of FORTRAN computer programme for estimating plant mating system heredity 47: 37-54.

Sahadevan, P.C. and Namboodri K.M.N. (1963). Natural crossing in rice. Natl. Acad. Proc. Nose. Sci. 0176 to 0185.

Saikat G., Kander, S., Narain, A.C. and N.C. Sahu (2015). Synthetic seeds production of medicinal plants. Acta Physical plantarum 37: 98.

Sai Prasad G.V.S. (2001). Artificial seeds and their applications. Resonance. May 2001. 39-47.

Saravanan, S. and Bhuvaneswaran, C. (2003). *Gloriosa superb* cultivation Adv. In plant Sci. 16: 23-28.

Saroj, P.L. (2012). Standardization for production of quality planting material of arid fruit p. 96-113. In Singh, H.P. 2012 Ed. Quality seed and planting material in Horticultural crops. IIIR Bangaluru.

Sarkar, S. *et al.*. (2015). Sunnhemp in India. ICAR Central Res. Institute for Jute and Allied Fibres Barrackpur. VV.B.

Sarjeet Singh *et al.* (2000). True Potato Seed Production. Central Potato Res. Institute Simla.

Santhiya, S., Kowshiga, K., Abisha, D., Sarankumar, C., Kavithamani, D., Selvi, B., and Senthil, N. (2020). SSR Marker-based DNA fingerprinting and morphological characterization for varietal identification in popular sorghum varieties of Tamil Nadu. Electronic Journal of Plant Breeding, 11: 521-527.

Saxena K B., D. Sharma, M. Vales (2018). Development and commercialisation of CMS pigeon pea hybrids. Plant Breeding. 41: 103-107.

Schonten, H.J., F.A. Krens and E. Jacobsen (2006). Do Cisgenetics plants warrant less stringent oversight. Nature Biotech, 24: 753.

Seema S. Yash Pal. S. Jhakeer, P. (2017). Quantification of colchicue in differenct parts of Gloriosa. Int. J. Chem. studies 5: 147-149.

Sekar S.N.C. and S.V.P. Sarathy (1975). Cytogenetics and Plant Breeding Revised Edition by N. Krishnaswamy and N. Hrishi. P Varadachary & Co Madras.

Senaratna T. (1992). Artificial seeds Biotech Adv. 10: 349-392.

Selvakumar P., Ravikesavan R., Thiaga G. Preetha, S. Boopathi N.M. (2010). Genetic purity analysis of cotton hybrid using SSR markers. Seed Sci Technology 38: 358-366.

Seetharam, A. Jeyarama gowda and B.H. Halaswamy (2003). Small Millets P 54-57. Chowdhury R.K. and S.K. Lal (Ed) 2003. Nucleus and Breeder Seed Production Manual NSP, IARI, New Delhi.

Senthil, N., Gayathri, V., Selva Babu, S., Madhumitha, K., Sarankumar, C., and Nirmalakumari A. (2020). Characterization of little millet (*Panicum sumatranse*) varieties using morphological discriptors and SSR based DNA fingerprinting. Journal of Phytology, 12: 29-34.

Senthil N.P., Tamilkumar, M. Raveendran R. Jerlin and R. Umarani (2016). Markers Technology for Plant Variety identification. TNAU Agri. Portal.

Sharma, A.K., Venkatachalam, E.P. and V. Kumar (2013). Potato minituber production during main and off crop season in high hills of N.W. Himalayas Potato J. 40: 29-37.

Sharma S., A. Shahzode, J.A. Jelxei and A. Selva (2013). Synthetic Seeds Techonology. Biotech. Advances 31: 186-207.

Singh I.D. (1992). Release of Biclonal seed stock in Tea. Tocklai Tea Research Inst.

Singh. H.P. *et al.* (Eds.) (2012). Quality seed and Planting Material in Horticultural Crops. SPH, IIHR, CHAI and NHB, I.I.H.R. Bengaluru. 2012 pp. 646.

Spillane A. *et al.* (2001). Apomixes in Agriculture – A quest for clonal seeds. Sexual Reproduction 14: 179-187.

Sripathy K.V., Jagadesh Hosamani, Ama Siddha Ballundagi and Prabhakar 2012.

Organic Seed Production. Environment and Ecology, 30: 102-105.

Singh N.N., S.B. Singh, S. Rakshi, Sudipta Basu and S. Vankatesh (2003). Maize P. 27-35 in R.K. Chowdhury and S.K. Lal (Ed) 2003. Nucleus and Breeder Seed Production Manual. N.S.P., I.A.R.I., New Delhi.

Sivolop, Y.M. (2013). Molecular Markers and Plant Breeding. Cytology and Genetics. 47: 188-195.

Singh R.S. and Singh M.N. (2016). Heterosis and inbreeding depresson in pigeonpea *Cajanus cajan*. Envt and Ecology. 34: 395-399.

Shoba N., S. Balakrishnan, P. Paramaguru and K. Vithya (2020). Genetic variability and character association studies in curry leaf (*Murraya koenigii*). Electronic Journal of Plant Breeding. 11: 694-697.

Sindhuja T. and Raja K. (2020). Establishment method, spacing and pinching effect on plant growth and seed yield in Lambs quarters (*Chenapodium album*. L). Madras agricultural J. 107: 153-160.

Sreenivasamoorthy, D. (2012). Hand pollination in Arka Sahon hybrid custard apple I.I.H.R. Bangaluru. Swetha Gautham, Singh, P. and Y. Shivhana 2011. Tinda-*Praecitrullus fistulosus*. A miraculours plant. Asian J. Pharm. Tech 2011. 1: 9-12.

Srinivasan C.S. (1996). Current status and future thrust areas of research on varietal improvement and horticultural aspects of coffee. J. Coffee Res., 26 (1): 1-16.

Sreenivasamurthy, D. (2012). Hand pollination in Arka Sahan custard Apple. IIHR, Report.

Srikant, G. Arpita Das and J. Banerigea (2017). Geographical distribution, botanical description self incompatability mechanisms of Stevia. Sugar Tec. 20: 1-20.

Sree Rangasamy, S.R. and P. Devasahayam (1972). Cytology of sex determination in palmyrah palm. (*Borassus flabellifer*. L) La. Cellule 69: 129-134.

Sree Rangasamy S.R. and C. Sridhar (1993). Multistage selection in coconut palm for genetic improvement. Adv. Coconut R-D 105-106. Oxford NBH Publn.

Swathi Madrasi, Sajeed Ahmed, G.N. Singh, Kanchakihli *et al.* (2010). *Stevia rebaudiana* A. Review Indian J. Natural. Products and Resource 1: 267-286.

Swetha Gantham, Singh P and Y.Shivhana 2011.Tinda. Praecitrulhis Fistulosus.A Miraculous plant. Asian J. pharma. Tech 1: 9-12.

Sundaram R.M., Naveenkumar, B., Biradar, S.K., Balachandran, S.M., Mishra, B., Ilyas Ahmed, M. and Sarma, N.P. (2008). Identification of informative SSR markers capable of distinguishing hybrid rice parental lines and their utilization in seed purity assessment. Euphytica, 163: 215- 224.

Subba Rao L.V., N. Shoba Rani B. Viswanath, B. Mishra and H. Prasad Rao (2003). Rice p. 18-26. In: R.K. Chowdhury and S. K. Lal (Ed) Nucleus and Breeder Seed Production Manual, N.S.P., I.A.R.I., New Delhi.

Suso *et al.* (1996). Reproductive Biology of *Vicia faba* – role of pollination conditions. Field Crop Res. 46: 81-91.

Sundararaj S.R. Agarwal. A. Tyagi R.K. (2010). Encapsulation for *in vitro* short term storage and exchange of ginger. *Zingiber officinalis* gernmplasm Sci. Hort 125: 761-766.

Steen P.C. Groot J.M., Vander Wolf, H. Jalink Cees, J. Langerak and Rund, W. (2004). Challenges for the production of high quality organic seeds. Seed Testing International, 127: 12-15.

Surabhi G.K. and Pattanaik S. (2015). Deciphering the genetic identity and fidelity of banana through intersimple sequence repeats fingerprinting. Horticultural Biotechnology Research, 1: 16- 22.

Sambandam, C.N. (1962). Natural cross polliantionin egg plant. Ecom. Botany 18: 128-131.

Singh, B.K. (2019). Khasi Batua Vegetable Sci. 45: 291-293.

Storey W.B. (1953). Genetics of papaya J. Heredity 44: 70-78.

Suso, M.J. Morrena, M.T. and F.M. Rodrigues, (1996). Reproductive biology of *Vicia Faba* – role of pollination conditions. Field Crops Research, 46: 81-91.

Samantaray S. *et al.* (2011). Identification of sex specific DNA markers in betelvine (*Piper betle* L.) J. Genetic Resources and crop evaluation 59: 645-653.

Suso, M.J. and Riode R. (2015). A crop pollination interplay approach to assessing seed production pattern in *Vicia faba* under two pollination environment Euphytica 201: 231-251.

Sundara, B. (1995). Causes of sugarcane vars. yield decline and techniques for rejuvenation. Annual Conf. Sugarcane Technologist. Association ND. P. 153-167.

Tamilkumar, P., Jerlin, R., Senthil, N., Ganesan, K.N., Jeevan, R.J., and Raveendran, M. (2009). Fingerprinting of rice hybrids and their parental lines using microsatellite markers and their utilization in genetic purity assessment of hybrid rice. Research Journal of Seed Science, 2: 40-47.

Tandon. J.P., Sharma. S.P. Sindhu. J.S., Yadava D.K. Prabhu K.V. and Yadava. O.P. (2015). Guidelines for testing crop varieties under All India coordinated crop improvement Projects. ICAR. New Delhi. PP. 249.

Tayler *et al.* (2001). *Morus sp.*/ mulberry/ CSRI Mysuru.

Thamburaj S. and Narendra singh (2001). Ed. Vegetables, Tuber Crops and Spices ICAR. New Delhi.

Thangaselvibai K.R. Sudha, T. Selvakumar and B. Ramachandran (2011). Nutmeg (*Myristica fragrance* Hutt), the tree spice. A review J. Agric. Renew 32: 283-293.

Thimmappiah, R. Ruthra G.T., J.R. Anil (2002). Micropropagation of Cashew (*Anacardium occidentale*) Sciences Horticulture. 92 : 177-182.

Thimmaraju. K.R. and Shivananda, M.C. (1993). Studies on Floral biology, pollination and fruit set in Jack fruit (*Artocarpus heterophyllus* L.) Proc. Internatl on Pollination in Tropics. Bangaluru. India. P. 44.

Tripathi, M. (2017). Synthetic seed Technology and its application – A Review. Int. J. Plant. Sci. 159: 962-978.

Trought T. (1930). Note on certain factors on vicirism and artificial pollination Egyptian. Cotton Gr. Rev. F. 13-18.

Tujeb Y and Oztakin G.P. (2017).Organic Seedling Production Acta – Horticulture 117: 1141-1148.

Tyson, K. and Amit Dhingra (2013). Rootstock – scion somatogenetic interaction in perennial composite plants. Plant cell Rep. 32: 1321-1327.

TNAU (Tamil Nadu Agricultural University) (i)TNAU Agri Portal- Horticulture (ii) Crop Production Techniques of Horticultural crops 2020.

UPASI Tea Research foundation (TRF) Valparai – 642 127 (2014). Planting Materials, Tea cultivation & Practices www.upasi tea research org.

UPOV- International Union for Protection of New varieties of Plants (1961). Publ No. 437(2019).

Vaidya K.R. (2000). Natural cross pollination in Rosella, *Hibiscus sabdariffa* L. (Malvaceae). Genet. Mol. Biol. 23: 667-669.

Vasanthi *et al.* (2011). *Stevia* Crop Research Hissar 41: 103-106 and 107-112.

Veerasamy R. Palsamy S.S. and R. Rathiasamy (1973). Natural cross pollination in *Cajanus cajan* and lablab niger. Madras Agric. J. 60: 1828.

Veera Raghava Thatham, S. Auxcilia N. Shoba and A. Veena (2005). Studies on Natural Cross Pollination in Tomato. Crop. Res. 29: 169-177.

Vinod K.K. (2016). Potential male gametophyte competition in isolated Biclonal seed orchard in Tea. Breeding for Quality improvement in plantation crops UPASI.org

Vijayan R. and V. Krishnasamy (2017). Organic seedling production in rice. National Seed Seminar ISST ICAQ-IART. New Delhi Page 214.

Virmani S.S. and D.S. Athwal (1977). Inheritance of floral characters influenced out crossing in rice Crop Science 14 : 350-353.

Virmani S.S., Sun Z.X., Mou, J.M. Jauhar A.G. A. Mao, C.X. (2003). Two line Hybrid Rice Breeding Manual. International Rice Research Institute, Manila Philippins.

Virginia Tech (2004). Biennial Vegetable. Horticulture 4764.

Vishnu Dhar, N.B. Singh, K.B. Wanjari, Farindra Singh and N.D. Majumdar (2003). Pigeon pea (P. 64-69) in R.K. Chowdhury and S.K. Lal (Ed) Nucleus and Breeder Seed Production Manual. NSP, I.A.R.I., New Delhi.

Viswanathan Reddy A., K. Arjara and Vijendra Singh (2003). Safflower page 93-97 in R.K. Chowdhury and S.K. Lal (Ed) 2003. Nucleus and Breeder Seed Production Manual. N.S.P., I.A.R.I., New Delhi.

Viswanathan Reddy A. and C. Hanumantha Rao (2003). Castor pp. 98-102. InL R.K. Chowdhury and S.K. Lal (Ed) 2003. Nucleus and Breeder Seed Production Manual N.S.P., I.A.R.I., New Delhi.

Verma, T.S. and Sharma, S.C. (2008). Producing Seeds of Biennial Vegetables of temperate region. ICAR. New Delhi.

Virmani, S.S. and D.S. Athwal (1977). Genetic variability in floral characteristics influencing out corssing in *Oryza sativa*. Crop Sci. 13: 66-67.

Yadav A.K. (2015). Organic Seed Production. National Center of Organic Farming. Ghaziabad, India.

Yadav R.K., Tomar, B.S. and B.K. Singh (2020). New bathua cultivars a weed to health. Indian Horticulture. 65: 124-126.

Yadavendra J.P., Parmaz J.P. and N.F. Melkania (2003). Lucerne P. 169-175. In: R.K. Chowdhury and S.K. Lal (Ed) 2003. Nucleus and Breeder Seed Production Manual. N.S.P., I.A.R.I., New Delhi.

Yadavah J.S. and K.H. Singh (2003). Rapeseed Mustard P. 111-119. In: R.K. Chowdhury and S.K. Lal (Ed) 2003 Nucleus and Breeder Seed Production Manual, N.S.P., I.A.R.I., New Delhi.

Yuang, L.P. (1987). Strategy concept of Hybrid rice breeding. Hybrid Paddy 1: 1-3.

Zhao, L., CAi, C.R., Mei, H.X. and Guo, W.Z. (2012). Screening microsatellite loci for identifying genome barcoding of Cotton cultivars Acta. Agronomica. Sinica 38: 1810-1817.

Zimmerman R.H. (1972). Juvenility and flowering in woody Plants - A review. Horticulture Sciences: 447-455.

Subject Index

A

Achene 238, 352, 372, 421, 529
Acid lime, sweet orange and mandarin orange 408
Aerial tubers 345
Agarose gel electrophoresis 104
Agathi 268, 287, 478
Ajawain 462, 483
Allelic barcode 94
Aloe vera 545
Amaranth 288, 290, 291
Amla 411, 412, 413
Andromonoecious 20, 438, 439, 459
Anise 464
Anthesis 5, 15, 110, 143, 160, 166, 176, 187, 193, 225, 238, 242, 258, 260, 268, 273, 301, 307, 310, 337, 340, 348, 359, 370, 387, 388, 390, 402, 405, 414, 433, 438, 442, 452, 453, 458, 466, 486, 493, 510, 515, 542, 552
Arecanut 6, 71, 466, 467, 468, 469
Ashgourd 293
Avaram 548

B

Banana 29, 33, 36, 44, 45, 47, 48, 49, 51, 52, 67, 97, 98, 100, 470, 472, 473, 474, 475
Bar coding 102
Barley 16, 75, 105, 106, 107
Barnyard millet 153, 154, 155
Beet root 299, 300, 371, 373
Ber 36, 414, 415, 416, 417
Berseem 18, 273, 274, 275, 276
Betelvine 476, 477, 478, 479, 480
Biochemical 27, 396, 438, 534, 502
Biparental inbreeding 27, 340
Bitter gourd 301
Black gram/Urd bean 163
Black pepper 51, 53, 481, 482
Bottle gourd 305, 386
Bred lines 85
Breeder Seed production 4-8, 31, 46, 69, 70, 72, 74, 76, 79-81, 83, 105-109, 113, 115-117, 119, 121, 123, 126, 129, 130, 133, 143-147, 149, 150, 152-154, 157, 164, 167, 169, 170, 172, 173, 177-180, 183, 185, 188, 191, 193, 194, 195, 197, 199, 201, 202, 204, 205, 207, 210, 216, 218, 219-221, 223, 226, 231, 234, 235, 239, 242, 243, 244, 247-250, 252, 253, 256, 261, 264, 267, 269, 272, 275, 280, 283, 287, 288, 291, 294, 296, 300, 303, 306, 310, 313, 317, 325, 329, 332, 334, 337, 339, 349, 354, 356, 357, 358, 360, 365, 368, 369, 378, 385, 393, 396, 399, 404, 407, 420, 437, 440, 444, 445, 450, 457, 460, 463, 465, 483, 485, 500, 503, 506, 509, 511, 514, 516, 518, 525, 531, 536, 542, 544, 549, 551, 554, 558
Breeding cycle 527
Brinjal 307
Broad bean 166, 167, 179
Bud pollination 310, 318, 349, 387, 542

C

Cabbage 18, 310, 311, 312, 313, 352
Cacao 589
Caraway 462, 483
Cardamom 484
Carrot 65, 315, 317
Cashew 33, 43, 485, 486, 487, 488
Cassia senna 550
Castor 17, 80, 198, 208, 209, 211, 212, 213, 214
Cauliflower 65, 100, 318, 319, 320, 321
Cereals 7, 71
Chickpea 17, 75, 168, 169
Chilli hot pepper 322
Chinese potato 330
Chow chow (Chayote) 327

Cleft grafting 429
Cleistogamous 159, 168, 255, 510
Cleistogamy 105
Climate changes 83
Clonal fidelity 54, 87, 94, 98
Clonal generations 378, 473
Clonal orchard 413, 414, 415, 451, 489, 534
Clonal seed orchard 342
Clonal seeds 9, 27, 28, 29, 30, 65, 69, 72, 399, 408
Clonality 9, 27, 30
Coconut 6, 67, 71, 287, 491, 492
Coffee 6, 71, 497, 498, 499, 500, 501
Colacasia 357, 400
Cole crops 319
Coleus 330331, 332, 357
Codominant markers 87, 100
Coriander 333
Cormels 344, 358, 400, 401
Corn 22, 29, 60, 109, 146
Cowpea 17, 75, 171, 172, 173, 174, 175
Cross Pollinated 5, 18, 20, 22, 25, 59, 61, 69, 76, 80, 81, 83, 106, 108, 145, 150, 166, 208, 217, 218, 219, 222, 224, 230, 245, 251, 261, 273, 278, 296, 299, 301, 307, 310, 316, 333, 359, 363, 370, 382, 386, 392, 405, 408, 418, 433, 438, 443, 451, 458, 467, 493, 497, 504, 515, 517, 520, 526, 531, 534
Cucumber 20, 335, 336, 337, 338
Cultivar identification 87
Cumin 462, 483, 502, 515, 516
Curry leaf 338, 339
Custard apple, sugar apple, sithapal 418
Cytoplasmic male sterility (GMS) 198

D

Daincha 61, 266
Descriptors 166, 174, 186, 206, 208, 222, 230, 235, 307, 310, 318, 322, 359, 402, 419, 498, 507
Deviating plant types 164, 252, 261, 311, 343, 369, 513
Diagnostic characteristics 70, 71, 81, 163, 199, 201, 216, 231, 240, 276, 280
Dill seed spice 504
Dioscorea 30, 98, 345, 357, 358
DNA Barcode 88, 92, 103
DNA finger printing 7, 90, 103
DNA markers 78, 88, 89, 92, 93, 100, 103, 583
Dolichos lab lab 186
Dormancy 9, 66, 227, 228, 312, 319, 327, 419, 420, 489, 527, 540, 552, 554
Doubled haploid 499
Drumstick 340, 343

E

Encapsulation 66, 67, 68
Elephant yam 344, 357, 358
Epicotyl grafting 430

F

Female sterility 28
Fennel 504, 507, 508, 509
Fenugreek 510, 511
Finger millet 154, 156, 157
Floral biology 5, 342, 429, 492
Food Security 2, 11
Foxtail millet 153, 158
French bean 178, 179, 180, 181, 193
Full vegetative grafts 41

G

GA 320
Genetic diversity 87, 92, 93, 340
Genetic fidelity 28, 29, 45, 49, 67, 97, 285, 451, 453, 474, 475
Genetic male sterility 126, 130, 197
Genetic purity 2, 5- 8, 13, 15, 20, 28, 29, 35, 36, 38, 40, 45, 49, 69, 70-74, 77, 78-81, 83, 85, 87, 90, 92-94, 100-103, 106, 109, 119, 126, 127, 130, 143, 149, 150, 152-154, 157, 164, 167, 173, 177, 180, 185, 192, 204, 216, 219, 220, 223, 224, 226, 227, 232-234, 236, 240, 243, 253-256, 262-265, 268, 269, 270, 280-283, 291, 294, 298, 303, 307, 310, 313, 317, 321, 325, 326, 330, 332, 334, 353, 356, 361, 367, 379, 385, 391, 393, 397, 399, 400, 440, 441, 446, 448, 454, 457, 463, 465, 479, 500, 501, 506, 511, 526
Genomics 11, 12
Gloriosa lily 552, 554
GOT 7, 46, 50, 71, 73-75, 78, 81, 85, 87, 99-102, 104, 111, 119, 121, 126, 130, 133, 143, 146, 154, 164, 167, 169, 173, 180, 183, 188, 196, 197, 207, 210, 213, 217, 219, 224, 229 232, 233, 234, 237, 242, 243, 248, 251, 254, 265, 274, 280, 283, 284, 292, 294, 297, 304, 309, 326, 332, 353, 356, 361, 364, 365, 369,

385, 386, 388, 400, 404, 440, 446, 447, 448, 456, 461, 475, 480, 506, 541, 542, 556
Grafting 6, 28, 30, 32, 33, 34, 39, 41, 42, 413, 415, 418, 423, 428, 429, 430, 434, 454, 455, 487, 488, 489, 491, 500, 517, 518, 527, 528, 533, 534
Grapevine 29, 92, 98, 422, 423
Greater yam 345
Grow out tests 72, 73, 85, 226, 457
Guar 181, 182, 183
Guava 33, 36, 92, 426
Guinea grass 276
Gynodioecious 420, 441, 442, 444, 448

H

HDP-High Density Planting 412, 420, 426, 435, 454
Horse gram 184, 185

I

In vitro grafting 6, 41, 42, 423
Inarching 428, 434, 454, 455
Inbreds 5, 71, 84, 93, 100, 126, 340
Inbreeding depression 5, 27, 28, 61, 76, 81, 83, 193, 222, 299, 316, 333, 340, 341, 382, 387, 438, 447, 494
Indian bean 186
Indian Minimum Seed Certification Standards 2, 8, 28, 35, 38, 49, 51, 73, 75, 77, 78,
Informal system 4, 78
Insect pollination 16, 70, 208, 235, 263, 327, 359
Isolation 5, 15, 18, 19-24, 39, 40, 49, 50, 60, 70, 76, 82, 85, 92, 105, 106, 108, 109, 111, 113, 116-119, 121, 123, 125, 126, 128, 129, 130, 133, 134, 136, 141-149, 152-154, 156, 157, 163, 164, 166, 168, 170, 172, 176, 179-182, 184, 185, 187, 188, 191, 193-195, 197, 199, 200, 202, 203, 205-210, 213, 216, 217, 218-220, 223-226, 229, 231-244, 246-248, 251-257, 259-261, 264, 265, 267, 269-274, 277, 279, 280, 282, 283, 287, 290, 291-293, 297, 299, 300, 302, 304-306, 308-313, 316, 317, 325, 326, 329, 334, 336, 337, 339, 341-343, 347, 349, 352, 353, 355, 356, 358, 360, 361, 363-369, 371, 373, 374, 377, 384-387, 389, 390, 391, 392, 404-406, 411, 414, 416, 419, 421, 423, 428, 429, 430, 434, 435, 437, 440, 441-445, 447, 448, 450, 451, 454, 456, 457, 460, 461, 463, 465, 468, 469, 479, 482, 484, 485, 490, 495, 496, 500, 501, 503, 504, 506, 509, 511, 514, 516, 518, 524, 525, 528, 529, 530, 535, 538, 540, 544, 546, 549, 550, 551, 553, 556, 558, 558
Ivy gourd 346

J

Jack fruit 36, 429, 430
Jute 17, 74, 255, 256, 257
Juvenility 30, 412

K

Kapok 258, 259
Kidney Bean 178
Knol khol 347, 348, 349
Kodo millet 153, 159

L

Labelling 72, 76, 100, 130, 166, 188, 191, 211, 223, 356, 407, 450
Lentil 190, 191, 192
Lettuce 63, 350, 351, 352, 353
Lima bean 192
Linseed 17, 25, 215, 216
Little millet 90, 153, 154, 160
Lucerne 18, 25, 278, 279, 280, 281

M

Maintainer line 121, 124, 127, 198, 209, 240, 249, 251, 363
Maintenance breeding 2, 6, 69, 70, 78, 81, 83, 84, 85, 302, 305, 324, 333, 336, 363, 377, 388, 392, 395, 459, 472, 513
Maize 16, 18, 22, 24, 25, 74, 75, 92, 108, 109, 110, 111, 565
Male sterile line 19, 125, 194, 195, 240, 253, 363, 369
Male sterility 16, 25, 121, 124, 125, 126, 127, 130, 135, 143, 194, 195, 197, 198, 221, 243, 249, 250, 251, 253, 363, 366, 459, 499
Manathakkali 354
Mango 34, 37, 67, 92, 431, 433, 434, 435
Mean ± 2SE 180, 231, 325
Mean ± CD 247
Mean ± SE 78
Mentha Mint 554, 555, 556
Mesta 260, 262

Micrografting 41, 42
Micropropagation 9, 29, 43, 45, 46, 49, 66, 72, 98, 284, 379, 381, 397, 398, 451, 474, 478, 485, 534, 536, 553
Microsatellite 88, 89, 90, 92, 95, 103
Microtubers 44, 379, 381, 553
Mini tuber 380
Molecular GOT 46, 81, 85, 87, 100, 101, 102, 104
Molecular markers 7, 29, 30, 54, 71, 87, 92, 93, 94, 97, 98, 100
Monitoring Team 51, 76, 85, 106, 109, 113, 119, 124, 126, 130, 143, 150, 152, 153, 157, 164, 167, 169, 172, 177, 180, 183, 185, 188, 192, 194, 195, 201, 204, 207, 210, 216218, 221, 224, 226, 232, 234, 236, 240, 242, 243, 244, 247, 251, 252, 253, 256, 261, 264, 268, 269, 272, 275, 280, 283, 288, 291, 294, 298, 303, 307, 310, 325, 330, 332, 334, 337, 339, 347, 353, 356, 358, 361, 365, 369, 375, 379, 385, 391, 393, 396, 400, 405, 411, 414, 416, 438, 440, 444, 446, 448, 453, 455, 457, 461, 463, 465, 473, 475, 479, 480, 482, 501, 503, 506, 509, 511, 515, 517, 525, 531, 538, 542, 547, 549, 551, 554, 559
Mulberry 43, 435, 436, 437, 438
Musk melon 438
Mustard 17, 100, 198, 217, 219, 221, 507

N

NAA 320, 324, 450
Natural cross pollination 15, 16, 18, 20, 21, 22, 23, 24, 25, 28, 81, 112, 156, 163, 199, 215, 235, 260, 270, 342
Nigella 515
Niger 222, 223, 224
Noni 556, 557, 558
Notified varieties 6
Nucellar embryos 408
Nucleus seed stage I 106, 115, 149, 206, 207
Nucleus seed stage II 106, 115, 149, 152, 207, 247, 334, 367
Nutmeg 517, 517
Nutritional security 1, 12, 13, 70

O

Oats 16, 75, 112, 113, 114, 115
Offtypes 123, 129, 243, 250, 270, 257
Okra bhendi 359
Onion 92, 198, 361, 362, 363, 364, 366, 367, 368, 369
OPV 108, 116, 121, 142, 143, 219, 238
Organic seedlings 61, 64
Organic seeds 61, 62, 63, 64
Outcrossing 15, 18, 19, 20, 21, 71, 93, 92, 100, 206, 225, 230

P

Paired cross 117, 118, 119, 195, 220, 201, 220, 221, 232, 240, 240, 249, 443, 446, 447
Palak 370, 371, 372, 373
Palmyra palm 519
Papaya 25, 98, 441, 442, 443, 445, 446, 447
PCR amplification 94, 97, 101, 104
Pearl millet 75, 100, 115, 116, 120, 121
Periwinkle 523, 524, 525
Pigeon pea 18, 100, 193, 197, 198, 204, 205
Pineapple 34, 37, 98, 449, 451
Pistillate line 209, 210, 212, 213, 214
Pointed gourd 374
Pollinators 16, 19, 20, 340, 419, 524
Polyembryonic 33, 34, 339, 408, 433, 434
Pomegranate 34, 451, 452
Potato 29, 30, 44, 45, 51, 56, 57, 62, 66, 71, 327, 330, 331, 350, 357, 376, 377, 378, 379, 380, 381, 394, 395, 396, 397
PPV & FRA (2001) 165, 285
Progeny rows 50, 79, 105, 106, 108, 109, 113, 116, 117, 118, 119, 123, 125, 128, 129, 133, 135, 137, 138, 142, 143, 144, 146, 149, 153, 156, 164, 166, 167, 168, 169, 170, 172, 176, 177, 179, 180, 182, 185, 188, 191, 194, 197, 202, 205, 206, 209, 210, 216, 218, 219,. 223, 225, 229, 231, 232, 233, 235, 237, 240, 242, 243, 244, 252, 253, 256, 257, 259, 261, 264, 265, 267, 269, 271, 272, 275, 276, 277, 279, 280, 288, 291, 292, 293, 298, 300, 301, 306, 309, 312, 317, 320, 324, 325, 326, 329, 331, 334, 337, 339, 343, 349, 352, 353, 355, 356, 358, 360, 361, 364, 365, 367, 374, 378, 386, 390, 393, 399, 404, 407, 416, 419, 428, 437, 440, 443, 445, 447, 448, 450, 454, 456, 457, 461, 473, 478, 480, 482, 495, 496, 500, 501, 505, 506, 511, 514, 525, 528, 528, 541, 543, 546, 549, 551, 553, 556, 558
Progeny testing 70, 78, 168, 492, 494
Proso millet 153, 161, 162
Prota.org 550

Protandrous 108, 222, 238, 270, 333, 362, 414, 462, 515
Protogynous 16, 115, 116, 310, 316, 318, 348, 405, 418, 453, 481, 531
Pumpkin and Squash 382
Purity testing 7, 87, 92, 93, 562

Q

Quality assurance 2, 6, 7, 80, 85, 93, 119, 130, 143, 164, 252, 260, 294, 313, 317, 321, 343, 353, 388, 407, 440, 447, 461

R

Radish 18, 386, 387, 388
RAPD 87, 89, 92, 97, 98, 99
AFLP 98, 342
Rapeseed 17, 198, 217, 221
Released varieties 5, 46, 70, 72, 79, 455, 543, 577, 563, 564, 566, 567,
Rice 18, 19, 20, 21, 62, 64, 66, 68, 89, 90, 91, 95, 100, 122, 123, 126, 127, 128, 129, 130, 131, 133, 134, 135, 136, 139, 141, 198, 268, 524, 561, 565, 566
Ridge gourd 304, 388, 389
Root nodules 269, 270
Root stock 5, 6, 9, 32, 33, 35, 39, 40, 41, 42, 43, 44, 66, 408, 409, 413, 414, 415, 416, 418, 419, 420, 423, 428, 429, 434, 435, 454, 455, 489, 490, 491, 527, 528, 533
Rubber 6, 41, 44, 71, 72, 513, 514, 526, 527, 528

S

Safflower 17, 198, 230, 231, 232, 233, 234
Sapota 34, 37, 453, 454
Scion 5, 6, 9, 32, 33, 35, 38, 39, 40, 41, 43, 44, 412, 413, 414, 415, 416, 418, 423, 428, 429, 434, 435, 455, 488, 489, 490, 491, 417, 527, 528, 533
Seed dormancy 227, 228
Seed purity 84, 88
Self Pollinated Crops 12, 20, 80, 83, 153, 163
Sembua 400
Sesamum 234, 235, 237
Sesbania 61, 266, 267, 268, 269, 287, 288, 476, 478
Sesbania rostrata 268, 269
Small Millets 153, 156
Snakegourd 392, 393
Somaclonal variation 51, 52, 54, 84, 98, 99
Somatic embryoids 9, 65, 66, 67, 68, 72, 100, 499
Sorghum 16, 74, 75, 90, 91, 100, 142, 143, 145, 146, 147, 148
Soybean 17, 18, 66, 74, 103, 206, 207
Spinach 294, 299, 370, 371, 372, 373
Sponge gourd 304, 388, 389, 390
SSR markers 88, 89, 90, 91, 92, 94, 96, 103, 535
Steckling 300, 416
Stem nodules 268, 269
Stem tuber 376
Stolon 376
Sugarcane 4, 30, 45, 46, 47, 51, 55, 66, 98, 100, 281, 282, 283, 284, 285
Sunflower 17, 100, 238, 239, 240, 241, 242, 243, 244
Sunn hemp 270, 271, 272
Sweet potato 29, 30, 66, 394, 395, 396, 397
Synthetic seeds 6, 9, 27, 65, 66, 67, 68, 72, 80, 100

T

Table examination 78, 274, 309, 349, 361, 404
Tamarind 37, 531, 532, 533, 534
Tapioca 30, 31, 397, 398, 399, 400
Taro 29, 30, 357, 358, 400, 401
Tassel 108, 109
Tea 6, 60, 71, 534, 535, 536, 537, 538
Temperature sensitive genic male sterility (TGMS) 198
Tinda 455, 456, 457
TNAU Agri portal 46
Tobacco 18, 262, 263, 264, 265, 327, 351
Tomato 18, 22, 23, 62, 95, 96, 402, 403, 404, 405
Trait purity 72, 81, 85
True 2, 3, 5, 6, 8, 9, 27, 31, 32, 39-41, 44-46, 4-69, 70-72, 76, 78, 79, 83-85, 94, 97, 98, 100, 105, 106, 108, 109, 111-113, 115-119, 121, 122, 123, 126, 128, 130, 133, 135, 136-138, 142-147, 149, 152-154, 156, 157, 163, 164, 166-168, 170, 172, 176-180, 182, 184, 185, 187, 190, 191, 194, 195, 197, 198-200, 202-210, 213, 216, 218, 219, 220, 223, 224, 226, 229, 231-238, 240, 242-244, 246-249, 251-

254, 256-259, 261, 263-267, 269, 270-275, 277, 279, 281-283, 285, 287-294, 296-306, 308-313, 316, 317, 319-321, 325, 326, 329-331, 334, 337, 339, 341, 343, 347, 349, 352, 353, 356-358, 360, 361-369, 374, 377-380, 384-387, 389, 391-393, 395-397, 399, 403-407, 410-412, 414- 416, 418-421, 423, 428, 429, 434, 435, 437-440, 442-446, 448, 451-456, 459-465, 468, 469, 472-474, 478-482, 484-487, 489, 500-506, 508, 509, 511, 513-518, 521, 524, 528, 530, 531, 534, 538, 541, 543-546, 548-551, 553, 554, 556, 558

Potato Seed (TPS) 380

Turmeric 66, 67, 539540, 541

Turnip 405, 406, 407

V

Variety identification 92

Vegetative grafts 41

Vernalisation 300, 312, 318, 319, 368

Vicinism 16, 70, 194, 223, 231, 252, 275, 280, 291, 334

Virus elimination 40, 42, 381, 398

Virus indexing 41, 45

Viviparous 9, 327, 329

W

Watermelon 18, 20, 457, 458, 459, 460, 461

Wheat 16, 74, 75, 106, 107, 148, 149, 150, 151152, 565

Wind pollination 22, 493

X

Xenogamy 342

Z

Zinger 66, 67, 542, 543